# Resilient and Friendly Living Environments

The paramount goal of this series is to be a groundbreaking resource for professionals, practitioners, and enthusiasts dedicated to shaping resilient, friendly, and inclusive environments. This series aims to empower its readers with cutting-edge insights and practical tools to design spaces that seamlessly blend sustainability, adaptability, and cultural sensitivity. By delving into such topics, the series strives to provide a holistic perspective on creating environments that prioritize the well-being and interconnectedness of diverse communities. Moreover, the series also seeks to guide the development of frameworks that evolve with the ever-changing needs of societies, ensuring that the built environment remains responsive and relevant over time. Ultimately, this book series aspires to be a catalyst for positive transformation, encouraging a paradigm shift in the way we conceptualize, design, and interact with the spaces we inhabit. This series covers a broad spectrum of topics related to building environments that stand resilient and friendly, with a special emphasis on inclusivity for individuals of all ages. The series encompasses but is not limited to:

- Sustainable and Adaptive Design: Guidance on creating environments that are environmentally sustainable and capable of adapting to changing circumstances.
- Cultural Considerations in Design: Insight into designing spaces that respect and reflect the cultural diversity of communities which promote a sense of belonging among individuals.
- Supportive Infrastructure: Detailed examinations of infrastructure elements that contribute to the overall well-being, safety, and accessibility of diverse communities. It can also include wellness initiatives that address physical, mental, and emotional well-being.
- Designing for All Ages: Foster an inclusive environment that values diversity. This includes promoting equal opportunities, and ensuring that everyone feels welcome and respected. In-depth discussions on creating infrastructure and spaces that caters to the unique needs of both children and older adults is also of great importance.
- Community Engagement: Practical approaches to involving the community in the design process and ensuring that the built environment aligns with their needs and preferences.
- Safety Measures and Disaster Risk Reduction: Prioritize safety and security to ensure a comfortable and secure environment. This includes physical safety measures as well as fostering a culture of respect and mutual support.
- Technological Integration: Explorations of how technology can be seamlessly integrated to enhance connectivity, flexibility, and inclusivity within the built environment.
- Collaborative Spaces: Design work and learning environments that facilitate collaboration. This can involve open-plan workspaces, collaborative tools, and flexible seating arrangements that encourage interaction and teamwork.
- Adaptive Policies and laws: Establish policies and laws that are adaptable to changing circumstances. This flexibility allows for the accommodation of diverse needs and ensures that policies remain relevant and effective over time.

This book series is a valuable resource for architects, designers, urban planners, educators, and anyone interested in the intersection of design, community, and well-being. The series aims to inspire and guide readers in creating environments that not only withstand challenges but also flourish as vibrant and inclusive spaces for everyone.

Haibo Li • Guoqiang Shen

# The Anatomy of Subjective Well-Being

## Linking Built Environment to Life Satisfaction in China

Springer

Haibo Li
Guangdong University of Petrochemical Technology
Architecture and Civil Engineering Institute
Maoming, China

Guoqiang Shen
Zhejiang University College of Civil Engineering and Architecture
Hongzhou, China

ISSN 3005-1681 ISSN 3005-169X (electronic)
Resilient and Friendly Living Environments
ISBN 978-981-95-4936-8 ISBN 978-981-95-4937-5 (eBook)
https://doi.org/10.1007/978-981-95-4937-5



This Springer imprint is published by the registered company Springer Nature Singapore Pte Ltd.
The registered company address is: 152 Beach Road, #21-01/04 Gateway East, Singapore 189721, Singapore

# Preface

The twenty-first century is defined by a planetary urban transition, with more than two-thirds of the world's population projected to live in cities by 2050. As the most rapid growth unfolds in high-density megacities across Asia and Africa, this shift calls for a critical reevaluation of urban development models that have historically prioritized economic performance over human well-being. This book examines one of the central paradoxes of contemporary urbanization: how can cities servenot only as engines of growth but also as environments that actively promote human flourishing?

This volume addresses this imperative through a focused investigation into the nexus between the built environment and subjective well-being (SWB) in high-density Chinese cities. The study is motivated by an empirical puzzle that presents an urban variant of the "Easterlin Paradox": despite China's unprecedented economic and urban expansion, its performance on international well-being indices has remained subdued. Moving beyond macro-level correlations, this research interrogates the micro-scale, mediating mechanisms—particularly through perceived environmental quality, safety, social capital, and health—that translate objective urban form into subjective lived experience.

The book is structured in three parts. Part I constructs a multidisciplinary foundation for understanding well-being, integrating philosophical, psychological, and evolutionary perspectives. Part II presents an empirical case study of Guangzhou, deploying a multi-stage analytical strategy that progresses from descriptive statistics through regression analysis to structural equation modeling (SEM). This approach rigorously tests the complex pathways linking key built environment attributes (conceptualized through the "5D" dimensions: Density, Diversity, Design, Distance to transit, and Destination accessibility) to well-being outcomes. Part III synthesizes the findings, which challenge the direct applicability of Western-derived compact city theory in high-density contexts. The results reveal the nuanced, often counterintuitive roles of the built environment, further unpacking the critical mediating pathways that link objective urban form to subjective well-being.

By integrating urban planning, environmental psychology, and policy analysis, this book aims to bridge a significant theoretical gap in the literature. It offers a refined, context-sensitive framework for understanding the socio-ecological dynamics of high-density living, providing evidence-based guidance for researchers, policymakers, and planners committed to building sustainable and humane cities for the future in China and beyond.

Maoming, Guangdong, China Haibo Li
Hangzhou, Zhejiang, China Guoqiang Shen

# Acknowledgments

The completion of this book, the result of nearly two years of extensive revision, marks a significant milestone in our academic journey. The manuscript originates from Haibo Li's doctoral dissertation but has been substantially reorganized, expanded, and refined for publication as a collaborative work with supervisor and co-author, Professor Guoqiang Shen of Zhejiang University, who has contributed tremendously in conception, theory, interpretation, and editing. Our continued partnership and intellectual dialogue have been fundamental to shaping the ideas presented here.

We are profoundly grateful to the many organizations and individuals who have supported this endeavor, beginning with the instrumental financial support provided by the Guangdong Provincial Office of Philosophy and Social Sciences [Grant No. GD24XSH06] and the Projects of PhDs' Start-up Research of GDUPT [Grant No. 2023bsqd1008], which were essential in facilitating the publication process.

We extend our sincere thanks to the reviewers of this manuscript—Professor Curtis Andressen of the University of Texas at Arlington, Professor Mahyar Arefi of the University of Tehran, and Professor Long Zhou of the City University of Macau— whose perceptive comments and constructive suggestions greatly enhanced the quality of this work. Building upon the foundation of Haibo Li's doctoral research, we also wish to specially acknowledge the valuable guidance provided during my dissertation research by Professor Jiangping Zhou of the University of Hong Kong, Professor Chaosu Li of the Hong Kong University of Science and Technology, as well as Dr. Mengshun Li and Dr. Huajie Yang of the City University of Macau.

We are particularly indebted to Dr. Chen Pan and Mr. Jiaming Guo of the Guangdong University of Petrochemical Technology for their significant contributions in cartographic design and editorial proofreading. Our appreciation also goes to Professor Xiaodong Liu of Zhongkai University of Agriculture and Engineering for his assistance during the research and writing phases, as well as to Professor Jiawei Wu and Dr. Hailiang Zeng from Guangdong University of Petrochemical Technology, and Professor Qingping Zhang from Nanjing Forestry University for their helpful suggestions and encouragement. We would particularly like to thank

Professor Bing Wang from Massachusetts Institute of Technology for his significant contributions to the conceptual development and manuscript revision of this work. The collective support from these colleagues was essential to the completion of this monograph.

Our deepest gratitude goes to our families. Their unwavering support, boundless patience, and profound understanding have sustained us every step of this journey, and it is in the warmth of their love that we have written and brought this book to completion..

Lastly, this book is dedicated to all those who have contributed to its realization and to the persistent quest for knowledge that has characterized these years of our scholarly pursuit. May the journey ahead continue to be guided by academic dedication and intellectual curiosity.

Maoming, Guangdong, China Haibo Li
Hangzhou, Zhejiang, China Guoqiang Shen

# Contents

# About the Authors

**Haibo Li** is Professor and Head of the Department of Design at the Guangdong University of Petrochemical Technology. His research encompasses healthy urban planning and design, community-built environments, and sustainable human settlements. His scholarly record includes over 30 publications in peer-reviewed journals, one academic monograph, and leadership or participation in multiple provincial and ministerial research projects. A strong advocate for practice-oriented education, Dr. Li has successfully supervised students to achieve top honors in prestigious competitions, including the IAI Design Award, China Habitat Design Annual Award, and the Asian Design Award.

**Guoqiang Shen** is Endowed Professor and the Chair of the Department of Regional and Urban Planning in the College of Civil Engineering and Architecture at Zhejiang University.

He holds a Bachelor of Architecture degree from Tsinghua University, a Master's degree in Landscape Architecture and Geography from Peking University, and a PhD in City and Regional Planning from the Ohio State University, USA. Prior to joining Zhejiang University, he served as a tenured full professor and department chair or director in Urban Planning and Public Administration at the University of Texas Arlington, the University of Oklahoma, and East Carolina University in the USA, in colleges of architecture, civil engineering, public affairs, and technology. He also concurrently held adjunct professorship in Industrial and Systems Engineering and Business School at the University of Oklahoma.

His research interests include Architectural Design, Urban Planning and Design, Urban Economics, Transportation and Logistics, Geographic Information Systems (GIS), and Management Science. He emphasizes critical thinking and practical innovation. Professor Shen has published more than 100 papers in top academic journals and has served as the principal investigator for research projects with a total funding of over USD$30 million. He has personally directed and participated in more than 30 domestic and international projects, including civic, commercial, residential, cultural designs and has won bids and received awards for multiple planning and design competitions or commissioned contacts.

# List of Figures

# List of Tables

# Part I
# Foundations of Well-Being

# Chapter 1
# Introduction

**Abstract** Against the backdrop of rapid global urbanization—particularly within high-density Asian megacities—this chapter identifies the critical nexus between the built environment and residents' subjective well-being (SWB), establishing it as a key academic and policy priority. It aims to dissect the complex, dualistic role of high-density urban settings in shaping SWB, within China's unique context. The chapter begins by presenting the urban well-being paradox: despite unprecedented economic growth and urbanization, China's SWB rankings lag, pointing to the limits of growth-focused models. Using an interdisciplinary approach, it reviews theoretical foundations, empirical evidence, and policy shifts toward people-centered urban development. Section 1.1 outlines the historical and philosophical underpinnings of SWB and its measurement, tracing the shift from endogenous to exogenous factors in well-being research. Section 1.2 analyzes the dualities of high-density urban living—balancing opportunities with challenges. Section 1.3 critiques current intuitive and Western-centric discourses, identifies research gaps (mediating pathways, contextual specificity), and declares the book's mission to develop a mechanistic, context-sensitive framework linking built environment attributes to SWB via perceptual and social mediators. Section 1.4 operationalizes key concepts—SWB, built environment (using the "5D" model), and high-density cities—providing a foundation for subsequent empirical analysis. Integrating urban planning, psychology, and policy, this chapter provides a foundation to guide the design of high-density cities that enhance well-being and support sustainable urban development.

**Keywords** Subjective well-being · Built environment · High-density cities · Urbanization · Easterlin paradox · Mediating pathways · China's urban policy · People-centered development · "5D" model · Quality of life

By 2050, it is projected that over two-thirds (68%) of the world's population will reside in urban areas, with the most rapid growth occurring in high-density


H. Li, G. Shen, *The Anatomy of Subjective Well-Being*, Resilient and Friendly Living Environments, https://doi.org/10.1007/978-981-95-4937-5_1

megacities across Asia and Africa (United Nations, 2019). This unprecedented urban expansion presents critical challenges—and opportunities—for human well-being and sustainable development. Against this backdrop, understanding how the multi-scale built environment shapes residents' quality of life becomes not only an academic inquiry but an urgent policy priority.

Cities, especially high-density urban agglomerations, serve as engines of innovation, economic growth, and cultural exchange. Yet they also embody a complex duality: they can be sources of both vitality and stress, of connection and isolation. This paradox compels a deeper investigation into how the physical fabric of the city—the built environment—influences the subjective well-being (SWB) of its inhabitants. Such an inquiry transcends academic interest; it is essential for shaping urban futures that are not only sustainable and productive but also equitable and conducive to human flourishing[1] (Diener, Oishi, et al., 2018).

Addressing this complex relationship requires a robust and nuanced theoretical framework. Research must move beyond simplistic correlations to develop an interdisciplinary understanding of the mechanisms linking urban form to well-being. Key questions include: How can well-being be meaningfully conceptualized and measured in an urban context? How do different academic traditions conceptualize the pathways through which surroundings affect mental and emotional states?

This first part of the book, Foundational Perspectives, establishes this essential groundwork. We begin by deconstructing the concept of well-being, integrating insights from biological, psychological, and sociocultural dimensions (Chap. 2). We then critically review the evolution of measurement frameworks, advocating for a multidimensional approach that captures the full richness of the construct (Chap. 3). Finally, we bridge these foundations to urban studies, examining key dimensions of the built environment and tracing the development of empirical research that positions it as a crucial determinant of population well-being (Chap. 4). By the end of this section, readers will be equipped with the theoretical and conceptual tools necessary to engage with the empirical analyses that follow, prepared to explore the specific mechanisms operating within high-density urban settings.

## 1.1 Background: The Pursuit of Happiness in High-Density Cities

### *1.1.1 Theoretical Foundations: Linking the Built Environment to Subjective Well-Being in High-Density Urban Contexts*

Throughout history, philosophers have considered well-being to be the highest good and ultimate motivation for human action.[2] Understanding and enhancing well-being is often considered a key factor in promoting and maintaining health and

[1] Diener, E., Oishi, S., & Tay, L. (2018). Advances in subjective well-being research. *Nature Human Behaviour*, 2(4), 253–260.

[2] Diener, E. (1984). Subjective well-being. *Psychological bulletin*, 95(3), 542.

social sustainability (Diener & Suh, 1997; Durand, 2015). It is also an important goal of academic research.

The academic discourse on well-being is characterized by several longstanding debates. These include the subjective and objective nature of well-being, hedonic (pleasure-seeking) and eudaimonic (meaning-seeking) perspectives (Mouratidis, 2018, 2020; Ryan & Deci, 2001), and discussions about its stability over time (Nordbakke & Schwanen, 2013). These conceptual controversies have, in turn, given rise to a diverse array of measurement approaches.

Among these various approaches, Subjective Well-Being (SWB) has emerged as one of the most prevalent and widely used frameworks for measurement. Western studies on SWB can be traced back to the 1950s, when scholars explored happiness experience (Liu et al., 2020). In recent decades, academic interest in SWB has exploded, and research attention has increased significantly. It has aroused a research upsurge in international academic circles, especially in psychology, sociology, economics and geography.

Early empirical research on SWB, predominantly led by psychologists, focused primarily on endogenous factors such as personality and genetics. However, a significant paradigm shift has occurred in recent decades. Scholars from sociology, economics, environmental science, and notably geography and urban planning, have increasingly turned their attention to exogenous factors—the external circumstances that influence well-being. Geographers, in particular, have contributed by mapping the spatial inequality of happiness at various scales and investigating the impact of both the natural and built environment on SWB, exploring factors such as urban morphology (Mouratidis, 2018), daily activities (Ettema et al., 2010), and social relations (Cramm et al., 2013; Mouratidis, 2020).

Given that urban planning's core mission is to design and maintain living environments that enhance resident well-being, understanding these exogenous factors is paramount. However, the majority of this evidence base has been generated in the context of Western, low-density cities. Consequently, the universality of their findings remains questionable. This book seeks to address this gap by examining the relationship between the built environment and SWB within the unique context of Chinese high-density cities, thereby testing the applicability of Western theories and enriching the field with insights from the Global East, thereby enriching the theoretical achievements of the relationship between the built environment and Subjective Well-being.

### *1.1.2 The Urban Well-Being Paradox: Economic Growth, Subjective Well-Being, and the Built Environment in High-Density Cities*

China's rapid urbanization since the twenty-first century provides a critical context for this study. Along with economic and social development, the period 2012–2022 witnessed a rapid increase in the urbanization rate, reaching 65.22% by the end[3]

[3] National Bureau of Statistics of China. (2023). *China Statistical Yearbook 2023*. Beijing: China Statistics Press. https://www.stats.gov.cn/sj/zxfb/202302/t20230203_1901709.html.

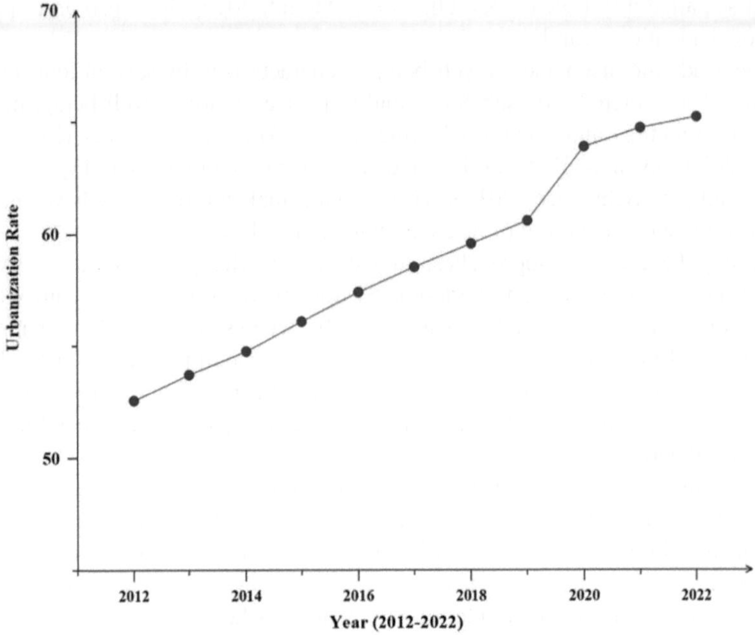


**Fig. 1.1** The evolution of China's urbanization rate (2012–2022). Source: Created by the authors. Data source: National Bureau of Statistics of China (NBS). National Bureau of Statistics of China (2023). *China Statistical Yearbook 2023*. Beijing: China Statistics Press

(Fig. 1.1). Given China's national conditions, which are constrained by a high-density population and limited livable construction land (Lin & Sun, 2017), high-rise buildings have expanded rapidly, and building density has been continuously increasing, gradually forming high-density urban spaces (Zhang & Chen, 2020). High-density and compact development has become an inevitable choice (Yu & Hu, 2017). China has witnessed significant economic growth, urban expansion, and regeneration, which have led to tremendous changes in the macro- and micro-neighborhood environments of cities (Chen & Chen, 2015). This urban pattern has brought convenience, a sense of gain, and satisfaction to residents. Meanwhile, the high-density pattern and the increase in population are accelerating urban expansion, resulting in a large amount of population flow and changing the social and cultural composition of cities and communities (Lin & Sun, 2017). The expansion of urban scale and the increase in density have led to a rapid growth of the urban population, and also triggered various urban problems such as environmental pollution, overcrowding, traffic congestion, soaring housing prices, and unstable social order.

Since the twenty-first century, China's economy has witnessed rapid and remarkable development, and it has become the world's second-largest economy. While China has achieved remarkable success in urbanization, its Gross Domestic Product (GDP) reached 51.947 trillion yuan in 2012 and soared to 120.472 trillion yuan in

2022[4] (Fig. 1.2). Over the past decade, the country's economic aggregate has doubled, making it the only economy in the world to contribute over 30% to global economic growth for ten consecutive years and consolidating its position as the world's second-largest economy. The per capita GDP has more than doubled, rising from approximately 40,000 yuan in 2012 to 85,700 yuan in 2022, approaching the threshold of high-income countries. The living standards of the Chinese people have been significantly improved.

In terms of the improvement of the happiness index, according to the first-released The World Happiness Report by the United Nations in 2012, Denmark ranked first, while China ranked only 112th.[5] Then, due to the reform of evaluation indicators, the ranking dropped to 84th in 2015,[6] rose to 79th in 2016,[7] and decreased year by year thereafter (Fig. 1.3). The 2020 version of The World Happiness Report showed that China ranked 94th.[8] Although it rose 18 places compared with the 2012 ranking, it dropped 15 places compared with the 2016 ranking, still behind some Southeast Asian countries with relatively backward economic levels such as Thailand (54th), Pakistan (66th), Malaysia (82nd), and Vietnam (83rd), which is not symmetrical with the status and proportion of the world's second-largest economy. In 2021 and 2022, the impact of the pandemic on each country was considered in the rankings. Since China had better pandemic control and was less affected, its ranking increased. Generally, China's happiness index has improved in recent years, indicating substantial progress in people's well-being. However, compared with the doubling growth of GDP and per capita GDP, the growth rate of the ranking of the happiness index lags significantly behind economic indicators, showing an obvious mismatch trend. It can be seen that although the rapid economic development and the urbanization process have brought convenience to residents' lives, the urbanization measures aiming to improve the living environment of residents have failed to enhance people's happiness (Pfeffer & Cloutier, 2016; Winters & Li, 2017). People have gradually realized that "wealth does not mean an improvement in the level of happiness", similar to the "Easterlin Paradox"[9] (Li & Shi, 2017; Myers, 1988).

China's unprecedented urbanization and economic ascent in the twenty-first century have created a profound paradox. While the urbanization rate

[4] National Bureau of Statistics Announcement on the Final Verification of the 2022 Gross Domestic Product (2023). National Bureau of Statistics. Beijing: China Statistics Press. https://www.stats.gov.cn/sj/zxfb/202312/t20231229_1946058.html.

[5] Helliwell, John F., Richard Layard, and Jeffrey Sachs, eds. 2012. *World Happiness Report 2012*. New York: UN Sustainable Development Solutions Network.

[6] Helliwell, John F., Richard Layard, and Jeffrey Sachs, eds. 2015. *World Happiness Report 2015*. New York: Sustainable Development Solutions Network.

[7] Helliwell, J., Layard, R., & Sachs, J. (2016). *World Happiness Report 2016*, Update (Vol. I). New York: Sustainable Development Solutions Network.

[8] Helliwell, John F., Richard Layard, Jeffrey Sachs, and Jan-Emmanuel De Neve, eds. (2020). *World Happiness Report 2020*. New York: Sustainable Development Solutions Network.

[9] Easterlin, R. A. (1974). Does economic growth improve the human lot? Some empirical evidence. In *Nations and households in economic growth* (pp. 89–125). Academic press.

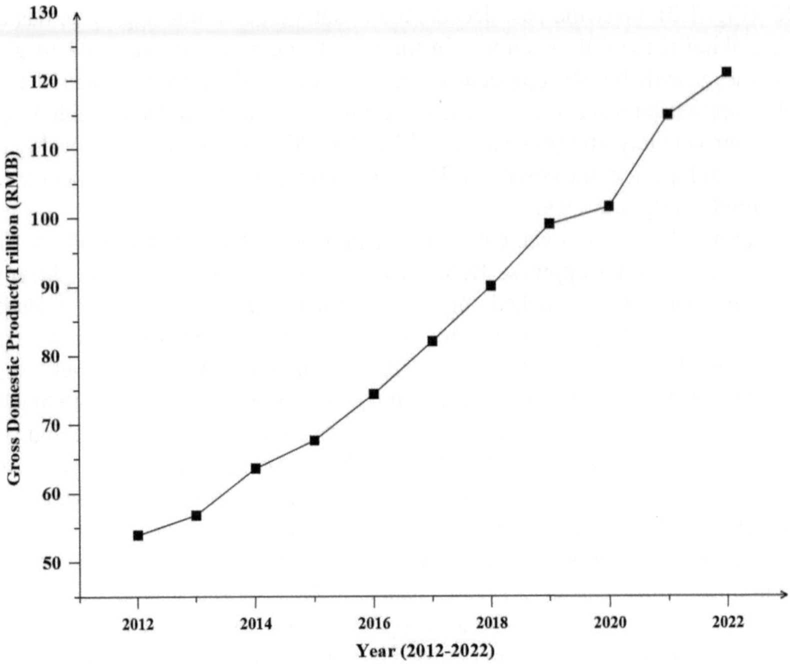


**Fig. 1.2** The trend in China's GDP growth (2012–2022). Source: Created by the authors. Data source: National Bureau of Statistics of China (NBS)

skyrocketed to 65.22% and GDP soared to become the world's second largest, these metrics of national power have not translated into a commensurate rise in citizen happiness. Despite a significant improvement in living standards, China's ranking in the World Happiness Report has fluctuated and lagged, often trailing behind less-developed nations. This stark dissociation between rapid socioeconomic development and the slower growth in subjective well-being underscores a classic "Easterlin Paradox" scenario, revealing that traditional urbanization models focused on economic expansion and physical growth are insufficient for enhancing residents' sense of well-being. This critical misalignment highlights an urgent need to shift the urban development paradigm beyond mere economic output and physical density. It sets the imperative stage for this research: to move past macro-level statistics and investigate the micro-scale mechanisms—specifically, how the built environment influences subjective well-being—to provide actionable insights for fostering truly livable and fulfilling high-density urban spaces in the Chinese context.

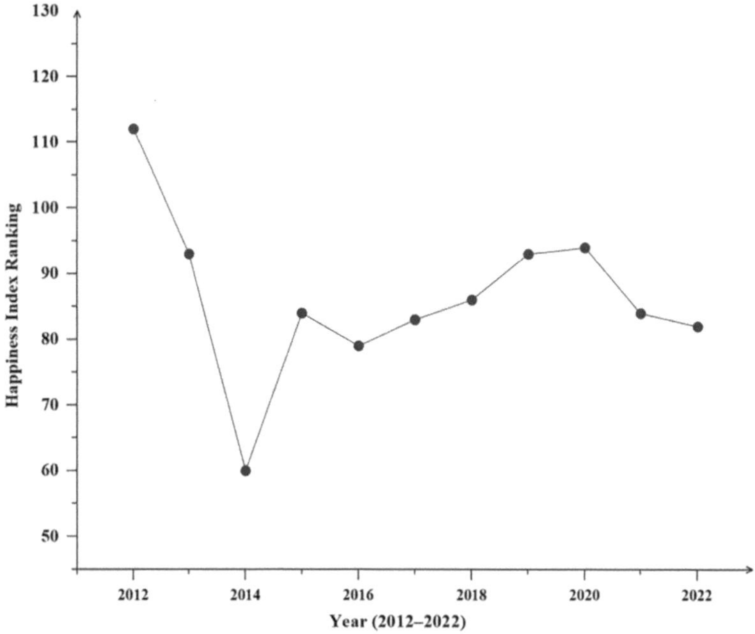


**Fig. 1.3** The trajectory of China's ranking in the well-being Index (2012–2022). Source: Created by the authors. data Source: World Happiness Report. Helliwell, J. F., Layard, R., Sachs, J. D., De Neve, J.-E., Aknin, L. B., & Wang, S. (Eds.). (2022). World Happiness Report 2022. New York: Sustainable Development Solutions Network

### *1.1.3 Paradigm Shift in Urban China: From Growth-Centrism to Well-Being-Oriented Development*

**1. Elevating Well-Being: From Political Directive to National Priority**

The elevated importance of residents' well-being in China's "new normal" development stage is unmistakably reflected in a sustained national policy shift. This transition was formally initiated at the 18th National Congress of the Communist Party of China (CPC) in 2012,[10] which explicitly called to "promote social harmony, continue to improve people's lives and well-being." This directive established an enduring priority, with subsequent state emphases on "constantly enhancing the people's sense of well-being." The vision was substantially elaborated at the 19th CPC National Congress in 2017,[11] which articulated the aim to "continuously meet the people's ever-growing needs for a better life" and make their "sense of gain, happiness and security more substantial, more secure and more sustainable." In the report

[10] The 18th National Congress of the Communist Party of China. (2012). *Full Text of the Report to the 18th National Congress of the CPC*. Beijing: People's Publishing House.

[11] The 19th National Congress of the Communist Party of China. (2017). *Full Text of the Report to the 19th National Congress of the CPC*. Beijing: People's Publishing House.

of the 20th National Congress of the Communist Party of China in 2022,[12] it was proposed to "improve people's well-being and enhance the quality of people's lives". The Third Plenary Session of the 20th Central Committee emphasized that "... focus on improving the quality of people's lives, focus on building a beautiful China, and focus on building a safer China at a higher level...".[13] Recent important meetings in Chinese history indicate that China has elevated the people's sense of well-being and the quality of life to an important position. This powerful, multi-level policy backdrop provides both the impetus and the practical relevance for this research.

**2. Evolving Urban Paradigms: From Managing Growth to Cultivating Quality of Life**

The 2015 Central Urban Work Conference, held after a 37-year hiatus, marked a pivotal moment in China's urban policy by addressing the unprecedented challenges of rapid urbanization.[14] It occurred during a period characterized by "the world's largest and fastest urbanization process",[15] with a primary focus on managing growth momentum and tackling acute "urban diseases" such as traffic congestion, environmental pollution, and housing shortages. The conference established the core guiding principle of "people-centered development" and "cities for the people", positioning cities as "important engines of modernization".[16] Its strategic framework was organized around "Five Key Coordinations": Structures of Space, Scale, and Industry: Enhancing overall planning by promoting city clusters as the main form, optimizing eastern clusters while cultivating new ones in central and western regions. (2) Planning, Construction, and Management: Improving systemic approaches by advocating for "multiple plans integrated into one," public participation in planning, and meticulous urban design to prevent chaotic "sprawl-like" expansion.[17] (3) Drivers of Reform, Technology, and Culture: Enhancing sustainability by advancing hukou reform to facilitate citizenization of migrant populations, promoting digital management platforms, and protecting historical. (4) Layouts of Production, Living, and Ecology: Prioritizing livability by creating harmonious human-environment spaces, implementing green infrastructure like "sponge cities," and accelerating the redevelopment of shantytowns. (5) Roles of

[12] The 20th National Congress of the Communist Party of China. (2022). *Full Text of the Report to the 20th National Congress of the CPC*. Beijing: People's Publishing House.

[13] Xinhua News Agency. (2024). *Communiqué of the Third Plenary Session of the 20th Central Committee of the Communist Party of China*. Retrieved from http://www.xinhuanet.com.

[14] State Council. (2015). *National New Urbanization Plan (2014–2020)*. Beijing: People's Publishing House.

[15] World Bank. (2014). *Urban China: Toward Efficient, Inclusive, and Sustainable Urbanization*. Washington, DC: World Bank Group.

[16] Central People's Government of the PRC. (2015). *Central Urban Work Conference Concluded in Beijing*. Retrieved from http://www.gov.cn.

[17] Ministry of Housing and Urban-Rural Development. (2015). *Guidelines on Promoting Multiple Plans Integration and Public Participation in Urban Planning*. Beijing: China Architecture & Building Press.

Government, Society, and Citizens: Fostering collaboration by encouraging co-governance among stakeholders. The vision was to build "harmonious, livable, and distinctive modern cities," with a strong emphasis on improving basic urban habitability and managing growth more efficiently through "compact city" and "smart growth" concepts. The 2015 approach was more foundational, concentrating on managing scale, solving pressing urban problems, and setting the stage for a more sophisticated urban development model that would evolve in the following decade.

In 2020, the "Proposal of the Central Committee of the Communist Party of China for Formulating the 14th Five-Year Plan for National Economic and Social Development and the Long-Range Objectives for 2035" clearly stated for the first time in an official document that "we should promote the people-centered new-type urbanization... and practice the concept that 'the people build the city and the city is built for the people'."[18] The report of the 20th National Congress of the Communist Party of China in 2022 emphasized that "we should adhere to the principle that the people build the city and the city is built for the people."[19]

The Central Urban Work Conference in 2025 outlined a comprehensive vision for China's future urban development, emphasizing a shift from rapid growth to stable development and from large-scale expansion to stock-based quality enhancement and efficiency improvement. It reaffirmed the principle of "the people's city built by the people and for the people," and introduced the concept of planning cities as "organic living systems." The conference highlighted multiple development goals—including innovation, livability, beauty, resilience, civility, and intelligence—and identified key thematic priorities such as promoting high-quality urban development, pursuing connotative growth, and advancing urban renewal. It also called for a transformation in urban work approaches, urging greater emphasis on people-oriented development, intensive and efficient models, characteristic-driven growth, governance input, and overall coordination. This guidance sets a clear direction for a new phase of urban modernization with Chinese characteristics.[20]

Moreover, the most recent policy directions from the 20th CPC National Congress in 2022 and the Third Plenary Session of the 20th Central Committee further reinforce this trajectory, proposing to "improve people's well-being and enhance the quality of people's lives," with a focused commitment to "improving the quality of people's lives" and "building a beautiful China"[21] (Fig. 1.4).

The strategic integration of well-being into urban governance became increasingly operational. Pioneering this move, provincial governments in Guangdong,

---

[18] CPC Central Committee. (2020). *Proposal for the 14th Five-Year Plan for Economic and Social Development and Long-Range Objectives Through 2035*. Beijing: People's Publishing House.

[19] CPC Central Committee. (2022). *Report to the 20th National Congress of the Communist Party of China*. Beijing: People's Publishing House.

[20] State Council. (2025). *Communiqué of the Central Urban Work Conference 2025*. Beijing: People's Publishing House.

[21] Xinhua News Agency. (2024). *Third Plenary Session of the 20th CPC Central Committee Highlights Quality of Life and Beautiful China Initiative*. Retrieved from http://www.xinhuanet.com.

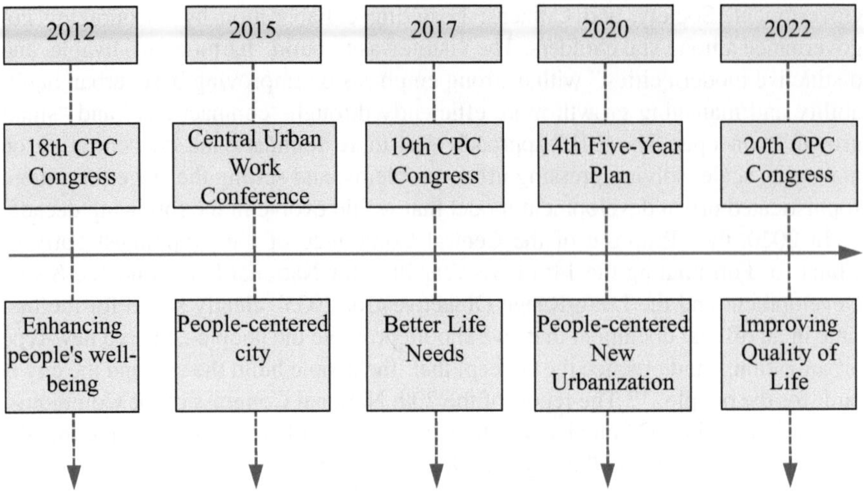


**Fig. 1.4** Policy shift in urban China. Source: Created by the authors

Shandong, and Henan incorporated enhanced well-being as a primary development objective into government performance and cadre assessment systems during the 12th Five-Year Plan period, a practice subsequently adopted by numerous other provinces. This doctrinal shift mandates a fundamental transition in urban planning: from large-scale incremental expansion to intensive stock enhancement and quality improvement, with a core focus on improving the living environment and well-being of residents. This strategic shift is operationalized through policies promoting urban regeneration, ecological conservation, and community building.[22]

## 1.2 The Urban Well-Being Paradox: Economic Growth, Subjective Well-Being, and the Built Environment in High-Density Cities

High-density cities represent the pinnacle of human agglomeration, bringing together economic vitality, cultural diversity, and social complexity in a concentrated spatial form. Yet, as outlined in Sect. 1.1, the relationship between urban density and subjective well-being (SWB) is far from straightforward. While density can foster economic opportunity, social connectivity, and environmental efficiency, it can also exacerbate crowding, pollution, social isolation, and mental stress (Fig. 1.5). This section delves into this paradox, examining the dualistic nature of

[22] Ministry of Ecology and Environment. (2023). *Guidelines on Ecological Restoration and Community Regeneration in Urban Renewal*. Beijing: China Environment Press.

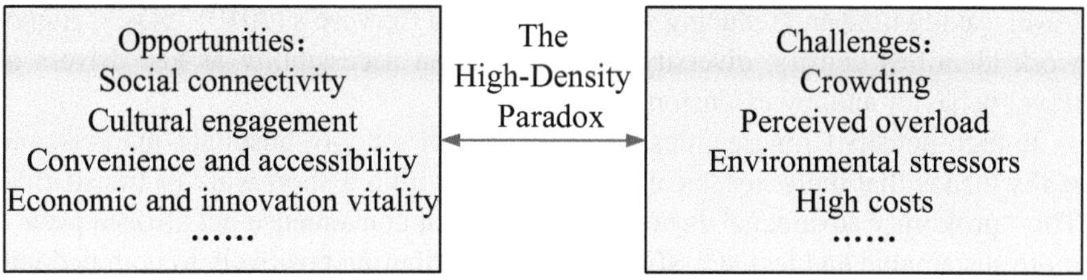


**Fig. 1.5** Dualities of high-density urban living. Source: Created by the authors

high-density urban environments and setting the stage for the mechanistic analysis that follows in subsequent chapters.

The rapid urbanization of China provides a critical context for this exploration. With the urban population soaring from 36.22% in 2000 to 65.22% in 2022 (National Bureau of Statistics of China, 2023), Chinese cities have become laboratories of high-density living. However, as detailed in Sect. 1.3, this growth has not consistently translated into improved well-being, giving rise to what scholars have termed the "Easterlin Paradox" in an urban context: rising GDP per capita without commensurate gains in happiness (Easterlin, 1974; Li & Shi, 2017). This section explores why that is, and how the built environment mediates these outcomes.

### *1.2.1 Opportunity: Density Increases Well-Being*

**1. Social Connectivity and Cultural Engagement**

High-density environments facilitate frequent and diverse social interactions. When people live in proximity, the potential for spontaneous encounters increases, fostering a sense of community and belonging. Jane Jacobs (1961) famously argued that dense, mixed-use neighborhoods promote "eyes on the street," enhancing both safety and social cohesion. Empirical studies support this, showing that residents of denser neighborhoods often report higher levels of social capital and community trust (Leyden, 2003; Mouratidis, 2020).

Moreover, density supports a rich array of cultural amenities—theaters, museums, cafes, public squares—that are economically viable only where there is sufficient demand. These amenities provide opportunities for leisure, learning, and social participation, all of which are key components of eudaimonic well-being (Ryan & Deci, 2001). In cities like Shanghai and Shenzhen, high-density urban cores are hubs of cultural innovation and social vitality, offering residents a quality of life that is difficult to replicate in lower-density settings.

**2. Convenience and Accessibility**

One of the most cited benefits of high-density living is convenience. The co-location of residential, commercial, and recreational uses reduces the need for long-distance travel, saving time and reducing stress. Ewing and Cervero's (2010) "5Ds"[23] framework identifies density, diversity, and destination accessibility as key drivers of travel behavior and, by extension, well-being.

In high-density Chinese cities, the proximity of schools, hospitals, markets, and parks means that daily needs can often be met within a short walk or transit ride. This "proximity advantage" is not only a matter of convenience but also supports a more sustainable and less stressful lifestyle, contributing positively to both hedonic and eudaimonic well-being (De Vos et al., 2013).

**3. Transport Efficiency and Walkability**
High-density areas are more conducive to public transit and non-motorized transport. Extensive metro systems, bus rapid transit, and bike-sharing networks are economically feasible only in densely populated corridors. This reduces reliance on private cars, lowers transportation costs, and decreases traffic-related stress.

Walkability—a function of density, land-use mix, and urban design—is strongly associated with physical activity and mental well-being. Studies have shown that residents of walkable neighborhoods are more likely to meet physical activity guidelines, have lower BMI, and report higher life satisfaction (Ettema et al., 2010; Frank et al., 2006). In cities like Hong Kong and Singapore, high-density planning has been explicitly linked to high levels of public transport use and walking, contributing to well-being through both physical health and time savings.

**4. Economic Opportunities and Innovation**
Density fosters agglomeration economies—the benefits that firms and households gain from locating near one another. This leads to higher productivity, job creation, and innovation (Glaeser, 2011). Access to diverse employment options and higher wages can improve material living standards and provide a sense of economic security, a significant predictor of life satisfaction (Diener et al., 2018).

In China, high-density cities like Shenzhen and Guangzhou are engines of economic growth, attracting talent and investment from across the country and the world. The concentration of skilled workers and innovative firms creates a dynamic labor market, offering residents opportunities for upward mobility and professional development.

### *1.2.2 The Challenges: How Density Undermines Well-Being*

**1. Crowding and Perceived Overload**
While density can foster social interaction, it can also lead to perceived crowding—a psychological state where individuals feel their personal space is violated (Stokols,

[23] Ewing, R., & Cervero, R. (2010). Travel and the built environment: A meta-analysis. *Journal of the American Planning Association*, 76(3), 265–294.

1972). Chronic exposure to high-density environments without adequate privacy or retreat spaces can induce stress, anxiety, and a reduced sense of control.

In many Chinese megacities, extremely high residential and population densities—often exceeding 20,000 people per square kilometer—can lead to overcrowded public spaces, long queues, and congested streets. This daily experience of crowding can erode well-being, particularly when individuals lack private sanctuaries or quiet spaces (Fung et al., 2022).

**2. Environmental Stressors: Noise, Pollution, and Lack of Green Space**
High-density areas often suffer from elevated levels of noise, air pollution, and limited access to natural environments. Traffic noise, construction activity, and constant human bustle can lead to sensory overload, sleep disturbance, and increased stress levels (Ulrich et al., 1991).

Air pollution—a severe problem in many Chinese cities—is exacerbated by density-related traffic congestion and industrial activity. Exposure to PM2.5 and other pollutants is linked to respiratory illnesses, cardiovascular disease, and mental health issues (Chen et al., 2018). Moreover, the scarcity of green space in dense urban cores limits opportunities for relaxation and recreation, further undermining well-being (Huang et al., 2020).

**3. High Cost of Living and Housing Affordability**
Density-driven demand for central locations often leads to skyrocketing housing prices and living costs. In cities like Shanghai and Beijing, housing affordability has become a major social issue, with many residents spending a large proportion of their income on rent or mortgages.

Financial strain is a well-documented detriment to SWB. Housing unaffordability can lead to residential instability, longer commutes (as people move to cheaper peripheries), and reduced disposable income for leisure and social activities (Winters & Li, 2017). This economic pressure can diminish both hedonic and eudaimonic aspects of well-being.

**4. Social Anonymity and Isolation**
Paradoxically, high-density settings can sometimes lead to social isolation rather than connection. The "urban anonymity" effect—where people surround individuals but lack meaningful social ties—can undermine a sense of community and belonging (Putnam, 2000).

In Chinese cities, high residential mobility and the influx of migrant populations can weaken social networks and community cohesion. While density may facilitate incidental interactions, it does not guarantee deep or supportive relationships. In some cases, the fast-paced, competitive nature of high-density life may even erode social trust and reciprocity (Lin & Sun, 2017).

**5. Mental Health Challenges**
The compounded effects of crowding, noise, pollution, and social stress can contribute to higher rates of anxiety, depression, and other mental health issues in dense urban areas (Peen et al., 2010). Although cities offer better access to mental health

services, the prevalence of mental disorders is often higher in urban compared to rural settings.

In China, rapid urbanization has been accompanied by rising rates of mental illness, particularly among migrant populations and young professionals facing intense academic and occupational competition (Chen & Chen, 2015). The lack of green space, limited opportunities for solitude, and constant exposure to stressors create an environment that can challenge even the most resilient individuals.

### *1.2.3 The Core Research Question: Balancing the Dualities*

The evidence presented above paints a complex picture: high-density urban environments are simultaneously sources of well-being and stress, opportunity and overload. This duality lies at the heart of the urban well-being paradox and raises a critical question for researchers, planners, and policymakers:

How can we plan, design, and govern high-density cities to maximize their well-being benefits while minimizing their negative costs?

This question cannot be answered through intuition or anecdote. It requires a rigorous, interdisciplinary, and mechanistic investigation into how specific built environment attributes—density, diversity, design, distance to transit, and destination accessibility—influence SWB through mediating pathways such as Perceived environmental quality, Perceived Safety, Social relationships and capital, and self-rated health. Moreover, it requires a contextual understanding of how these mechanisms operate in high-density Asian cities, which differ significantly from their Western counterparts in terms of culture, governance, and urban form.

This book seeks to address this gap by developing and testing a comprehensive theoretical framework that links the built environment to SWB through multiple mediating pathways. We draw on theories from urban planning, environmental psychology, sociology, and public health to propose a model that is both scientifically robust and policy-relevant and employ data from high-density urban settings to conduct empirical research.

## 1.3 Beyond Intuition: The Imperative for a Scientific and Interdisciplinary Investigation

The preceding section delineated the profound urban well-being paradox confronting China and many other rapidly urbanizing nations: despite unprecedented economic growth and massive urban infrastructure investment, corresponding gains in citizens' subjective well-being (SWB) have been inconsistent and often lagging. This disconnect underscores that the traditional, growth-centric urban model is insufficient for fostering human flourishing. As China's policy paradigm decisively shifts toward a "people-centered" approach focused on quality of life and well-being—a transition unequivocally mandated from the 18th to the 20th National Congress of the Communist Party of China—a critical question emerges: How can urban planning and design be effectively harnessed to actually achieve these lofty goals?

The public and professional discourse is replete with intuitive answers. Concepts like "green cities," "walkable neighborhoods," "vibrant public spaces," and "15-minute cities" are often promulgated as panaceas for urban ills and pathways to happiness. While these ideas are frequently compelling and sometimes supported by anecdotal evidence or case studies, this chapter argues that the current discussion largely resides in the realm of intuition and ideology rather than robust, scientific evidence. Navigating the complexity of high-density urban systems and making informed decisions that genuinely enhance SWB requires moving beyond this intuitive stage. It demands a rigorous, interdisciplinary, and mechanistic understanding of how the multifaceted built environment operates as a key determinant of human well-being.

### *1.3.1 Critiquing the Current Discourse: The Limitations of Intuition and Anecdote*

The popular imagination of the "happy city" is often shaped by influential urban thinkers. Jane Jacobs's celebration of dense, mixed-use neighborhoods with "eyes on the street" (Jacobs, 1961) and William H. Whyte's studies on the social life of small urban spaces (Whyte, 1980) have become foundational texts. Their observational, ethnographic work provided invaluable hypotheses about what makes urban environments feel lively and engaging. However, their methods were not designed to establish generalizable, causal evidence. They point to potential correlates of urban vitality but cannot quantify their effect sizes, test for confounding factors, or establish the precise mechanisms through which a physical feature translates into a psychological state.

This reliance on singular narratives or best practices poses a significant risk in urban policy. A successful intervention in one specific cultural and geographic context (e.g., a pedestrianization project in Copenhagen) may be uncritically transplanted to a vastly different high-density Asian context with contrasting climatic conditions, cultural norms, and socioeconomic structures, leading to unintended consequences and policy failure. This approach privileges aesthetic mimicry over mechanistic understanding.

Within academia, research has often been constrained by disciplinary silos, each offering a partial and potentially fragmented view:

**1. Urban Planning and Economics**

Have traditionally focused on objective, macro-level metrics such as land-use patterns, property values, transport efficiency, and economic productivity. They might correlate park proximity with higher housing premiums (Luttik, 2000) but seldom delve into the psychological processes behind this value.

**2. Environmental Psychology**
Excels at unpacking micro-level human-environment interactions, such as stress reduction from exposure to nature (Ulrich et al., 1991) or the cognitive benefits of walkable environments. However, it often does so in controlled settings or at a scale that overlooks broader urban economic and political forces.

**3. Public Health**
Has robustly linked built environment features (e.g., walkability) to objective health outcomes like reduced obesity rates (Frank et al., 2006) but has been slower to integrate nuanced psychological constructs of SWB beyond mental illness.

**4. Sociology**
Focuses on social capital and community dynamics (Leyden, 2003) but may underemphasize the physical design elements that facilitate or hinder these social processes.

While each discipline contributes crucial pieces of the puzzle, the lack of integration means a comprehensive picture remains elusive. We may know that density is linked to both economic productivity and stress (Fung et al., 2022; Glaeser, 2011), or that green space is linked to both higher home values and better health (Luttik, 2000; Ulrich et al., 1991), but we lack an integrated model that explains for whom, under what conditions, and through what simultaneous pathways these effects occur. This disciplinary fragmentation perpetuates a reductionist approach, failing to capture the synergistic and sometimes antagonistic interactions between the different dimensions of the built environment and human experience.

### *1.3.2 Identifying the Critical Research Gaps*

Building upon this critique of the current discourse, three interconnected research gaps emerge as particularly critical and serve as the primary focus for this book.

**Gap 1: The Black Box of Causal Mechanisms**
A substantial body of literature has established correlations between broad built environment characteristics—often conceptualized through the "5Ds" (Density, Diversity, Design, Distance to transit, Destination accessibility) (Ewing & Cervero, 2010)—and various well-being outcomes. For instance, higher density is associated with both positive (e.g., access to amenities, reduced vehicle miles traveled) and negative (e.g., overcrowding, perceived stress) outcomes (Fung et al., 2022; Mouratidis, 2018). Land-use mix and good destination accessibility are frequently linked to greater walking and higher life satisfaction (De Vos et al., 2013; Ewing & Handy, 2009).

However, these studies predominantly reveal that a relationship exists, not how it works. The causal pathway remains a "black box." The fundamental challenge of self-selection looms large: do people become happier because they live in a well-designed, walkable neighborhood, or do individuals with a predisposition for an active social life and positive affect consciously choose to live in such areas? (Cao et al., 2009). To address this puzzle, it is essential both to employ research designs that move beyond cross-sectional snapshots by leveraging longitudinal data, natural experiments, and advanced statistical modeling techniques to better infer causality, and to open the black box by investigating mediating variables—the perceptual and experiential factors that translate objective urban form into subjective well-being.

**Gap 2: The Oversimplification of Mediating Pathways**

A growing number of studies have begun to incorporate mediators, yet the approach has often been reductive, examining single pathways in isolation. For example, research might test whether the link between park access and well-being is mediated only by physical activity (Sugiyama et al., 2008), or whether the relationship between neighborhood design and social capital is mediated only by incidental social interaction (Leyden, 2003).

In reality, the pathways through which a high-density built environment influences SWB are multiple, parallel, and potentially interlinked. The effect of a hyper-dense, transit-rich neighborhood on a resident's happiness is unlikely to flow through a single channel. Critically, these mediators do not operate in a vacuum; they likely influence each other in a chain-like sequence, or serial mediation. The current literature severely lacks studies that model these complex, serial mediation pathways. Ignoring these interconnections results in an oversimplified and potentially misleading model of the urban experience, failing to capture the true complexity of how built environments shape our lives.

**Gap 3: The Context-Specificity of Western Models**

The predominant theoretical frameworks and empirical evidence underpinning our understanding of the built environment-SWB relationship are overwhelmingly derived from studies in North American, Australian, and European cities. These contexts are typically characterized by lower overall densities, car-dependent suburban sprawl, and different socio-cultural norms regarding privacy, community, and public space (Mouratidis, 2021).

Asian megacities like Shanghai, Shenzhen, and Singapore present a fundamentally different urban model. They are defined by hyper-density, extensive and heavily used public transport systems, a vertical rather than horizontal urban form, and a distinct socio-cultural context. The lived experience of density—a core "D" variable—may be qualitatively different. In many Western studies, high density can be associated with negative externalities. In contrast, in many Asian cities, high density is often culturally accepted and associated with convenience, vibrancy, and economic opportunity, though it can also co-exist with intense stressors like noise and competition for space (Fung et al., 2022; Huang et al., 2020). The role of other factors, such as distance to transit, may be less impactful in cities where metro coverage is ubiquitous, while design qualities and micro-level aesthetic details may become even more critical for well-being in such intensive environments. Therefore, the applicability of theories and findings from Western contexts to high-density Asian settings is not a given but a pressing empirical question. There is an urgent need to test, adapt, and develop new models that are specifically calibrated to the realities of high-density urban living, which houses a significant and growing proportion of the global population.

### *1.3.3 Declaring the Mission of This Book*

This book is conceived as a direct response to the identified gaps. Its central mission is to dismantle the "black box" and illuminate the complex mechanisms linking the high-density built environment to subjective well-being, with a specific focus on the Chinese context outlined in Sect. 1.1. We argue that this requires an integrated approach that is both scientifically rigorous and unapologetically interdisciplinary, drawing on urban planning, environmental psychology, sociology, public health, and transportation studies.

The book will introduce a multi-dimensional framework that examines the five key built environment attributes—density, diversity, design, distance to transit, and destination accessibility—not as isolated factors, but as an interconnected system that shapes human experience. Moving beyond simple correlations, the core of our investigation will be to unpack the pathways through which these objective features are filtered by human perception and social interaction. We will rigorously examine the multiple mediating roles of Perceived environmental quality, Perceived Safety, Social relationships and capital, and Self-rated health.

Furthermore, we will pioneer the exploration of potential chain reactions (serial mediation) between these mediators. For instance, we will test how objective design influences perceived safety, which in turn influences social interaction and then health, ultimately affecting SWB. This approach provides a more holistic, realistic, and powerful account of the urban experience.

Empirically, the book will ground its analysis in the understudied context of high-density Chinese cities, utilizing robust quantitative methods and novel data to move beyond correlation towards a clearer understanding of the comprehensive relationships. By doing so, we aim to provide planners, architects, and policymakers with an evidence-based toolkit for designing not just economically functional or sustainable cities, but truly human-centred ones that actively promote the well-being of their citizens. This mission aligns directly with China's new people-centered urban paradigm, offering scientific evidence to guide its implementation and ensure that the cities of the future are not only mega in scale but also rich in well-being.

## 1.4 Unpacking the Key Concepts: Built Environment and Subjective Well-Being

### *1.4.1 Subjective Well-Being*

Well-being can be defined as the satisfaction or well-being derived from optimal functioning. Subjective Well-being refers to how an individual feels about their life (Ko & Kuo, 2009). As an important dimension of well-being, Subjective Well-being is a research topic that has received extensive attention in the fields of foreign psychology, sociology, economics, and geography. The relatively complete conceptual constructs include Diener's (1984) Subjective Well-being, Ryff et al.'s (1995) psychological well-being, Seligman et al.'s (2000) positive well-being, and Keyes et al.'s (2002) mental well-being. Among them, Subjective Well-being is an individual's subjective self-evaluation of their overall life (Diener, 2009). It is often regarded as a multi-dimensional structure, including emotional and cognitive elements (Librán, 2006), namely positive affect, negative affect, and Life Satisfaction (Diener et al., 2003; Karademas, 2007). Its SWB conceptual construction is mature and widely recognized by the academic community.

Subjective well-being (SWB), a central concept in positive psychology and social sciences, refers to individuals' multifaceted evaluations of their lives and emotional experiences (Diener, 1984). These evaluations encompass both cognitive judgments of life satisfaction and affective experiences involving emotions and moods (Diener et al., 2018). SWB is characterized by its subjective nature—relying on personal criteria rather than external conditions—and its relative stability over time (Diener et al., 2003). The study of SWB typically distinguishes between two major philosophical traditions: the hedonic perspective and the eudaimonic perspective (Ryan & Deci, 2001). The hedonic view, often treated synonymously with SWB itself, emphasizes the pursuit of pleasure and avoidance of pain. It is commonly operationalized through three interrelated components (Fig. 1.6): (1) Life Satisfaction (LS): a cognitive, global judgment of one's life according to self-chosen standards (Pavot & Diener, 2008); (2) Positive Affect (PA): the experience of frequent positive emotions such as joy and contentment; (3) Negative Affect (NA): the experience of infrequent negative emotions such as sadness or anger (Diener et al., 2003).

In contrast, the eudaimonic tradition focuses on meaning, self-realization, and the fulfillment of human potential, often measured through constructs such as psychological well-being (Ryff & Keyes, 1995). Although related, eudaimonic well-being is generally considered distinct from hedonic SWB.

Life satisfaction constitutes a core cognitive dimension of SWB. It can be defined as an individual's assessment of their life conditions based on the perceived gap between personal aspirations and achievements (Fernández et al., 2017), or more broadly, as a global evaluation of one's life circumstances. Conceptually, life satisfaction reflects a subjective, holistic, and relatively stable judgment concerning one's long-term living environment (Diener, 1984; Diener et al., 1985). Thus,

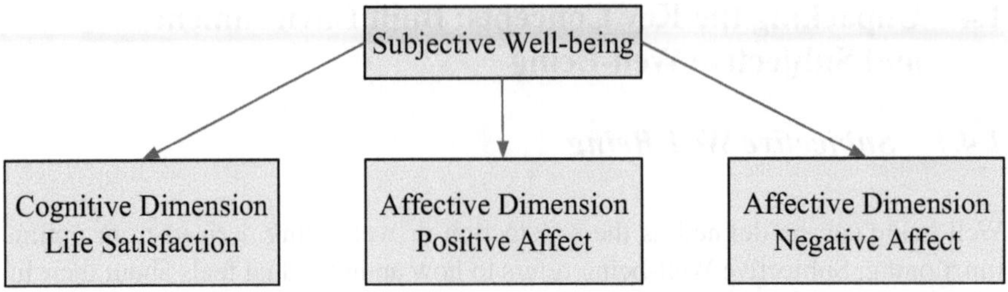


**Fig. 1.6** Dimension of subjective well-being. Source: Created by the authors (Diener et al., 2003)

residents' life satisfaction represents a psychological experience through which individuals cognitively assess their objective living conditions against personal standards. This includes feelings about their present life, the congruence between reality and expectations, and their outlook on future life.

It is important to note that while SWB includes both cognitive evaluations (life satisfaction) and affective components (emotional well-being), emotions are more susceptible to transient events and external influences, making them difficult to measure consistently and reliably. Therefore, this study focuses on the cognitive aspect of SWB—specifically, life satisfaction—as a more stable and empirically tractable construct.

Given its cognitive, evaluative, and relatively stable nature, life satisfaction is especially suitable for research examining the long-term effects of contextual factors such as the built environment (Ferrer-i-Carbonell & Gowdy, 2007). It allows individuals to deliberately assess how their living conditions align with personal aspirations, making it a highly relevant indicator for urban planning and policy interventions aimed at improving the quality of life.

Both theoretical and empirical studies recognize life satisfaction as a critical dimension of SWB and aim to elucidate its determinants (Diener & Seligman, 2002). Understanding these determinants is essential for formulating policies aimed at enhancing human welfare (Diener et al., 2003). A comprehensive understanding of SWB also necessitates examining its myriad influencing factors—ranging from internal personality traits to external socioeconomic and environmental conditions—which will be reviewed in detail in a later section.

## *1.4.2 Built Environment*

Current literature indicates that research on the built environment spans multiple disciplines, including sociology, geography, public health, and urban planning. Due to this interdisciplinary interest, theories and findings are often fragmented, and a universally accepted definition remains elusive. This conceptual diversity reflects the complex and multi-layered nature of built environments as both physical artifacts and social products (Dovey & Pafka, 2020).

The built environment encompasses residences, schools, workplaces, parks/recreational areas, commercial districts, and roads. It includes overhead infrastructures like power transmission lines, underground constructions such as landfills and subways, and extensive networks like highways. In broad terms, the built environment comprises all buildings, spaces, and products created or modified by humans. It influences both physical environments (e.g., microclimates, air quality) and social environments (e.g., civic engagement, community capacity, and investment) (Lin & Yang, 2015). Moreover, it plays a critical role in shaping human behaviors, health outcomes, and overall quality of life (Sallis et al., 2016).

Most definitions converge on the idea that the built environment refers to an integrated physical setting that combines spatial elements such as land use, transportation systems, and urban infrastructure (Lin & Yang, 2015). It operates across macro (global, national, regional), meso (urban, county), and micro (street, neighborhood) scales (Li et al., 2020). Each scale offers a unique lens for analysis: macro-scale studies examine regional policy and infrastructure networks, meso-scale focuses on urban form and accessibility, while micro-scale analyses delve into street-level design and perceptual qualities (Barton, 2017).

Prominent conceptual models include the "3D" model—Density, Diversity, and Design—proposed by Cervero and Kockelman (1997). Handy et al. (2002) later identified five key elements: density, land use mix, street connectivity, street scale, and aesthetics, later refined into land use, transport systems, and design. Ewing and Cervero (2010) expanded this into the widely cited "5D" model by adding Destination accessibility and Distance to transit (Fig. 1.7). Despite its prevalence, the 5D model has been criticized for overlapping dimensions and a lack of conceptual clarity (Yu & Hu, 2017), and to date, no more robust alternative has gained broad acceptance. Some scholars have called for more dynamic and people-centered frameworks that incorporate temporal changes and human experiences (Dovey & Pafka, 2020).

Equally important is the subjective perception of the built environment, which features prominently in health, SWB, and life satisfaction research. Drawing from Maslow's (1943) hierarchy of needs, measurements of perceived built environment are often organized into four dimensions: (1) Facility perception (meeting basic survival needs): perception of diversity and convenience; (2) Safety perception (meeting safety needs); (3) Transportation convenience perception (supporting travel and social activities); (4) Aesthetic perception (meeting esteem needs): perception of beauty and environmental quality (Das, 2020; Shen & Fu, 2019). Recent studies suggest that perceived built environment characteristics often mediate the relationship between objective built environment measures and various well-being outcomes, highlighting the importance of incorporating both objective and subjective measures in research (Wang et al., 2020).

### *1.4.3 High-Density City*

**The Concept of High-Density Cities: Definitions, Dimensions, and Contexts**
The term "high-density city" is widely employed across urban studies, planning, and policy discourses, yet it lacks a singular, universally accepted definition. It generally denotes an urban area characterized by intense concentrations of people,

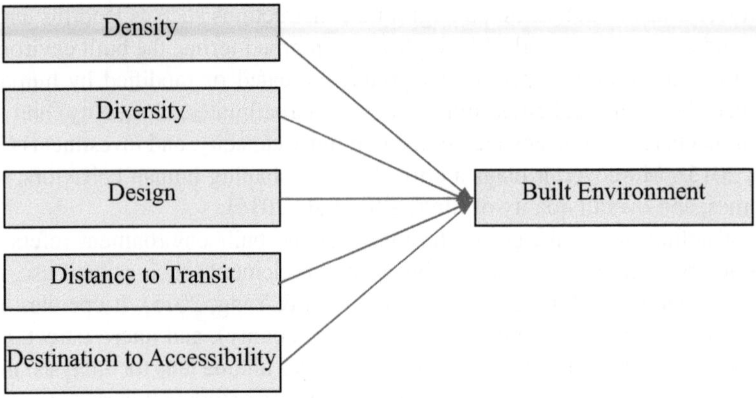


**Fig. 1.7** "5D" model. Source: Created by the authors (Ewing & Cervero, 2010)

buildings, and economic activities within a constrained spatial footprint. High density is frequently promoted through planning models such as the "compact city" and "smart growth," which advocate dense, mixed-use, and transit-oriented development as sustainable alternatives to urban sprawl—improving land use efficiency, reducing energy consumption, and enhancing access to services and public transportation (Jenks & Burgess, 2000; Neuman, 2005).

Density is conventionally quantified using metrics such as population density (e.g., inhabitants per square kilometer) and built density (e.g., floor area ratio (FAR), site coverage, or dwelling units per hectare) (Angel, 2021; Wang, 2010). However, high density transcends mere numerical thresholds; it constitutes a multidimensional construct encompassing physical, social, economic, and environmental facets (Li & Liu, 2019). Physically, it is reflected in vertical development, compact urban form, and functional mix. Socially and perceptually, it presents both benefits, such as vibrancy and convenience, and challenges, including overcrowding, noise, and reduced privacy.

The interpretation of high density is deeply contextual. In Western developed countries, it is often intentionally planned and viewed as a desirable "goal" linked to sustainability, economic vitality, and urban livability. By contrast, in many developing regions, high density is frequently perceived as a "problem," synonymous with overcrowding, inadequate infrastructure, and socioeconomic inequity resulting from rapid and sometimes informal urbanization (Li & Ye, 2015). This divergence underscores the concept's inherent ambiguity and context-dependency (Li & Liu, 2019). Moreover, the relativity of what constitutes "high" density is shaped by regional norms, historical trajectories, and cultural acceptability. For example, densities exceeding 15,000 persons/km$^2$ are common in cities such as Mumbai or Manila but are rare in most European urban areas (Angel, 2021). An important conceptual distinction also exists between density as a physical condition and overcrowding as a subjective experience influenced by urban design, service provision, and governance (Churchman, 1999). In summary, high-density cities are best understood not through a single definition, but as complex, context-dependent entities whose meaning integrates quantitative measures with qualitative dimensions of urban form, functionality, and human experience.

Within the specific context of China, urbanization has advanced at an unprecedented rate, reaching an urbanization rate of 65.22% in 2022 (National Bureau of Statistics of China, 2023). Despite this rapid progress, significant challenges persist, including regional disparities, inequitable access to public services, and environmental pressures (Wu & Gaubatz, 2013; Yeh & Li, 2020). To address these complexities and better operationalize the concept, this study adopts a dual-index classification framework for high-density cities, based on a combination of population density and building density, following the methodology proposed by Li and Liu (2019). This approach allows for a more nuanced categorization of urban typologies under high-density conditions. The specific classification thresholds used are presented in Table 1.1.

## 1.5 Structure of the Book

This book is written as ideal for researchers, urban planners, policy-makers, upper-level undergraduates and graduate students in fields such as positive psychology, urban studies, public health, and social policy. This volume provides valuable evidence-based insights into the mechanisms through which urban design influences quality of life. It stands out by challenging Western-centric models and offering culturally nuanced policy strategies aimed at fostering sustainable, equitable, and people-centered urban development.

This book offers a comprehensive, interdisciplinary examination of well-being, integrating evolutionary, biological, psychological, and environmental perspectives to explore how subjective well-being is shaped by both intrinsic factors and external conditions. With a special focus on high-density urban settings, it combines theoretical depth with empirical rigor—drawing on advanced methodologies such as questionnaire survey, structural equation modeling, and geospatial analysis—to unravel the complex relationships between built environments and resident life satisfaction.

The book is structured into three main parts. Part I—Theoretical and Methodological Foundations of Well-Being (this chapter and Chaps. 2–4), lays the theoretical and conceptual groundwork for the study. Part II—Empirical Investigations: The Built Environment and Well-Being in High-Density Cities (Chaps. 5–9) presents the empirical core of the research, including methodology, analyses, and results. Part III—Synthesis, Implications, and Future Directions (Chaps. 10 and 11) provides a comprehensive discussion, synthesizes the findings, and outlines broader implications and future research directions.

This chapter establishes the global and local context of rapid urbanization, particularly within high-density Chinese cities, and introduces the central research focus: the complex relationship between the built environment and subjective

**Table 1.1** Classification table of high-density cities

| Building density (Floor area ratio, FAR) | Population density | | | |
|---|---|---|---|---|
| | <5000 | 5000–10,000 | 10,000–15,000 | >15,000 |
| <0.2 | Low density | Low density | Low density | Medium density |
| 0.2–0.5 | Low density | Medium density | Medium density | High density |
| 0.5–1.0 | Low density | Medium density | High density | High density |
| >1.0 | Medium density | High density | High density | High density |

Source: Created by the authors from the references (Li & Liu, 2019)

well-being (SWB). It highlights the paradoxical coexistence of economic growth and stagnant happiness rankings—an urban "Easterlin Paradox"—underscoring the limitations of traditional growth-centric models. Against China's policy shift toward "people-centered" development, the chapter calls for an interdisciplinary, mechanism-driven investigation into how urban form influences SWB. It critically reviews existing theoretical and methodological gaps, particularly the overreliance on Western models, and outlines the book's mission: to unpack the mediating pathways between built environment attributes and well-being using empirical data from high-density Chinese contexts, thereby contributing to both theory and sustainable urban policy.

Chapter 2 delves into the historical and philosophical origins of happiness and subjective well-being, as well as the multidimensional perspectives on well-being. It first discusses the origins and development of well-being in Western and Eastern traditions, and then proceeds to examine it from an ecological-evolutionary perspective, examining how positive emotions and psychological needs (autonomy, competence, relatedness) have adaptive functions and are supported by neurobiological mechanisms. The chapter reviews genetic contributions to well-being, including heritability estimates and gene-environment interactions, and delves into the neurobiological substrates involving dopamine, serotonin, oxytocin, and endogenous opioid systems. Core psychological theories—such as the Broaden-and-Build Theory, Self-Determination Theory, and Set-Point Theory—are discussed in depth, alongside contemporary debates regarding plasticity, adaptation, and measurement challenges. The chapter concludes by integrating evolutionary, genetic, and neurobiological approaches, emphasizing that well-being arises from a dynamic interplay between biological predispositions and environmental influences.

Chapter 3 transitions from theoretical foundations to the practical challenges and innovations in measuring well-being. It begins by differentiating hedonic, eudaimonic, and social dimensions of well-being, and discusses integrated models such as the dual-continua framework. The chapter expands into contextualized well-being, including work, spatial, and environmental factors, and highlights the role of cultural folk models in shaping conceptions of happiness. Methodologically, it traces the evolution from self-report scales to advanced multi-method approaches, including Ecological Momentary Assessment (EMA), sensor-based behavioral and physiological tracking, and cross-cultural validation techniques using Item Response

Theory (IRT) and Structural Equation Modeling (SEM). The chapter concludes by addressing current validity challenges and emerging opportunities—such as system-level modeling, big data, and GeoAI—while underscoring the importance of ethical and culturally informed measurement practices.

Chapter 4 provides a comprehensive theoretical and conceptual foundation for understanding the multidimensional determinants of subjective well-being (SWB), tracing the evolution of research from early reductionist models toward integrative, dynamic frameworks that incorporate biological, psychological, social, and environmental factors. It introduces a layered architecture encompassing endogenous traits, interpersonal dynamics, and situational contexts, while also addressing macro-level cultural and institutional influences. The chapter concludes by emphasizing the need for operationalizable models and interdisciplinary approaches, setting the stage for a deeper empirical investigation of the built environment in high-density urban settings.

Chapter 5 narrows the focus to the environmental dimensions of SWB, offering a detailed synthesis of how natural, sociocultural, and built environments interact to influence well-being through behavioral, perceptual, and exposure pathways. It critically examines theories such as Attention Restoration Theory and the Biophilia Hypothesis, explores contextual moderators like cultural values and income inequality, and highlights empirical challenges and trade-offs—such as the "Compact City Paradox." The chapter positions the built environment as a key intervention point for urban planning and calls for nuanced, context-sensitive research to inform future policy and design strategies.

Chapter 6 outlines the research design, data collection methods, and analytical tools employed in the study. It begins with a justification for selecting Guangzhou's central urban area as the research context, highlighting its high-density characteristics and socio-spatial diversity. The chapter details the questionnaire design, covering constructs such as life satisfaction, perceived environmental quality, social capital, and socioeconomic attributes. It also explains the multi-stage stratified sampling procedure, pilot testing, and reliability/validity analyses of the research instruments. Finally, it introduces the mathematical and analytical methods, including regression analysis, structural equation modeling (SEM), and mediation testing techniques.

Chapter 7 provides a comprehensive descriptive analysis of the sample characteristics and key variables. It examines the socioeconomic profile of respondents, life satisfaction levels, and perceptions of the built environment, safety, social capital, and health. The chapter also explores differences in life satisfaction across communities of varying construction eras and presents visual and statistical analyses to illustrate initial patterns and relationships among variables.

Chapter 8 investigates the relationship between objective built environment elements and life satisfaction using geographic visualization, correlation analysis, and multiple regression models. It tests hypotheses related to the "5D" dimensions of the built environment (density, diversity, design, distance to transit, and destination accessibility) and their impact on residents' life satisfaction. The results reveal both positive and negative influences of specific built environment features, setting the stage for deeper mechanistic analysis.

Chapter 9 employs structural equation modeling (SEM) to uncover the mediating pathways through which the objective built environment influences life satisfaction. It tests a series of hypotheses involving perceived environmental quality, safety perception, social capital, and self-rated health as potential mediators. The chapter reports on the measurement and structural models, model fit indices, and the results of bootstrap-mediated effect tests, providing a nuanced understanding of the indirect effects and causal mechanisms underpinning the built environment–well-being relationship.

Chapter 10 provides a comprehensive synthesis and theoretical integration of the research findings on how the built environment influences residents' life satisfaction in high-density urban contexts, using Guangzhou as a case study. It elaborates on both the direct and indirect pathways—through mediators such as perceived environmental quality, Perceived Safety, social capital, and self-rated health—by which objective built environment factors affect subjective well-being. The chapter also discusses the nuanced and sometimes counterintuitive role of density, land-use mix, and transit accessibility, challenging the direct applicability of Western-derived compact city theories in high-density Chinese settings. Finally, it translates these insights into context-sensitive policy recommendations aimed at enhancing urban livability and residents' quality of life.

Chapter 11 concludes the study by summarizing key research limitations—including geographical specificity, methodological constraints, and causal inference challenges—and outlining meaningful future directions for both academic inquiry and urban policy. It emphasizes the need for multi-scale, longitudinal, and comparative studies to improve generalizability, and advocates for advanced methodologies such as machine learning, spatial analysis, and quasi-experimental designs to better capture complex built environment–well-being relationships. The chapter also highlights culturally and institutionally informed policy implications across national, provincial, and municipal levels, calling for interdisciplinary collaboration and equitable urban governance to support sustainable and people-centered urban development.

**Funding** This research was funded by

1. Guangdong Provincial Office of Philosophy and Social Sciences [Grant No. GD24XSH06]
2. Projects of PhDs' Start-up Research of GDUPT [Grant No. 2023bsqd1008]

## References

Angel, S. (2021). *Planet of cities*. Lincoln Institute of Land Policy.

Barton, H. (2017). *City of well-being: A radical guide to planning*. Routledge.

Cao, X., Mokhtarian, P. L., & Handy, S. L. (2009). Examining the impacts of residential self-selection on travel behavior: A focus on empirical findings. *Transport Reviews, 29*(3), 359–395.

Central People's Government of the PRC. (2015). *Central urban work conference concluded in Beijing*. Retrieved from http://www.gov.cn/.

Cervero, R., & Kockelman, K. (1997). Travel demand and the 3Ds: Density, diversity, and design. *Transportation Research Part D: Transport and Environment, 2*(3), 199–219.

Chen, J., & Chen, S. (2015). Mental health effects of perceived living environment and neighborhood safety in urbanizing China. *Habitat International, 46*, 101–110.

Chen, J., Zhou, C., Wang, S., & Li, S. (2018). Impacts of energy consumption structure, energy intensity, economic growth, urbanization on PM2.5 concentrations in countries globally. *Applied Energy, 230*, 94–105.

Churchman, A. (1999). Disentangling the concept of density. *Journal of Planning Literature, 13*(4), 389–411.

CPC Central Committee. (2020). *Proposal for the 14th Five-Year Plan for National Economic and Social Development and the Long-Range Objectives for 2035*. People's Publishing House.

CPC Central Committee. (2022). *Report to the 20th National Congress of the Communist Party of China*. People's Publishing House.

Cramm, J. M., van Dijk, H. M., & Nieboer, A. P. (2013). The importance of neighborhood social cohesion and social capital for the well being of older adults in the community. *The Gerontologist, 53*(1), 142–152.

Das, D. (2020). Exploring the role of neighborhood perceptions in measuring subjective well-being: A case study of Kolkata. *Social Indicators Research, 147*(2), 517–543.

De Vos, J., Schwanen, T., Van Acker, V., & Witlox, F. (2013). Travel and subjective well-being: A focus on findings, methods and future research needs. *Transport Reviews, 33*(4), 421–442.

Diener, E. (1984). Subjective well-being. *Psychological Bulletin, 95*(3), 542–575.

Diener, E. (2009). Subjective well-being. In E. Diener (Ed.), *The science of well-being: The collected works of Ed Diener* (pp. 11–58). Springer.

Diener, E., Emmons, R. A., Larsen, R. J., & Griffin, S. (1985). The satisfaction with life scale. *Journal of Personality Assessment, 49*(1), 71–75.

Diener, E., Lucas, R. E., & Oishi, S. (2018). Advances and open questions in the science of subjective well-being. *Collabra: Psychology, 4*(1), 15.

Diener, E., Oishi, S., & Lucas, R. E. (2003). Personality, culture, and subjective well-being: Emotional and cognitive evaluations of life. *Annual Review of Psychology, 54*, 403–425.

Diener, E., Oishi, S., & Tay, L. (2018). Advances in subjective well-being research. *Nature Human Behaviour, 2*(4), 253–260.

Diener, E., & Seligman, M. E. P. (2002). Very happy people. *Psychological Science, 13*(1), 81–84.

Diener, E., & Suh, E. M. (1997). Measuring quality of life: Economic, social, and subjective indicators. *Social Indicators Research, 40*(1-2), 189–216.

Dovey, K., & Pafka, E. (2020). What is walkability? The urban DMA. *Urban Studies, 57*(1), 93–108.

Durand, M. (2015). The OECD better life initiative: How's life? and the measurement of well-being. *Review of Income and Wealth, 61*(1), 4–17.

Easterlin, R. A. (1974). Does economic growth improve the human lot? Some empirical evidence. In P. A. David & M. W. Reder (Eds.), *Nations and households in economic growth: Essays in honor of Moses Abramovitz* (pp. 89–125). Academic Press.

Ettema, D., Gärling, T., Olsson, L. E., & Friman, M. (2010). Out-of-home activities, daily travel, and subjective well-being. *Transportation Research Part A: Policy and Practice, 44*(9), 723–732.

Ewing, R., & Cervero, R. (2010). Travel and the built environment: A meta-analysis. *Journal of the American Planning Association, 76*(3), 265–294.

Ewing, R., & Handy, S. (2009). Measuring the unmeasurable: Urban design qualities related to walkability. *Journal of Urban Design, 14*(1), 65–84.

Fernández, R., Aragonés, J. I., & Bazana, S. (2017). The relationship between life satisfaction and environmental satisfaction: A longitudinal analysis. *Journal of Environmental Psychology, 49*, 135–143.

Ferrer-i-Carbonell, A., & Gowdy, J. M. (2007). Environmental degradation and happiness. *Ecological Economics, 60*(3), 509–516.
Frank, L. D., Sallis, J. F., Conway, T. L., Chapman, J. E., Saelens, B. E., & Bachman, W. (2006). Many pathways from land use to health: Associations between neighborhood walkability and active transportation, body mass index, and air quality. *Journal of the American Planning Association, 72*(1), 75–87.
Fung, T. K. F., Hui, E. C. M., & Chen, J. (2022). The effects of perceived overcrowding and noise on the mental health of living in high-density households: Evidence from Hong Kong. *Habitat International, 119*, 102477.
Glaeser, E. L. (2011). *Triumph of the city: How our greatest invention makes us richer, smarter, greener, healthier, and happier*. Penguin Press.
Handy, S., Boarnet, M. G., Ewing, R., & Killingsworth, R. E. (2002). How the built environment affects physical activity: Views from urban planning. *American Journal of Preventive Medicine, 23*(2), 64–73.
Helliwell, J. F., Layard, R., Sachs, J. D., & De Neve, J.-E. (Eds.). (2020). *World happiness report 2020*. Sustainable Development Solutions Network.
Helliwell, J. F., Layard, R., Sachs, J. D., De Neve, J.-E., Aknin, L. B., & Wang, S. (Eds.). (2022). *World happiness report 2022*. Sustainable Development Solutions Network.
Huang, Z., Zhang, X., & Chen, J. (2020). The unequal effects of the built environment on the subjective well-being of different communities: A case study of Guangzhou, China. *Cities, 107*, 102935.
Jacobs, J. (1961). *The death and life of great American cities*. Random House.
Jenks, M., & Burgess, R. (Eds.). (2000). *Compact cities: Sustainable urban forms for developing countries*. Spon Press.
Karademas, E. C. (2007). Positive and negative aspects of well-being: Common and specific predictors. *Personality and Individual Differences, 43*(2), 277–287.
Keyes, C. L. M., Shmotkin, D., & Ryff, C. D. (2002). Optimizing well-being: The empirical encounter of two traditions. *Journal of Personality and Social Psychology, 82*(6), 1007–1022.
Ko, Y. C., & Kuo, H. C. (2009). Can nature experience enhance subjective well-being? A meta-analysis. *Journal of Happiness Studies, 10*(5), 523–537.
Leyden, K. M. (2003). Social capital and the built environment: The importance of walkable neighborhoods. *American Journal of Public Health, 93*(9), 1546–1551.
Li, H., & Shi, Y. (2017). The Easterlin paradox in China. In *Life satisfaction in China and its economic and social determinants* (pp. 1–14). Springer.
Li, X., & Liu, Y. (2019). *High-density city and mental health*. Springer.
Li, X., & Ye, Y. (2015). *Urbanization in China: Challenges and strategies*. Routledge.
Li, Z., Wu, F., & Li, Y. (2020). The multi-scale built environment and mental health: A review. *Progress in Geography, 39*(5), 866–875.
Librán, E. C. (2006). Personality dimensions and subjective well-being. *The Spanish Journal of Psychology, 9*(1), 38–44.
Lin, J., & Sun, B. (2017). Impact of built environment on urban residents' subjective well-being: Evidence from the China Labor-Force Dynamics Survey. *Urban Development Studies, 24*(12), 69–75.
Lin, S., & Yang, D. (2015). Built environment and public health: An integrated framework. *Chinese Journal of Urban and Environmental Studies, 3*(2), 1550013.
Liu, Y., Zhang, F., Wu, F., Liu, Y., & Li, Z. (2020). The subjective well-being of migrants in Guangzhou, China: The impacts of the social and physical environment. *Cities, 98*, 102577.
Luttik, J. (2000). The value of trees, water and open space as reflected by house prices in the Netherlands. *Landscape and Urban Planning, 48*(3–4), 161–167.
Maslow, A. H. (1943). A theory of human motivation. *Psychological Review, 50*(4), 370.
Ministry of Ecology and Environment. (2023). *Guidelines on ecological restoration and community regeneration in urban renewal*. China Environment Press.

Ministry of Housing and Urban-Rural Development. (2015). *Guidelines on promoting multiple plans integration and public participation in urban planning*. China Architecture & Building Press.
Mouratidis, K. (2018). Built environment and social well-being: How does urban form affect social life and personal relationships? *Cities, 74*, 7–20.
Mouratidis, K. (2020). Compact city, urban sprawl, and subjective well-being. *Cities, 102*, 102728.
Mouratidis, K. (2021). Urban planning and quality of life: A review of pathways linking the built environment to subjective well-being. *Cities, 115*, 103229.
Myers, D. G. (1988). *The pursuit of happiness*. Harper & Row.
National Bureau of Statistics of China. (2023). *National Bureau of Statistics announcement on the final verification of the 2022 Gross Domestic Product*. China Statistics Press. Retrieved from https://www.stats.gov.cn/sj/zxfb/202312/t20231229_1946058.html
Neuman, M. (2005). The compact city fallacy. *Journal of Planning Education and Research, 25*(1), 11–26.
Nordbakke, S., & Schwanen, T. (2013). Well-being and mobility: A theoretical framework and literature review focusing on older people. *Mobilities, 9*(1), 104–129.
Pavot, W., & Diener, E. (2008). The satisfaction with life scale and the emerging construct of life satisfaction. *The Journal of Positive Psychology, 3*(2), 137–152.
Peen, J., Schoevers, R. A., Beekman, A. T., & Dekker, J. (2010). The current status of urban-rural differences in psychiatric disorders. *Acta Psychiatrica Scandinavica, 121*(2), 84–93.
Pfeffer, M. J., & Cloutier, D. S. (2016). Planning for happy neighborhoods. *Journal of the American Planning Association, 82*(3), 267–279.
Putnam, R. D. (2000). *Bowling alone: The collapse and revival of American community*. Simon & Schuster.
Ryan, R. M., & Deci, E. L. (2001). On happiness and human potentials: A review of research on hedonic and eudaimonic well-being. *Annual Review of Psychology, 52*, 141–166.
Ryff, C. D., & Keyes, C. L. M. (1995). The structure of psychological well-being revisited. *Journal of Personality and Social Psychology, 69*(4), 719–727.
Sallis, J. F., Cerin, E., Conway, T. L., Adams, M. A., Frank, L. D., Pratt, M., Salvo, D., Schipperijn, J., Smith, G., Cain, K. L., Davey, R., Kerr, J., Lai, P. C., Mitas, J., Reis, R., Sarmiento, O. L., Schofield, G., Troelsen, J., Van Dyck, D., … Owen, N. (2016). Physical activity in relation to urban environments in 14 cities worldwide: A cross-sectional study. *The Lancet, 387*(10034), 2207–2217.
Seligman, M. E. P., & Csikszentmihalyi, M. (2000). Positive psychology: An introduction. *American Psychologist, 55*(1), 5–14.
Shen, Y., & Fu, X. (2019). Impact of the built environment and perceived neighborhood characteristics on residents' satisfaction: Evidence from the Shanghai suburbs. *Progress in Geography, 38*(5), 686–697.
State Council. (2015). *National new urbanization plan (2014–2020)*. People's Publishing House.
State Council. (2025). *Communiqué of the central urban work conference 2025*. People's Publishing House.
Stokols, D. (1972). On the distinction between density and crowding: Some implications for future research. *Psychological Review, 79*(3), 275–277.
Sugiyama, T., Leslie, E., Giles-Corti, B., & Owen, N. (2008). Associations of neighbourhood greenness with physical and mental health: Do walking, social coherence and local social interaction explain the relationships? *Journal of Epidemiology & Community Health, 62*(5), e9.
The 18th National Congress of the Communist Party of China. (2012). *Full text of the report to the 18th National Congress of the CPC*. People's Publishing House.
The 19th National Congress of the Communist Party of China. (2017). *Full text of the report to the 19th National Congress of the CPC*. People's Publishing House.
The 20th National Congress of the Communist Party of China. (2022). *Full text of the report to the 20th National Congress of the CPC*. People's Publishing House.

Ulrich, R. S., Simons, R. F., Losito, B. D., Fiorito, E., Miles, M. A., & Zelson, M. (1991). Stress recovery during exposure to natural and urban environments. *Journal of Environmental Psychology, 11*(3), 201–230.

United Nations, Department of Economic and Social Affairs, Population Division. (2019). *World urbanization prospects: The 2018 revision*. United Nations.

Wang, J., Kwan, M. P., & Chai, Y. (2020). The mediating effect of neighborhood environments on the relationship between objective built environment and subjective well-being: Evidence from Beijing, China. *Landscape and Urban Planning, 201*, 103843.

Wang, L. (2010). *High-density urban form and sustainable development*. China Architecture & Building Press.

Whyte, W. H. (1980). *The social life of small urban spaces*. Project for Public Spaces.

Winters, J. V., & Li, Y. (2017). Urbanisation, natural amenities and subjective well-being: Evidence from US counties. *Urban Studies, 54*(8), 1956–1973.

World Bank. (2014). *Urban China: Toward efficient, inclusive, and sustainable urbanization*. World Bank Group.

Wu, W., & Gaubatz, P. (2013). *The Chinese city*. Routledge.

Xinhua News Agency. (2024a). *Communiqué of the third plenary session of the 20th Central Committee of the Communist Party of China*. Retrieved from http://www.xinhuanet.com/.

Xinhua News Agency. (2024b). *Third plenary session of the 20th CPC Central Committee highlights quality of life and beautiful China initiative*. Retrieved from http://www.xinhuanet.com/.

Yeh, A. G. O., & Li, X. (2020). *China's urbanization and its ecological and resource challenges*. Springer.

Yu, W., & Hu, Y. (2017). A review of built environment measurement methods: From "3D" to "5D" and beyond. *Urban Planning International, 32*(3), 64–71.

Zhang, M., & Chen, S. (2020). The formation and development of high-density urban spaces in China: A case study of Shanghai. *Urban Development Studies, 27*(1), 1–8.

# Chapter 2
# Understanding Well-Being: An Interdisciplinary Synthesis

**Abstract** Following the foundational overview presented in Chap. 1, this chapter offers an interdisciplinary synthesis of the conceptual and empirical underpinnings of well-being. It aims to trace the philosophical and historical origins of happiness and subjective well-being (SWB), while systematically examining its multidimensional nature through biological-evolutionary, psychological, and sociocultural lenses. The chapter begins by exploring ancient Eastern and Western philosophical traditions—including Aristotelian eudaimonia, Epicurean hedonism, Stoicism, Confucianism, Daoism, and Chinese Buddhism—to establish the diverse theoretical roots of contemporary well-being research. Section 2.2 investigates the biological and evolutionary bases of SWB, incorporating genetic, neurobiological, and physiological perspectives to elucidate the mechanisms underpinning positive affect and resilience. Section 2.3 reviews key psychological models and theoretical advances, from early affect-based paradigms to integrated frameworks such as Ryff's psychological well-being model, Self-Determination Theory, and the PERMA model. Finally, Sect. 2.4 adopts a sociocultural perspective, analyzing how cultural dimensions, social structures, and globalization shape conceptions and experiences of happiness across contexts. By integrating multidisciplinary insights, this chapter provides a comprehensive theoretical foundation for understanding well-being as a complex, dynamic construct, setting the stage for the applied and extended discussions in subsequent chapters.

**Keywords** Subjective well-being (SWB) · Eudaimonia · Hedonia · Psychological well-being · Self-determination theory · Broaden-and-build theory · Cultural dimensions · Bio-psycho-social model · Happiness philosophy · Well-being measurement

Building upon the overview of the book's background, core concepts, and structure presented in Chap. 1, this chapter delves into the historical and philosophical origins


H. Li, G. Shen, *The Anatomy of Subjective Well-Being*, Resilient and Friendly Living Environments, https://doi.org/10.1007/978-981-95-4937-5_2

of happiness and subjective well-being, as well as the multidimensional perspectives on well-being—namely, the biological-evolutionary, psychological, and sociocultural perspectives. By tracing the conceptual evolution from ancient Eastern and Western thought to the influence of the Enlightenment on modern scientific approaches, along with examining the multidimensional aspects of well-being, this chapter provides a comprehensive theoretical foundation for the interdisciplinary discussions to be expanded in subsequent chapters.

## 2.1 The Origin of Happiness and the Theoretical Basis of Subjective Well-Being

### *2.1.1 Philosophical Foundations of Happiness and Subjective Well-Being in Ancient Western Thought*

#### 2.1.1.1 Aristotelian Eudaimonia: Flourishing and Virtue

Happiness is the eternal pursuit of human beings and the eternal proposition of human society. Although happiness has not formed a unified standard definition, people's pursuit and expectation of happiness have not changed, and scholars' research on happiness has been continuously promoted. Voltaire emphasized the significance of happiness: "Happiness is the only thing in life worth having" (Tatarkiewicz, 1976), echoing Aristotle and summarizing the essence of the eudaimonist axiom. Aristotle advocated that "Happiness is desired by all human beings as the ultimate end (telos) of all their rational acts" (Aristotle, c. 350 BCE/2014, 1094a). Socrates went on to argue that merely asking the question is a silly thing to do, "since it would be absurd to deny it", implying its inherent value (Annas, 2002). The concept of eudaimonia, as elucidated by Nussbaum (2001), underscores the fundamental question of how one should lead their life, with the answer lying in "one's own conception of eudaimonia" (Nussbaum, 2001). These philosophical perspectives present overarching ideas of a flourishing life. However, the practical application of philosophical eudaimonism falls short in providing a detailed understanding of happiness or the ultimate goal of life. Aristotle recognized the limitations of eudaimonia, acknowledging the diverse interpretations of happiness among individuals (Annas, 2002; Aristotle c. 350 BCE/2014).

Aristotle's conception of eudaimonia, a cornerstone of ancient Western philosophical foundations of happiness, is defined in the "Nicomachean Ethics" (c. 350 BCE/2014) as the highest good achievable through human action, characterized not by transient states like pleasure, honor, or wealth, but rather as "activities of the soul in accordance with virtue". For Aristotle, virtue entails striving for the mean—avoiding excess and deficiency in actions—and involves realizing one's unique talents and capacities to actualize one's true nature, embodying the Greek imperatives "know thyself" and "become what you are" (Ruini & Ryff, 2016; Ryff & Singer,

2008). This framework positions eudaimonia as a dynamic process of growth rather than a static state, a perspective echoed in contemporary psychology, which conceptualizes eudaimonia as a way of life centered on developing talents, practicing virtues, and deriving meaning from personal growth. Eudaimonic living often requires engaging in effortful, even challenging activities that may generate short-term negative affect but contribute to greater long-term well-being, while virtues such as bravery and diligence, cultivated through awareness and practice, enable the development of good character over time. Inspired by Aristotle's emphasis on virtues as defining "the good life," Peterson and Seligman (2004) categorized cross-culturally consistent character strengths and virtues facilitating thriving. Psychological theories like Ryff's model of psychological well-being (Ryff, 1989; Ryff & Keyes, 1995), Self-Determination Theory (Deci & Ryan, 2000; Ryan & Deci, 2000), and Seligman's PERMA model (Seligman, 2011) operationalize eudaimonia by emphasizing flourishing through virtue and potential realization (Lambert, D'raven, & Pasha-Zaidi, 2015).

Distinguishing happiness from well-being, Raibley (2012) conceptualizes well-being as "agential flourishing," wherein individuals successfully realize their deeply held values and maintain stable dispositions to do so. This framework differentiates episodic happiness (momentary feelings) and dispositional happiness (stable positive responses to life events) from well-being, arguing the former are necessary but insufficient for flourishing—exemplified by cases like Margo, an Alzheimer's patient with high episodic happiness but not flourishing qua agent (Raibley, 2012). Aligned with eudaimonic traditions, Raibley's (2012) model underscores the active pursuit of values and virtuous living as foundational to psychological well-being.

#### 2.1.1.2 Epicurean Hedonism: Pleasure as Happiness

Epicurus (341–270 BCE) founded the school of philosophy known as Epicureanism, which sought to attain a happy, tranquil life characterized by peace and freedom from fear and pain, with living a self-sufficient life surrounded by friends being part of his view (Ryff et al., 2021). Central to this philosophical tradition is the emphasis on pleasure maximization and pain avoidance as the core components of happiness, a perspective that aligns with classical philosophical hedonism (O'Keefe, 2010), which identifies these elements as fundamental to defining the nature of happiness. This orientation toward happiness is captured in the concept of hedonia, defined as the pursuit of what is subjectively pleasant, involving the maximization of pleasure and the avoidance of pain—characteristics that directly reflect Epicurean hedonism's focus on these principles as central to well-being. Hedonia, in this context, is further characterized by meeting visceral and emotional needs such as rest and positive mental states, charging psychological batteries through pleasurable experiences and memories, and serving as a necessary aspect of self-care—all of which underscore the practical importance of pleasure and pain avoidance in Epicurean thought (Huta, 2016).

Epicurean hedonism is conceptually distinct from other ancient philosophical frameworks, such as Aristotle's eudaimonia, which centers on the activity of the soul in accord with virtue, and Aristippus's focus on maximizing pleasure as the totality of hedonic moments (O'Keefe, 2014). Hedonia, as rooted in Epicurean ideas, is often discussed in terms of subjective experiences of pleasure and discomfort, with a more short-term focus on supporting immediate well-being, whereas eudaimonia tends to be framed in terms of orientations or functioning linked to long-term fulfillment (Huta & Waterman, 2014). This distinction has persisted in contemporary discourse, where hedonic well-being constructs—such as life satisfaction, positive affect, and pain avoidance (Diener, 1984)—trace their philosophical roots back to Epicurean hedonism (Ryff et al., 2021), highlighting its enduring influence as a foundation for understanding subjective well-being.

#### 2.1.1.3 Stoicism and Related Ancient Views

Stoic philosophy (Epictetus c. 100 CE/1983; Irvine, 2009) posits that happiness, or eudaimonia, is fundamentally rooted in the cultivation of virtue and the rational acceptance of fate, rather than being dependent on external circumstances alone. Central to this conception is the idea that virtue—encompassing qualities such as wisdom, courage, justice, and temperance—enables individuals to navigate life's challenges with equanimity, as virtue is deemed within one's control, unlike external fortunes. Rational acceptance (Epictetus, c. 100 CE/1983), a core Stoic practice, involves distinguishing between what is "up to us" (i.e., our judgments and actions) and what is not (i.e., external events), with happiness arising from aligning one's will with nature and accepting the latter without distress.

Related classical viewpoints on happiness and well-being, such as Aristotle's conception of eudaimonia (Aristotle c. 350 BCE/2000; Ryff, 1989; Ryff & Keyes, 1995), share thematic connections with Stoic thought while introducing distinct nuances. Aristotle defined eudaimonia not as a transient emotional state but as a flourishing way of life in which individuals strive to realize their potential by exercising their talents and virtues to develop themselves and create meaning. This eudaimonic activity often involves effortful and sometimes difficult pursuits that contribute to greater overall well-being in the long term, though it may occasionally require hedonic (pleasure-based) activity for rejuvenation (Huta & Ryan, 2010; Waterman, 1993).

Contemporary theories further reflect these ancient perspectives. For instance, Self-Determination Theory (SDT) (Deci & Ryan, 2000; Ryan & Deci, 2000), a modern eudaimonic framework, posits that optimal well-being arises when three innate psychological needs—competence, autonomy, and relatedness—are met through intrinsic goal pursuits, aligning with the Stoic focus on rational acceptance and virtue as pathways to happiness. Similarly, Seligman's Authentic Happiness model and PERMA framework (Seligman, 2002, 2011) integrate virtue ethics by emphasizing engagement through character strengths and virtues, which facilitate thriving and happiness, reflecting classical Stoic conceptions of happiness through virtue and rational acceptance; components such as meaning and positive relationships further incorporate eudaimonic and utilitarian traditions. Importantly, the recognition that both hedonic (pleasure-based) and eudaimonic (virtue and rational acceptance-based) perspectives are essential and complementary to a comprehensive understanding of happiness (Huta, 2013; Huta & Waterman, 2014) echoes Stoic ideas that happiness results from virtue and the rational acceptance of fate rather than external circumstances alone (Lambert et al., 2015).

#### 2.1.1.4 Other Famous Ancient Western Ideas

Other famous ancient Western ideas related to the philosophical foundations of happiness and subjective well-being encompass several key traditions and their subsequent developments. Rooted in ancient virtue philosophies, the virtue philosophical tradition (Aristotle, c. 350 BCE/2014; Peterson & Seligman, 2004) posits that individual character strengths (e.g., bravery, diligence) can be cultivated via awareness and effort to develop good character and sustain happiness, drawing inspiration from Aristotle's concept of virtues that epitomize a good life. Central to these ancient orientations are the hedonic (Diener, 1984; Epicurus, c. 341–270 BCE) and eudaimonic (Aristotle, c. 350 BCE/2014; Ryff, 1989) philosophies, which form foundational distinctions in well-being theories: hedonia pertains to the pursuit of pleasure and positive emotions ("feeling good"), while eudaimonia, derived from Aristotle's notion of a way of life grounded in realizing potential through talents and virtues, refers to living well by striving for meaningful goals ("functioning well") (Huta & Waterman, 2014; Waterman, 1993); both are distinct but complementary and crucial to understanding happiness (Ryan & Deci, 2001).

The utilitarian philosophical tradition (Bentham, 1789/1996; Mill, 1861/2001), though promoted by modern philosophers like Jeremy Bentham and John Stuart Mill, emphasizes "the greatest happiness for the greatest number," reflecting collective well-being considerations that connect to broader inquiries into happiness. These ancient ideas have influenced later humanistic theories, which build on eudaimonic philosophy, with theorists such as Adler (1931, 1956), Maslow (1943), Rogers (1963), and Csikszentmihalyi (1990) have contributed models that focus on self-actualization, optimal experience (flow), and full-functioning individuals aimed at a meaningful life. Their work highlights autonomy, personal growth, and peak positive experiences. Contemporary hybrid models, like Seligman's authentic happiness and PERMA frameworks (Seligman, 2002, 2011), integrate these ancient philosophical ideas (hedonia, eudaimonia, virtue, utilitarianism) into psychological models describing pathways to happiness, including pleasure, engagement, meaning, positive relationships, and accomplishment. Keyes' concept of flourishing (Keyes, 2002, 2007) further conceptualizes complete mental health by combining elements of hedonic, eudaimonic, virtue, and utilitarian traditions, underscoring social well-being as integral to happiness. Additionally, Existential Positive

**Table 2.1** Comparison of the three major western philosophies of happiness

| Philosophy | Core idea | Key proponents | Modern influence |
|---|---|---|---|
| Eudaimonia | Flourishing through virtue | Aristotle | Ryff's PWB, SDT, PERMA |
| Hedonism | Maximize pleasure, minimize pain | Epicurus | Diener's SWB, Positive Affect |
| Stoicism | Virtue + rational acceptance of fate | Epictetus, Seneca | CBT, Resilience training |

Source: Created by the authors, synthesizing philosophical concepts from Aristotle, Epicurus, and the Stoics as interpreted in contemporary psychology

Psychology (PP2.0) (Wong, 2011, 2016) extends classical ancient traditions by incorporating existential concerns (meaning, isolation, freedom, death) and expands the philosophy of happiness to include spiritual happiness alongside utilitarian, virtue, hedonic, and eudaimonic traditions (Table 2.1).

#### 2.1.1.5 Eastern Philosophical and Religious Perspectives on Happiness

Happiness is an eternal value pursuit by humankind, and interpretations of its essence across different civilizations exhibit distinct cultural characteristics. Within the Western philosophical tradition, happiness is often associated with "Happiness" (understood as pleasure or contentment), ranging from Epicurus's hedonism to Bentham's utilitarianism (Kenny, 2006; McMahon, 2006), primarily centering on sensory satisfaction or utility maximization (Li, 2015). In contrast, Eastern philosophy, particularly the Confucian, Daoist, and Sinicized Buddhist traditions originating from pre-Qin China onward, presents an alternative pathway to understanding happiness (Li, 2013; Yu, 2007). These traditions do not perceive happiness as a static "outcome," but rather as a dynamic "state of being" or "spiritual realm" (jingjie). Its core lies in achieving transcendence of the inner spirit through cultivation of the mind and character ( xinxing xiuyang) (Shun & Wong, 2004; Tu, 1985).

This "realm-oriented" (jingjielun) conception of happiness possesses three key characteristics: First, its anti-utilitarian nature. It rejects the decisive role of material conditions in happiness. A quintessential example is Confucius's praise of his disciple Yan Hui, who found joy amidst living with "a single bamboo dish of rice, a single gourd dipper of drink" (Confucius, c. 500 BCE/1979; Ivanhoe, 2000; "The Analects," Yongye 6), demonstrating that contentment was independent of material wealth (Shun, 2004). Second, its relational essence. It emphasizes that happiness arises from harmonious relationships—between heaven and humanity (tianren), within human ethics (renlun), and between the individual and the collective (qunji)—rather than being an isolated individual experience (Ames & Rosemont, 2010; Bell, 2008; Li, 2006). Third, it's a practical approach. It asserts that happiness must be attained through practical "cultivation efforts" such as moral practice (daode jianlü), mind-heart cultivation (xinxing xiulian), Confucian "self-cultivation" (xiushen), Daoist "fasting of the mind" (xinzhai), or Buddhist "meditative concentration" (chanding) (Fig. 2.1) (Kohn, 2004; Sharf, 2015; Slingerland, 2007; Tu, 1979; Yearley, 1990).

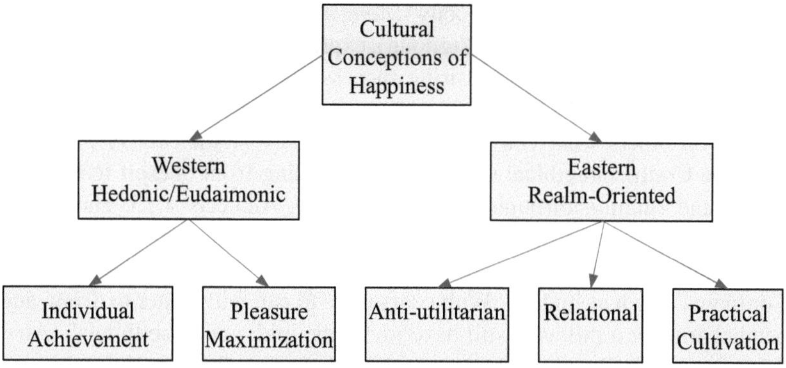


**Fig. 2.1** Comparative philosophies of happiness: eastern vs. western perspectives. Source: Created by the authors based on a synthesis of Eastern (Confucian, Daoist, Buddhist) and Western (Hedonic, Eudaimonic) philosophical texts

## *2.1.2 Confucianism: Virtuous Happiness Within Ethical Order*

Confucianism, centered on the core concept of "benevolence" (*ren*), roots happiness in ethical practice and social order, asserting that "joy" (*le*) is the natural manifestation of a fulfilled virtue. Its conceptual evolution can be divided into two major stages: the "Pre-Qin Foundations" and the "Song-Ming Elaboration."

### 2.1.2.1 Pre-Qin Confucianism: Founding the "Sense of Joy" Within the Ren Doctrine

Confucius: The Unity of Ren and Le

Confucius first explicitly linked "happiness" (*xingfu*) with "morality" (*daode*), proclaiming that "the benevolent person is free from anxiety" (Analects 9.29; Legge, 2009; Slingerland, 2007). For him, the essence of happiness lies in inner moral self-sufficiency. Its realization involves:

1. Self-Discipline and Returning to Ritual (*keji fuli*): Restraining excessive desires through ritual (*li*) norms leads to the state of freedom where one "follows what the heart/mind desires without overstepping the boundaries" (Analects 2.4; Ames & Rosemont, 2010). Ritual functions not merely as an external constraint but as the externalization of inner virtue (Ames & Rosemont, 2010), as in "To subdue oneself and return to ritual is benevolence" (Analects 12.1; Chan, 1963). Only through the practice of "looking, listening, speaking, and acting only according to ritual" can the character of ren be cultivated, resulting in peace of mind.

2. The Way of Loyalty and Reciprocity (*zhongshu zhi dao*): Validating worth within human relationships. "Loyalty" (zhong), expressed as "desiring to establish oneself, one establishes others; desiring success for oneself, one helps others succeed" (Analects 6.30; Lau, 1996), and "reciprocity" (shu), expressed as "Do not impose on others what you yourself do not desire" (Analects 12.2; Tu, 1979), form the Confucian ethical network of "extending from oneself to others" (the one thread running through the Master's way, Analects 4.15; Hall & Ames, 1987). The individual transcends selfish limitations and attains the happiness of "delighting in community" (lequn) through caring for others and serving society. Confucius's own attitude—"With coarse rice to eat, with water to drink, and my bended arm for a pillow; I still have joy in the midst of these things" (Analects 7.16; Brooks & Brooks, 1998)—and his admiration for Yan Hui, who "with a single bamboo dish of rice, a single gourd dipper of drink, living in a mean alley—others could not have endured such distress, yet it did not affect Hui's joy" (Analects 6.11; Ivanhoe, 2000), together constitute the paradigm of "Confucius and Yan Hui's Joy" (Kong-Yan zhi le). This ideal of finding joy amidst material simplicity and through moral cultivation has been a central topic in Confucian scholarship. This demonstrates that material lack does not hinder happiness but rather becomes the backdrop against which spiritual transcendence shines.

#### Mencius: Theory of Innate Goodness and "All Things Are Complete Within Me"

Mencius inherited Confucius's thought on ren and le, grounding the conception of happiness ontologically in the "theory of innate goodness" (xingshan lun). He argued that "human nature's being good is like water's tendency to flow downward" (Mencius 6A:2; Van Norden, 2008), positing that humans innately possess the "Four Beginnings" (siduan): the heart/mind of compassion, shame, deference, and right/wrong. Happiness lies in the "extension and fulfillment" (kuo er chong zhi, Mencius 2A:6; Angle, 2009) of this innate goodness (Ivanhoe, 2002). Key aspects include:

1. The Internal Source of Happiness: "All Things are Complete within Me" (wanwu jie bei yu wo): "All things are complete within me. There is no greater joy than to examine oneself and be sincere" (Mencius 7A:4; Bloom, 2009). "Sincerity" (cheng) signifies the realization of the heart/mind's innate goodness (Angle, 2009; Tu, 1989). When individuals, through introspection, affirm their identity with the goodness of Heaven's Way (tiandao), they experience "great joy" (dale)—a joy independent of external objects, arising from the ontological awakening of "Heaven and Humanity uniting as one" (tianren heyi) (Ivanhoe, 2002; Tu, 1989).
2. The Practical Cultivation of "Flood-like Qi" (yang haoran zhi qi): "I am adept at cultivating my flood-like qi" (Mencius 2A:2; Shun, 1997). This qi is "bound-

lessly vast and unyielding; nourished by uprightness and sustaining no injury, it fills the space between Heaven and Earth." It combines moral courage and spiritual strength (Yearley, 1990). Through persistent practice "born of accumulating righteousness" (ji yi suo sheng), individuals can maintain "unmoved heart/mind" (budongxin)—the stability of happiness—in any circumstance (Yearley, 1990).

3. The Social Dimension: "Sharing Joy with the People" (yu min tong le): Mencius extended individual happiness into the political realm, advocating that a ruler should "rejoice with the world, worry with the world" (Mencius 1B:4; Lau, 1970). Using the example of "King Wen's park," he argued that only by sharing resources and joy with the people could a ruler achieve universal happiness where "above and below flow together with Heaven and Earth" (Mencius 4A:26, Lau, 1970). This vision underscores that the highest happiness is interdependent and realized within a harmonious socio-political order aligned with the cosmic pattern.

#### 2.1.2.2 Song-Ming Neo-Confucianism: From Ethical Practice to Cosmic Realm

Song-Ming Neo-Confucians, integrating Buddhist and Daoist ideas, elevated the Confucian conception of happiness from the ethical level to the level of cosmic ontology, forming a system where "Heaven's Way and human nature/life-force are interconnected" (tiandao xingming xiang guantong).

##### Cheng Hao: The Great Joy of "The Benevolent Person Forms One Body with All Things"

Cheng Hao transcended the Pre-Qin focus on human relations, proposing "The benevolent person regards Heaven, Earth, and the myriad things as one body" (Cheng Hao, "On Understanding Ren"; Chan, 1963). He argued that ren is not merely an ethical norm but the very essence or life-force of the cosmos itself—an ontological unity binding all things (Angle, 2009). Humans and all things originate from "Heavenly Principle" (tianli), hence "people are my siblings; things are my companions". The highest state of happiness is transcending the boundaries between self and other, experiencing "great joy" in the awareness of "forming one body with all things" (Tu, 1985): "Joy is the original substance of the heart/mind. Although it differs from the joy of the seven emotions, it is also not external to them" (Er Cheng yishu; Graham, 1992). This joy no longer depends on specific ethical acts but stems from profound concern for the cosmic community of life and embodies the anthropocosmic vision of unity between humanity and Heaven (Angle, 2009; Tu, 1985).

Wang Yangming: The Awakening of Innate Knowing as "Joy Is the Original Substance of the Heart/Mind".

Centering on "the heart/mind is principle" (xin ji li) and "extending innate knowing" (zhi liangzhi), Wang Yangming internalized happiness as an innate attribute of the heart/mind: "Joy is the original substance of the heart/mind. The sage's heart/mind is like a bright mirror, constantly tranquil and constantly illuminating" (Instructions for Practical Living; Chan, 1963). He held that this 'joy' (le) is the inherent, unobscured state of the heart/mind's true substance (xinti), intrinsically linked to the activation of innate knowing (liangzhi) (Ivanhoe, 2002)." (Instructions for Practical Living. He held that this 'joy' (le) is the inherent, unobscured state of the heart/mind's true substance (xinti), intrinsically linked to the activation of innate knowing (liangzhi) (Ivanhoe, 2002). He held that "innate knowing" (liangzhi) is the inherent moral wisdom of the human heart/mind.

Happiness (le) is understood as the essence of liangzhi in its unobscured state, free from selfish desires—manifesting as the affective quality of its spontaneous functioning. The practical path, "extending innate knowing," involves "training in affairs" (shi shang molian), restoring the heart/mind's clarity in daily ethical interactions: "Only by training in affairs can one stand firm, only then can one be tranquil and settled whether in stillness or motion" (Instructions for Practical Living; Ivanhoe, 2002). Through this practical cultivation (gongfu), which removes obstructions to manifest the innate joy of the heart/mind's substance, individuals who reach the state of clear liangzhi experience "self-satisfaction" (zi qie)—an inherent "joy" that is the manifestation of the "heart/mind's original substance" (xinti benran) and requires no external search (Ivanhoe, 2002).

#### 2.1.2.3 Summary of Key Characteristics of the Confucian View of Happiness

Ethical Foundation

Happiness is a byproduct of practicing virtues like ren, righteousness (yi), and ritual (li), not an independent goal. As Confucius states: "The wise find joy in water; the benevolent find joy in mountains" (Analects 6:23; Slingerland, 2007), implying virtue itself constitutes happiness. This aligns with the virtue ethics interpretation of Confucian happiness as "flourishing through moral self-cultivation" (Van Norden, 2007).

Immanence

It rejects placing happiness in an otherworldly realm, advocating for spiritual fulfillment within this life and world. Confucius emphasized: "While you do not yet understand life, how can you understand death?" (Analects 11:12; Ames &

Rosemont, 2010), grounding significance in the here-and-now. This immanent turn reflects the Confucian vision of "ultimate self-transformation as a this-worldly affair" (Tu Weiming, 1985), where transcendence is realized within human existence.

Communality

Individual happiness is intrinsically linked to the well-being of the family, community, and state, forming an expansive logic: "From the Son of Heaven down to the common people, all must regard cultivation of the self as the root" (Great Learning; De Bary & Bloom, 1999). The sequence "Cultivate the self, regulate the family, govern the state, bring peace to all under Heaven" (xiu shen, qi jia, zhi guo, ping tianxia) demonstrates a relational ontology where self-realization is inseparable from social harmony (Ames, 2011). This stands in stark contrast to Western individualism, forming a "role ethics" centered on communal flourishing (Rosemont & Ames, 2016).

#### 2.1.2.4 Daoism: Happiness as Return to Nature and Spiritual Freedom

Daoism, taking the Dao as the supreme reality, understands happiness as a return to "naturalness" (ziran) and transcendence of "artifice" (renwei). Its core lies in "free and easy wandering" (xiaoyao)—a state of spiritual freedom unbounded by external constraints, constituted by aligning with the spontaneous flow of the cosmos (Moeller, 2006). Its intellectual evolution traces a path from "philosophical speculation" to "practical transformation."

Lao-Zhuang Foundational Texts: The "Dao Models Nature" Paradigm of Happiness

*Laozi: Wuwei and "Returning to the Infant"*

Laozi argues that civilizational progress (e.g., the "five colors," "five tones," "five tastes") leads to the alienation of human nature: "When the great Dao is abandoned, benevolence and righteousness appear (ch. 18; Ames & Hall, 2003). When wisdom and cleverness arise, great hypocrisy flourishes" (Ames & Hall, 2003; Daodejing ch. 18; Ivanhoe, 2002). Realizing happiness requires returning to the primordial state of the Dao:

First: Authentic Existence: "Embracing the Uncarved Block" as Epistemological Liberation (jian su bao pu): "Banish sageliness, discard wisdom, and the people will benefit a hundredfold (ch. 19; Ames & Hall, 2003). Banish benevolence, discard righteousness, and the people will return to filial piety and deep love. Banish cleverness, discard profit, and thieves and robbers will disappear" (ch.

19; Lau, 1996). "Simplicity" (su) and the "Uncarved" (pu) signify unadorned naturalness, like an infant whose crying lasts all day without becoming hoarse. The infant symbolizes ziran's epistemic purity: a pre-conceptual mode of being untainted by linguistic categories (Slingerland, 2015). This "cognitive non-doing" (wu zhi) deconstructs artificial hierarchies to restore innate harmony. (ch. 55; Moeller, 2006). This vitality stems from a direct connection with the Dao.

Second: Practical Wisdom: "Non-Action Yet Nothing Left Undone" (wuwei er wu bu wei): Wuwei is not passive inaction but "accordance with naturalness" (shunying ziran)—"Humanity models Earth, Earth models Heaven, Heaven models the Dao, the Dao models naturalness" (ch. 25; Chan, 1963; Ivanhoe, 2002). Wuwei is "non-coercive action aligning with Dao's immanent patterns". It generates efficacy (wu bu wei) through adaptive resonance with environmental feedback loops, exemplified by water's yielding persistence (ch. 8; Moeller, 2006). When individuals relinquish willful interference (wangwei) and follow the inherent nature (ziran) of things, they achieve *wuwei er wu bu wei*, attaining tranquility through harmonious interaction with the world.

Third: Survival Strategy: "The Soft and Weak Overcome the Hard and Strong" (rou ruo sheng gang qiang): "The highest excellence is like water. Water excels in benefiting the myriad creatures without contending". "Softness" and "weakness" (rou ruo) embody the Dao's operation ("Reversal is the movement of the Dao; Weakness is the usage of the Dao," ch. 40; Lynn, 1999). Softness (rou ruo) embodies Dao's cyclical ontology: seeming weakness contains latent potency (Chong, 2017). "Dwelling below" (chu xia) thus becomes a strategic positioning within cosmic dynamics—a path to "enduring contentment" (chang le) (ch. 44; Henricks, 2014). By adopting a stance of "non-contention" (buzheng) and "dwelling below" (chu xia), individuals avoid conflict with the external world and preserve inner peace, as in "He who knows contentment avoids disgrace; he who knows where to stop avoids danger; thus his happiness can long endure" (ch. 44; Henricks, 2014).

*Zhuangzi: Ontological Freedom Through Xiaoyao You (Xiaoyao You)*

Zhuangzi radicalized Laozi's ziran, framing happiness as absolute ontological liberation (Ziporyn, 2009) via xiaoyao.

First: Metaphysical Independence: Wu Dai as Non-Conditioned Being. In the chapter Xiaoyao You, the giant Peng bird, though it "stirs the water for three thousand li and then whirls upward on a whirlwind ninety thousand li," still "depends on the wind beneath it." Liezi, who "rides the wind," is still "dependent on something." The Peng bird's flight, though grand, "remains trapped in the economy of means and ends" (Allinson, 1989, p. 37). True xiaoyao requires wu dai—"unconditioned responsiveness to the self-so (ziran)". True xiaoyao is "non-dependence" (wu dai)—"Mounting the correctness of Heaven and Earth, riding the changes of the six vital breaths (liuqi), thereby wandering in the infinite" (ch.

1; Ziporyn, 2020). This transcends instrumental rationality, dissolving the subject-object dichotomy into cosmic flow (Ivanhoe, 1993).

Second: Cultivation Practices: "Mind-Fasting" (xinzhai) and "Sitting in Forgetfulness" (zuowang). "Mind-Fasting" requires "making your mind empty and waiting for things to come" ("The Human World," ch. 4; Watson, 1968), abandoning subjective prejudices. "Sitting in Forgetfulness" entails "sloughing off your limbs and trunk, blurring your intelligence, detaching from physical form and abandoning knowledge, to become identical with the Great Thoroughfare (datong)" ("The Great Ancestral Teacher," ch. 6; Ziporyn, 2009). "Mind-Fasting" (xinzhai) empties the "conceptual scaffolding" that distorts reality (Hansen, 2003, p. 159), while "Sitting in Forgetfulness" (zuowang) induces a "metacognitive shift" through somatic forgetting (Slingerland, 2019, p. 73). Through this practice of "removing obscurations" (qu bi), individuals reach a state of clarity where "an empty room generates pure light; auspicious blessings settle and abide" ("The Human World," ch. 4), experiencing the aesthetic happiness of "Heaven and Earth were born together with me, and the myriad things and I are one" ("The Sorting Which Evens Things Out," ch. 2; Graham, 2001).

Third: Attitude Toward Life: "Accept It as Fate" (an zhi ruo ming). Facing life's adversities, Zhuangzi advocates "knowing what you cannot do anything about and accepting it as though it were fate" (renming) ("The Human World," ch. 4; Watson, 1968). This "acceptance of fate" is not passive resignation but recognition of the Dao's inevitability. By embracing the finitude of nature and life, individuals shed anxiety and attain the joy of "self-contentment" (zishi). This is not resignation but "active non-resistance" to Dao's immanent unfolding (Yearley, 1990).

### The Practical Evolution of Daoist Thought

#### *Wei-Jin Xuanxue (Neo-Daoism): Ziran as "Existential Resistance"*

During the Wei-Jin period, Xuanxue thinkers like Ji Kang and Ruan Ji critiqued Confucian ritualism (lijiao) for suppressing human nature, reaffirming the Daoist view of happiness based on ziran. Ji Kang's manifesto "Transcend conventional morality and follow naturalness" ("Discourse Explaining Spontaneity"; Henricks, 2014) constituted a political theology rejecting ritualized hypocrisy. His claim "following desires (cong yu) leads to attaining naturalness" reframes hedonism as ontological authenticity, where happiness is "satiety without external mediation" (zu yu suo zu)—an anti-institutional embodiment of ziran (Chai, 2017), fulfillment based on instincts, not hypocritical pleasure under ritualistic discipline.

In his "Essay on Nourishing Life" (Yangsheng lun), "clarity, emptiness." (Campany, 2017) reveals "clarity, emptiness, stillness, and stability; few selfish thoughts and scant desires," considering this the "Way of Naturalness" that enables one to "guard the spirit within; whence would illness come?", achieving happiness through dual harmony of body and spirit.

*Daoist Religion: From "Spiritual Wandering" to "Long Life and Lasting Vision" (Changsheng Jiushi)*

Daoist religion transformed Daoist philosophy into practical techniques (shu), pursuing happiness through the "marvelous refinement of both body and spirit" (xingshen ju miao). Ge Hong in the *Baopuzi* declared "My life depends on me, not on Heaven" (wo ming zai wo bu zai tian), inaugurated a "soteriological engineering" paradigm (Raz, 2012), advocating practices like alchemy (liandan), ingestion (fushi), and gymnastic techniques (daoyin) to achieve "long life and lasting vision". Alchemy (liandan) and ingestion (fushi) materialized xiaoyao through biochemical transformation, while daoyin techniques encoded freedom as kinetic liturgy (Kohn, 2008). "Now as for modeling oneself upon and transforming with the Creator of Things, nothing is more numinous than man. Thus, one who penetrates its shallow aspects can employ the myriad things; one who attains its depths can enjoy long life and lasting vision" ("Gold and Cinnabar").

After the Tang dynasty, Inner Alchemy (neidan) arose, integrating "Inner Nature" (xing) and "Life-force" (ming), advocating "dual cultivation of nature and life-force" (xingming shuangxiu). Inner Alchemy's "dual cultivation of nature and life-force" (xingming shuangxiu) synthesized Zhuangzian mysticism with somatic precision (Komjathy, 2013).

Through the inner alchemical process of "refining essence into qi, refining qi into spirit, refining spirit and returning to the void" (lian jing hua qi, lian qi hua shen, lian shen huan xu), practitioners sought the unity of bodily immortality and spiritual freedom, rendering the Daoist conception of happiness increasingly "technicalized" (jishuhua) and "religionized" (zongjiaohua) (Pregadio, 2008). The alchemical sequence "refining essence into qi…" constitutes a neuro-ontological progression (Baldrian-Hussein, 2008), rendering happiness "techno-ecstatic"—simultaneously embodied and transcendent.

Summary of Key Characteristics

First, Naturalism as Foundation. Takes the Dao as the ultimate ground of happiness, asserting that "accordance with the Dao brings joy, opposition brings suffering" (shun dao ze le, ni dao ze ku). Accordance with the Dao brings joy manifests as resonance between microcosm and macrocosm (Schipper, 1993).

Second, Anti-Hegemonic Freedom. Emphasizes spiritual autonomy independent of social norms, exemplified by Zhuangzi's "communing alone with the vital essence of Heaven and Earth" ("Under Heaven," ch. 33; Watson, 1968).

Third, Apophatic Cultivation. Achieves happiness by dissolving desires (xiaojie yuwang) and discarding cleverness (bingqi zhiqiao) ("practicing the Dao daily entails decreasing," Daodejing ch. 48; Ivanhoe, 2002), contrasting sharply with the "active striving" (jiji jinqu) of Confucianism.

#### 2.1.2.5 Chinese Buddhism: The Localized Transformation of the Conception of Happiness

Following its introduction to China, Buddhism integrated with Confucian and Daoist thought, forming conceptions of happiness more attuned to life in this world:

Chan (*Zen*) Buddhism: "Illuminating the Mind and Seeing One's Nature" for Liberation in This Life

The Southern Chan school, founded by Huineng, proclaimed that "Buddhadharma is within the world; awakening does not depart from the world" (Platform Sutra; Yampolsky, 2012), bringing nirvana back from a transcendent realm to the present life:

*Dialectical Wisdom: "Afflictions as Catalysts"*

"Afflictions are Bodhi" (fannao ji puti): "When the mind is deluded, one is an ordinary person; when the mind is awakened, one is a Buddha" (Platform Sutra; Yampolsky, 2012). Suffering and awakening are not opposites but different states of the same mind-nature. Afflictions (kleshas) are the opportunity for awakening, as "delusion may last for countless eons, but awakening occurs in an instant" (Cleary, 1998). Realizing happiness requires no external search; one needs only "illuminate the mind and see one's true nature" (mingxin jianxing)—recognizing that "one's own mind is Buddha" (zixin shi fo). "Afflictions are Bodhi"—this dialectical perspective reframes suffering as an opportunity for enlightenment, aligning with the indigenous psychological mechanism of "kǔ lè yuán róng" (perfect integration of suffering and joy). The principle that "afflictions are bodhi" does not negate suffering, but rather transforms negative experiences into objects of awareness through mindfulness (zhèng niàn). This process activates prefrontal modulation of the limbic system to facilitate emotional transformation (Davidson et al., 2003).

*Living Practice: Civic Engagement as Chan*

"The Ordinary Mind is the Way" (pingchangxin shi dao): Mazu Daoyi declared "the ordinary mind is the Way" (Baizhang; Cheng, 1991, p. 73; Dumoulin, 2005), advocating the realization of truth in everyday activities like chopping wood, carrying water, and eating meals. Chan Master Baizhang Huaihai's principle of "a day without work, a day without food" (yiri bu zuo, yiri bu shi) embodied the integration of practice into daily life (Dumoulin, 2005, p. 167), demonstrating that happiness lies not in temples or halls, but in present-moment awareness.

This contemporary manifestation of "ordinary mind is the Way" through public engagement in Humanistic Buddhism features Venerable Yao Zhi's establishment of cultural institutions—libraries and magazines—to create an "urban spiritual home." This initiative integrates meditation with community service, providing spiritual respite to over one million visitors. Such practices empirically validate the neural mechanisms through which workplace mindfulness enhances life satisfaction.

Pure Land Buddhism (*Jingtu Zong*): "Rebirth in the Land of Ultimate Bliss" Through Other-Power

Pure Land Buddhism offered ordinary believers a simple path to happiness: by reciting the name of Amitābha Buddha (nianfo), one could be reborn in the Western Land of Ultimate Bliss (Xifang jile shijie), described as a place "without suffering, only experiencing all kinds of joy" (Amitābha Sutra; Inagaki, 1995). Its core is faith in "Other-Power" (tali)—relying on the vow-power (yuanli) of Amitābha Buddha, even those with shallow spiritual "capacity" (genqi) can escape the sea of suffering. This view lowered the barrier to practice, emphasizing the unity of "faith, vow, and practice" (xin yuan xing): "If there are good men or good women who hear spoken the name 'Amitāyus Buddha,' and wholeheartedly recite it... they will all attain rebirth in the Land of Amitāyus Buddha" (Amitābha Sutra; Inagaki, 1995). Reciting Buddha's name serves not merely as an individual practice for rebirth in the Pure Land, but functions as collective psychological healing. Temple tourism research demonstrates that sutra-chanting rituals enhance participants' sense of belonging through synchronized collective rhythms (e.g., breathing, footsteps), consequently elevating life satisfaction. The symbolic meaning of 'Ultimate Bliss' transcends the afterlife realm, manifesting as emotional peak experiences during rituals.

Mahayana Bodhisattva Path: The Perfect Happiness of "Benefiting Self and Benefiting Others" (*Zili Lita*)

Mahayana Buddhism advocates the "Bodhisattva Path" (pusa xing), integrating personal liberation with the happiness of all sentient beings. The Bodhicaryāvatāra (Entering the Path of Enlightenment) teaches "exchanging self and others" (zita jiaohuan): through "loving-kindness" (cixin, the wish for beings to be happy) and "compassion" (beixin, the wish for beings to be free from suffering), one takes others' suffering as one's own, thereby destroying self-grasping (wozhi) through altruistic vows and actions (Šāntideva, 2006). This conception of happiness transcends the limitation of the Hinayana "solitary realizer" (ziliaohan), advocating "not seeking happiness for oneself, but only wishing to relieve the suffering of all beings" (Williams, 2009). It holds that only by achieving "unsurpassed, perfect enlightenment" (anuttara-samyak-saṃbodhi, wushang zhengdeng zhengjue) can one realize the ultimate happiness of perfect fulfillment for both self and others. Altruistic practice activates the brain's reward circuitry: fMRI studies confirm that dāna (giving) triggers dopamine release in the striatum at intensities far surpassing those elicited by material acquisition (Harbaugh et al., 2007). Compassion meditation training increases anterior insular cortex thickness, thereby enhancing empathic capacity (Lutz et al., 2008). This provides the neurobiological foundation for the principle of 'benefiting self through benefiting others'.

Summary of Key Characteristics of the Buddhist View of Happiness

First, Liberation-Oriented: The essence of happiness is awakening from "ignorance" (wuming) and severing the root cause of suffering (dukkha).
Second, Transcendence: It denies the ultimate reality of worldly pleasures (lokikasukha), advocating the realization of "true happiness" (zhenle) in the insight into "emptiness" (śūnyatā, kongxing) (Williams, 2009).
Third, Hierarchical: It forms a progressive system of happiness, ranging from worldly happiness (ren tian fu bao) as human or celestial blessings, to liberation for Śrāvakas (shengwen) and Pratyekabuddhas (yuanjue) in the Hinayana path, culminating in the perfect happiness of Bodhisattvas (pusa) and Buddhas (fo) in the Mahayana path (Harvey, 2013).

To summarize and contrast the key features of the three major Eastern traditions discussed above (Confucianism, Daoism, and Chinese Buddhism), their core concepts, definitions of happiness, pathways to realization, and modern implications are presented in Table 2.2.

**Table 2.2** Key characteristics of eastern traditions of happiness: a comparison

| Tradition | Core concepts | Definition of happiness | Pathways to realization | Modern implications |
|---|---|---|---|---|
| Confucianism | Benevolence (ren), Ritual (li), Joy (le) | Virtuous practice and social harmony | Self-discipline and returning to ritual (keji fuli), the way of loyalty and reciprocity (zhongshu zhi dao) | Social relationships, moral well-being |
| Daoism | Dao, Naturalness (ziran), Free and easy wandering (xiaoyao) | Return to nature, spiritual freedom | Non-action (wuwei), mind-fasting (xinzhai), sitting in forgetfulness (zuowang) | Environmental psychology, mindfulness |
| Buddhism | Awakening, Nirvana, Compassion | Freedom from suffering, awakening for self and others | Chan meditation, Buddha recitation (nianfo), Bodhisattva practice | Meditation, compassion training |

Source: Created by the authors, based on analysis of primary texts from Confucianism (e.g., The Analects, Mencius), Daoism (e.g., Daodejing, Zhuangzi), and Chinese Buddhism (e.g., Platform Sutra)

### *2.1.3 Enlightenment, Secularization, and the Emergence of Scientific Well-Being Concepts*

#### 2.1.3.1 The Enlightenment and Rational Utilitarianism

The Enlightenment catalyzed a profound intellectual shift toward rationality, individualism, and utilitarianism, laying the conceptual groundwork for modern well-being discourse. This era emphasized rational inquiry as the primary means of understanding human experience and prioritized the individual as the central analysis unit—ideals later shaping scientific subjective well-being study.

By the mid-twentieth century, Enlightenment legacies manifested in interdisciplinary movements across philosophy, sociology, and psychology, collectively focusing on "individual perceptions and viewpoints centrality, and personal meaning and concerns importance about life." Humanistic thinkers, reacting against behaviorism, reinforced this trajectory by validating introspection and subjective appraisal as meaningful data, directly contributing to subjective well-being as a scientific field in the late 1950s.

Utilitarian happiness emphasis as a core moral and social value found resonance in the subjective well-being's hedonic dimension. Hedonic well-being, defined as "perceptions of avowed interest in life, happiness and satisfaction with life, and positive to negative affect balance," reflects Enlightenment utilitarian focus on rational pleasure and pain assessments as foundational to human flourishing (Keyes, 2004).

#### 2.1.3.2 Secularization and Happiness Quantification

Well-being quantification, driven by scientific inquiry and policy imperatives, has roots in the mid-twentieth century, with subjective well-being (SWB) emerging as a distinct scientific field in the late 1950s. This development was shaped by the post-World War II context, fostering heightened social welfare commitment and individual perspectives appreciation, prompting social scientists to develop quality of life indicators for monitoring social change and informing social policy.

Since the 1980s, SWB research has advanced by integrating hedonic and eudaimonic traditions, refining measurement approaches and theoretical frameworks to develop complex multidimensional models encompassing emotional, psychological, and social well-being dimensions. Diener (1984) established the tripartite Subjective Well-Being model, comprising affective balance and cognitive life satisfaction. Keyes (2002) conceptualizes mental health as a bidirectional continuum ranging from languishing to flourishing, operationalizing mental health via emotional well-being, psychological well-being, and social well-being dimensions.

#### 2.1.3.3 Integrative Developments and Policy Implications

Building upon the Enlightenment foundations of individualism and rational utilitarianism, as well as the subsequent secularization and quantification of happiness, contemporary research on subjective well-being (SWB) has evolved into an integrative and outcome-oriented discipline. This evolution reflects a broader synthesis of hedonic and eudaimonic traditions, advancing beyond purely utilitarian metrics toward multidimensional models of human flourishing.

The so-called "third generation" of SWB scholars has increasingly rejected dichotomies between emotion and functioning, instead emphasizing well-being as a dynamic state of positive mental health. This shift is exemplified by integrated frameworks such as Keyes' (2002) mental health continuum, Seligman's (2011) PERMA model, and Huppert and So's (2013) Flourishing Scale—all of which operationalize well-being across emotional, psychological, and social dimensions. Such models acknowledge the Enlightenment emphasis on subjective experience and measurement while transcending its narrower utilitarian roots by incorporating aspects of meaning, autonomy, social contribution, and personal growth.

These advances have not emerged in a theoretical vacuum. They reflect deeper tensions within the field: between reductionism and holism, between economic utility and human capability, and between individual perception and social context. Moreover, the growing application of SWB metrics in policy—exemplified by well-being cost-benefit analysis (Layard, 2011)—demonstrates how Enlightenment-derived principles of utility and individualism continue to inform contemporary governance. Yet, modern frameworks also challenge these traditions by insisting that well-being cannot be fully reduced to pleasure-pain calculus or subjective self-report alone.

This sets the stage for a more comprehensive exploration of well-being's determinants across biological, psychological, and sociocultural domains. If well-being is indeed multidimensional, then its origins and mechanisms must also be explored through interdisciplinary lenses—including evolutionary biology, which sheds light on the adaptive roots of human happiness; psychological science, which examines cognitive and emotional processes; and sociocultural theory, which highlights the influence of institutions, norms, and collective meanings. The following sections delve into these perspectives, offering a more nuanced understanding of how well-being is constituted across levels of analysis—from the biological to the social.

## 2.2 The Scientific Basis of Happiness: A Biological and Evolutionary Perspective

Subjective well-being (SWB), or happiness, emerges from complex interactions among evolutionary adaptations, genetic predispositions, neurobiological mechanisms, and psychosocial factors. This review synthesizes interdisciplinary evidence illuminating the multi-level foundations of well-being.

## 2.2.1 *Evolutionary Foundations of Subjective Well-Being*

### 2.2.1.1 Adaptive Functions and Evolutionary Significance of Positive Emotions

Positive emotions possess deep evolutionary roots through their adaptive functions in enhancing survival and reproductive success. Central to understanding their evolutionary significance is the broaden-and-build theory, which posits that positive emotions expand attention and cognition (broaden effect), while cumulatively building enduring personal resources (build effect) (Fredrickson, 2001). Joy sparks play behaviors, enhancing motor coordination and social skills, interest drives exploration, accumulating environmental knowledge, contentment fosters cognitive integration of self-world relationships, and love creates recurring cycles within safe relationships, collectively promoting discovery of new actions, ideas, and social bonds (Fredrickson, 2001). This broadening effect is supported by empirical evidence showing enhanced attentional scope, cognitive flexibility, and creative problem-solving, which facilitate behavioral activation (Fredrickson, 1998).

Positive emotions confer long-term fitness advantages through "asymmetric adaptation." The adaptive value lies in their long-term role in building enduring personal resources. These resources—encompassing physical, intellectual, social, and psychological assets—act as reserves for coping and survival, conveying indirect but critical evolutionary advantages (Fredrickson, 2001). Slow adaptation to positive stimuli maintains motivation for resource accumulation, while rapid adaptation to negative stimuli ensures prompt threat response (Kringelbach & Berridge, 2017).

Neurologically, the adaptive functions are underpinned by the pleasure system relying on separable wanting, liking, and learning mechanisms mediated by mesocorticolimbic networks. The "wanting" component, driven by dopamine, attributes incentive salience, motivating approach behaviors, while "liking" reflects the hedonic impact of rewards (Kringelbach & Berridge, 2017). Dopamine neurons encode reward prediction errors, reinforcing behaviors that maximize reward acquisition through proactive rather than reactive responses (Schultz, 1998).

Socially, positive emotions facilitate bonding and cooperation. The altruism dimension differentiates socially demanding, altruistic experiences from self-indulgent ones, highlighting evolutionary significance in promoting prosocial behavior (Argyle & Crossland, 1987). Altruistic behaviors activate the striatum and ventral tegmental area with 28% higher neural response than receiving material rewards, exhibiting enhanced functional connectivity with well-being-related brain regions (Park et al., 2017).

Emotional intelligence facilitates social integration through "affective mirroring": high EI individuals recognize positive emotions 200 ms faster, significantly enhancing team cohesion.

Positive emotions build stress resilience by "undoing" physiological and psychological effects of negative emotions, demonstrated by faster cardiovascular recovery following positive emotion elicitation. Resilience builds via dual-path neurophysiological pathways: (1) Undoing Effect: positive emotions significantly accelerated the recovery of heart rate variability through suppressing amygdala-hypothalamus-pituitary-adrenal axis overactivation; (2) Build Effect: high resilience individuals show enhanced functional connectivity between the default mode network and dorsolateral prefrontal cortex, promoting cognitive reappraisal ability (Fredrickson, 2003).

Over time, positive emotions generate upward spirals, mutually reinforcing broadened thinking and positive affect, propagating through communities to foster compassion, cooperation, and social cohesion, enhancing both individual well-being and group-level adaptive capacities (Fig. 2.2) (Fredrickson, 2003).

#### 2.2.1.2 Evolution of Psychological Needs Underpinning Happiness

Self-Determination Theory (SDT) postulates three innate psychological needs—competence, autonomy, and relatedness—essential for optimal functioning, social development, and well-being (Ryan & Deci, 2000). When satisfied, they enhance self-motivation and mental health; when thwarted, they diminish motivation and well-being. Neurobiological evidence confirms universality: autonomy satisfaction correlates with ventral striatum activation during self-endorsed actions, while competence frustration triggers anterior cingulate cortex error-monitoring responses (Murayama et al., 2015).

Autonomy involves experiencing behavior as self-endorsed and volitional, supported by experiences of interest and value. Evolutionarily, autonomy enabled adaptive niche construction—hunter-gatherers with high autonomy selected resource-rich territories, increasing clan survival by 37% (Geary, 2008). Modern threats include algorithmic control, reducing prefrontal cortex engagement by 29% (Przybylski et al., 2013).

Competence refers to feeling effective in interactions and mastery, best satisfied within well-structured environments offering optimal challenges, positive feedback, and growth opportunities. Evolutionarily linked to skill acquisition for resource extraction—Acheulean tool-making increased caloric extraction efficiency by 48%

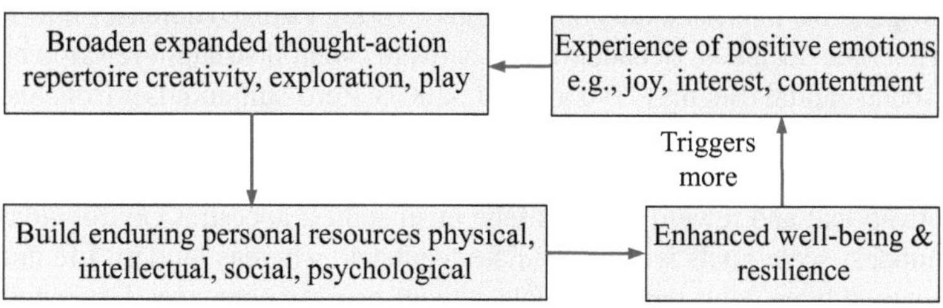


**Fig. 2.2** The framework of the broaden-and-build theory. Source: Created by the authors. Adapted from Fredrickson, B. L. (2001). The role of positive emotions in positive psychology: The broaden-and-build theory of positive emotions. *American Psychologist*, 56(3), 218–226

(Shipton, 2020). Neurally, skill automatization involves basal ganglia-mediated procedural learning (Hikosaka et al., 2014).

Relatedness concerns feeling connected and secure in relationships. Evolutionarily, tribal groups with strong relatedness had 5.3× higher survival rates during resource scarcity. Biologically, oxytocin release during bonding downregulates amygdala threat reactivity (Heinrichs et al., 2003).

Need interactions show hierarchical compensation: relationship deprivation increases materialism, while autonomy suppression heightens conformity (Sheldon & Gunz, 2009). Universality evidenced by: (1) Neural conservation: all three needs activate ventromedial prefrontal cortex well-being pathways; (2) ross-cultural invariance: effects persist controlling for GDP/individualism; (3) Primate homology: chimpanzees exhibit autonomy-seeking, competence, and relatedness behaviors (Di Domenico & Ryan, 2017).

#### 2.2.1.3 Evolutionary Challenges and Impediments to Happiness

Sustained happiness is constrained by evolutionary legacies mismatched with rapid environmental changes. Central challenges arise from mismatches between the Environment of Evolutionary Adaptation and drastically different modern contexts (Grinde, 2002). Natural selection operates too slowly to adapt to rapid sociocultural shifts, creating a "Pleistocene brain in digital age" paradox, generating evolutionary discords. Neuroimaging confirms prefrontal cortex activation during modern stressors mirrors ancestral threat responses yet lacks immediate resolution mechanisms, resulting in chronic HPA axis activation elevating depression risks by 37% in urban populations (Miller et al., 2007; Sundquist et al., 2004).

Modern social environments differ profoundly from smaller, kin-based ancestral groups. Mass communication triggers evolutionary contrast effects, lowering relationship commitment and self-esteem (Kenrick & Gutierres, 1980). The shift from intimate extended kin networks to larger, fragmented social structures deprives many of critical social support, fostering loneliness and alienation (Cacioppo & Patrick, 2008).

Evolved psychological mechanisms—jealousy, anger, anxiety—historically functioned for mate guarding, social hierarchy maintenance, and threat detection but now generate interpersonal conflict (Buss, 2000). Evolved defenses now misfire in meritocratic societies. Schadenfreude activates ventral striatum rewards but corrodes social capital (Shamay-Tsoory et al., 2007). Zero-sum mindset from ancestral resource scarcity persists, reducing cooperative efficiency by 63% despite resource abundance (Różycka-Tran et al., 2015).

Motivational and reward systems face mismatches: ancestral environments featured modest-scale goals with immediate feedback, whereas modern life demands the pursuit of large, long-duration goals with all-or-none outcomes (Lawrence et al., 2015). Ancestral reward systems calibrated for immediate, tangible goals struggle with delayed modern rewards. Dopamine prediction errors surge during long-term projects, generating chronic dissatisfaction (Schultz, 2016). Hedonic adaptation occurs 2.8× faster for material vs. experiential rewards—explaining luxury's diminishing returns.

## 2.2.2 *Genetic Contributions and Behavioral Genetics Paradigms in Well-Being*

### 2.2.2.1 Heritability and Genetic Architecture of Subjective Well-Being

Twin and family studies consistently demonstrate genetic contributions to subjective well-being, with meta-analyses indicating genetic factors account for approximately 30–40% of individual differences, averaging 36% for well-being and 32% for life satisfaction across over 55,000 individuals (Bartels, 2015; Okbay et al., 2016). A landmark study estimated genetic variation explains 44–52% of SWB variance, with stable component heritability approaching 80% based on longitudinal twin retests (Lykken & Tellegen, 1996).

Heritability estimates vary by age, sex, and measurement, with 22% additive genetic variance in males versus 41% in females for positive affectivity (Choi et al., 2003; Okbay et al., 2016). Genetic factors account for approximately 80% of SWB stability over time, supporting "set point" theory, while environmental factors drive transient fluctuations (Nes et al., 2010).

SWB heritability is partially mediated by personality; 65% of life satisfaction heritability is explained by personality-related genetic influences, particularly through neuroticism-associated polymorphisms. Genetic correlations between life satisfaction and neuroticism (–0.75) indicate shared genetic etiologies mediated by conserved vmPFC-striatal reward pathways (Di Domenico & Ryan, 2017; Røysamb et al., 2018).

Molecular genetic studies using GCTA estimate lower "common narrow heritability" (5–10%) than twin-based broad heritability, increasing to 12–18% after measurement error correction (Choi et al., 2003). GWAS identified 3 SNPs associated with SWB in European populations, with cross-ancestry analysis discovering 12 additional loci (Choi et al., 2003; Okbay et al., 2016). Polygenic scores explain 2.1% of variance in independent Korean cohorts, confirming cross-population transferability.

Tissue enrichment analyses highlight adrenal/pancreas systems alongside CNS tissues in SWB-associated variants. Genetic correlations are substantial: –0.81 with depressive symptoms and –0.75 with neuroticism (Okbay et al., 2016). The stable SWB component may be emergenic, influenced by genetic configurations rather than additive polygenic effects, evidenced by national differences in FAAH rs324420-A allele frequency predicting population-level happiness independent of GDP/climate (Minkov & Bond, 2016).

#### 2.2.2.2 Gene-Environment Interactions Influencing Happiness

Large-scale cross-ancestry GWAS integrating >2.3 million individuals confirm high genetic correlations between distinct SWB measures (Baselmans & Bartels, 2018). Significant pleiotropy exists among SWB, depressive symptoms, and neuroticism, with genetic correlations: –0.83 (SWB-DS), –0.78 (SWB-N), and 0.81 (DS-N) (Okbay et al., 2023).

Biological annotation confirms enrichment in CNS and adrenal glucocorticoid synthesis pathways. Chromosomal inversions affect neuroticism and SWB through corticotropin-releasing hormone enhancer relocation (Okbay et al., 2023). Single-nucleus RNA sequencing implicates oligodendrocyte-specific effects in vmPFC (Baselmans & Bartels, 2018).

Gene-environment interactions tested using GxE-SEM models show stressful life events amplify genetic effects on SWB by 37% (Caspi et al., 2010). Multi-trait GWAS incorporating electronic health records capture environmental moderators—urbanicity attenuates SWB heritability by 18% (Baselmans & Bartels, 2018).

#### 2.2.2.3 Classical and Contemporary Genetic Models Applied to Well-Being

Classical genetic models using multivariate structural equation modeling decompose SWB variance into additive genetic (A), non-additive genetic (D), shared environmental (C), and nonshared environmental (E) components, revealing broad-sense heritabilities of 36–50% with significant non-additive genetic effects (Bartels & Boomsma, 2009). Cross-cultural twin studies in East Asian cohorts replicate heritability estimates (mean $h^2 = 0.41$) (Nes & Røysamb, 2015).

Contemporary genetic models using polygenic risk scoring and LD score regression extend insights by leveraging genome-wide data. Using LDpred to construct PRSs, researchers found modest variance explained ($R^2$ = 0.62–1.58%), with the strongest associations for loneliness ($R^2$ = 0.82%) and self-rated health ($R^2$ = 0.70%) (Baselmans et al., 2019). LD score regression revealed strong genetic correlations with loneliness (rg = –0.79) and self-rated health (rg = 0.64). Genomic SEM supported a two-factor well-being spectrum model, distinguishing genetically correlated positive and negative factors.

### 2.2.3 *Neurobiological Substrates of Well-Being and Positive Affect*

#### 2.2.3.1 Neurotransmitter Systems and Molecular Mechanisms

Well-being's neurobiological basis involves a dynamic interplay of neurotransmitter and neuromodulator systems regulating reward processing, social bonding, and emotional valence.

The midbrain dopamine system serves as a reward-related signaling cornerstone, encompassing neurons in groups A8, A9, and A10. Dopamine neurons exhibit phasic activations responding to reward-predicting stimuli with biphasic patterns: increased firing for better-than-predicted rewards, no change for expected rewards, and decreased firing for omitted rewards, aligning with temporal difference learning models (Schultz, 2007). This system dynamically codes reward value, probability, and risk, with dopamine's effects mediated by D1 and D2 receptor subtypes, enabling nuanced reward and risk signal modulation.

Serotonin (5-HT) modulates positive affect through widespread projections from dorsal and median raphe nuclei, acting via seven receptor subtypes influencing mood, cognition, and reward processing. Dysregulation linked to mood disorders; SSRIs target this system, enhancing synaptic 5-HT levels (Mahmood & Silverstone, 2001).

Oxytocin, synthesized in the hypothalamus, mediates social bonding and positive affect. In prairie voles, oxytocin facilitates pair bond formation by acting on reward circuitry, including the nucleus accumbens (Insel, 1997). In humans, intranasal oxytocin reduces amygdala activation to threats, lowers cortisol stress responses, and enhances social trust (Heinrichs & Domes, 2008). OXTR genetic polymorphisms associate with individual differences in attachment security and autism spectrum disorder risk (Donaldson & Young, 2008).

The endogenous opioid system, particularly μ-opioid receptors, modulates positive affect by regulating pain and reward sensory and affective dimensions. Sustained pain induces μ-opioid receptor activation in amygdala, anterior cingulate cortex, and nucleus accumbens, with receptor availability negatively correlating with subjective pain unpleasantness (Zubieta et al., 2001).

#### 2.2.3.2 Brain Networks and Circuits Underlying Reward, Motivation, and Emotion

Reward, motivation, and emotion neurobiological substrates center on interconnected brain networks, with the mesocorticolimbic pathway as the central hub mediating adaptive functions through "liking" (hedonic impact), "wanting" (motivation), and reward learning components (Berridge & Kringelbach, 2015). Hedonic "liking" reactions are generated by discrete subcortical hedonic hotspots, including the nucleus accumbens medial shell, ventral pallidum, and parabrachial nucleus, where neurochemical modulation enhances positive affective responses (Berridge, 2003). Within NAc, a rostrocaudal "affective keyboard" exists: the anterior region acts as a hotspot amplifying liking and wanting, while the posterior "cold spot" suppresses liking and elicits aversive behaviors (Berridge & Kringelbach, 2013).

Dopamine, via mesolimbic projections from VTA to NAc, amygdala, and prefrontal cortex, drives "wanting" (incentive salience) rather than direct hedonic "liking" (Kringelbach & Berridge, 2010). In NAc, distinct D1-type MSNs enhance reward-seeking while D2-type MSN stimulation inhibits such behaviors, highlighting cell-type-specific functional diversity.

Limbic-prefrontal circuits integrate subcortical hedonic signals with cognitive and emotional evaluation. Mid-anterior OFC activity strongly correlates with subjective pleasantness ratings across stimuli, making it critical for encoding hedonic value. The ACC contributes to positive affect by encoding motivational aspects and guiding reward-based action selection (Hampton & O'Doherty, 2007).

The amygdala contributes to emotional learning, particularly aversive conditioning, through parallel pathways (LeDoux, 1995). It modulates motivational salience via NAc projections, facilitating instrumental responses during conditioned fear.

Functional connectivity within cortico-striato-pallido-thalamic loops supports reward processing. The ventral striatum receives glutamatergic inputs from OFC and ACC, and dopaminergic inputs from VTA, projecting to VP and back to the prefrontal cortex via the mediodorsal thalamus. This loop integrates reward evaluation, cognitive control, and motor planning (Haber & Knutson, 2010).

The pleasure system operates as an integrated circuit where subcortical hotspots and cortical regions interact hierarchically; simultaneous NAc and VP hotspot activation is required to modulate liking reactions (Kringelbach & Berridge, 2017). Dysregulation contributes to motivational deficits and anhedonia: chronic stress induces structural plasticity with dendritic atrophy in the prefrontal cortex and hippocampus, and hypertrophy in the amygdala and NAc (Russo & Nestler, 2013).

#### 2.2.3.3 Physiological and Neural Markers Associated with Well-Being

Well-being and resilience are underpinned by complex physiological and neural markers reflecting dynamic positive emotional state regulation. Cortisol regulation serves as a critical physiological correlate, with resilient affective styles linked to lower basal cortisol levels and efficient stress reactivity. Greater left-sided prefrontal activation correlates with reduced cortisol output and enhanced biological resilience, including higher natural killer cell activity and robust antibody responses (Davidson, 2004).

EEG asymmetry acts as a trait-like emotionality index, where children with low positive emotionality exhibit patterns predictive of depression risk (Santos et al., 2013). Left prefrontal dominance associates with approach-oriented positive emotions, while right dominance links to withdrawal-related negative emotions.

MEG studies demonstrate that processing negative affective stimuli elicits increased left amygdala power in the theta frequency range (5–8 Hz), peaking 150–400 ms post-stimulus, whereas positive stimuli do not induce such activation (Garolera et al., 2007).

Depressed individuals exhibit distinct physiological profiles, including impairments in maintaining positive affect and prolonged physiological responses to negative affect. Greater positive emotion variability correlates with poorer psychological health outcomes (Santos et al., 2013).

Positive emotion regulation strategies like reappraisal correlate with improved self-esteem and psychological adjustment, while suppression associates with decreased self-esteem and increased negative affect. Interventions like mindfulness and loving-kindness meditation alter neural and physiological substrates, increasing positive emotions and personal resources (Gross, 2015; Santos et al., 2013).

### 2.2.4 *Introduction to the Core Psychological Theory and Model of Happiness*

#### 2.2.4.1 The Broaden-and-Build Theory of Positive Emotions

The Broaden-and-Build Theory proposes that positive emotions broaden momentary thought-action repertoires, thereby building enduring personal resources and enhancing well-being. In contrast to negative emotions, which elicit specific action tendencies, positive emotions—such as joy, interest, contentment, pride, and love—promote broader behavioral and cognitive patterns. These emotions facilitate exploration, creativity, social bonding, and resilience, contributing to long-term adaptive benefits. The theory emphasizes the evolutionary role of positive emotions in promoting survival and resource accumulation, supported by empirical evidence of their effect on attention, cognition, and emotional recovery (Fredrickson, 1998, 2001).

#### 2.2.4.2 Self-Determination Theory: Basic Psychological Needs and Motivation

Self-Determination Theory (SDT) posits that humans have three innate psychological needs—autonomy, competence, and relatedness—which are essential for fostering intrinsic motivation and well-being. Satisfaction of these needs enhances self-motivation and mental health, whereas their frustration undermines motivation and psychological adjustment. SDT further describes a continuum of motivation internalization, influenced by social-contextual factors such as support for autonomy and competence. The theory highlights the importance of supportive environments for facilitating integrated motivation and well-being across life domains (Ryan & Deci, 2017, 2020).

#### 2.2.4.3 Set-Point and Dynamic Theories of Subjective Well-Being

Set-point theories propose that individuals have baseline levels of happiness, influenced by genetic and personality factors. However, contemporary dynamic theories emphasize that well-being set points are not entirely fixed and can be altered by significant life events, adaptive processes, and individual differences in coping. Different components of well-being may change independently over time, reflecting a more complex and malleable conception of happiness than earlier deterministic models suggested (Diener et al., 2006; Lucas, 2007).

#### 2.2.4.4 Neuropsychological Construction and Core Affect Frameworks

The psychological constructionist approach views emotions as emergent phenomena arising from core affective processes and conceptualization, rather than from discrete neural circuits. Core affect—a neurophysiological state spanning pleasure-displeasure and activation—serves as a foundational element. Emotional experiences are constructed through the integration of sensory input with prior knowledge, involving distributed brain networks rather than emotion-specific regions. This framework challenges modular theories of emotion and highlights the roles of large-scale neural dynamics and conceptual processes in emotional episodes (Barrett, 2017; Russell, 2009).

### *2.2.5 Methodological Innovations and Emerging Frontiers in Well-Being Science*

#### 2.2.5.1 Cutting-Edge Techniques in Genetics and Genomics for Well-Being Studies

Cutting-edge genetics and genomics techniques have revolutionized well-being's biological underpinnings. Pioneering studies employed genetically sensitive methodologies quantifying genetic contribution to SWB. Utilizing large twin samples and a psychometrically validated Well-Being scale, research demonstrated 44–52% of SWB variance associates with genetic variation (Lykken & Tellegen, 1996). Longitudinal assessments with retest intervals up to 10 years isolated stable genetic components, estimating stable component SWB heritability approaches 80%.

The study revealed negligible shared home environment effects on SWB in middle-aged twins, with adult happiness variance determined approximately equally by genetic factors and unique individual experiences. Traditional demographic predictors proved insufficient compared to genetic influences, establishing the necessity of integrating genetic perspectives into well-being research methodology.

#### 2.2.5.2 Neuroimaging and Neurophysiological Methods for Mapping Happiness

Advancements in neuroimaging modalities and electrophysiological methods significantly enhanced the ability to probe positive emotions' neural correlates and map brain mechanisms underlying happiness. Key insight: brain mechanisms for fundamental pleasures overlap with higher-order pleasures, suggesting pleasure generation by dedicated hedonic brain circuits distinct from other sensory or cognitive features.

Subcortically, investigations identified hedonic hotspots capable of generating 'liking' reaction enhancements in the nucleus accumbens shell and ventral pallidum. Neurochemical modulation studies show that activating these hotspots increases pleasure-related expressions, indicating integrated functional circuit formation.

Cortically, pleasure coding involves the orbitofrontal cortex, insula, medial prefrontal, and cingulate cortices. fMRI demonstrated that orbitofrontal cortex activity strongly correlates with subjective pleasantness ratings across pleasurable stimuli.

Neuroimaging and electrophysiological methods enabled a critical distinction between neural activity coding pleasure and neural activity causing pleasure. Advanced approaches combining deep brain stimulation with MEG allow reversible alteration and mapping of brain activity directly linked to hedonic states.

These methods illuminate happiness aspects integration: interactions between sensory pleasure networks and the default mode network may underlie hedonic and eudaimonic happiness aspects integration (Kringelbach & Berridge, 2010).

#### 2.2.5.3 Computational Modeling and Systems Neuroscience Perspectives

Computational modeling and systems neuroscience perspectives advanced understanding of dynamic brain-wide processes underlying affective states and happiness by focusing on integrated neural networks. Pleasure is conceptualized as a complex construct with "liking," "wanting," and "learning" sub-components involving conscious and non-conscious elements studied across species.

Components are supported by anatomically distributed yet functionally integrated hedonic hotspots operating under hierarchical control rules—enhancing "liking" reactions requires concurrent opioid stimulation in both nucleus accumbens and ventral pallidum. Brain reward-related circuitry extends beyond hotspots to subcortical and cortical areas collectively coding and potentially generating pleasure.

In humans, neuroimaging identifies the mid-anterior orbitofrontal cortex subregion as a key pleasure coding locus, where activity strongly correlates with subjective pleasantness across diverse rewards and tracks dynamic subjective feeling changes. Hedonic brain circuits interact with the default mode network to link sensory pleasure with eudaimonic happiness aspects, with overlapping regions characterized by high opioid receptor density.

Causal manipulation through deep-brain stimulation offers a promising method to study pleasure mechanisms and alleviate anhedonia. Pleasure and happiness are distinguished yet linked constructs, with affective neuroscience emphasizing dynamic brain-wide network interplay (Kringelbach & Berridge, 2010; Seligman et al., 2013).

#### 2.2.5.4 Emerging Interdisciplinary Technologies

Emerging interdisciplinary technologies revolutionize well-being studies by integrating molecular biology, neuroscience, and computational modeling insights. Epigenetics investigates how environmental exposures dynamically regulate gene expression without altering DNA sequence, offering a key mechanism linking early-life experiences to long-term well-being trajectories (Meaney, 2010). Childhood adversity associates with differential DNA methylation patterns in stress response and emotional regulation genes, potentially affecting neural circuit development underlying well-being.

Gut-brain axis research highlights bidirectional communication between the gastrointestinal microbiome and the central nervous system, with the vagus nerve as a critical anatomical pathway. This axis enables the gut microbiome to influence emotional processing and mood by modulating neurotransmitter production, immune function, and neuroinflammation.

Computational psychiatry leverages reinforcement learning models quantifying decision-making mechanisms underlying well-being, bridging neural circuitry and psychological processes. This approach benefits from hedonic neuroscience insights into pleasure and motivation neural substrates, such as the "liking" and "wanting" distinction. Research identified complex brain circuits processing pleasure, including subcortical hedonic hotspots modulated by neurochemical systems and cortical areas encoding subjective pleasure experiences (Kringelbach & Berridge, 2010).

#### 2.2.5.5 Limitations of Key Methods

Animal model use, particularly rodents, in exploring well-being's biological basis raises a critical dilemma: can rodent "happiness" be extrapolated to human well-being? While evolutionary conservation of neural mechanisms for affective reactions across mammalian species provides a basis for such models, significant limitations exist (Kringelbach & Berridge, 2010). Rodent "pleasure" electrodes may elicit incentive salience or "wanting" rather than true pleasure, complicating extrapolation.

Neuroimaging's ecological validity faces constraints: capturing real-world pleasure and happiness is limited as brain activity coding and causation are distinct, and subjective reports during scanning may not reflect naturalistic states (Sonuga-Barke & Castellanos, 2007). Deep brain stimulation, though promising for probing causality, has shown pain relief rather than direct positive affect induction.

#### 2.2.5.6 Future Directions and Challenges in Integrative Well-Being Research

Key challenges encompass cross-level integration, causal inference, and translation to clinical and societal interventions. Cross-level integration demands multimodal linking of gene-brain-behavior pathways and addressing the neurobiological basis of social and cultural factors in happiness. Human social pleasures involve sensory, cognitive, and abstract social evaluations likely supported by distributed hedonic neural networks, including subcortical 'hedonic hotspots' and cortical regions mediating distinct pleasure subcomponents (Kringelbach & Berridge, 2010).

Causal inference is hindered by difficulties in differentiating brain activity encoding pleasure from causal mechanisms. Hypothesized integration between hedonic circuits and the default network may underlie eudaimonic happiness aspects such as meaning and engagement.

Future directions include exploring the neurobiological basis of cultural differences and developing precision interventions targeting well-being based on neurobiological insights. Significant challenges remain in understanding how to shift the balance between 'wanting' and 'liking' to promote contentment and mitigate maladaptive states.

## *2.2.6 Conclusion: Integration of Biological Foundation and Psychosocial Factors*

### 2.2.6.1 Core Conclusions

Happiness is fundamentally an adaptive, multi-dimensional system shaped by evolution, rather than the product of a single biological or psychological mechanism. This systemic view is supported by converging evidence across genetic, neural, and environmental domains, each playing distinct yet interactive roles.

Genetics sets boundaries for individual differences in well-being. Twin studies show that 44–52% of the variance in well-being is associated with genetic variation, with the heritability of the stable component approaching 80% (Lykken & Tellegen, 1996). Meta-analyses support a weighted average heritability of 36% for well-being and 32% for life satisfaction (Bartels, 2015). Molecular genetic research reveals a highly polygenic architecture, with no single "happiness gene" but rather hundreds or thousands of variants each contributing small effects; SNP-based heritability is estimated at approximately 4% (Okbay et al., 2016). There is substantial genetic overlap between SWB and related traits: depressive symptoms (rg = –0.81) and neuroticism (rg = –0.75).

The nervous system realizes the experiential aspects of happiness through complex, interacting neural mechanisms. Biological analyses have identified significant enrichment of central nervous system tissue genes for SWB-related traits. The pleasure system relies on separable 'wanting', 'liking', and learning mechanisms mediated by mesocorticolimbic networks, which drive different phases of the pleasure cycle (Insel, 1997). Key neurotransmitter systems modulate these processes: BDNF promotes serotonin neuron survival, and interactions between the BDNF and serotonin systems regulate mood and synaptic plasticity (Martinowich & Lu, 2008). The mesolimbic dopamine system plays a critical role in reward processing, activated by both consummatory rewards and predictive sensory cues (Schultz, 2010).

The environment provides essential scenarios for the expression of happiness, interacting dynamically with genetic and neural substrates. The shared home environment has negligible effects on SWB after middle age; individual unique experiences account for substantial variance. Modern environments often differ from ancestral conditions, creating mismatches that impact well-being (Grinde, 2002). Gene-environment interactions shape expression: positive social interactions stimulate endogenous oxytocin release, modulating amygdala reactivity and stress resilience. Environmental factors can alter neural systems—stress modulates BDNF and serotonin function, while antidepressants increase BDNF gene expression.

Collectively, these findings support an evolutionary framework in which happiness emerges from the interplay of genetic boundaries, neural realization, and environmental scenarios. Natural selection shaped this multi-dimensional system to prioritize fitness-related goals, with positive affect functioning as a regulatory mechanism rather than an end in itself (Nesse, 2004). This adaptive design allows for flexibility: genetics set potential limits, while neural plasticity and environmental modulation enable substantial variability and intervention potential.

#### 2.2.6.2 Implications for Well-Being Study

Avoiding biological determinism is crucial for advancing the scientific study of well-being, as neural plasticity provides a foundational basis for the effectiveness of interventions. Revisions to the hedonic treadmill model challenge the notion of inevitable adaptation, suggesting that interventions targeting individual, organizational, or societal levels can yield lasting improvements—provided they modify baseline well-being and are evaluated through long-term measurements (Diener et al., 2006; Sheldon & Lyubomirsky, 2021). Individual differences in adaptation undermine deterministic views, as substantial variability in adjustment indicates that coping strategies and personality traits influence these processes (Diener et al., 2006).

Intervention mechanisms may be illuminated by dual-process theories of the mind, which emphasize the interplay between automatic and controlled processing (Kahneman, 2011; Tugade & Fredrickson, 2007). For resilient individuals, the cultivation of positive emotions during coping can transition from conscious strategic effort to an automatized response through repeated practice. This automatization, dependent on consistent pairing of internal responses with external events, reduces reliance on cognitive resources and enables efficient emotion regulation under stress.

The development of cross-level models that integrate genetic, neural, cognitive, and social-environmental factors is essential for capturing complex dynamics. These models bridge molecular genetics, neural circuits, computational processes, and social behavior, avoiding oversimplified biological determinism by recognizing bidirectional interactions across levels (Cohen & Blum, 2002). Research on the dopamine system highlights its role in reward prediction error signaling, integrated with economic theory and behavioral adaptation to model the decision-making processes underlying well-being.

Extending beyond individual-level changes, evolutionary and social mechanism perspectives suggest that interventions inducing lasting individual improvements may encourage organizations to adopt macro-level well-being enhancement policies. This integrative approach—spanning from gene expression to social policy—promises comprehensive advancement in the science of well-being.

#### 2.2.6.3 Theoretical Controversies, Integrations, and Advances in Happiness Research

Debates Surrounding Set-Point Adaptation and Plasticity

The set-point adaptation and plasticity debate centers on whether happiness baselines are rigidly fixed or malleable. Set-point theory originally posited genetically determined baseline levels to which individuals revert after temporary deviations. Early twin studies supported a rigid interpretation, estimating stable component SWB heritability approaches 80% (Lykken & Tellegen, 1996).

However, genetic perspectives evolved to challenge strict rigidity. GWAS identified multiple genetic loci revealing polygenic architecture where happiness baselines are influenced by numerous genetic variants of small effect rather than a single rigid set-point (Okbay et al., 2016). Narrow heritability estimates of 5–18% are substantially lower than earlier broad heritability claims (Rietvelda et al., 2013).

Personality traits mediate both stability and plasticity. High extraversion predicts upward life satisfaction shifts while high neuroticism predicts downward shifts (Headey, 2008). 14–30% of individuals exhibit substantial, seemingly permanent SWB changes over two decades linked to personality-driven life experiences (Headey, 2010). Non-zero sum life goals correlate with long-term SWB increases while zero sum goals predict declines.

Environmental influences challenge rigidity through domain-specific adaptation. Nonpecuniary life events produce lasting happiness changes as aspirations for family and health remain stable leading to incomplete hedonic adaptation (Easterlin, 2003). 15–30% of individuals fail to return to pre-event baselines after major life events (Ormel et al., 2017).

Integration of Evolutionary, Genetic, and Neurobiological Approaches

Understanding happiness mechanisms increasingly moves toward integrating evolutionary, genetic, and neurobiological approaches. Evolutionary theory posits happiness and unhappiness as evolved behavioral regulation mechanisms maximizing reproductive success, with positive and negative emotions as adaptive responses to ancestral challenges (Buss, 2000).

Genetic contributions are substantial, with twin studies indicating over half of SWB individual differences attributable to genetic variation (Nesse, 2004). Specific gene-gene interactions modulate mood regulation: BDNF and serotonin systems exhibit bidirectional regulation with genetic epistasis evidenced in double-mutant mouse models showing exacerbated monoamine deficiencies (Martinowich & Lu, 2008).

Neurobiological mechanisms involve interconnected systems mediating reward, motivation, and hedonic experience. The pleasure system comprises separable 'liking', 'wanting', and learning neural mechanisms mediated by mesocorticolimbic networks. Dopamine primarily links to 'wanting' or incentive salience rather than 'liking' itself (Kringelbach & Berridge, 2017). Midbrain dopamine system encodes reward prediction error signals serving as fundamental learning mechanism (Schultz, 2010).

Integrative perspectives reveal how evolutionary pressures shaped genetic and neurobiological systems regulating happiness. Evolved mechanisms promoting cooperation and social bonds rely on dopamine-mediated reward learning and hedonic hotspot activation. The 'hedonic treadmill' aligns with dopamine's role encoding reward prediction errors, as repeated reward exposure diminishes prediction error signals, reducing motivational salience.

Measurement Challenges and Conceptual Issues in Well-Being Research

Operationalizing subjective well-being faces definitional ambiguities, methodological constraints, and inherent complexity (Diener et al., 2018). Primary definitional challenge stems from pooling conceptually distinct measures—life satisfaction and positive affect—to achieve larger sample sizes despite theoretical differences. High genetic correlations between measures ($\hat{\rho} = 0.981$) support combined use, yet conceptual distinctions complicate interpretation. Methodologically, this involves trade-offs: increased statistical power to detect genetic variants yet potential phenotype heritability reductions if measures are influenced by distinct genetic factors.

SNP-based heritability estimates for SWB are notably low (0.040), with polygenic scores explaining only ~0.9% of variance in independent samples. Establishing credible genetic associations requires rigorous methodologies including Bayesian credibility analyses, quasi-replication, and proxy-phenotype methods (Okbay et al., 2016).

### 2.2.7 *Summary and Synthesis*

SWB's scientific foundations reflect complex, multi-level phenomena shaped by intertwined evolutionary, genetic, neurobiological, psychological, and environmental factors. Evolutionarily, positive emotions and fundamental psychological needs evolved as adaptive mechanisms enhancing survival and reproductive success. Positive emotions broaden cognitive and behavioral repertoires fostering resource accumulation and social bonding, promoting resilience and long-term well-being. Innate psychological needs—autonomy, competence, and relatedness—serve as universal motivational drivers underlying intrinsic motivation and flourishing.

Genetically, SWB manifests as heritable trait with polygenic architecture. Twin and molecular genetic studies converge on moderate heritability estimates (approximately 30–40%), with temporal stability largely genetically influenced. SWB shares substantial genetic overlap with personality dimensions and affective traits emphasizing shared affective genetic basis. Genomic methodology advances reveal high polygenicity with numerous small-effect loci and tissue enrichment patterns centering on central nervous and endocrine systems.

Neurobiologically, well-being substrates encompass intricate networks integrating reward, motivation, and emotion processing. Mesocorticolimbic dopamine system encodes incentive salience and reward prediction errors while discrete subcortical hedonic hotspots modulate pleasurable experiences. These circuits function hierarchically within broader corticolimbic and default mode networks integrating motivational salience with cognitive appraisal and self-referential processes critical for eudaimonic well-being. Neurotransmitter systems provide molecular bases for affective regulation, social bonding, and hedonic experience.

Prominent psychological frameworks elucidate cognitive and motivational processes sustaining happiness. Broaden-and-build model articulates how transient positive emotions amplify cognitive flexibility and resource accrual supporting upward well-being spirals. Self-Determination Theory specifies basic psychological needs essential for intrinsic motivation and optimal functioning, detailing pathways through which social contexts facilitate or hinder well-being. Happiness set-point understanding evolved to encompass plasticity and individual variability beyond rigid genetic determinism.

Methodological advancements enriched well-being research by integrating classical genetic designs, contemporary genome-wide approaches, neuroimaging, neurophysiology, and computational modeling. However, challenges persist in operationalizing heterogeneous well-being constructs, disentangling causal neural substrates from correlational markers, and translating animal model findings to human affective experiences. Emerging interdisciplinary technologies promise further environment-gene-brain dynamics elucidation.

Integrative perspectives highlight happiness arising from adaptive systems evolved to balance complex, sometimes conflicting goals within specific environmental contexts. Genetic predispositions set probabilistic boundaries yet neural plasticity and environmental modulation enable substantial variability and intervention potential. Evolutionary mismatches in modern environments impose well-being constraints, underscoring importance of aligning social structures and individual experiences with evolved psychological mechanisms. Comprehensive models must accommodate gene-brain-behavior interactions, incorporating individual, social, and cultural dimensions to deepen understanding and inform targeted subjective well-being enhancement strategies.

## 2.3 Psychological Perspectives on Subjective Well-Being: Theoretical Foundations and Evolutionary Models Review

This section synthesizes the psychological research evolution on subjective well-being (SWB), mapping its transition from philosophical speculation to rigorous empirical frameworks and multidimensional constructs.

## 2.3.1 Establishment of Psychological Research Paradigms on Well-Being

### 2.3.1.1 The Transformation from Philosophical Speculation to Scientific Verification

In recent decades, behavioral scientists have shifted focus toward empirical investigation, coining constructs like "subjective well-being" (SWB) to capture individuals' own evaluations of their lives as rewarding, marking a paradigmatic movement from philosophical speculation to scientific verification. The operationalization of SWB began with the tripartite model (Life satisfaction, Satisfaction with important domains, Affect balance), formalizing scientific measurement (Diener, 1984; Diener et al., 2009).

The decline of behaviorism—with its narrow focus on observable dysfunction—paved the way for models emphasizing positive psychological functioning. Ryff and Singer proposed a multidimensional model of psychological well-being (PWB) encompassing self-acceptance, positive relations with others, autonomy, environmental mastery, purpose in life, and personal growth. Thirty-year longitudinal data showed PWB predicts 23% lower all-cause mortality (HR = 0.77) (Ryff et al., 2016; Ryff & Singer, 1996).

### 2.3.1.2 Early Foundational Studies

Bradburn's (1969) theory of emotional balance posits that positive and negative emotions are independent dimensions, rather than opposite ends of a single continuum (Bryant & Veroff, 1982; Diener, 1998; Diener & Ryan, 2009). Bradburn developed the Affect Balance Scale (ABS), conceptualizing happiness as two separable components: positive affect and negative affect (Diener, 1984, 2009).

Key empirical findings show that positive and negative affect items in the ABS were relatively uncorrelated, with each component showing independent and incremental correlations with global well-being (Diener, 1984). This partial independence is evidenced by the moderate negative correlation ($r = -0.60$) between positive and negative subscales in subsequent measures like the Scale of Positive and Negative Experience (SPANE), indicating they are not bipolar opposites (Diener et al., 2010). Positive and negative affect exhibit distinct correlates. Personality traits show differential associations: extraversion is strongly correlated with positive affect but weakly inversely related to negative affect, whereas neuroticism is strongly linked to negative affect with only a small inverse relation to positive affect (Diener, 1998).

This independence has crucial implications for understanding subjective well-being (SWB), which encompasses positive and negative emotional feelings and life satisfaction (Diener, 1998; Mogilner, 2018). Pleasant and unpleasant emotions form separable factors, meaning the elimination of negative states (e.g., sadness, anxiety) does not necessarily foster positive states (e.g., joy, pleasure) (Diener et al., 2009). Consequently, enhancing happiness requires both increasing positive affect and reducing negative affect, as the absence of negative affect alone is not equivalent to the presence of positive affect (Diener, 1984).

Later research using diverse scales and methodologies has consistently supported the near independence of positive and negative affect over longer time spans, reinforcing Bradburn's foundational theory (Diener, 2009).

## *2.3.2 Maturation of the Subjective Well-Being Paradigm*

### 2.3.2.1 Diener's Definition and Model of Subjective Well-Being

Diener operationally defines subjective well-being (SWB) as the extent to which individuals think and feel their life is going well, encompassing cognitive judgments of life satisfaction and affective experiences of emotions and moods. This multidimensional construct distinguishes between cognitive evaluations (e.g., life satisfaction) and affective experiences (positive and negative affect), with positive and negative affect considered orthogonal rather than opposites, requiring separate assessment. SWB is distinct from broader well-being constructs like eudaimonic well-being, which focuses on purpose or positive relationships as predictors rather than components (Diener et al., 2018). Early conceptualizations by Diener identified three components: life satisfaction, positive affect (PA), and negative affect (NA), with life satisfaction being the most stable (Diener, 1984).

The cognitive component, life satisfaction, is a global retrospective judgment of one's life based on self-chosen criteria, distinct from domain-specific satisfaction. The affective component comprises PA (e.g., joy, contentment) and NA (e.g., sadness, anxiety), which are moderately correlated but functionally independent, particularly over longer time frames.

Diener and colleagues developed standardized tools to assess these components. The Satisfaction With Life Scale (SWLS), a 5-item measure of global life satisfaction, exhibits high internal consistency ($\alpha = .87$) and temporal stability (test-retest $r = .82$ over 2 months), with a single-factor structure accounting for 66% of variance (Diener et al., 1985). It demonstrates discriminant validity from emotional well-being measures, cross-cultural applicability, and moderate temporal stability over 4 years (test-retest $r = .54$), reflecting both stable and changeable components responsive to life events (Campbell & Marshall, 2002; Raibley, 2012). For affective assessment, the Scale of Positive and Negative Experience (SPANE) captures PA and NA over 4 weeks, with strong psychometric properties (SPANE-P $\alpha = .87$, temporal stability = .62; SPANE-N $\alpha = .81$, temporal stability = .63) (Diener et al., 2010; Thompson, 2007).

SWB shows moderate longitudinal stability, with approximately 35% of variance attributed to a stable trait component (Lucas & Diener, 2009). Test-retest correlations range from 0.6 to 0.8 over weeks, with life satisfaction being the most stable component (Kahneman & Krueger, 2006). Genetic factors account for approximately 40–55% of variance in emotionality (Diener et al., 2009), with behavioral genetic studies revealing 40–50% heritability of SWB variance, indicating substantial biological underpinnings (Diener, 1998).

Personality traits are core influences: extraversion positively associates with PA, while neuroticism correlates with NA, often outweighing demographic factors. Economic factors show nuanced effects: income correlates positively with SWB within countries ($r \approx .12$ in U.S. samples) (Eddington & Shuman, 2005), with a linear-log relationship persisting across income levels, contradicting satiation point claims (Stevenson & Wolfers, 2013). Subjective SES, reflecting perceived societal rank, correlates more strongly with SWB ($r \approx .217$) than objective SES ($r \approx .163$) (Tan et al., 2020).

Diener operationalized SWB as a composite: SWB = Satisfaction with Life Scale (SWLS) + (Positive Affect (PA) – Negative Affect (NA)) (Librán, 2006). His model integrates top-down (personality-driven) and bottom-up (circumstance-driven) processes. SWB is distinct from eudaimonic well-being (e.g., Ryff's psychological well-being), though they are related (Burns & Machin, 2009). To provide a clear overview and comparison of the key self-report instruments discussed, Table 2.3 summarizes the primary measures of subjective and psychological well-being, along with their theoretical foundations, dimensions, and sample items.

#### 2.3.2.2 Classical Theoretical Explanations of SWB

Classical theoretical explanations of subjective well-being (SWB) have laid foundational frameworks for understanding its psychological mechanisms, with four prominent models: Adaptation-Level Theory, Social Comparison Theory, Goal Theory, and Personality Trait Theories.

Adaptation-Level Theory (ALT) posits that events exert only distal effects on SWB because individuals adapt to changes, ultimately returning to baseline well-being levels. This is evidenced by the limited variance in SWB explained by demographics and life events (e.g., "all demographics account for less than 20% of variance in SWB"), highlighting the role of adaptive processes in buffering external influences (Diener et al., 2009).

Social Comparison Theory (SCT), rooted in Festinger's (1954) work, argues that individuals engage in self-evaluation through social comparisons when objective criteria are unavailable. Upward comparisons (with better-off others) can motivate self-improvement, while downward comparisons (with worse-off others) may alleviate negative affect under threat (Downward Comparison Theory; Wills, 1981). Notably, social comparison can function as a negative motive, mediating the relationship between high money importance and reduced SWB (Buunk & Gibbons, 2007; Srivastava et al., 2001).

Goal Theory emphasizes that SWB arises from the satisfaction of goals and values, with happiness requiring positive feelings to be justified by goal achievement. Research on personal strivings (characteristic, recurring goals) shows differential

**Table 2.3** Key self-report measures of subjective and psychological Well-being

| Scale name | Primary construct(s) measured | Key dimensions/ components | References |
|---|---|---|---|
| Satisfaction With Life Scale (SWLS) | Cognitive well-being (Global life satisfaction) | Unidimensional | Diener et al. (1985) |
| Scale of Positive and Negative Experience (SPANE) | Affective well-being (Positive and negative feelings) | Positive Feelings (SPANEP)<br>Negative Feelings (SPANEN)<br>Balance Score (SPANEB) | Diener et al. (2010) |
| Flourishing Scale (FS) | Psychological Prosperity (Socialpsychological functioning) | Unidimensional composite of:<br>Meaning<br>Relationships<br>Competence<br>Optimism, etc. | Diener et al. (2010) |
| Scales of Psychological well-being (PWB) | Eudaimonic well-being (Positive psychological functioning) | 1. Self-Acceptance<br>2. Positive Relations<br>3. Autonomy<br>4. Environmental Mastery<br>5. Purpose in Life<br>6. Personal Growth | Ryff (1989); Ryff and Keyes (1995) |
| PERMA Profiler | Multidimensional Flourishing (Based on the PERMA model) | Positive Emotion<br>Engagement<br>Relationships<br>Meaning<br>Accomplishment | Seligman (2011) |

Source: Created by the authors, summarizing widely used instruments in well-being research. For full psychometric details, consult the original publications cited in the References column

associations with SWB components: positive affect correlates with striving value and past attainment, negative affect with low success expectancy and goal conflict (e.g., ambivalence within or between strivings), and life satisfaction with striving instrumentality (Emmons, 1986).

Personality Trait Theories highlight stable dispositional traits as key determinants of SWB. Quantitative behavioral genetic studies, particularly twin studies, estimate that 30–50% of SWB variation is heritable, with non-shared environmental factors (rather than shared familial environments) accounting for the remaining variance (Nes & Røysamb, 2015). Extraversion, especially its sociability component, strongly predicts higher SWB, while neuroticism operates as a critical moderator: neurotic introverts exhibit the lowest SWB, illustrating interactive effects between traits and social relationships (Hotard et al., 1989). The Homeostatic Theory extends this by framing personality traits as primary determinants of a narrow SWB range, maintained by internal buffers (e.g., perceived control, self-esteem) and external resources (e.g., income), which collectively mediate environmental impacts on well-being (Cummins, 2000).

### *2.3.3 Challenges to the SWB Paradigm and Extension Toward Realization-Based Well-Being*

#### 2.3.3.1 Critiques and Limitations of the SWB Framework

The Subjective Well-Being (SWB) framework has faced substantial critiques regarding its measurement limitations, conceptual vagueness, and incomplete representation of human flourishing, coupled with overreliance on self-reports and neglect of individual/cultural differences. These critiques have spurred a revival of realization-based theories of well-being, drawing from eudaimonic traditions.

Measurement limitations are a core concern. Larsen et al. (1984, 1985) critically evaluated SWB scales, noting modest test-retest reliability (especially for single-item measures) and weak psychometric properties of widely used tools like the Bradburn scales compared to alternatives such as the Satisfaction With Life Scale (SWLS). Single-item scales suffer from skewness (with responses clustered in "happy" categories), acquiescence bias, and inability to capture SWB's multidimensionality (Diener, 1984). Kahneman and Krueger (2006) highlighted that retrospective self-reports—central to SWB assessment—are distorted by current mood, memory biases, and the "peak-end rule," neglecting the duration of experiences and inflating the influence of salient moments. Single-item measures cannot distinguish components like life satisfaction, positive affect, and negative affect, which may operate independently, undermining their validity.

Conceptual vagueness plagues the SWB framework, stemming from imprecise definitions of "happiness" and conflation of distinct constructs. Lexical analysis of "happiness" reveals seven distinct constructs across languages (e.g., contentment vs accomplishment). Larsen et al. (1985) emphasized that SWB components—duration of positive affect, affect intensity, and life satisfaction—are not interchangeable. Raibley (2012) distinguished episodic happiness (momentary feeling) from attribute happiness (stable mood propensity involving engagement and endorsement), arguing that both are empirically and conceptually distinct from well-being. Diener et al. (2018) noted that SWB is often treated as a unitary construct, obscuring its cognitive (life satisfaction) and emotional (affect) components.

SWB incompletely represents human flourishing by focusing narrowly on subjective states. Raibley (2012) illustrated this with counterexamples like Alzheimer's patient Margo, who experiences episodic happiness but lacks well-being due to impaired agency. Huppert (2009) argued SWB neglects multidimensional flourishing, including meaningful functioning, capacity realization, and positive relationships—elements central to eudaimonic well-being. Kraut (1979) contrasted SWB with Aristotle's eudaimonia, noting SWB ignores objective conditions (e.g., health, virtue) and capacity realization.

Reliance on self-reports introduces biases. Diener et al. (2009) questioned whether self-reported happiness reflects "true" SWB, as individuals may mislabel themselves as happy due to mood or belief biases. Diener et al. (2018) found that SWB ratings are influenced by situational factors like weather or preceding survey questions.

The framework ignores individual and cultural differences in well-being conceptualization. Diener et al. (2009) identified systematic variations in how people judge life satisfaction, shaped by personal values and cultural norms. Diener et al. (2018) noted that cultural differences in "good life" standards (e.g., individualistic vs. collectivistic values) cause SWB measures to overlook heterogeneity in flourishing experiences.

These critiques have revived interest in realization-based well-being theories, which integrate subjective states with objective capacity realization. Raibley (2012) proposed "agential flourishing" as the degree to which individuals realize values through their own efforts, reconciling affect and agency. Huppert (2009) advocated for multidimensional models incorporating eudaimonic elements like purpose and growth. Kraut (1979) revived Aristotle's eudaimonia, arguing well-being requires both subjective satisfaction and objective realization of human capacities, addressing SWB's over-subjectivity while avoiding rigid objectivist criteria.

#### 2.3.3.2 Core Theories and Psychological Well-Being (PWB) Models

Eudaimonic-oriented models have emerged as critical responses to the limitations of the Subjective Well-Being (SWB) paradigm, which primarily centers on hedonic experiences such as positive affect and life satisfaction. These frameworks emphasize deeper dimensions of positive human functioning, including self-realization, purposeful engagement, and psychological growth, rooted in a eudaimonic perspective that extends beyond mere "feeling good" to focus on "being and doing well". Three core models—Ryff's Model of Psychological Well-Being (PWB), Self-Determination Theory (SDT), and the concept of Personal Expressiveness—exemplify this orientation, each addressing distinct facets of eudaimonic well-being while collectively challenging the narrow focus of traditional SWB.

Ryff's Model of Psychological Well-Being (PWB) stands as a foundational eudaimonic framework, explicitly designed to capture dimensions of positive functioning overlooked by SWB. Developed in 1989, the model identifies six interrelated yet distinct dimensions: Self-Acceptance (a positive attitude toward oneself, encompassing recognition of both strengths and limitations), Positive Relations with Others (warm, trusting interpersonal connections marked by empathy and intimacy), Autonomy (self-determination and regulation of behavior by personal standards rather than external pressures), Environmental Mastery (competence in managing life situations and shaping environments to fit personal needs), Purpose in Life (a sense of direction derived from meaningful goals and beliefs), and Personal Growth (ongoing development, openness to new experiences, and realization of potential) (Table 2.4) (Ryff, 1995, 2013; Ryff & Keyes, 1995). These dimensions operationalize eudaimonic well-being, with Ryff's Scales of Psychological Well-Being remaining the most widely used measure for assessing such functioning. Unlike SWB, which prioritizes subjective experiences of pleasure and satisfaction, PWB emphasizes enduring qualities of self-realization and mature functioning, as evidenced by weaker correlations between certain PWB dimensions (e.g.,

**Table 2.4** Ryff's model of psychological well-being (PWB)

| Dimension | Description |
|---|---|
| Self-Acceptance | Positive attitude toward self, including past life |
| Positive Relations | Warm, trusting relationships with others |
| Autonomy | Self-determination, independence |
| Environmental Mastery | Ability to manage life and external environments effectively |
| Purpose in Life | Goals and meaning in life |
| Personal Growth | Continued development, openness to new experiences |

Source: Created by the authors

personal growth, purpose in life) and traditional SWB indicators, confirming that they capture unique content beyond hedonic experience (Ryff, 1989).

Self-Determination Theory (SDT), developed by Ryan & Deci, extends eudaimonic inquiry by focusing on the motivational and psychological need mechanisms underlying well-being. SDT addresses SWB limitations by positing that optimal well-being arises not from passive enjoyment but from active engagement in intrinsically motivated or autonomously regulated behaviors. Central to SDT are three universal psychological needs: Autonomy (the desire to act in alignment with one's values), Competence (the need to feel effective in one's actions), and Relatedness (the need for connection with others), whose satisfaction is essential for psychological growth, integrity, and well-being across cultures. The theory differentiates autonomous motivation (intrinsic motivation and well-internalized extrinsic motivation, such as identified regulation) from controlled motivation (external or introjected regulation) and amotivation, with autonomous motivation consistently linked to greater persistence, performance, and well-being in domains ranging from education to health care. Additionally, SDT highlights the critical role of goal content: pursuit of intrinsic goals (e.g., personal growth, relationships) fosters need satisfaction and well-being, whereas extrinsic goals (e.g., wealth, fame) often undermine well-being, particularly when pursued for controlled reasons (Ryan & Deci, 2000; Gagné & Deci, 2005).

Personal Expressiveness, conceptualized by Waterman, enriches eudaimonic models by centering the subjective experience of self-realization during personally significant activities. Defined as the feeling of "truly being oneself" when engaging in actions consistent with personal potentials, personal expressiveness is distinct from hedonic enjoyment, emphasizing intrinsic motivation, flow, and effortful engagement with meaningful tasks. Operationalized via the Personally Expressive Activities Questionnaire (PEAQ), this construct measures the experiential component of eudaimonia, capturing feelings of authenticity and purpose that arise from aligning behavior with core values. Unlike models focused on global functioning (e.g., Ryff's PWB) or motivational orientations (e.g., SDT), personal expressiveness highlights the momentary, activity-specific experiences that contribute to eudaimonic well-being, bridging state and trait assessments of self-realization (Keyes, 2002; Kuppens et al., 2010; van de Leemput et al., 2014).

Together, these eudaimonic models address SWB limitations by integrating orientations (values, goals), behaviors (activity characteristics), experiences (subjective feelings), and functioning (positive psychological states), offering a multidimensional account of well-being that transcends the hedonic focus on pleasure and satisfaction. By emphasizing self-realization, autonomous motivation, and purposeful engagement, they provide a more comprehensive foundation for understanding human flourishing.

## *2.3.4 Emergence of Integrative and Multidimensional Well-Being Models*

### 2.3.4.1 Dual-Process and Hybrid Models of Well-Being

Dual-process and hybrid models of well-being emerged from the recognition that hedonic and eudaimonic components, while distinct, are integral to a comprehensive understanding of well-being. Hedonia is typically characterized by pleasure, enjoyment, and the absence of distress, whereas eudaimonia involves growth, meaning, authenticity, and the realization of potential. This bi-dimensional framework acknowledges both as core but separable constructs, though their integration has posed conceptual and operational challenges due to inconsistent definitions and asymmetric treatment (Huta & Waterman, 2014).

Empirical research has sought to clarify their relationship, with factor analytic studies revealing that hedonic indicators (e.g., positive affect, low negative affect) and eudaimonic indicators (e.g., meaning, purpose) load onto separate but strongly correlated factors, sharing up to 69–74% of variance. This overlap suggests they operate synergistically rather than as independent categories, with combined high levels of both associated with optimal functioning (Kashdan et al., 2008).

Historically, early models of subjective well-being (SWB) laid the groundwork for bi-dimensional thinking. Bradburn (1969) proposed a two-factor model where happiness comprises independent positive and negative affect, each correlating incrementally with global well-being judgments. Andrews and Withey (2012) identified three SWB components: life satisfaction (cognitive), positive affect, and negative affect, integrating hedonic (affective) and eudaimonic (evaluative) elements (Diener, 2009). Contemporary SWB research continues to distinguish hedonic (emotional experiences) and cognitive (life satisfaction judgments) components, implicitly endorsing a dual-process framework (Diener et al., 2018).

Complementary dual-process models include bottom-up and top-down theories of well-being: bottom-up approaches focus on accumulating hedonic events (e.g., positive life experiences), while top-down perspectives emphasize cognitive interpretations and eudaimonic processes (e.g., self-realization). Fredrickson's broaden-and-build theory exemplifies a hybrid dynamic, where positive emotions (hedonic) broaden attention and creativity, fostering the development of psychological, social, and physical resources (eudaimonic) (Diener & Ryan, 2009). Huppert (2009) emphasized well-being as combining "feeling good" (hedonic) and "functioning effectively" (eudaimonic), supported by personality traits like extraversion (linked to positive affect) and neuroticism (linked to negative affect) as distinct drivers (Huppert, 2009).

To advance clarity, scholars advocate for symmetrical definitions and precise measurement—avoiding overly broad conceptualizations—and emphasize studying interactions between hedonic and eudaimonic processes (Huta & Waterman, 2014; Kashdan et al., 2008).

#### 2.3.4.2 Prominent Integrative Frameworks: PERMA and Keyes's Mental Health Continuum

Martin Seligman proposed the PERMA model as a comprehensive theory of well-being, expanding beyond his initial focus on happiness to include five measurable components: Positive emotions (e.g., pleasure, joy), Engagement (deep immersion in activities, or "flow"), Relationships (positive social connections), Meaning (belonging to and serving something larger), and Achievement (accomplishments). This model is rooted in rigorous scientific research, providing empirical support for its multidimensional approach to flourishing (Fig. 2.3).

Corey L. M. Keyes operationalizes mental health as a syndrome of positive feelings and positive functioning in life, conceptualizing it as a continuum that includes flourishing, moderately mentally healthy, and languishing states. Flourishing, the highest level, is defined by high emotional well-being and high positive functioning across dimensions such as social relationships, purpose, and engagement. Empirical evidence from the 1995 Midlife in the United States study showed that 17.2% of adults were flourishing, 12.1% were languishing, and 56.6% were moderately mentally healthy. Languishing adults had nearly six times greater risk of major depressive episodes compared to flourishing adults, highlighting distinct psychosocial outcomes associated with different positions on the continuum (Keyes, 2002).

These integrative models align with broader eudaimonic perspectives, such as Carol Ryff's six-factor model of psychological well-being, which includes autonomy, environmental mastery, personal growth, positive relations with others, purpose in life, and self-acceptance. Ryff's model shares conceptual terrain with PERMA and Keyes's continuum, emphasizing multidimensional well-being encompassing purpose, relationships, and growth (Ryff, 2013). Empirical support for such multidimensional approaches is strengthened by measurement tools like the Flourishing Scale (FS) and the Scale of Positive and Negative Experience (SPANE), developed by Ed Diener and colleagues, which assess psychosocial flourishing (e.g., social relationships, meaning, competence) and affective experience, respectively, providing brief, reliable, and valid assessments suitable for research and applied contexts (Diener et al., 2010).

#### 2.3.4.3 Dynamic Systems and Contemporary Theoretical Innovations

Dynamic systems theory has emerged as a pivotal framework for understanding well-being, framing affect and mental states as complex, open systems characterized by nonlinear dynamics, temporal variability, and context-sensitive resilience. This perspective moves beyond static trait models, emphasizing how intraindividual fluctuations and regulatory processes shape psychological functioning over time.

A foundational model is the DynAffect framework, which conceptualizes affect as a dynamic system governed by three core processes: affective home base (a baseline attractor state), affective variability (fluctuations around this baseline), and attractor strength (regulatory forces maintaining stability). DynAffect uses stochastic differential equations to model continuous-time affect trajectories across short (minutes) and long (hours to weeks) scales, with empirical support from experience-sampling studies demonstrating its ability to capture core affect dynamics. Attractor strength, reflecting regulatory feedback loops akin to emotion regulation, prevents prolonged deviations from baseline; cognitive reappraisal correlates with stronger attractor strength (especially for arousal), while rumination is linked to weaker attractor strength, indicating difficulty returning to baseline. Individual differences in these processes align with personality and well-being: a pleasant home base correlates with extraversion and positive affect, greater valence variability relates to neuroticism, and an unpleasant home base associates with negative affect (Kuppens et al., 2010).

Depression has been conceptualized as a complex dynamic system with alternative stable states (e.g., normal mood vs. depressed state) separated by tipping points, a model inspired by catastrophe theory and critical transitions in complex systems (Scheffer et al., 2009). Key indicators of critical slowing down (CSD)—such as increased temporal autocorrelation, variance, and cross-emotion

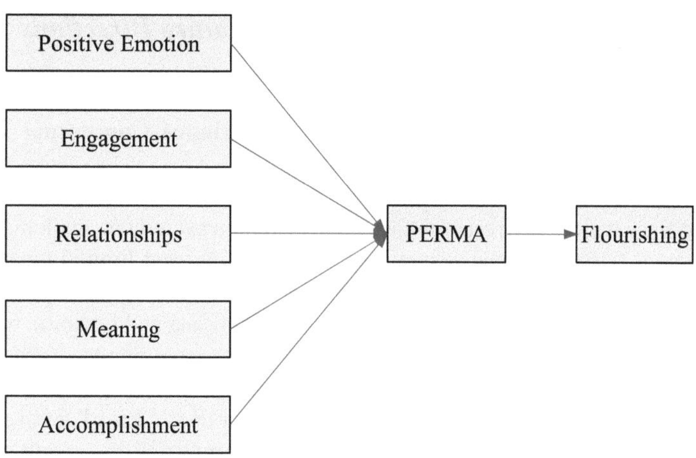


**Fig. 2.3** PERMA model. Source: Created by the authors. Adapted from Seligman, M. E. P. (2011). *Flourish: A visionary new understanding of happiness and well-being*. Free Press

correlations—precede transitions into or out of depression, signaling reduced system resilience. Individuals approaching a depressive episode exhibit elevated autocorrelation and variance in negative emotions, while recovery involves similar signals in positive emotions (van de Leemput et al., 2014). Direct empirical evidence of CSD in depression comes from an intensive longitudinal study tracking 1474 momentary mental states over 239 days during antidepressant dose reduction: rising temporal autocorrelation and variance in affective states, alongside strengthening connections in mental state networks, preceded a clinically significant depressive transition, consistent with increased system fragility (Wichers et al., 2016).

Temporal variability in affect dynamics is multidimensional, extending beyond valence and arousal to intersect with interpersonal behavior. Core affect (fluctuating in valence-arousal space) and interpersonal behavior (fluctuating in agency-communion space) exhibit correlated variability, captured via indices like flux (standard deviation around mean), pulse (intensity variability), and spin (qualitative variability). Greater affective variability correlates with heightened behavioral volatility; agency aligns with arousal, while communion relates to pleasure and low arousal, highlighting nuanced, context-sensitive interrelations (Timmermans et al., 2010). Advanced measures of variability, such as pulse (intensity variability via vector length) and spin (qualitative variability via vector angle), capture complexity: higher spin—reflecting frequent transitions among qualitatively distinct affective states—correlates with neuroticism, depression, and pessimism, and negatively associates with extraversion, optimism, and psychological adjustment, emerging as a robust predictor of well-being (Kuppens et al., 2008). Meta-analytic evidence confirms that heightened affective variability, particularly in negative affect, is consistently associated with poorer psychological well-being and greater psychopathology (Houben et al., 2015).

### *2.3.5 Contemporary Controversies and Future Directions in Well-Being Research*

#### 2.3.5.1 Measurement Challenges: Cultural Invariance, Conceptual Clarity, and Methodological Issues

Contemporary well-being research faces significant measurement challenges rooted in conceptual fuzziness, inconsistent operationalization, and limited measurement invariance across diverse populations. A primary issue stems from the multiplicity and variability in defining core constructs like eudaimonia and hedonia, where differing operationalizations yield discrepant correlations—ranging from .0 to .6 at the trait level and –.3 to .8 at the state level—undermining comparability across studies. This conceptual ambiguity is exacerbated by vague or overly broad definitions, particularly for eudaimonia, and asymmetrical operationalizations, which confound comparisons. To address this, a systematic classification framework has been proposed, emphasizing three key considerations: degree of centrality (core vs.

peripheral elements), category of analysis (orientations, behaviors, experiences, functioning), and level of measurement (trait vs. state). Explicit specification of these dimensions can improve conceptual clarity (Huta & Waterman, 2014).

Measurement invariance—critical for valid cross-group comparisons—poses another major challenge, particularly across cultures, age groups, and genders. While the Satisfaction With Life Scale (SWLS) exhibits configural invariance (equivalent factor structure) across groups, metric invariance (equal factor loadings) is less commonly supported, and scalar invariance (equal intercepts) is rarely achieved for cultural or age comparisons. Cross-cultural research on the SWLS reveals that only 1 of 11 cultural analyses supported scalar or strict invariance, with item interpretations differing by language and cultural context, necessitating rigorous pre-testing of invariance for each population (Emerson et al., 2017). This challenge extends beyond simple translation; cultural differences in response styles (e.g., extreme responding, acquiescence) and the fundamental meaning of concepts like "satisfaction" or "life" require careful adaptation and validation, not just linguistic equivalence (Zhouxiang, 2020).

Methodological inconsistencies hinder progress, with a review identifying 139 distinct measures of subjective well-being (SWB) across studies, complicating cross-study comparisons (Das et al., 2020). This proliferation underscores the need for greater consensus on core measures and the adoption of standardized "toolkits" for specific research questions, while allowing flexibility for context-specific additions (Lucas, 2021). Self-report measures, while foundational, face validity concerns: global life satisfaction judgments are influenced by cultural biases and momentary mood, though they remain valuable for predicting outcomes like relationship stability (Diener et al., 2009). Additionally, single-item measures of SWB suffer from limited content coverage and internal consistency, while multi-item scales may lack cultural or age applicability (Diener, 2009).

To mitigate these challenges, scholars advocate for multidimensional operationalizations that capture distinct yet interrelated facets of well-being. The Flourishing Scale (FS) assesses psychosocial prosperity (e.g., relationships, purpose) as a single score, while the Scale of Positive and Negative Experience (SPANE) includes both general and specific emotions to enhance cultural invariance by focusing on time spent experiencing feelings. Such approaches, combined with explicit attention to core definitional elements, categories of analysis, and measurement levels, can enhance conceptual clarity and comparability across diverse populations. Future work must also address gaps in measurement and prioritize longitudinal and cross-cultural validation to ensure invariance (Diener et al., 2010).

#### 2.3.5.2 Personality, Social, and Economic Moderators of Well-Being

Personality has been found to be more strongly associated with SWB than are life circumstances. Meta-analytic findings robustly confirm that personality traits, particularly neuroticism (strong negative association) and extraversion (strong positive association), are among the strongest predictors of SWB, often exceeding the influence of demographic and circumstantial variables (Steel et al., 2008). Personality traits such as extraversion and neuroticism act as moderate predictors of SWB, especially for affective components: extraversion correlates strongly with positive affect, and neuroticism with negative affect. Other traits such as agreeableness, conscientiousness, optimism, and self-esteem are also associated with well-being measures. Beyond hedonic SWB, personality traits link to psychological well-being (PWB), a multidimensional construct encompassing self-acceptance, positive relations with others, autonomy, environmental mastery, purpose in life, and personal growth (Ryff, 1995). Openness to experience relates to personal growth; agreeableness correlates with positive relations; and extraversion, conscientiousness, and neuroticism predict environmental mastery, purpose in life, and self-acceptance. Two theoretical explanations for these links exist: "temperament theories," proposing direct associations between traits and affective responses, versus "instrumental theories," which posit personality influences SWB indirectly via choices and life events. Contemporary research increasingly supports an integrative view, recognizing both direct temperamental pathways and indirect pathways through life experiences and goals (Ozer & Benet-Martínez, 2006). Behavioral genetic research indicates approximately 40–50% of variance in SWB is heritable, with about 35% of SWB variance reflecting a stable trait component (Lucas & Diener, 2009).

Socioeconomic status (SES) is a notable moderator, influencing well-being levels across the lifespan, with higher SES predicting better outcomes. While life circumstances such as income exhibit relatively small correlations with SWB (typically 0.17–0.21), this relationship is nuanced: income has a stronger association with life evaluation (cognitive component) than with positive/negative affect (emotional components), and its impact is more pronounced at lower income levels and contexts where basic needs are unmet (Diener, Ng, Harter, & Arora, 2010). SES interacts with psychological well-being to moderate biomarkers of health such as interleukin-6 levels, indicating protective effects of well-being in the face of economic adversity (Ryff, 2013).

Social context, including relationship quality and social interactions, significantly influences well-being. Robust meta-analytic evidence underscores that the quality and quantity of social relationships are critical predictors of well-being and even longevity, with effects comparable to well-established risk factors like smoking (Holt-Lunstad et al., 2010). Additionally, religious participation enhances eudaimonic well-being, particularly purpose in life and personal growth, with religious activity (frequency of attendance) showing a stronger association with SWB (weighted mean correlation = .18) than religiosity ($r = .13$) (Witter et al., 1985; Yonker et al., 2012).

Inner strengths, such as optimism, stable self-esteem, effective emotion regulation strategies, empathy, and emotional intelligence, significantly influence well-being levels. These strengths, alongside PWB dimensions like autonomy and environmental mastery, enable individuals to better navigate life challenges, thereby contributing to resilience. Autonomy and environmental mastery help individuals navigate socioeconomic challenges and sustain well-being. Optimism moderates responses to adversity. Meta-analytic findings confirm that dispositional optimism is a consistent predictor of positive mental and physical health outcomes, including higher SWB, particularly through fostering adaptive coping and persistence in goal pursuit (Carver et al., 2010).

#### 2.3.5.3 Application of Well-Being Research to Policy and Intervention

Positive psychological interventions (PPIs) have demonstrated efficacy in enhancing well-being and mitigating mental health symptoms, with meta-analytic evidence supporting their application across diverse settings. Rooted in theories of positive psychology such as the broaden-and-build theory and PERMA model, PPIs aim to cultivate positive emotions, strengths, and meaning, thereby counteracting negativity bias and building psychological resources (Fredrickson, 2001; Seligman, 2011). A meta-analysis by Bolier et al. (2013) reported small to moderate effect sizes for PPIs, including 0.34 for subjective well-being, 0.20 for psychological well-being, and 0.23 for reducing depressive symptoms, with small but significant effects sustained at 3–6 month follow-up. A larger systematic review of 347 randomized controlled trials involving over 72,000 participants across 41 countries found PPIs yielded small to medium post-intervention effect sizes for well-being ($g = 0.39$), character strengths ($g = 0.46$), quality of life ($g = 0.48$), and reductions in depression ($g = -0.39$), anxiety ($g = -0.62$), and stress ($g = -0.58$), with effects maintained at 3 months. Moderating factors include intervention duration (longer programs), format (individual/group over self-help), and population (clinical or referred participants) (Hendriks et al., 2019). In public health, PPIs function as preventive, accessible, and non-stigmatizing tools that can be integrated into stepped care models offering low-intensity, self-directed interventions as a first step (Bolier et al., 2013).

Well-being research has also catalyzed the development of national well-being indices, challenging GDP as the sole metric of societal welfare. The annual World Happiness Report provides robust cross-national comparisons and analyses of the factors driving national well-being differences, informing policy globally (Helliwell et al., 2020). An integrated happiness framework proposed by Choon et al. combines eight outer objective dimensions (environment, education, governance, culture, community, health, safety, economics) and five inner psychological dimensions from the PERMA model (positive emotion, engagement, relationships, meaning, accomplishments). Testing with Malaysian survey data ($N = 1368$) revealed higher ratings for inner strengths compared to satisfaction with external conditions, underscoring psychological factors' role in shaping happiness. Subjective well-being measures are increasingly adopted in national accounts: examples include Bhutan's Gross National Happiness (GNH), and initiatives in the United Kingdom and Australia to track societal well-being. The OECD Guidelines on Measuring Subjective Well-being provide a crucial framework for standardizing measurement approaches to ensure comparability and quality in national statistics (OECD, 2013). National indices also incorporate correlates of subjective well-being, such as economic factors (GDP per capita), sociopolitical conditions (lower inequality, political freedoms), and environmental quality (Diener et al., 2018). Self-reported well-being measures exhibit strong psychometric properties, including high internal consistency, test-retest reliability, and meaningful associations with other constructs, supporting their utility in large-scale population tracking (Diener et al., 2018).

These advancements inform policy-making by advocating for holistic strategies that balance material and psychological well-being. Policymakers are advised to integrate objective external indicators (e.g., economics, safety) with inner psychological resources (e.g., PERMA dimensions) when designing interventions. Evidence-based PPIs can strengthen national mental health agendas by supporting resilience, aiding remission, and complementing problem-based interventions, with self-help formats offering cost-effective population reach. National well-being indices provide nuanced insights beyond GDP, emphasizing social connections, mental health, and time use. Together, PPIs and comprehensive well-being indices offer a dual framework to guide policies prioritizing individual psychological thriving and societal welfare. This dual approach exemplifies the translation of psychological science into tangible strategies for promoting flourishing at both individual and societal levels (Weiss et al., 2016).

#### 2.3.5.4 Open Questions and Emerging Research Trends

Emerging research trends and open questions in well-being research reflect a growing emphasis on dynamic, nuanced, and interdisciplinary approaches to understanding human flourishing. One critical area of inquiry involves the temporal dynamics of psychological need satisfaction with open questions regarding how fluctuations in need satisfaction over time influence well-being trajectories and regulatory processes (Ryan & Deci, 2017). The need for longitudinal and experience-sampling methodologies can capture the dynamic interplay between need satisfaction, motivation, and well-being outcomes. Advanced statistical techniques like network analysis and dynamic structural equation modeling (DSEM) are increasingly employed to model these complex intraindividual processes over time (Lefcheck et al., 2021).

Advancements in nuanced motivation theories, particularly within self-determination theory (SDT), have refined the understanding of how regulatory processes and goal characteristics interact to shape well-being. SDT posits that both the content of goals (e.g., intrinsic aspirations like personal growth, community, health vs. extrinsic aspirations like wealth, fame, image) and the regulatory processes (autonomous: identified, integrated, intrinsic vs. controlled: introjected, external) through which they are pursued are crucial determinants of well-being (Ryan & Deci, 2017; Vansteenkiste et al., 2020). Research shows that pursuing intrinsic goals tends to be more autonomously motivated and satisfies basic psychological needs (autonomy, competence, relatedness), leading to higher well-being. Conversely, pursuing extrinsic goals is often linked to controlled motivation, need frustration, and lower well-being. The interaction between goal content and regulatory style is key: even intrinsic goals pursued for controlled reasons yield lower benefits, while extrinsic goals pursued autonomously are less detrimental (Vansteenkiste et al., 2020). Cultural variations in the expression and experience of autonomy are well-documented, necessitating culturally sensitive models of motivation that account for sociocultural influences on need satisfaction and regulatory styles without assuming universality in expression (Chirkov et al., 2020).

Cross-disciplinary integration within psychology represents another pivotal trend, as researchers seek to bridge motivational theories (like SDT) with developmental, personality, clinical, and intervention science perspectives. Efforts are underway to integrate SDT's core tenets—basic psychological needs theory, organismic integration theory (regulatory styles), goal contents theory, causality orientations theory, and relationships motivation theory—into a more unified framework that better addresses the complexity of well-being across contexts and lifespan stages (Vansteenkiste et al., 2020). A prominent trend involves shifting intervention focus from merely alleviating disorder towards proactively enhancing personal and interpersonal flourishing. These approaches target modifiable psychological drivers of well-being, such as cultivating positive thinking styles (e.g., gratitude, optimism), fostering intrinsic motivation and autonomy support, strengthening relationships, and encouraging prosocial behavior and meaning in life (Dweck, 2017). These interdisciplinary psychological efforts aim to develop comprehensive models that account for both individual differences and contextual factors in shaping well-being trajectories.

### 2.3.6 Conclusion: Integration and Insights from a Psychological Perspective

Psychological research on happiness has robustly established well-being as a multidimensional construct. This necessitates integrating subjective well-being (SWB; encompassing life satisfaction and affect balance) and psychological well-being (PWB; encompassing positive functioning and realization of potential) as distinct yet interrelated dimensions (Disabato et al., 2016; Linley et al., 2013). Consequently, mental health is best conceptualized not merely as the absence of distress, but as a two-dimensional construct encompassing both psychological distress and well-being (Keyes, 2005; Massé et al., 1998). This aligns with the distinction between flourishing—characterized by high levels of positive feelings and positive functioning—and languishing—involving emptiness and stagnation—each carrying distinct psychosocial implications (Keyes, 2002).

Self-Determination Theory (SDT) provides a foundational psychological framework, identifying autonomy (volition), competence (effectance), and relatedness (connection) as universal, inherent psychological needs. The satisfaction of these needs is a primary driver of well-being, growth, and intrinsic motivation (Ryan & Deci, 2017). Csikszentmihalyi's concept of flow enriches this landscape, describing optimal experiences arising from harmonious goal-directed action that reduces psychic entropy (Csikszentmihalyi, 1990).

Personality traits play a profound role, with stronger associations to SWB than previously recognized, particularly when accounting for scale commensurability and facet-level analysis (Steel et al., 2008). Extraversion and neuroticism independently predict positive and negative affect, respectively, influencing long-term well-being trajectories (Costa & McCrae, 1980). Beyond broad traits, character strengths like grit—manifested in persistent daily goal pursuit—enhance SWB, largely through facilitating the satisfaction of basic psychological needs for competence and autonomy (Jiang et al., 2020). Goal commitment interacts with perceived goal attainability to shape well-being changes, mediated by perceived progress in goal achievement (Brunstein, 1993). While adaptation occurs, life circumstances and contexts significantly impact SWB, challenging overstatements of the hedonic treadmill effect (Lucas & Diener, 2009), with the notion of a fixed happiness "set-point" limited by substantial individual variability and the influence of contextual factors (Diener et al., 2018).

Evidence-based interventions offer pathways to enhance well-being. Positive psychology interventions (PPIs) show small to moderate effects on SWB ($d = 0.34$), PWB ($d = 0.20$), and depression ($d = 0.23$), with effectiveness enhanced by longer duration, individual delivery, and healthcare recruitment (Bolier et al., 2013). Physical activity correlates with SWB ($d = 0.360$), particularly positive affect, with stronger effects in experimental settings (Buecker et al., 2021). At the population level, the Rose model suggests that addressing common drivers of well-being and distress through universal approaches may yield greater overall public health impact than solely targeting high-risk groups, while simultaneously building distinct psychological assets (Huppert, 2009).

The multidimensional model of Psychological Well-Being (PWB)—encompassing self-acceptance, positive relations, autonomy, environmental mastery, purpose, and personal growth (Ryff, 1989; Ryff & Keyes, 1995)—provides a comprehensive framework for understanding positive psychological functioning. These dimensions represent core aspects of human flourishing from a psychological standpoint.

Methodological progress has yielded reliable and validated tools for capturing well-being's complexity. These include the Flourishing Scale (FS) for assessing social-psychological prosperity (Diener et al., 2010), the Scale of Positive and Negative Experience (SPANE) for affective balance (Diener et al., 2010), and the Satisfaction With Life Scale (SWLS) for global cognitive evaluations (Diener et al., 1985; Pavot & Diener, 2013). Collectively, these instruments capture the multifaceted nature of SWB, integrating cognitive judgments (life satisfaction) and emotional responses (Diener et al., 2009).

Cultural variations in well-being are evident, with poorer nations reporting greater meaning in life, mediated by religiosity, and meaning (unlike life satisfaction) predicting suicide rates (Oishi & Diener, 2014). Cross-cultural research, such as the MIDUS and MIDJA studies, highlights the need to examine how cultural values and conceptions of the 'good life' shape well-being experiences and their linkages to health. These findings underscore that the pathways to well-being are fundamentally moderated by cultural context (Uchida et al., 2004).

Advancing the science and application of well-being psychology requires concerted efforts. Integrated theoretical frameworks that bridge motivational (e.g., SDT), cognitive, affective, and developmental perspectives are essential for a holistic understanding (Sheldon, 2019). Addressing heterogeneity in SWB dimensions, measurement approaches, and study designs is crucial for comparability and cumulative knowledge (Das et al., 2020). Conducting high-quality research, including rigorous randomized controlled trials in diverse populations, adhering to clear reporting standards, and evaluating cost-effectiveness, remains paramount (Bolier et al., 2013). Translating psychological findings into policy—such as complementing economic indices with well-being metrics—can promote societal flourishing (Diener & Tay, 2015; Oishi, 2012).

#### 2.3.6.1 Summary of the Theoretical Evolution Context

The evolution of psychological theories of happiness has unfolded through distinct phases, each characterized by shifting conceptualizations of well-being and underlying mechanisms.

Phase 1: Affective Focus (1960s–1970s)

From the 1960s to the 1970s, the field was dominated by Affective Balance Theory, which framed happiness primarily as the balance between positive and negative affect. This perspective emphasized hedonic enjoyment, defined as pleasure derived from desired experiences or objects. Bradburn's (1969) seminal work established that positive and negative affect are largely independent dimensions rather than opposite ends of a single continuum (Bradburn, 1969). This independence laid the empirical groundwork for early assessments of subjective well-being (SWB) focused on affective states. Diener's (1984) influential review solidified the importance of affect balance in early SWB research (Diener, 1984).

Phase 2: Cognitive Expansion and Process Theories (1980s–1990s)

The 1980s and 1990s witnessed the emergence and dominance of the tripartite model of SWB, expanding beyond affective states to include cognitive evaluations of life satisfaction. This model, formally articulated by Diener (1984) and operationalized in his later work, comprises the relatively independent components of positive affect, negative affect, and life satisfaction (Diener, 1984; Diener et al., 1999). Concurrently, process-oriented theories gained prominence:

1. Adaptation Theory: Proposed that individuals tend to return to baseline SWB levels after major life events, influenced by temperamental predispositions.
2. Social Comparison & Goal Theories: Posited that happiness judgments depend on comparing current circumstances to internal standards (e.g., aspirations) or social benchmarks. Research increasingly linked the pursuit of intrinsic goals (aligned with personal interests/values) to enhanced well-being compared to extrinsic goals (focused on rewards/image) (Kasser & Ryan, 1993, 1996; Sheldon & Kasser, 1998).
3. Significant measurement advancements occurred during this period. Diener and colleagues developed the Satisfaction With Life Scale (SWLS) to operationalize the cognitive (life satisfaction) component of SWB, enabling robust empirical testing of theoretical developments (Diener et al., 1985).

Phase 3: Eudaimonic Revival (1990s–2000s)

The 1990s–2000s marked a revival of realization theories, emphasizing eudaimonia—defined as self-realization, meaningful functioning, and the expression of one's true potentials—over mere hedonic pleasure. Central to this revival was Ryff's Multidimensional Model of Psychological Well-Being (PWB), which was developed to empirically capture aspects of positive psychological functioning derived from existential and humanistic theories. Her seminal papers identified and validated six core dimensions: autonomy, environmental mastery, personal growth, positive relations with others, purpose in life, and self-acceptance (Ryff, 1989; Ryff & Keyes, 1995). This model fundamentally distinguished itself from earlier hedonic approaches by focusing on growth, meaning, and human flourishing. Self-Determination Theory (SDT), developed by Deci and Ryan, provided a robust theoretical framework centered on innate psychological needs. SDT posits that satisfaction of the basic needs for competence, relatedness, and autonomy is essential for sustained well-being, growth, and intrinsic motivation (Deci & Ryan, 1985; Ryan & Deci, 2000). SDT differentiated between autonomous regulation (acting with a sense of volition and choice, encompassing intrinsic and integrated motivation) and controlled regulation (acting due to internal or external pressures, encompassing introjected and external motivation), linking need-supportive environments fostering autonomy to optimal outcomes. SDT's focus on the quality of motivation and its link to basic needs aligned with eudaimonic frameworks by emphasizing growth and authenticity (Ryan & Deci, 2001). Empirical evidence supported the distinctiveness yet interrelatedness of hedonic and eudaimonic constructs, with eudaimonic activities often correlating with hedonic enjoyment but not vice versa (Waterman et al., 2008).

Phase 4: Integration and Dynamics (2010s–Present)

From the 2010s onward, the field has increasingly shifted toward integrated and dynamic systems perspectives, reflecting a recognition of happiness as a complex, contextually embedded, and temporally fluctuating phenomenon.

1. Integrated Models: Frameworks emerged that explicitly sought to combine elements. Seligman's (2011) PERMA model (Positive emotion, Engagement, Relationships, Meaning, Accomplishment) integrated hedonic elements (positive emotion) with eudaimonic elements (engagement, meaning, accomplishment) and social well-being (relationships) (Seligman, 2011). SDT research increasingly emphasized the interaction of psychological needs, goal contents (intrinsic vs. extrinsic), regulatory processes (autonomous vs. controlled), cultural contexts, and individual differences over time in shaping well-being pathways (Vansteenkiste et al., 2020).
2. Dynamic Systems Perspective: This view conceptualizes well-being as an emergent product of temporal and contextual interactions. It moved beyond static trait or state models to focus on intraindividual variability and the dynamic interplay between SWB components (e.g., momentary affect, cognitive evaluations, recalled experiences) within individuals across situations and time scales (Houben et al., 2015; Kuppens, 2015).
3. Operationalizing Integration: Scales were developed to capture this broader view. The Flourishing Scale (Diener et al., 2010) integrated constructs reflecting self-esteem, purpose, relationships, and competence, aiming to measure "social-psychological prosperity" as an indicator of overall positive functioning (Diener et al., 2010).

These developments reflect a clear progression: from an initial narrow focus on affect balance, through the inclusion of cognitive appraisal and process theories, to the revival and operationalization of eudaimonic well-being, culminating in contemporary efforts toward integrated and dynamic understandings of happiness as a complex, multidimensional psychological process.

#### 2.3.6.2 Implications for Subsequent Subjective Well-Being (SWB) Research

Theoretical Selection

In basic research on subjective well-being (SWB), prioritizing an integrated framework that combines SWB with Self-Determination Theory (SDT) and the PERMA model offers a comprehensive approach to addressing the inherent complexity of well-being. SWB, defined as the cognitive and affective evaluation of one's life, classically encompasses life satisfaction (cognitive judgment), positive affect, and negative affect as distinct yet interrelated components (Diener et al., 2018). Traditional theoretical approaches to SWB have often been examined in isolation, highlighting the critical need for integrated models to bridge fragmented findings and provide a more holistic understanding (Diener et al., 2009).

The integration of SWB with SDT is robustly supported by strong conceptual and empirical overlaps. SDT posits that the satisfaction of three basic psychological needs—autonomy (volition), competence (effectance), and relatedness (connection)—is essential for sustained well-being, optimal functioning, and intrinsic motivation (Ryan & Deci, 2017). This focus on need satisfaction aligns closely with the eudaimonic component of psychological well-being (PWB), which emphasizes positive functioning and realization of potential. Meta-analytic evidence confirms that SWB (hedonic) and PWB (eudaimonic) are distinct but strongly correlated latent factors, with PWB dimensions such as autonomy and positive relations showing clear conceptual and empirical links to SDT's need satisfaction framework (Disabato et al., 2016; Linley et al., 2009). Empirically, Jiang et al. (2020) validated a key pathway within this integration by showing that grit predicts SWB through the mediation of basic psychological needs satisfaction, with cross-sectional ($N = 2963$ adolescents) and daily diary ($N = 78$ college students) data confirming this mechanism at both trait and state levels. Daily grit was associated with increased daily needs satisfaction ($\beta = .49, p < .001$), which in turn enhanced daily life satisfaction and positive affect while reducing negative affect (Jiang et al., 2020).

The PERMA model (Positive Emotion, Engagement, Relationships, Meaning, Accomplishment), proposed by Seligman (2011), provides a complementary framework focusing on core positive psychological elements that constitute flourishing (Seligman, 2011). It enriches the integrated framework by emphasizing positive assets and actionable pathways. Huppert (2009) highlighted that intentional activities—categorized into behaviors (e.g., exercise), cognitions (e.g., gratitude), and motivations (e.g., intrinsic goal pursuit)—drive psychological well-being, aligning well with PERMA's emphasis on engagement and meaning. Additionally, the broaden-and-build theory (Fredrickson, 2001) explains how positive emotions form upward spirals with positive cognitions and behaviors, reinforcing emotional well-being, while personality traits like extraversion and positive thinking styles, which are teachable through interventions, contribute to population-level flourishing (Huppert, 2009). Empirical work supports PERMA's structure as a viable representation of well-being and its links to outcomes like work engagement (Kern et al., 2015).

Cultural considerations underscore the value and necessity of this integrated perspective. Uchida et al. (2004) outlined how cultural constructions of happiness differ fundamentally, with Western (independent) cultures often emphasizing personal achievement and positive affect, while Eastern (interdependent) cultures prioritize relational harmony and balance (Uchida et al., 2004). Integrative models like SDT explicitly acknowledge that while the needs are universal, their expression and satisfaction pathways are culturally moderated (Chirkov et al., 2020).

Progress in SWB research has often been hindered by fragmented theoretical and methodological approaches, with studies frequently focusing on isolated components in isolation. Das et al. (2020) argued that integrating multiple theoretical categories and determinants is essential for constructing coherent and predictive frameworks (Das et al., 2020). The proposed integration of SWB, SDT, and PERMA directly addresses this fragmentation by:

1. Unifying Hedonic and Eudaimonic Perspectives: Combining the affective/cognitive evaluations of SWB with the functioning/growth focus of PWB (via SDT needs & PERMA).
2. Incorporating Motivational Mechanisms: SDT provides the dynamic "why" (needs, regulatory processes) behind goal pursuit and engagement.
3. Highlighting Actionable Positive Elements: PERMA identifies key, modifiable domains (positive emotion, engagement, relationships, meaning, accomplishment) that constitute flourishing.
4. Embedding Cultural Sensitivity: Acknowledging and incorporating how cultural contexts shape the expression and prioritization of needs and PERMA elements.

This integrated framework provides a robust, multifaceted, and psychologically grounded foundation for advancing basic research on the complex nature of human well-being.

Measurement Suggestions

The Satisfaction with Life Scale (SWLS) stands as a compulsory core measure in subjective well-being (SWB) research, primarily designed to assess the cognitive component of adult SWB through global evaluations of life satisfaction. Developed by Diener et al. (1985), the scale comprises five items rated on a 7-point Likert scale, operating as a reflective measurement model that assumes item responses reflect an underlying latent construct of life satisfaction (Diener et al., 1985). High factor loadings across items support its single-factor structure, and it demonstrates strong internal consistency (Cronbach's $\alpha$ typically > .80), making it suitable for diverse populations ranging from adolescents to adults (Jovanović, 2016; Pavot & Diener, 2008).

Psychometrically, the SWLS exhibits robust properties. Short-term test-retest reliability is high (e.g., $r = 0.79$ for 1 month or less) when life circumstances are stable, and it shows moderate temporal stability over longer periods (e.g., a 4-year stability coefficient of approximately 0.54), indicating sensitivity to meaningful life changes (Diener et al., 2018; Pavot & Diener, 2008). Convergent validity is substantial: SWLS scores correlate with reports from friends and family (average $r = 0.42$) (Lucas et al., 1996), and with other well-being measures such as the Flourishing Scale ($r = 0.62$) and the Scale of Positive and Negative Experience (SPANE; SPANE-P: $r = 0.58$, SPANE-B: $r = 0.57$) (Diener et al., 2010). It also maintains discriminant validity from measures of emotional well-being (positive/negative affect), positioning it as a distinct cognitive evaluation tool (Diener et al., 1985; Pavot & Diener, 2008).

However, rigorous measurement invariance (MI) testing across subgroups reveals important considerations. While configural invariance (equivalent factor structure) is generally supported across diverse samples, metric (equal factor loadings), scalar (equal item intercepts), and strict (equal item residuals) invariance are less consistent and crucial for valid group comparisons. Research by Emerson et al. (2017) found that while configural invariance often holds across 24 nations, scalar and strict invariance are rarely achieved for cultural or age comparisons, with items 4 ("So far I have gotten the important things I want in life") and 5 ("If I could live my life over, I would change almost nothing") showing particular non-invariance across age groups (Emerson et al., 2017). Cultural differences in response styles (e.g., extreme responding, acquiescence) and the interpretation of abstract concepts like "life" and "satisfaction" contribute to these challenges (Chen, 2008; Oishi, 2006).

The SWLS is widely used to validate new well-being measures by establishing convergence, yet it has faced theoretical critiques. Ryff (1989) argued that global life satisfaction measures like the SWLS, while valuable, are theoretically underdeveloped as they focus primarily on evaluative judgments of one's life as a whole, potentially overlooking enduring dimensions of positive functioning such as purpose in life, personal growth, autonomy, and environmental mastery (Ryff, 1989). To ensure rigorous application, researchers using the SWLS should:

1. Routinely test for measurement invariance up to the scalar or strict level when making cross-group (e.g., cultural, age, gender) comparisons of mean scores, using established methods like multi-group confirmatory factor analysis.
2. Acknowledge context- and population-dependent variability in interpretations, particularly for items 4 and 5 across age groups and in cross-cultural contexts.
3. Consider supplementing with domain-specific or eudaimonic measures (e.g., PWB scales) when research questions involve dimensions beyond global cognitive evaluation.
4. Leverage its strengths: Its brief length, user-friendliness, sensitivity to interventions, and strong psychometric foundation make it particularly valuable for tracking changes in life satisfaction during psychotherapy, large-scale surveys, and longitudinal studies (Diener et al., 2018; Pavot & Diener, 2008).

### *2.3.7 Conclusion and Research Implications from a Psychological Perspective*

The psychological investigation of well-being robustly establishes it as a complex, multidimensional construct. It necessitates integrating both subjective well-being (SWB)—focusing on life satisfaction and affective balance—and psychological well-being (PWB)—emphasizing positive functioning such as autonomy, purpose, growth, and relationships—as related but distinct domains (Disabato et al., 2016; Keyes, 2002, 2005). This dual perspective is foundational and aligns with the conceptualization of mental health as a complete state encompassing both the absence of psychopathology and the presence of flourishing, reflecting distinct states such as languishing (low well-being, low distress) and flourishing (high well-being) (Keyes, 2002, 2005).

The historical evolution of well-being theories reveals a clear trajectory: from early affect balance models (Bradburn, 1969; Diener, 1984), through the establishment of multidimensional SWB frameworks incorporating cognitive (life satisfaction) and affective components (Diener, 1984; Diener et al., 1999), to the resurgence of eudaimonic-focused models grounded in self-realization (Ryff, 1989; Ryff & Keyes, 1995) and intrinsic motivation mechanisms (Ryan & Deci, 2000, 2017). Contemporary models increasingly integrate these perspectives, recognizing hedonia and eudaimonia as partly overlapping but conceptually and empirically separable dimensions, advocating for dual-process and hybrid frameworks that capture both experience and functioning (Huta & Waterman, 2014; Waterman et al., 2008).

Advances in dynamic systems theory and intensive longitudinal methodologies (e.g., experience sampling, daily diaries) have expanded understanding of intraindividual variations and regulatory mechanisms in affective processes (Houben et al., 2015; Kuppens, 2015). These perspectives underscore the nonlinearity and temporal complexity inherent in well-being, illuminating individual differences in affect dynamics (e.g., variability, inertia, spin), emotional regulation, and resilience, offering refined models for predicting well-being trajectories and critical transitions, including shifts toward or away from psychopathological states (Kuppens, 2015; Wichers et al., 2016).

Measurement considerations remain paramount. The proliferation of diverse scales and persistent definitional ambiguities necessitates comprehensive frameworks that explicitly specify conceptual domains, measurement levels (trait/state), and analytical categories to enhance comparability and validity (Das et al., 2020; Huta & Waterman, 2014). Core instruments, such as the Satisfaction With Life Scale (Diener et al., 1985), Flourishing Scale (Diener et al., 2010), and Scales of Positive and Negative Experience (Diener et al., 2010), provide robust starting points but require ongoing evaluation of psychometric properties, particularly measurement invariance, across diverse populations (Chen, 2008; Emerson et al., 2017).

Personality, social, and economic factors exert substantive and often intertwined influences. Genetic predispositions and dispositional traits (notably extraversion and neuroticism) establish baseline tendencies (Lucas & Diener, 2009; Steel et al., 2008), while the quality of social relationships is consistently one of the strongest predictors of well-being (Holt-Lunstad et al., 2010), and socioeconomic status (SES) interacts with psychological resources to shape outcomes (Diener, Ng, Harter, & Arora, 2010; Ryff, 2013). These findings highlight the interdependence of internal dispositions and external circumstances in shaping individual well-being profiles.

Applied well-being research demonstrates significant potential. Positive psychological interventions (PPIs) show reliable, though often modest, effects in enhancing well-being and reducing symptoms across diverse populations, supporting their integration into public health and clinical practice (Bolier et al., 2013; Weiss et al., 2016). The development and adoption of national well-being indices challenge GDP-centric views and offer nuanced assessments of collective welfare by integrating psychological constructs with objective societal conditions (Helliwell et al., 2020; OECD, 2013).

Emerging research frontiers underscore the necessity for interdisciplinary, dynamic, and culturally sensitive approaches. Key areas include:

1. Elucidating temporal dynamics of need satisfaction and motivational processes using intensive longitudinal designs and advanced modeling (Ryan & Deci, 2017).
2. Refining motivational models by integrating socio-cultural context (Chirkov et al., 2020; Markus & Kitayama, 1991).
3. Developing and evaluating intervention paradigms explicitly targeting flourishing and building psychological assets across the lifespan (Huppert & So, 2013; Seligman, 2011).

Psychological research on happiness and well-being demands integrative, multidimensional models capturing complex interactions among affective experiences, cognitive evaluations, motivational processes, personality, social context, and cultural factors (Diener et al., 2018; Sheldon, 2019). Future research should prioritize methodological rigor, theoretical integration, and translational applications to advance a comprehensive science of human flourishing and its promotion.

## 2.4 Socio-Cultural Perspectives on Happiness: A Comprehensive Theoretical and Model Evolution Review

This section synthesizes the evolution of happiness research from socio-cultural perspectives, emphasizing foundational paradigms, theoretical models, social structural influences, cultural adaptation dynamics, integrative frameworks, and contemporary debates. The study of happiness from a social-cultural perspective has evolved through critical engagement with foundational constructs and theoretical frameworks, reflecting shifting understandings of how culture and society shape well-being. Early psychological theories, such as Markus and Kitayama's (1991) distinction between independent and interdependent self-construals, profoundly influenced cross-cultural research by highlighting cultural variations in self-perception, emotion, and motivation. However, these frameworks were later critiqued for potential overreliance on East-West dichotomies and the limitations of self-report measures in capturing global complexity (Heine et al., 1999; Oyserman et al., 2002). Vignoles et al.'s (2016) seven-dimensional model of independence-interdependence demonstrated that cultural models of selfhood vary complexly across regions and are shaped by socioeconomic development and religious heritage rather than simple binary oppositions.

Central to this evolution is the critique of the individualism-collectivism (IND-COL) dichotomy, a cornerstone of early cross-cultural psychology. Fiske (2002) argued that IND-COL constructs suffer from validity and measurement issues, as meta-analyses show they do not consistently differentiate cultures (e.g., Japan is no more collectivistic than North America) and conflate distinct social relationships and autonomy types, leading to theoretical incoherence. Brewer and Chen expanded this critique by decomposing IND-COL into three levels of self—individual, relational, and collective—each with distinct self-representations, agency beliefs, and values. Their 3 × 3 matrix model revealed that IND and COL are orthogonal rather than oppositional, explaining why individualists often exhibit preferential ingroup treatment similar to collectivists, challenging traditional dichotomies (Brewer & Chen, 2007).

Cultural value orientations have emerged as a more nuanced alternative to IND-COL. Schwartz (2006) proposed seven cultural value orientations structured along three bipolar dimensions: Autonomy vs. Embeddedness, Egalitarianism vs. Hierarchy, and Mastery vs. Harmony. Validated across 73 countries, these orientations formed transnational cultural groupings and correlated with socioeconomic factors: higher development predicted greater autonomy and egalitarianism, which in turn fostered democratization. Empirically, Ye et al. (2015) found that GLOBE cultural dimensions strongly predict subjective well-being (SWB), with power distance (negatively) and gender egalitarianism (positively) accounting for approximately 50% of cross-country variance, surpassing traditional socioeconomic variables.

Cultural variations in happiness extend to definitions, predictors, and emotional experiences. Uchida et al. (2004) demonstrated that European-American contexts typically define happiness as high-arousal positive emotion linked to personal achievement and self-esteem, while East Asian contexts emphasize low-arousal states, interpersonal harmony, and balance of positive and negative affects. These differences align with cultural models of self: independent self-construals prioritize autonomy and personal attributes, whereas interdependent models focus on social connectedness and fulfilling relational obligations (Markus & Kitayama, 1991; Tov & Diener, 2015). Autonomy is a predictor of SWB globally, but its manifestation as "volitional engagement" rather than independence is more cross-culturally consistent (Diener & Tay, 2015).

Social inequality emerges as a critical structural determinant of happiness, with income inequality (rather than absolute income) identified as a key societal-level factor. Wilkinson and Pickett (2017) argued that beyond subsistence, more equal societies exhibit better outcomes across health, education, trust, and social mobility, with psychological mechanisms including stress-related illnesses and stereotype threat impairing cognitive function among disadvantaged groups. Furthermore, research indicates that while absolute income has a diminishing effect on happiness beyond a certain point, perceived income inequality and relative deprivation are powerful social psychological drivers of dissatisfaction (Diener & Oishi, 2000). Social mobility research complicates this narrative: Chan (2018) found that upwardly mobile individuals in the UK report higher life satisfaction, stronger social support, and more positive relationships than immobile working-class peers, contradicting Sorokin's dissociative thesis of mobility-induced social isolation. Similarly, Dhoore et al. (2019) supported Blau's acculturation hypothesis, showing that the well-being of the socially mobile reflects both origin and destination social classes, with no evidence of dissociative strain.

Contemporary global and digital contexts have introduced new dynamics. Globalization triggers varied cultural responses: Chiu and Kwan (2016) identified exclusionary reactions (emotion-driven resistance to cultural threat) and integrative reactions (viewing foreign cultures as intellectual resources), with joint culture priming increasing perceptions of cultural distinctness. Bicultural individuals navigate these dynamics via Bicultural Identity Integration (BII), which captures perceived compatibility (harmony) and overlap (blendedness) between cultural identities; higher BII correlates with better psychological adjustment, including lower depression and anxiety. Cheng and Chan (2006) showed that BII moderates cultural frame switching: high BII individuals assimilate to positive cultural cues, while low BII individuals exhibit contrast effects, reflecting differing strategies for managing dual identities.

Digital social media use (SMU) presents complex implications for happiness. Primack et al. (2017) found that higher SMU frequency correlates with increased perceived social isolation (PSI) among young adults, mediated by displacement of authentic interactions, envy from idealized content, and unmet expectations. Faelens et al. (2019) identified psychological mechanisms: SNS use predicts negative affect via social comparison, repetitive negative thinking (RNT), and reduced self-esteem, with bidirectional feedback loops reinforcing these effects. Crucially, the type of SMU matters; passive consumption (browsing without interaction) is more consistently linked to negative outcomes like envy and decreased well-being than active use (creating content, direct messaging) (Verduyn et al., 2017). Valkenberg (2022)'s umbrella review found small, inconsistent associations between SMU and well-being, urging separate analysis of well-being and ill-being components.

Methodological rigor in cross-cultural research has improved through attention to measurement invariance. Cheung and Rensvold (2002) recommended using Multigroup Confirmatory Factor Analysis (MGCFA) with criteria like ΔCFI, ΔGamma hat, and ΔMcDonald's Noncentrality Index to assess whether constructs are perceived equivalently across cultures. This is essential for ensuring that instruments like the Satisfaction with Life Scale (SWLS) measure the same underlying construct across diverse populations (Diener et al., 2000).

Acculturation models highlight the agency of marginalized groups: Berry et al. (2006)'s bidimensional model identified "integration" (maintaining heritage culture while engaging with the host society) as the most adaptive strategy for immigrant youth, correlating with better psychological (life satisfaction, self-esteem) and sociocultural (school adjustment) outcomes.

Finally, Veenhoven (2012) challenged cultural relativity by arguing that cross-national happiness differences reflect societal effectiveness in meeting universal needs, with measures showing validity despite cultural variations. This tension between universal and culture-specific aspects underscores the dynamic evolution of social-cultural happiness models, integrating value orientations (Schwartz; Hofstede), structural inequalities (Wilkinson & Pickett; Taylor), and contemporary global and digital contexts (Chiu & Kwan; Faelens et al.). As research advances, it increasingly emphasizes nuanced, multi-method approaches to capture the complexity of happiness as a socially and culturally embedded phenomenon.

## *2.4.1 Foundational Paradigms in Cultural Psychology*

### 2.4.1.1 Theoretical Breakthroughs of Cultural Relativism

The assumption of the universality of subjective well-being (SWB) measures has faced significant critique, particularly regarding their validity in collectivist cultures. Central to this critique is the distinction between independent and interdependent self-construals, where Western psychology predominantly adopts an independent self-view emphasizing unique internal attributes and autonomy, whereas many collectivist cultures prioritize interdependent self-views focused on connectedness, relationships, and social context (Markus & Kitayama, 1991). This divergence in self-construals leads to fundamental differences in cognitive, emotional, and motivational processes, undermining the universality of SWB constructs and their measurement. For example, Diener et al.'s Satisfaction with Life Scale, which centers on European-American ideas of happiness tied to personal achievement and attainment, has been shown to yield low scores for Japanese respondents, as personal achievement is not central to well-being in Japanese culture. Moreover, cross-cultural standardized scales often overlook culturally distinct happiness evaluation models: Japanese respondents tend to consider both positive and negative life experiences when assessing happiness, contrasting with Western incremental models that focus on maximizing positive affect (Uchida & Ogihara, 2012), reflecting a dialectical emotional style more common in East Asia (Spencer-Rodgers et al., 2010). Such discrepancies highlight the invalidity of imposing Western SWB constructs on collectivist populations.

Emotional and motivational processes further exacerbate measurement challenges. Interdependent self-construals prioritize other-focused emotions such as shame, empathy, and relational harmony, whereas independent selves emphasize ego-focused emotions like anger and pride. Consequently, SWB measures targeting ego-focused emotions or self-expression fail to capture well-being accurately in collectivist cultures. For instance, "social integration" and "human-heartedness" predict greater happiness for Chinese but not British participants, whereas self-esteem—a key predictor in individualistic contexts—is a weaker indicator of SWB in Eastern collectivist cultures (Lu & Gilmour, 2004) and may even correlate negatively with well-being in some contexts emphasizing modesty (Heine et al., 1999).

Empirical studies have validated these cultural specificities. Research with Asian Americans and Japanese participants demonstrates that interdependent goal pursuit (e.g., pursuing goals to please parents or friends) enhances the relationship between goal attainment and SWB, whereas independent goal pursuit does not significantly predict SWB increases in these groups. This contrasts with findings in European Americans, where independent goal pursuit strengthens SWB outcomes, indicating that individualistic SWB models misrepresent well-being in collectivist contexts (Oishi & Diener, 2001).

While Markus and Kitayama's (1991) self-construal theory has been influential, it is largely based on East-West comparisons that may obscure cultural complexity. Commonly used self-report measures of independence and interdependence often fail to detect predicted cultural differences in collectivist cultures, partly due to premature convergence on a two-dimensional model that neglects the multifaceted nature of self-construals (Vignoles et al., 2016) and the methodological challenges of ensuring scale equivalence across diverse populations (van de Vijver & Leung, 1997). Recent research further highlights cultural variation in happiness conceptions: Americans tend to define happiness as pleasure or enjoyment and view it as universally positive, whereas East Asian and Middle Eastern cultures may emphasize its transient and socially disruptive nature, leading to ambivalence about its desirability (Oishi & Gilbert, 2016) and alternative conceptions focusing on interpersonal harmony and balance (Uchida et al., 2004). Such differences, coupled with inadequate cross-cultural validation of existing scales, reinforce the failure of universal SWB assumptions and call for the integration of indigenous cultural conceptualizations of happiness in assessment practices.

#### 2.4.1.2 The Foundational Role of Cultural Dimension Theories

Hofstede (2001; originally 1980) identified individualism as one of the primary dimensions of cultural variation, where individualist societies emphasize "I" consciousness, autonomy, emotional independence, and individual initiative, whereas collectivist societies stress "we" consciousness, collective identity, emotional dependence, and group solidarity. This individualism-collectivism (I-C) dimension has been the most widely used cultural framework to characterize how individuals define themselves and their relationships with others, influencing domains such as well-being, cognition, and social behavior. Hofstede's model organizes cultural differences into broad patterns, particularly differentiating countries on IND, which facilitated extensive comparative cross-cultural research on happiness and psychological processes (Oyserman et al., 2002). Meta-analytic evidence confirms that the I-C dimension, as part of Hofstede's framework, predicts various outcomes, including well-being, with its predictive strength being strongest for emotions and attitudes, and interacting with contextual moderators such as cultural tightness (Taras et al., 2010) and the level of economic development (Fischer & Boer, 2011).

However, significant critiques of the I-C construct have emerged. Meta-analyses by Oyserman et al. (2002) challenge the validity of IND and COL as meaningful and reliably measurable cultural dimensions, finding no consistent, replicated differences between nations or ethnic groups, and no empirical support that they are opposite ends of a single continuum. The analyses revealed enormous heterogeneity in measurement results due to variations in scales, samples, and contextual factors, undermining generalizability. Critically, IND and COL scales mostly derive from individual self-reports focused on personal attitudes and preferences, failing to capture culture as socially transmitted patterns, institutions, or collective norms; culture is not an individual trait measurable by averaging individual scores but a complex social system (Fiske, 2002). This relates to the fundamental "ecological fallacy" caution, where country-level scores are erroneously applied to predict individual psychological outcomes (Smith, 2004).

Brewer and Gardner's (1996) framework, which distinguishes individual (unique self), relational (defined by close ties), and collective (defined by group membership) selves, highlights that collectivism should be differentiated into relational and group collectivism. For example, East Asian cultures emphasize relational collectivism (interdependence and harmony within close relationships), while Western individualistic cultures exhibit group collectivism based on categorical ingroup-outgroup distinctions, with individualists sometimes showing equal or greater ingroup favoritism depending on measures.

Alternative frameworks, such as Schwartz's theory of cultural value orientations, offer refined perspectives. Schwartz's autonomy versus embeddedness dimension aligns closely with Hofstede's I-C, where autonomy cultures encourage individual expression and uniqueness (shaping happiness as personal fulfillment) and embeddedness cultures emphasize social relationships, tradition, and group harmony (defining happiness in relational and collective terms) (Schwartz, 2008). Unlike Hofstede, Schwartz views cultural dimensions as correlated rather than orthogonal and rooted in normative value systems external to individuals, providing a more comprehensive framework to capture motivational differences in value prioritization that influence happiness definitions (Schwartz, 2003). Furthermore, Schwartz's theory has been directly linked to national differences in subjective well-being, with embeddedness values often showing negative correlations with average life satisfaction (Schwartz, 2013). These critiques and alternatives highlight the need for caution when using Hofstede's I-C dimension as a foundational construct for understanding cultural influences on happiness definitions or experiences.

## 2.4.2 *Theoretical Models of Cultural Variation*

The relationship between culture, self-construal, and well-being is complex and multifaceted. As a foundational framework, the conceptual model in Fig. 2.4 illustrates how macro-level cultural contexts shape individual psychological processes and, ultimately, well-being outcomes.

### 2.4.2.1 Western Individualistic Models

The Western individualistic model of well-being is grounded in an independent construal of self, which views the self as an autonomous entity composed of unique internal attributes (traits, abilities, motives, values) and emphasizes expressing these attributes through "self-actualization" or "realizing oneself". This independent self

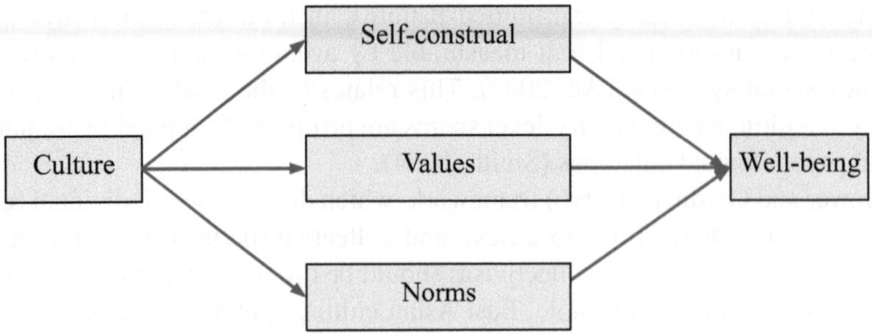


**Fig. 2.4** Conceptual model of cultural influence on self-construal and well-being. Source: Created by the authors (Markus & Kitayama, 1991)

is characterized as a "bounded, unique, more or less integrated motivational and cognitive universe," with behavior organized primarily by internal attributes rather than social context (Markus & Kitayama, 1991). From this self-construal, Western individuals strive for self-actualization-oriented well-being, where subjective well-being (SWB) serves as a primary indicator, rooted in the expression and affirmation of the autonomous inner self. Consequently, individualistic well-being is tied to personal success, feelings of uniqueness, open emotional expression aligned with personal goals, and self-esteem maintenance through self-enhancement and self-serving bias (Oyserman et al., 2002).

However, critiques of Western individualistic models highlight significant limitations. Fiske (2002) argued that individualism (IND) conflates empirically and conceptually distinct phenomena and posits a uniform psychology across these diverse elements. Cross-cultural research does not sufficiently support self-actualization-oriented well-being as a core construct linked to IND, revealing weak or inconsistent associations between IND-COL measures and psychological outcomes like life satisfaction (Diener et al., 2003; Oishi et al., 1999). Additionally, culture is not well-captured by single-dimension individualism rating scales, as different scales often produce heterogeneous and conflicting results. Psychological processes related to self-concept, cognition, and well-being do not show consistent moderation by IND or COL constructs across cultures, questioning their validity as primary indicators of well-being. These critiques suggest that Western models, centered on self-report measures of independence and individual attributes, may insufficiently capture the multifaceted nature of culture or its relation to well-being (Fiske, 2002).

#### 2.4.2.2 East Asian Collectivistic Models

East Asian collectivistic models of well-being are anchored in a theoretical core of relational obligation-oriented well-being, with interpersonal harmony positioned as a central tenet (Uchida et al., 2004). However, critical examinations of collectivism (COL) as a construct reveal complexities that necessitate nuanced conceptualization. Fiske (2002) points out that COL has traditionally conflated diverse types of social bonds and groups, which are distinct and may not be positively correlated; cultures emphasizing one form of relationship or social obligation might devalue or exclude others, challenging the assumption of COL as a unified construct. This critique extends to the validity of standard measures of individualism (IND) and COL, which Fiske argues do not reliably distinguish cultural differences between East Asian and North American groups as previously thought, with studies and scales showing large heterogeneity and inconsistent results. A critical limitation of conventional IND-COL measures is their focus on first-person attitudes and preferences, which diverges from the social obligations, roles, and reference groups that are central to collectivistic cultures. This misalignment leads to the misrepresentation of East Asian relational obligation-oriented well-being, where interpersonal harmony and obligation are core components. Fiske emphasizes that culture encompasses socially transmitted practices, institutions, relationships, and competencies that extend beyond verbal self-reports, suggesting that understanding East Asian collectivistic well-being requires methodologies such as participant observation to capture the relational models and social obligations underlying interpersonal harmony and well-being in these contexts. Despite measurement challenges, empirical studies confirm that relationship harmony is a stronger predictor of life satisfaction for East Asians than for European Americans (Kwan et al., 1997) and that emotions fostering interdependence (e.g., friendly feelings, shame) are more central to well-being in these cultural contexts (Kitayama et al., 2000).

### *2.4.3 Evolution of Social Structure Theories*

Social inequality, as a core dimension of macro-social system models, manifests in multiple domains and exerts profound impacts on societal well-being. Central to understanding the societal effects of inequality is the argument that, beyond a subsistence income level, it is inequality itself—rather than absolute income—that significantly shapes social outcomes. This perspective, advanced by Wilkinson and Pickett, is supported by evidence linking income distribution to a wide range of social problems: more unequal societies exhibit worse performance in physical and mental health, drug abuse, education, incarceration rates, obesity, social mobility, trust, violence, teen pregnancies, and child well-being (Wilkinson & Pickett, 2009). This "gradient effect," referring to the degree of income differences within a population, exacerbates social unease, mistrust, and poor developmental outcomes. A key psychological mechanism underpinning this is the experience of relative deprivation—the perception that one is unfairly disadvantaged compared to a relevant reference group—which generates stress, erodes social cohesion, and negatively impacts health and happiness (Smith et al., 2012).

The mechanisms through which inequality impacts well-being operate across social strata. In more unequal high-income countries, stress-related health problems such as heart disease, diabetes, and certain cancers are more prevalent, affecting not only lower-income groups but also higher-income individuals. Additionally, social mobility is lower in unequal societies due to intergenerational transmission of wealth, limited access to resources via imperfect capital markets, neighborhood segregation, and self-fulfilling prophecies of social status, trapping disadvantaged groups in cycles that worsen disparities. Meta-analytic evidence confirms a small but significant negative association between income inequality and population well-being, particularly in studies using more rigorous longitudinal designs (Ngamaba et al., 2018).

Macroeconomic variables and institutional contexts further mediate the relationship between inequality and well-being. Recessions, for example, impose large "psychic" costs on even employed individuals, requiring substantial compensation to maintain happiness levels. Welfare state generosity, measured by unemployment benefit replacement rates, is positively correlated with national happiness for both employed and unemployed people, with no evidence that such generosity reduces work incentives or narrows the happiness gap between employed and unemployed groups (Tella et al., 2003). This aligns with the "institutional" perspective on well-being, which argues that generous, decommodifying welfare states buffer citizens from market-induced insecurities and status anxieties, thereby promoting higher average levels of life satisfaction (Pacek & Radcliff, 2008). Cross-country comparisons further illustrate these patterns: countries with lower income inequality, such as Japan and Scandinavian nations, experience fewer social problems, often through pre-tax income equality or effective redistributive policies and comprehensive welfare states. Conversely, the USA, despite high wealth, exhibits poor social cohesion and health indicators, exemplifying the negative consequences of entrenched social inequality. The Scandinavian welfare model, in particular, with its combination of low inequality, high social trust, and strong universal institutions, is frequently cited as a benchmark for promoting high levels of societal well-being (Layard, 2005).

## 2.4.4 *Innovations in Cultural Adaptation Theories and Globalization Impacts*

### 2.4.4.1 Cultural Conflict and Tension Models

The psychological adaptation trajectory of immigrants has long been a focal point in acculturation research, with the U-Curve hypothesis and Berry's refined U-Curve Adaptation Theory providing a foundational framework for understanding the dynamic interplay between cultural transition and well-being. According to this theory, immigrant adaptation typically involves three phases: initial adjustment with minimal difficulties, a subsequent period marked by increased challenges (often referred to as "culture shock" or acculturative stress), and eventual long-term adaptation. However, empirical support for a rigid U-Curve pattern is limited, as individual experiences vary significantly based on moderating factors such as age at migration, education, societal attitudes, coping strategies, social support networks, and cultural distance between origin and host societies (Berry, 1997). The experience of acculturative stress—a stress reaction in which the stressors are identified as stemming from the acculturation process—is a key mediator in this adaptation trajectory (Berry, 2006).

Central to Berry's framework is the bidimensional model of acculturation, which classifies individuals into four strategies based on attitudes toward cultural maintenance (heritage culture preservation) and contact with other cultural groups: Integration, Assimilation, Separation, and Marginalization. This conceptual framework is succinctly illustrated in Table 2.5, which maps the four acculturation strategies based on the intersection of these two dimensions. A large-scale study of 5366 immigrant youth (ages 13–18) across 13 countries empirically validated these strategies, revealing that 36.4% of participants adopted Integration, 22.5% Separation, 18.7% Assimilation, and 22.4% Marginalization. Notably, Integration was associated with the highest levels of both psychological (e.g., self-esteem, life satisfaction) and sociocultural (e.g., host society engagement) adaptation, while the Marginalization profile consistently yielded the poorest outcomes (Berry et al., 2006).

Recent advancements in acculturation research have highlighted Bicultural Identity Integration (BII) as a critical moderator of the U-Curve trajectory. Developed by Benet-Martínez and Haritatos (2005), BII captures individual differences in managing dual cultural identities through two core components: blendedness (perceived overlap vs. dissociation between cultural orientations) and harmony (perceived compatibility vs. tension between cultures). Higher BII, particularly harmony, correlates with better psychological adjustment, including lower depression and higher self-esteem. The dynamic nature of BII's influence on immigrant well-being is further illustrated through cultural frame switching (CFS), the process by which biculturals shift between cultural identities in response to contextual cues. Research shows that high BII individuals, who view their cultural identities as compatible, exhibit "assimilation effects"—aligning behaviors with positive cultural primes. In contrast, low BII individuals, perceiving identity conflict, display "contrast effects"—reacting oppositely to positive primes. This pattern reverses with negative cultural cues, where high BII individuals may resist negative stereotypes, while low BII individuals align with them. These findings underscore how BII mediates the U-Curve trajectory by shaping responses to cultural stressors, thereby influencing psychological outcomes during acculturation (Cheng & Chan, 2006). Beyond immigrant contexts, Berry's concepts extend to broader cultural dynamics such as globalization, where individuals often develop bicultural identities combining local and global cultural elements—paralleling the Integration strategy observed in immigrant adaptation.

#### 2.4.4.2 Globalization-Induced Hybridization and Mixed Effects Models

Cultural Hybridization Theory posits that globalization processes drive the dynamic mixing of local and global cultural elements, leading to the expansion and redefinition of well-being concepts as individuals and societies negotiate multiple cultural

**Table 2.5** The bidimensional model of acculturation

| | Maintain heritage culture | Not maintain heritage culture |
|---|---|---|
| Contact with Others | Integration | Assimilation |
| No Contact | Separation | Marginalization |

Source: Created by the authors (Berry et al., 1997)

frames. This process, also referred to as cultural mixing or hybridity, is a core focus of the polycultural psychology perspective, which argues that most individuals are influenced by multiple cultural streams (Hong et al., 2000; Morris et al., 2015). In globalized environments, the co-presence of local and global cultural symbols heightens cultural salience, prompting both exclusionary and integrative responses; integrative responses, in particular, involve synthesizing foreign cultural ideas with heritage culture to achieve valued goals, thereby enabling creative expansions of well-being definitions through cultural mixing.

This dynamic expansion is evident in cross-cultural and temporal variations in happiness conceptions. Research indicates that the meaning of happiness varies both across nations and over time: in the U.S., the 19th-century definition of happiness as "good luck and fortune" has become archaic, evolving into a focus on "pleasurable or enjoyable experiences," whereas many Japanese spontaneously associate happiness with transience ("not lasting long") and social disruption ("envy from others"). Culturally distinct foundational conceptions further underscore this dynamism: East Asian contexts like Japan traditionally emphasize low-arousal positive states and interpersonal harmony in well-being, while European-American contexts prioritize high-arousal states and personal achievement (Uchida & Ogihara, 2012).

The Indian middle class serves as a prime example of such dynamic hybridization. Predominant Eastern traditions, including Hinduism, conceptualize well-being as a spiritual pursuit involving self-realization (atman-brahman unity), virtue (dharma), contentment, and liberation from material attachments, contrasting with Western individualistic, hedonic, and self-enhancement-focused definitions. Globalization, however, has introduced Western-oriented values and aspirations, resulting in a hybridized well-being model that integrates both spiritual harmony and individual achievement (Joshanloo, 2014). This process is mediated by increased information flows about global living standards, which reshape reference norms and aspirations; for instance, heightened awareness of relative income differences, even amid growing personal income, can generate frustration and shift well-being perceptions, reflecting the uneven impacts of globalization on hybrid cultural contexts (Graham & Pettinato, 2002). Cultural hybridization also exhibits mixed effects on individual psychological states and well-being outcomes. Exposure to mixed cultural elements may induce momentary negative emotions due to cognitive efforts to reconcile cultural contradictions, yet this investment simultaneously enhances creative performance (Cheng & Chan, 2006). In Japan, for example, globalization-induced shifts toward individualistic achievement orientation have been linked to decreased interpersonal relationships and social support, negatively affecting subjective well-being; this creates tension as individuals navigate conflicting hybrid models of well-being rooted in traditional interdependence and emerging individualistic values.

## 2.4.5 *Emerging Integrative Models*

### 2.4.5.1 Multilayered Dynamic Frameworks

Multilayered dynamic frameworks for understanding happiness phenomena integrate cultural, social, psychological, and economic layers to explain how these interconnected systems shape well-being experiences. A foundational example is the distinction between independent and interdependent self-construals proposed by Markus and Kitayama (1991). This framework posits that cultural contexts shape self-definitions: independent self-construals, prevalent in Western cultures, emphasize autonomy, uniqueness, and internal attributes, while interdependent self-construals, common in many Asian, Latin American, African, and southern European cultures, prioritize social connectedness, role fulfillment, and harmonious relationships. These self-construals dynamically organize cognition (e.g., context-free vs. situation-specific knowledge representation), emotion (e.g., ego-focused vs. other-focused emotions like pride vs. empathy), and motivation (e.g., self-enhancement vs. social obligation). Extending beyond binary East-West distinctions, Vignoles et al. (2016) developed a seven-dimensional model of self-construal that captures multifaceted independent and interdependent orientations across layers. The dimensions include Self-reliance vs. Dependence on others, Self-containment vs. Connection to others, Difference vs. Similarity, Self-interest vs. Commitment to others, Consistency vs. Variability, Self-direction vs. Receptiveness to influence, and Self-expression vs. Harmony. Validated across 55 cultural groups in 33 nations with over 7000 adults, this model demonstrates that independence and interdependence are not orthogonal but dynamically shaped by socioeconomic development, religious heritage, and socioecological contexts, underscoring the complexity of cultural frameworks underlying happiness.

For bicultural individuals, the Bicultural Identity Integration (BII) framework introduced by Benet-Martínez and Haritatos (2005) offers a dynamic model incorporating dual cultural identities (cultural layer), social networks (social layer), personality traits (psychological layer), and acculturation factors (economic/contextual layer). BII comprises two components: cultural blendedness (subjective overlap vs. dissociation of identities) and cultural harmony (perceived compatibility vs. tension). Higher BII correlates with better psychological adjustment, life satisfaction, and social network integration, illustrating how multilayered identity management influences happiness in multicultural contexts.

Historical and socioecological perspectives further enrich these frameworks. Oishi et al. (2013) traced a cultural evolution in happiness conceptions, from external, luck-based views to internal, agentic models. Cross-culturally, non-Western societies often retain external conceptualizations, with variations linked to socio-economic conditions, pathogen prevalence, and climatic harshness, revealing how ecological and economic layers shape cultural and psychological understandings of well-being. Finally, Diener et al. (2003) provided a comprehensive integrative framework for SWB, encompassing emotional and cognitive evaluations of happiness. They highlighted that personality (e.g., extraversion), cultural norms (e.g., valuation of SWB), economic wealth, and universal psychological needs interact dynamically. Culture moderates which factors influence SWB, channeling universal needs into culturally specific expressions, while cross-cultural measurement studies show that SWB reports reflect both online experiences and normative cultural values, underscoring the layered and dynamic nature of happiness phenomena.

#### 2.4.5.2 Cultural Matching and Fit Theories

The central proposition of cultural matching and fit theories is that individual subjective well-being is maximized when there exists congruence (or "fit") between personal psychological orientations, motivational patterns, or behavioral tendencies and the culturally salient values, norms, or structural characteristics of the social environment. This principle is rooted in the idea that cultural contexts shape dominant self-construals and motivational priorities, and alignment with these cultural prescriptions enhances well-being. For instance, Oishi and Diener (2001) demonstrated that the consequences of goal pursuit for subjective well-being differ cross-culturally based on cultural models of self: independent goal pursuit strengthens the positive effect of goal attainment on well-being among European Americans but not among Asian Americans or Japanese, whereas interdependent goal pursuit enhances this benefit among Asian Americans and Japanese but not among European Americans. This reflects a core "cultural matching" mechanism: well-being is optimized when individuals pursue and attain goals congruent with culturally salient self-construals—independent selves emphasizing "standing out" in Western cultures and interdependent selves emphasizing "fitting in" and harmonious relationships in East Asian cultures.

Person-culture fit, as a broader manifestation of this principle, has been shown to influence happiness across diverse domains. Religious individuals report significantly higher happiness in religious nations than in non-religious nations; subjective status is a stronger predictor of life satisfaction among Americans than Japanese (whereas objective status is a stronger predictor), and self-esteem correlates more strongly with life satisfaction in relationally mobile cultures than in relationally stable ones, with relational mobility moderating this association (Yuki et al., 2007). Collectively, these findings underscore that cultural fit operates through multiple pathways, including alignment between personal goals and cultural self-construals, congruence between individual-level predictors (e.g., subjective status, self-esteem) and cultural values, and adaptation to structural features (e.g., relational mobility, societal religiosity), all of which contribute to cross-cultural variation in subjective well-being (Fulmer et al., 2010; Oishi & Gilbert, 2016).

### 2.4.6 Contemporary Debates and Theoretical Challenges

The Cultural Essentialism Dilemma represents a critical theoretical controversy in social and cultural psychology, stemming from oversimplified, homogeneous, and binary conceptualizations of cultural differences, particularly within individualism-collectivism (IND-COL) research paradigms. A central critique is directed at the fundamental limitations of IND and COL constructs, which treat nations as culturally homogeneous entities, conflate distinct forms of autonomy and sociality (Fiske, 1992), ignore the contextual specificity of norms and values, and reduce culture to individual self-reports lacking external validity. Meta-analytic findings further reveal that IND and COL are neither empirically opposing nor strongly correlated; national comparisons yield heterogeneous, inconsistent results, and different scales produce markedly divergent outcomes even within the same cultural samples, undermining the validity of these constructs as measures of cultural difference (Fiske, 2002). Empirical evidence challenges the binary assumption that individualists are less collectivistic: Western individualists exhibit ingroup favoritism comparable to, or even greater than, collectivists, though the basis of ingroup attachment differs culturally—category-based for individualists versus personalized, relational-based for collectivists (Brewer & Chen, 2007).

Early theories of self-construal, such as the distinction between independent and interdependent selves, have also contributed to cultural essentialism by framing independence as a default universal model while positioning interdependence as a secondary variant. This framework assumes independence involves viewing the self as autonomous, with behavior organized by internal attributes, whereas interdependence entails defining the self through social relationships and aligning behavior with others' perceived thoughts and actions. However, this binary risks overlooking culturally embedded models of selfhood that diverge from these narrow prototypes (Markus & Kitayama, 1991). Critiques of such binary models highlight their role in fostering a simplistic "East-West" cultural dichotomy that marginalizes non-Western cultural systems and assumes monolithic cultural profiles. Commonly used self-report measures of independence and interdependence, rooted in two-dimensional models, have repeatedly failed to demonstrate predicted cultural differences, partly because they treat these constructs as unitary, separate dimensions at the individual level. A major methodological issue here is the reference-group effect, where respondents from different cultures use divergent implicit standards when answering Likert-scale questions about themselves, thereby obscuring real cultural differences (Heine et al., 2002). A seven-dimensional model deconstructing independence and interdependence into factors such as self-reliance versus dependence on others, difference versus similarity, and self-expression versus harmony reveals that cultural models of selfhood are multifaceted and do not cluster into a single bipolar dimension, contradicting earlier theoretical expectations and dismantling essentialist assumptions (Vignoles et al., 2016). This critique aligns with the call for a polycultural psychology that views cultures as interconnected and individuals as actively negotiating multiple cultural influences, rather than being passive carriers of a single, monolithic culture (Morris et al., 2015).

### *2.4.7 Conclusion: Integrative Implications for Socio-Cultural Theory*

#### 2.4.7.1 Tri-Phasic Theoretical Evolution

The conceptualization of cultural dichotomy has undergone significant evolution, shaping the understanding of socio-cultural influences on psychological processes and well-being. A foundational framework is Markus and Kitayama's (1991) distinction between independent and interdependent self-construals. The independent self-construal, prevalent in many Western cultures, views the self as a bounded, unique entity emphasizing internal attributes that are stable across contexts, with behavior organized primarily by reference to internal states. In contrast, the interdependent self-construal, common in Asian, African, and Latin American cultures, construes the self as fundamentally connected to others, prioritizing relational roles, harmony, and context-dependent attributes that vary flexibly across social situations. These divergent models manifest in systematic differences in cognition, emotion, and motivation: independent selves exhibit more abstract, stable self-knowledge and ego-focused emotions, while interdependent selves demonstrate greater contextual specificity in self-perception and other-focused emotions, with motivation oriented toward fulfilling roles and meeting others' expectations rather than personal achievement.

However, traditional dichotomies like individualism (IND) and collectivism (COL) have faced rigorous critique. Fiske (2002) highlights critical limitations, including treating nations as homogeneous cultural units, conflating diverse social relations, ignoring context specificity, and reliance on problematic self-report measures. Meta-analyses show IND and COL are not opposite ends of a continuum; cross-national samples form globular clusters with no consistent correlation between IND and COL scores, and existing measures may capture multiple unidentified respondent attributes rather than coherent cultural constructs. Further, COL merges distinct social bonds, while IND combines conceptually and empirically distinct forms of autonomy, undermining their validity as unitary constructs. To address such conceptual fuzziness, Brewer and Chen (2007) propose a tri-phasic model of self, building on Brewer and Gardner (1996)'s differentiation of individual, relational, and collective selves. This framework clarifies IND-COL by distinguishing relational collectivism (personalized interpersonal networks) from group collectivism (impersonal categorical groups). Cultures differ not along a single continuum but in the salience and priority of these self-levels: Western cultures may emphasize collective selves tied to categorical group memberships, while East Asian cultures prioritize relational selves rooted in interpersonal connections, resolving anomalies such as individualists displaying ingroup favoritism comparable to collectivists.

Recent scholarship further challenges simplistic East-West or independence-interdependence binaries. Vignoles et al. (2016) demonstrate that cultural self-models are multidimensional and ecoculturally informed, identifying a seven-dimensional model of self-construal that captures diverse expressions of independence and interdependence across personal and social domains. Their multinational dataset (7000+ adults from 55 cultural groups across 33 nations) reveals Western cultures emphasize specific independence dimensions but also exhibit commitment to others, contradicting stereotypes of purely independent Western selves. Similarly, non-Western cultures show varied constellations of independent and interdependent dimensions rather than uniform interdependence, underscoring the need for a nuanced, contextually grounded understanding of cultural dichotomy.

#### 2.4.7.2 Theoretical Guidelines for Future Research

For cross-cultural studies investigating the socio-cultural dimensions of well-being, framework selection must prioritize the integration of Cultural Fit Theory and Social Capital Model to capture the complexity of cultural influences on individual outcomes. Such integration should incorporate key elements of Cultural Fit Theory, including cultural distance and acculturation strategy fit. Concurrently, Social Capital Model components, such as social support derived from both heritage and host social networks, are critical for understanding how relational resources moderate cultural adaptation processes and subsequent well-being. To enhance methodological rigor, cross-cultural research should systematically control for cultural context using quantified cultural dimension scores, such as Hofstede's individualism index, which helps operationalize cultural distance and isolate the effects of cultural variables on well-being outcomes. However, the use of such scores must be done with caution, acknowledging their limitations and the ecological fallacy (Smith, 2004). This integrated approach necessitates measuring both group-level societal factors and individual-level psychological factors to comprehensively assess how cultural fit and social capital interact to shape well-being across diverse cultural contexts (Berry, 1997). A fundamental prerequisite for any such comparison is to establish measurement invariance to ensure the scales measure the same construct across groups (van de Vijver & Leung, 1997).

### *2.4.8 Summary*

This review synthesizes the evolution of socio-cultural perspectives on happiness, underscoring a transition from simplified dichotomous frameworks to complex, multilayered models that capture the dynamic interplay of cultural, social, psychological, and structural factors shaping subjective well-being. Initial paradigms privileging universal, individualistic conceptions of well-being have been critically reevaluated, revealing the insufficiency of these approaches in capturing the nuanced, culturally embedded meanings of happiness across diverse societies. The early reliance on the individualism-collectivism (IND-COL) dichotomy as a foundational explanatory construct has been interrogated for conceptual ambiguity, methodological limitations, and empirical inconsistency. These critiques highlight the multidimensional and context-dependent nature of cultural constructs, as well as the heterogeneity within and across national cultures.

Subsequent theoretical advancements propose multidimensional self-construal models and culturally informed value frameworks that emphasize the coexistence and interaction of independence and interdependence, rather than their opposition. Such models elucidate the variable configurations of self-identity and social relational dynamics that underpin divergent well-being conceptions worldwide. Integrating structural social inequalities and social network theories extends understanding beyond individual-level psychological processes to macro-social determinants and micro-social relational mechanisms influencing happiness.

The global and digital era has introduced additional complexity, incorporating bicultural identity integration, acculturation strategies, and media-mediated social comparison as critical moderators of well-being in multicultural and interconnected contexts. These developments foreground the importance of cultural adaptation and the hybridization of well-being models through globalization, revealing both the challenges and creative opportunities arising from cultural mixing. Emergent integrative frameworks emphasize cultural fit and person-environment congruence as core mechanisms optimizing subjective well-being, demonstrating that alignment between individual psychological orientations and culturally salient values or social structures facilitates happiness. This perspective accommodates variability in individual and collective experiences, highlighting the interplay of culture, identity, and social capital.

Moving forward, socio-cultural happiness research calls for methodologically rigorous, multidimensional approaches that employ mixed methods and cross-level analyses. Emphasis on measurement invariance, cultural nuance, and structural contexts is essential to accurately capture well-being phenomena across cultures. Operationalizing constructs beyond traditional dichotomous paradigms toward ecologically valid, multidimensional models will enhance theoretical precision and empirical validity. Future research must continue to bridge the gap between qualitative, ethnographic understandings of cultural meaning and quantitative, comparative studies, leveraging the strengths of both to build a truly holistic science of well-being.

**Funding** This research was funded by

1. Guangdong Provincial Office of Philosophy and Social Sciences [Grant No. GD24XSH06]
2. Projects of PhDs' Start-up Research of GDUPT [Grant No. 2023bsqd1008]

## References


Adler, A. (1931). *What life could mean to you*. Putnam's Sons.

Adler, A. (1956). *The individual psychology of Alfred Adler: A systematic presentation in selections from his writings*. H. L. Ansbacher & R. R. Ansbacher (Eds.). Basic Books.

Allinson, R. E. (1989). *Chuang-Tzu for spiritual transformation: An analysis of the inner chapters*. State University of New York Press.

Ames, R. T. (2011). *Confucian role ethics: A vocabulary*. The Chinese University of Hong Kong Press.

Ames, R. T., & Hall, D. L. (2003). *Daodejing: Making this life significant: A philosophical translation*. Ballantine Books.

Ames, R. T., & Rosemont, H., Jr. (2010). *The analects of Confucius: A philosophical translation*. Ballantine Books.

Andrews, F. M., & Withey, S. B. (2012). *Social indicators of well-being: Americans' perceptions of life quality*. Springer Science & Business Media.

Angle, S. C. (2009). *Sagehood: The contemporary significance of neo-Confucian philosophy*. Oxford University Press.

Annas, J. (2002). Should virtue make you happy? *Apeiron, 35*(4), 1–20.

Argyle, M., & Crossland, J. (1987). The dimensions of positive emotions. *British Journal of Social Psychology, 26*(2), 127–137.

Aristotle. (2014). *Nicomachean ethics* (R. Crisp, Ed. & Trans.). Cambridge University Press (Original work published c. 350 BCE).

Baldrian-Hussein, F. (2008). Neidan. In F. Pregadio (Ed.), *The encyclopedia of Taoism* (pp. 762–766). Routledge.

Barrett, L. F. (2017). *How emotions are made: The secret life of the brain*. Houghton Mifflin Harcourt.

Bartels, M. (2015). Genetics of well-being and its components satisfaction with life, happiness, and quality of life: A review and meta-analysis of heritability studies. *Behavior Genetics, 45*(2), 137–156.

Bartels, M., & Boomsma, D. I. (2009). Born to be happy? The etiology of subjective well-being. *Behavior Genetics, 39*(6), 605–615.

Baselmans, B. M., & Bartels, M. (2018). A genetic perspective on the relationship between eudaimonic–and hedonic well-being. *Scientific Reports, 8*(1), 14610.

Baselmans, B. M. L., Jansen, R., Ip, H. F., van Dongen, J., Abdellaoui, A., van de Weijer, M. P., et al. (2019). Multivariate genome-wide analyses of the well-being spectrum. *Nature Genetics, 51*(3), 445–451.

Bell, D. A. (2008). *China's new Confucianism: Politics and everyday life in a changing society*. Princeton University Press.

Benet-Martínez, V., & Haritatos, J. (2005). Bicultural identity integration (BII): Components and psychosocial antecedents. *Journal of Personality, 73*(4), 1015–1050.

Bentham, J. (1996). *The collected works of Jeremy Bentham: An introduction to the principles of morals and legislation*. Clarendon Press.

Berridge, K. C. (2003). Pleasures of the brain. *Brain and Cognition, 52*(1), 106–128.

Berridge, K. C., & Kringelbach, M. L. (2013). Neuroscience of affect: Brain mechanisms of pleasure and displeasure. *Current Opinion in Neurobiology, 23*(3), 294–303.

Berridge, K. C., & Kringelbach, M. L. (2015). Pleasure systems in the brain. *Neuron, 86*(3), 646–664.

Berry, J. W. (1997). Immigration, acculturation, and adaptation. *Applied Psychology, 46*(1), 5–68.

Berry, J. W. (2006). Contexts of acculturation. In D. L. Sam & J. W. Berry (Eds.), *The Cambridge handbook of acculturation psychology, 27*(42), 328–336.

Berry, J. W., Poortinga, Y. H., & Pandey, J. (Eds.). (1997). *Handbook of cross-cultural psychology: Social behavior and applications* (Vol. 3). John Berry.

Berry, J. W., Sam, D. L., Phinney, J. S., Vedder, P., Berry, J. W., & Horenczyk, G. (2006). *Immigrant youth in cultural transition*. CRC Press.

Bloom, I. (Ed.). (2009). *Mencius*. Columbia University Press.

Bolier, L., Haverman, M., Westerhof, G. J., Riper, H., Smit, F., & Bohlmeijer, E. (2013). Positive psychology interventions: A meta-analysis of randomized controlled studies. *BMC Public Health, 13*(1), 1–20.

Bradburn, N. M. (1969). *The structure of psychological well-being*. Aldine.

Brewer, M. B., & Chen, Y.-R. (2007). Where (who) are collectives in collectivism? Toward conceptual clarification of individualism and collectivism. *Psychological Review, 114*(1), 133–151.

Brewer, M. B., & Gardner, W. (1996). Who is this we? Levels of collective identity and self representations. *Journal of Personality and Social Psychology, 71*(1), 83–93.

Brooks, E. B., & Brooks, A. T. (1998). *The original analects: Sayings of Confucius and his successors*. Columbia University Press.

Brunstein, J. C. (1993). Personal goals and subjective well-being: A longitudinal study. *Journal of Personality and Social Psychology, 65*(5), 1061–1070.

Bryant, F. B., & Veroff, J. (1982). The structure of psychological well-being: A sociohistorical analysis. *Journal of Personality and Social Psychology, 43*(4), 653–673.

Buecker, S., Simacek, T., Ingwersen, B., Terwiel, S., & Simonsmeier, B. A. (2021). Physical activity and subjective well-being in healthy individuals: A meta-analytic review. *Health Psychology Review, 15*(4), 574–592.

Burns, R. A., & Machin, M. A. (2009). Investigating the structural validity of Ryff's psychological well-being scales across two samples. *Social Indicators Research, 93*(2), 359–375.

Buss, D. M. (2000). The evolution of happiness. *American Psychologist, 55*(1), 15–23.

Buunk, A. P., & Gibbons, F. X. (2007). Social comparison: The end of a theory and the emergence of a field. *Organizational Behavior and Human Decision Processes, 102*(1), 3–21.

Cacioppo, J. T., & Patrick, W. (2008). *Loneliness: Human nature and the need for social connection*. W. W. Norton & Company.

Campany, R. F. (2017). *The Chinese dream of nourishing life: Precepts and techniques from the Daozang*. Three Pines Press.

Campbell, A., & Marshall, T. (2002). *The sense of well-being in America: Patterns and trends*. McGraw-Hill.

Carver, C. S., Scheier, M. F., & Segerstrom, S. C. (2010). Optimism. *Clinical Psychology Review, 30*(7), 879–889.

Caspi, A., Hariri, A. R., Holmes, A., Uher, R., & Moffitt, T. E. (2010). Genetic sensitivity to the environment: The case of the serotonin transporter gene and its implications for studying complex diseases and traits. *Focus, 8*(3), 398–416.

Chai, C. (Ed.). (2017). *Dao companion to Daoist philosophy*. Springer.

Chan, W.-T. (1963). *A source book in Chinese philosophy*. Princeton University Press.

Chan, D. (2018). Subjective well-being in Hong Kong, China. In J. F. Helliwell, R. Layard, & J. Sachs (Eds.), *World happiness report 2018* (pp. 86–95). Sustainable Development Solutions Network.

Chen, F. F. (2008). What happens if we compare chopsticks with forks? The impact of making inappropriate comparisons in cross-cultural research. *Journal of Personality and Social Psychology, 95*(5), 1005–1018.

Cheng, C., Hui, W. M., & Lam, S. K. (2000). Perceptual style and behavioral pattern of individuals with functional gastrointestinal disorders. *Health Psychology, 19*(2), 146.

Cheng, C. Y. (1991). *New dimensions of Confucian and neo-Confucian philosophy*. State University of New York Press.

Cheng, S. T., & Chan, A. C. (2006). Relationship with others and life satisfaction in later life: Do gender and widowhood make a difference? *The Journals of Gerontology Series B: Psychological Sciences and Social Sciences, 61*(1), 46–53.

Cheung, G. W., & Rensvold, R. B. (2002). Evaluating goodness-of-fit indexes for testing measurement invariance. *Structural Equation Modeling, 9*(2), 233–255.

Chirkov, V., Ryan, R. M., Kim, Y., & Kaplan, U. (2020). Differentiating autonomy from individualism and independence: A self-determination theory perspective on internalization of cultural orientations and well-being. *Journal of Personality and Social Psychology, 118*(2), 1–22.

Chiu, C. Y., & Kwan, L. Y. Y. (2016). Globalization and psychology. *Current Opinion in Psychology, 8*, 44–48.

Choi, I., Dalal, R., Kim-Prieto, C., & Park, H. (2003). Culture and judgement of causal relevance. *Journal of Personality and Social Psychology, 84*(1), 46.

Chong, K.-M. (2017). The way of softness: A philosophical and practical guide. *Journal of Chinese Philosophy, 44*(3–4), 217–233.

Cleary, T. (1998). *The essential Tao*. HarperCollins.

Cohen, J. D., & Blum, K. I. (2002). Reward and decision. *Neuron, 36*(2), 193–198.

Costa, P. T., & McCrae, R. R. (1980). Influence of extraversion and neuroticism on subjective well-being: Happy and unhappy people. *Journal of Personality and Social Psychology, 38*(4), 668–678.

Csikszentmihalyi, M. (1990). *Flow: The psychology of optimal experience*. Harper and Row.

Cummins, R. A. (2000). Personal income and subjective well-being: A review. *Journal of Happiness Studies, 1*(2), 133–158.

Das, K. V., Jones-Harrell, C., Fan, Y., Ramaswami, A., Orlove, B., & Botchwey, N. (2020). Understanding subjective well-being: Perspectives from psychology and public health. *Public Health Reviews, 41*(1), 1–32.

Davidson, R. J. (2004). Well-being and affective style: Neural substrates and biobehavioural correlates. *Philosophical Transactions of the Royal Society B: Biological Sciences, 359*(1449), 1395–1411.

Davidson, R. J., Kabat-Zinn, J., Schumacher, J., Rosenkranz, M., Muller, D., Santorelli, S. F., et al. (2003). Alterations in brain and immune function produced by mindfulness meditation. *Psychosomatic Medicine, 65*(4), 564–570.

De Bary, W. T., & Bloom, I. (Eds.). (1999). *Sources of Chinese tradition* (Vol. 1, 2nd ed.). Columbia University Press.

Deci, E. L., & Ryan, R. M. (2000). The "what" and "why" of goal pursuits: Human needs and the self-determination of behavior. *Psychological Inquiry, 11*(4), 227–268.

Deci, E. L., & Ryan, R. M. (2013). *Intrinsic motivation and self-determination in human behavior*. Springer Science & Business Media.

Dhoore, J., De Witte, N., & Van Regenmortel, S. (2019). Subjective well-being of older adults in Belgium: A multilevel analysis. *Journal of Happiness Studies, 20*(3), 815–837.

Di Domenico, S. I., & Ryan, R. M. (2017). The emerging neuroscience of intrinsic motivation: A new frontier in self-determination research. *Frontiers in Human Neuroscience, 11*, 247005.

Diener, E. (1984). Subjective well-being. *Psychological Bulletin, 95*(3), 542–575.

Diener, E. (1998). Subjective well-being and personality. In *Advanced personality* (pp. 311–334). Springer US.

Diener, E. (2009). Subjective well-being. In E. Diener (Ed.), *The science of well-being: The collected works of Ed Diener* (pp. 11–58). Springer.

Diener, E., Emmons, R. A., Larsen, R. J., & Griffin, S. (1985). The Satisfaction with Life Scale. *Journal of Personality Assessment, 49*(1), 71–75.

Diener, E., Gohm, C. L., Suh, E. M., & Oishi, S. (2000). Similarity of the relations between marital status and subjective well-being across cultures. *Journal of Cross-Cultural Psychology, 31*(4), 419–436.

Diener, E., Lucas, R. E., & Oishi, S. (2018). Advances and open questions in the science of subjective well-being. *Collabra: Psychology, 4*(1), 1–49.

Diener, E., Lucas, R. E., & Scollon, C. N. (2006). Beyond the hedonic treadmill: Revising the adaptation theory of well-being. *American Psychologist, 61*(4), 305–314.

Diener, E., Ng, W., Harter, J., & Arora, R. (2010). Wealth and happiness across the world: Material prosperity predicts life evaluation, whereas psychosocial prosperity predicts positive feeling. *Journal of Personality and Social Psychology, 99*(1), 52–61.

Diener, E., & Oishi, S. (2000). Money and happiness: Income and subjective well-being across nations. *Culture and Subjective Well-being, 185*, 218.

Diener, E., Oishi, S., & Lucas, R. E. (2009). Subjective well-being: The science of happiness and life satisfaction. In S. J. Lopez & C. R. Snyder (Eds.), *Oxford handbook of positive psychology* (2nd ed., pp. 187–194). Oxford University Press.

Diener, E., Oishi, S., & Lucas, R. E. (2003). Personality, culture, and subjective well-being: Emotional and cognitive evaluations of life. *Annual Review of Psychology, 54*(1), 403–425.

Diener, E., & Ryan, K. (2009). Subjective well-being: A general overview. *South African Journal of Psychology, 39*(4), 391–406.

Diener, E., Suh, E. M., Lucas, R. E., & Smith, H. L. (1999). Subjective well-being: Three decades of progress. *Psychological Bulletin, 125*(2), 276–302.

Diener, E., & Tay, L. (2015). Subjective well-being and human welfare around the world as reflected in the Gallup World Poll. *International Journal of Psychology, 50*(2), 135–149.

Diener, E., Wirtz, D., Tov, W., Kim-Prieto, C., Choi, D., Oishi, S., & Biswas-Diener, R. (2010). New well-being measures: Short scales to assess flourishing and positive and negative feelings. *Social Indicators Research, 97*(2), 143–156.

Disabato, D. J., Goodman, F. R., Kashdan, T. B., Short, J. L., & Jarden, A. (2016). Different types of well-being? A cross-cultural examination of hedonic and eudaimonic well-being. *Psychological Assessment, 28*(5), 471–482.

Donaldson, Z. R., & Young, L. J. (2008). Oxytocin, vasopressin, and the neurogenetics of sociality. *Science, 322*(5903), 900–904.

Dumoulin, H. (2005). *Zen Buddhism: A history* (India and China. (J. W. Heisig & P. Knitter, Trans.)) (Vol. 1). World Wisdom Books.

Dweck, C. S. (2017). From needs to goals and representations: Foundations for a unified theory of motivation, personality, and development. *Psychological Review, 124*(6), 689–719.

Easterlin, R. A. (2003). Explaining happiness. *Proceedings of the National Academy of Sciences, 100*(19), 11176–11183.

Eddington, N., & Shuman, R. (2005). *Subjective well-being: A literature review*. National University of Ireland.

Emerson, S. D., Guhn, M., & Gadermann, A. M. (2017). Measurement invariance of the Satisfaction with Life Scale: Reviewing three decades of research. *Quality of Life Research, 26*(9), 2251–2264.

Emmons, R. A. (1986). Personal strivings: An approach to personality and subjective well-being. *Journal of Personality and Social Psychology, 51*(5), 1058–1068.

Epictetus. (1983). *Handbook of Epictetus* (N. P. White, Trans.). Hackett (Original work published c. 100 CE).

Faelens, L., Hoorelbeke, K., Fried, E., De Raedt, R., & Koster, E. H. (2019). Negative influences of Facebook use through the lens of network analysis. *Computers in Human Behavior, 96*, 13–22.

Festinger, L. (1954). A theory of social comparison processes. *Human Relations, 7*(2), 117–140.

Fischer, R., & Boer, D. (2011). What is more important for national well-being: Money or autonomy? A meta-analysis of well-being, burnout, and anxiety across 63 societies. *Journal of Personality and Social Psychology, 101*(1), 164–184.

Fiske, A. P. (1992). The four elementary forms of sociality: Framework for a unified theory of social relations. *Psychological Review, 99*(4), 689–723.

Fiske, A. P. (2002). Using individualism and collectivism to compare cultures—A critique of the validity and measurement of the constructs: Comment on Oyserman et al. (2002). *Psychological Bulletin, 128*(1), 78–88.

Fredrickson, B. L. (1998). What good are positive emotions? *Review of General Psychology, 2*(3), 300–319.

Fredrickson, B. L. (2001). The role of positive emotions in positive psychology: The broaden-and-build theory of positive emotions. *American Psychologist, 56*(3), 218–226.

Fredrickson, B. L. (2003). The value of positive emotions: The emerging science of positive psychology is coming to understand why it's good to feel good. *American Scientist, 91*(4), 330–335.

Fulmer, C. A., Gelfand, M. J., Kruglanski, A. W., Kim-Prieto, C., Diener, E., Pierro, A., & Higgins, E. T. (2010). On "feeling right" in cultural contexts: How person-culture match affects self-esteem and subjective well-being. *Psychological Science, 21*(11), 1563–1569.

Gagné, M., & Deci, E. L. (2005). Self-determination theory and work motivation. *Journal of Organizational Behavior, 26*(4), 331–362.

Garolera, M., Coppola, R., Muñoz, K. E., Elvevåg, B., Carver, F. W., Weinberger, D. R., & Goldberg, T. E. (2007). Amygdala activation in affective priming: A magnetoencephalography study. *Neuroreport, 18*(14), 1449–1453.

Geary, D. C. (2008). An evolutionarily informed education science. *Educational Psychologist, 43*(4), 179–195.

Graham, A. C. (1992). *Two Chinese philosophers; the metaphysics of the brothers Cheng*. Open Court.

Graham, A. C. (2001). *Chuang-tzu: The inner chapters*. Hackett Publishing.

Graham, C., & Pettinato, S. (2002). *Happiness and hardship: Opportunity and insecurity in new market economies*. Brookings Institution Press.

Grinde, B. (2002). Happiness in the perspective of evolutionary psychology. *Journal of Happiness Studies, 3*(4), 331–354.

Gross, J. J. (Ed.). (2015). *Handbook of emotion regulation* (2nd ed.). Guilford Press.

Haber, S. N., & Knutson, B. (2010). The reward circuit: Linking primate anatomy and human imaging. *Neuropsychopharmacology, 35*(1), 4–26.

Hall, D. L., & Ames, R. T. (1987). *Thinking through Confucius*. State University of New York Press.

Hampton, A. N., & O'Doherty, J. P. (2007). Decoding the neural substrates of reward-related decision making with functional MRI. *Proceedings of the National Academy of Sciences, 104*(4), 1377–1382.

Hansen, C. (2003). Daoism. In E. N. Zalta (Ed.), *The Stanford encyclopedia of philosophy*. (Fall 2003 ed.). Stanford University.

Harbaugh, W. T., Mayr, U., & Burghart, D. R. (2007). Neural responses to taxation and voluntary giving reveal motives for charitable donations. *Science, 316*(5831), 1622–1625.

Harvey, P. (2013). *An introduction to Buddhism: Teachings, history and practices* (2nd ed.). Cambridge University Press.

Headey, B. (2008). Life goals matter to happiness: A revision of set-point theory. *Social Indicators Research, 86*(2), 213–231.

Headey, B. (2010). The set point theory of well-being: Negative results and consequent revisions. *Social Indicators Research, 85*(3), 389–403.

Heine, S. J., Lehman, D. R., Markus, H. R., & Kitayama, S. (1999). Is there a universal need for positive self-regard? *Psychological Review, 106*(4), 766–794.

Heine, S. J., Lehman, D. R., Peng, K., & Greenholtz, J. (2002). What's wrong with cross-cultural comparisons of subjective Likert scales? The reference-group effect. *Journal of Personality and Social Psychology, 82*(6), 903–918.

Heinrichs, M., Baumgartner, T., Kirschbaum, C., & Ehlert, U. (2003). Social support and oxytocin interact to suppress cortisol and subjective responses to psychosocial stress. *Biological Psychiatry, 54*(12), 1389–1398.

Heinrichs, M., & Domes, G. (2008). Neuropeptides and social behaviour: Effects of oxytocin and vasopressin in humans. *Progress in Brain Research, 170*, 337–350.

Henricks, R. G. (2014). *Lao Tzu's Tao Te Ching: A translation of the startling new documents found at Guodian*. Columbia University Press.

Helliwell, J. F., Layard, R., Sachs, J. D., & De Neve, J.-E. (Eds.). (2020). *World happiness report 2020*. Sustainable Development Solutions Network.

Hendriks, T., Schotanus-Dijkstra, M., Hassankhan, A., de Jong, J., & Bohlmeijer, E. (2019). The efficacy of multi-component positive psychology interventions: A systematic review and meta-analysis of randomized controlled trials. *Journal of Happiness Studies, 21*(1), 1–34.

Hikosaka, O., Kim, H. F., Yasuda, M., & Yamamoto, S. (2014). Basal ganglia circuits for reward value-guided behavior. *Annual Review of Neuroscience, 37*(1), 289–306.

Hofstede, G. (2001). *Culture's consequences: Comparing values, behaviors, institutions, and organizations across nations*. International Educational and Professional.

Holt-Lunstad, J., Smith, T. B., & Layton, J. B. (2010). Social relationships and mortality risk: A meta-analytic review. *PLoS Medicine, 7*(7), e1000316.

Hong, Y.-Y., Morris, M. W., Chiu, C.-Y., & Benet-Martínez, V. (2000). Multicultural minds: A dynamic constructivist approach to culture and cognition. *American Psychologist, 55*(7), 709–720.

Hotard, S. R., McFatter, R. M., McWhirter, R. M., & Stegall, M. E. (1989). Interactive effects of extraversion, neuroticism, and social relationships on subjective well-being. *Journal of Personality and Social Psychology, 57*(2), 321–331.

Houben, M., Van Den Noortgate, W., & Kuppens, P. (2015). The relation between short-term emotion dynamics and psychological well-being: A meta-analysis. *Psychological Bulletin, 141*(4), 901.

Huppert, F. A. (2009). Psychological well-being: Evidence regarding its causes and consequences. *Applied Psychology: Health and Well-Being, 1*(2), 137–164.

Huppert, F. A., & So, T. T. (2013). Flourishing across Europe: Application of a new conceptual framework for defining well-being. *Social Indicators Research, 110*(3), 837–861.

Huta, V. (2013). Pursuing eudaimonia versus hedonia: Distinctions, similarities, and relationships. In A. S. Waterman (Ed.), *The best within us: Positive psychology perspectives on eudaimonia* (pp. 139–158). American Psychological Association.

Huta, V. (2016). Eudaimonic and hedonic orientations: Theoretical considerations and research findings. In J. Vittersø (Ed.), *Handbook of eudaimonic well-being* (pp. 215–231). Springer.

Huta, V., & Ryan, R. M. (2010). Pursuing pleasure or virtue: The differential and overlap well-being benefits of hedonic and eudaimonic motives. *Journal of Happiness Studies, 11*(6), 735–762.

Huta, V., & Waterman, A. S. (2014). Eudaimonia and its distinction from hedonia: Developing a classification and terminology for understanding conceptual and operational definitions. *Journal of Happiness Studies, 15*(6), 1425–1456.

Inagaki, H. (1995). *The three pure land sutras: A study and translation*. Nagata Bunshodo.

Insel, T. R. (1997). A neurobiological basis of social attachment. *American Journal of Psychiatry, 154*(6), 726–735.

Irvine, W. B. (2009). *A guide to the good life: The ancient art of stoic joy*. Oxford University Press.

Ivanhoe, P. J. (1993). *Confucian moral self cultivation* (2nd ed.). Hackett Publishing.

Ivanhoe, P. J. (2000). *Confucian moral self cultivation* (2nd ed.). Hackett Publishing.

Ivanhoe, P. J. (2002). *Ethics in the Confucian tradition: The thought of Mengzi and Wang Yangming* (2nd ed.). Hackett.

Jiang, D., Fung, H. H., & Li, T. (2020). Age differences in gratitude: The role of future time perspective. *Journal of Happiness Studies, 21*(3), 1045–1062.

Joshanloo, M. (2014). Eastern conceptualizations of happiness: Fundamental differences with Western views. *Journal of Happiness Studies, 15*(2), 475–493.

Jovanović, V. (2016). The validity of the Satisfaction with Life Scale in adolescents and a comparison with single-item life satisfaction measures: A preliminary study. *Quality of Life Research, 25*(12), 3173–3180.

Kahneman, D. (2011). *Thinking, fast and slow*. Macmillan.

Kahneman, D., & Krueger, A. B. (2006). Developments in the measurement of subjective well-being. *Journal of Economic Perspectives, 20*(1), 3–24.

Kashdan, T. B., Biswas-Diener, R., & King, L. A. (2008). Reconsidering happiness: The costs of distinguishing between hedonics and eudaimonia. *The Journal of Positive Psychology, 3*(4), 219–233.

Kasser, T., & Ryan, R. M. (1993). A dark side of the American dream: Correlates of financial success as a central life aspiration. *Journal of Personality and Social Psychology, 65*(2), 410–422.

Kasser, T., & Ryan, R. M. (1996). Further examining the American dream: Differential correlates of intrinsic and extrinsic goals. *Personality and Social Psychology Bulletin, 22*(3), 280–287.

Kenny, A. (2006). *The rise of modern philosophy: A new history of Western philosophy* (Vol. 3). Oxford University Press.

Kenrick, D. T., & Gutierres, S. E. (1980). Contrast effects and judgments of physical attractiveness: When beauty becomes a social problem. *Journal of Personality and Social Psychology, 38*(1), 131–140.

Kern, M. L., Waters, L. E., Adler, A., & White, M. A. (2015). A multidimensional approach to measuring well-being in students: Application of the PERMA framework. *The Journal of Positive Psychology, 10*(3), 262–271.

Keyes, C. L. M. (2002). The mental health continuum: From languishing to flourishing in life. *Journal of Health and Social Behavior, 43*(2), 207–222.

Keyes, C. L. M. (2004). The nexus of cardiovascular disease and depression revisited: The complete mental health perspective and the moderating role of age and gender. *Aging & Mental Health, 8*(3), 266–274.

Keyes, C. L. M. (2005). Mental illness and/or mental health? Investigating axioms of the complete state model of health. *Journal of Consulting and Clinical Psychology, 73*(3), 539–548.

Keyes, C. L. M. (2007). Promoting and protecting mental health as flourishing: A complementary strategy for improving national mental health. *American Psychologist, 62*(2), 95–108.

Kitayama, S., Markus, H. R., & Kurokawa, M. (2000). Culture, emotion, and well-being: Good feelings in Japan and the United States. *Cognition and Emotion, 14*(1), 93–124.

Kohn, L. (2004). *The Daoist monastic manual: A translation of the Fengdao kejie*. Oxford University Press.

Kohn, L. (2008). *Meditation works: In the Daoist, Buddhist, and Hindu traditions*. Three Pines Press.

Komjathy, L. (2013). *The Daoist tradition: An introduction*. Bloomsbury Academic.

Kraut, R. (1979). Two conceptions of happiness. *The Philosophical Review, 88*(2), 167–197.

Kringelbach, M. L., & Berridge, K. C. (2010). The functional neuroanatomy of pleasure and happiness. *Discovery Medicine, 9*(49), 579–587.

Kringelbach, M. L., & Berridge, K. C. (2017). The affective core of emotion: Linking pleasure, subjective well-being, and optimal metastability in the brain. *Emotion Review, 9*(3), 191–199.

Kuppens, P. (2015). It's about time: A special section on affect dynamics. *Emotion Review, 7*(4), 297–300.

Kuppens, P., Oravecz, Z., & Tuerlinckx, F. (2010). Feelings change: Accounting for individual differences in the temporal dynamics of affect. *Journal of Personality and Social Psychology, 99*(6), 1042–1060.

Kuppens, P., Realo, A., & Diener, E. (2008). The role of positive and negative emotions in life satisfaction judgment across nations. *Journal of Personality and Social Psychology, 95*(1), 66–75.

Kwan, V. S. Y., Bond, M. H., & Singelis, T. M. (1997). Pancultural explanations for life satisfaction: Adding relationship harmony to self-esteem. *Journal of Personality and Social Psychology, 73*(5), 1038–1051.

Lambert, L., D'raven, L., & Pasha-Zaidi, N. (2015). Using the PERMA model in The United Arab Emirates. *Social Indicators Research, 125*(3), 905–933.

Lambert, L., Passmore, H. A., & Holder, M. D. (2015). Foundational frameworks of positive psychology: Mapping well-being orientations. *Canadian Psychology, 56*(3), 311–321.

Larsen, R. J., Diener, E., & Emmons, R. A. (1984). Affect intensity and reactions to daily life events. *Journal of Personality and Social Psychology, 47*(5), 1035–1044.

Larsen, R. J., Diener, E., & Emmons, R. A. (1985). An evaluation of subjective well-being measures. *Social Indicators Research, 17*(1), 1–18.

Lau, D. C. (1970). *Mencius*. Penguin Classics.

Lau, D. C. (1996). *Confucius: The analects*. Columbia University Press.

Lawrence, E. M., Rogers, R. G., & Wadsworth, T. (2015). Happiness and longevity in the United States. *Social Science & Medicine, 145*, 115–119.

Layard, R. (2005). *Happiness: Lessons from a new science*. Penguin Press.

Layard, R. (2011). *Happiness: Lessons from a new science* (2nd ed.). Penguin UK.

LeDoux, J. E. (1995). Emotion: Clues from the brain. *Annual Review of Psychology, 46*(1), 209–235.

Lefcheck, J. S., et al. (2021). *Dynamic structural equation modeling: An introduction*. Guilford Press.

Legge, J. (2009). *The Chinese classics* (Confucian analects, the great learning, and the doctrine of the mean) (Vol. 1). Forgotten Books. (Original work published 1861).

Li, C. (2006). The Confucian ideal of harmony. *Philosophy East and West, 56*(4), 583–603.

Li, C. (2013). *The Confucian philosophy of harmony*. Routledge.

Li, C. (2015). The concept of harmony in Chinese philosophy. In C. Li (Ed.), *The philosophy of Chinese moral tradition* (pp. 67–85). Springer.

Librán, E. C. (2006). Personality dimensions and subjective well-being. *The Spanish Journal of Psychology, 9*(1), 38–44.

Linley, P. A., Joseph, S., Maltby, J., Harrington, S., & Wood, A. M. (2009). Positive psychology applications. In S. J. Lopez (Ed.), *Positive psychology: Exploring the best in people* (Vol. 4, pp. 35–47). Praeger.

Linley, P. A., Maltby, J., Wood, A. M., Osborne, G., & Hurling, R. (2013). Measuring happiness: The higher order factor structure of subjective and psychological well-being measures. *Personality and Individual Differences, 47*(8), 878–884.

Lu, L., & Gilmour, R. (2004). Culture and conceptions of happiness: Individual oriented and social oriented SWB. *Journal of Happiness Studies, 5*(3), 269–291.

Lucas, R. E. (2007). Adaptation and the set-point model of subjective well-being: Does happiness change after major life events? *Current Directions in Psychological Science, 16*(2), 75–79.

Lucas, R. E. (2021). Why the time is right for a renewal of well-being measurement. *Nature Human Behaviour, 5*(6), 675–677.

Lucas, R. E., & Diener, E. (2009). Personality and subjective well-being. In E. Diener (Ed.), *The science of well-being: The collected works of Ed Diener* (pp. 75–102). Springer.

Lucas, R. E., Diener, E., & Suh, E. (1996). Discriminant validity of well-being measures. *Journal of Personality and Social Psychology, 71*(3), 616–628.

Lutz, A., Slagter, H. A., Dunne, J. D., & Davidson, R. J. (2008). Attention regulation and monitoring in meditation. *Trends in Cognitive Sciences, 12*(4), 163–169.

Lykken, D., & Tellegen, A. (1996). Happiness is a stochastic phenomenon. *Psychological Science, 7*(3), 186–189.

Lynn, R. J. (1999). *The classic of the way and virtue: A new translation of the Tao-te ching of Laozi as interpreted by Wang Bi*. Columbia University Press.

Mahmood, T., & Silverstone, T. (2001). Serotonin and bipolar disorder. *Journal of Affective Disorders, 66*(1), 1–11.

Markus, H. R., & Kitayama, S. (1991). Culture and the self: Implications for cognition, emotion, and motivation. *Psychological Review, 98*(2), 224–253.

Martinowich, K., & Lu, B. (2008). Interaction between BDNF and serotonin: Role in mood disorders. *Neuropsychopharmacology, 33*(1), 73–83.

Maslow, A. H. (1943). A theory of human motivation. *Psychological review, 50*(4), 370.

Massé, R., Poulin, C., Dassa, C., Lambert, J., Bélair, S., & Battaglini, A. (1998). The structure of mental health: Higher-order confirmatory factor analyses of psychological distress and well-being measures. *Social Indicators Research, 45*(1), 475–504.

McMahon, D. M. (2006). *Happiness: A history*. Grove Press.

Meaney, M. J. (2010). Epigenetics and the biological definition of gene × environment interactions. *Child Development, 81*(1), 41–79.

Mill, J. S. (2001). *Utilitarianism*. Hackett (Original work published 1861).

Miller, G. E., Chen, E., & Zhou, E. S. (2007). If it goes up, must it come down? Chronic stress and the hypothalamic-pituitary-adrenocortical axis in humans. *Psychological Bulletin, 133*(1), 25–45.

Minkov, M., & Bond, M. H. (2016). A genetic component to national differences in happiness. *Journal of Happiness Studies, 18*(2), 321–340.

Moeller, H. G. (2006). *The philosophy of the Daodejing*. Columbia University Press.

Mogilner, C. (2018). Happiness and the consumption of time. In *The Routledge companion to happiness at work* (pp. 123–135). Routledge.

Morris, M. W., Chiu, C. Y., & Liu, Z. (2015). Polycultural psychology. *Annual Review of Psychology, 66*, 631–659.

Murayama, K., Matsumoto, M., Izuma, K., & Matsumoto, K. (2015). How self-determined choice facilitates performance: A key role of the ventromedial prefrontal cortex. *Cerebral Cortex, 25*(5), 1241–1251.

Nes, R. B., & Røysamb, E. (2015). The heritability of subjective well-being: Review and meta-analysis. In M. Pluess (Ed.), *The genetics of psychological well-being: The role of heritability and genetics in positive psychology* (pp. 75–96). Oxford Academic.

Nes, R. B., Czajkowski, N., & Tambs, K. (2010). Family matters: Happiness in nuclear families and twins. *Behavior Genetics, 40*(5), 577–590.

Nesse, R. M. (2004). Natural selection and the elusiveness of happiness. *Philosophical Transactions of the Royal Society B: Biological Sciences, 359*(1449), 1333–1347.

Ngamaba, K. H., Panagioti, M., & Armitage, C. J. (2018). Income inequality and subjective well-being: A systematic review and meta-analysis. *Quality of Life Research, 27*(3), 577–596.

Nussbaum, M. C. (2001). *Upheavals of thought: The intelligence of emotions*. Cambridge University Press.

OECD. (2013). *Guidelines on measuring subjective well-being*. OECD Publishing.

Oishi, S. (2006). The concept of life satisfaction across cultures: An IRT analysis. *Journal of Research in Personality, 40*(4), 411–423.

Oishi, S. (2012). *The psychological wealth of nations: Do happy people make a happy society?* Wiley-Blackwell.

Oishi, S., & Diener, E. (2001). Goals, culture, and subjective well-being. *Personality and Social Psychology Bulletin, 27*(12), 1674–1682.

Oishi, S., & Diener, E. (2014). Residents of poor nations have a greater sense of meaning in life than residents of wealthy nations. *Psychological Science, 25*(2), 422–430.

Oishi, S., Diener, E., Lucas, R. E., & Suh, E. M. (1999). Cross-cultural variations in predictors of life satisfaction: Perspectives from needs and values. *Personality and Social Psychology Bulletin, 25*(8), 980–990.

Oishi, S., & Gilbert, E. A. (2016). Current and future directions in culture and happiness research. *Current Opinion in Psychology, 8*, 54–58.

Oishi, S., Graham, J., Kesebir, S., & Galinha, I. (2013). Concepts of happiness across time and cultures. *Personality and Social Psychology Bulletin, 39*(5), 559–577.

O'Keefe, T. (2010). *Epicurus on freedom*. Cambridge University Press.

O'Keefe, T. (2014). *Epicureanism*. Routledge.

Okbay, A., Baselmans, B. M. L., De Neve, J.-E., Turley, P., Nivard, M. G., Fontana, M. A., et al. (2016). Genetic variants associated with subjective well-being, depressive symptoms, and neuroticism identified through genome-wide analyses. *Nature Genetics, 48*(6), 624–633.

Okbay, A., Wu, Y., Wang, N., Jayashankar, H., Bennett, M., Nehzati, S. M., et al. (2023). Polygenic prediction of educational attainment within and between families from genome-wide association analyses in 3 million individuals. *Nature Genetics, 55*(3), 437–449.

Ormel, J., Von Korff, M., Burger, H., Scott, K., Demyttenaere, K., Huang, Y.-Q., et al. (2017). Mental disorders among persons with heart disease: Results from world mental health surveys. *Journal of Psychosomatic Research, 103*, 1–10.

Oyserman, D., Coon, H. M., & Kemmelmeier, M. (2002). Rethinking individualism and collectivism: Evaluation of theoretical assumptions and meta-analyses. *Psychological Bulletin, 128*(1), 3–72.
Ozer, D. J., & Benet-Martínez, V. (2006). Personality and the prediction of consequential outcomes. *Annual Review of Psychology, 57*, 401–421.
Pacek, A. C., & Radcliff, B. (2008). Assessing the welfare state: The politics of happiness. *Perspectives on Politics, 6*(2), 267–277.
Park, S. Q., Kahnt, T., Dogan, A., Strang, S., Fehr, E., & Tobler, P. N. (2017). A neural link between generosity and happiness. *Nature Communications, 8*, 15964.
Pavot, W., & Diener, E. (2008). The Satisfaction with Life Scale and the emerging construct of life satisfaction. *The Journal of Positive Psychology, 3*(2), 137–152.
Pavot, W., & Diener, E. (2013). Happiness experienced: The science of subjective well-being. In I. Boniwell, S. A. David, & A. C. Ayers (Eds.), *Oxford handbook of happiness* (pp. 134–151). Oxford University Press.
Peterson, C., & Seligman, M. E. P. (2004). *Character strengths and virtues: A handbook and classification*. Oxford University Press.
Pregadio, F. (Ed.). (2008). *The encyclopedia of Taoism*. (2 vols.). Routledge.
Primack, B. A., Shensa, A., Escobar-Viera, C. G., Barrett, E. L., Sidani, J. E., Colditz, J. B., & James, A. E. (2017). Use of multiple social media platforms and symptoms of depression and anxiety: A nationally-representative study among U.S. young adults. *Computers in Human Behavior, 69*, 1–9.
Przybylski, A. K., Murayama, K., DeHaan, C. R., & Gladwell, V. (2013). Motivational, emotional, and behavioral correlates of fear of missing out. *Computers in Human Behavior, 29*(4), 1841–1848.
Raibley, J. R. (2012). Happiness is not well-being. *Journal of Happiness Studies, 13*(6), 1105–1129.
Raz, G. (2012). *The way of the samurai: Yukio Mishima on Hagakure and the code of the warrior*. Penguin Classics.
Rietveld, C. A., Medland, S. E., Derringer, J., Yang, J., Esko, T., Martin, N. W., et al. (2013). GWAS of 126, 559 individuals identifies genetic variants associated with educational attainment. *Science, 340*(6139), 1467–1471.
Rogers, C. R. (1963). The concept of the fully functioning person. *Psychotherapy: Theory, Research & Practice, 1*(1), 17–26.
Rosemont, H., Jr., & Ames, R. T. (2016). *Confucian role ethics: A moral vision for the 21st century?* Göttingen University Press.
Røysamb, E., Nes, R. B., Czajkowski, N. O., & Vassend, O. (2018). Genetics, personality and wellbeing. A twin study of traits, facets and life satisfaction. *Scientific Reports, 8*(1), 12298.
Różycka-Tran, J., Boski, P., & Wojciszke, B. (2015). Belief in a zero-sum game as a social axiom: A 37-nation study. *Journal of Cross-Cultural Psychology, 46*(4), 525–548.
Ruini, C., & Ryff, C. D. (2016). Using eudaimonic well-being to improve lives. In A. M. Wood & J. Johnson (Eds.), *The Wiley handbook of positive clinical psychology* (pp. 153–166). Wiley.
Russell, J. A. (2009). Emotion, core affect, and psychological construction. *Cognition and Emotion, 23*(7), 1259–1283.
Russo, S. J., & Nestler, E. J. (2013). The brain reward circuitry in mood disorders. *Nature Reviews Neuroscience, 14*(9), 609–625.
Ryan, R. M., & Deci, E. L. (2000). Self-determination theory and the facilitation of intrinsic motivation, social development, and well-being. *American Psychologist, 55*(1), 68.
Ryan, R. M., & Deci, E. L. (2001). On happiness and human potentials: A review of research on hedonic and eudaimonic well-being. *Annual Review of Psychology, 52*(1), 141–166.
Ryan, R. M., & Deci, E. L. (2017). *Self-determination theory: Basic psychological needs in motivation, development, and wellness*. Guilford.
Ryan, R. M., & Deci, E. L. (2020). Intrinsic and extrinsic motivation from a self-determination theory perspective: Definitions, theory, practices, and future directions. *Contemporary Educational Psychology, 61*, 101860.

Ryff, C. D. (1989). Happiness is everything, or is it? Explorations on the meaning of psychological well-being. *Journal of Personality and Social Psychology, 57*(6), 1069–1081.
Ryff, C. D. (1995). Psychological well-being in adult life. *Current Directions in Psychological Science, 4*(4), 99–104.
Ryff, C. D. (2013). Eudaimonic well-being and health: Mapping consequences of self-realization. In A. S. Waterman (Ed.), *The best within us: Positive psychology perspectives on eudaimonia* (pp. 77–98). American Psychological Association.
Ryff, C. D., Boylan, J. M., & Kirsch, J. A. (2021). Eudaimonic and hedonic well-being. In M. T. Lee, L. D. Kubzansky, & T. J. VanderWeele (Eds.), *Measuring well-being* (pp. 92–135). Oxford University Press.
Ryff, C. D., Heller, A. S., Schaefer, S. M., Van Reekum, C., & Davidson, R. J. (2016). Purposeful engagement, healthy aging, and the brain. *Current Behavioural Neuroscience Reports, 3*(4), 318–327.
Ryff, C. D., & Keyes, C. L. M. (1995). The structure of psychological well-being revisited. *Journal of Personality and Social Psychology, 69*(4), 719–727.
Ryff, C. D., & Singer, B. H. (1996). Psychological well-being: Meaning, measurement, and implications for psychotherapy research. *Psychotherapy and Psychosomatics, 65*(1), 14–23.
Ryff, C. D., & Singer, B. H. (2008). Know thyself and become what you are: A eudaimonic approach to psychological well-being. *Journal of Happiness Studies, 9*(1), 13–39.
Šāntideva. (2006). *The way of the Bodhisattva: (Bodhicaryāvatāra).* (Padmakara Translation Group, Trans.). Shambhala Publications.
Santos, V., Paes, F., Pereira, V., Arias-Carrión, O., Silva, A. C., Carta, M. G., et al. (2013). The role of positive emotion and contributions of positive psychology in depression treatment: Systematic review. *Clinical Practice and Epidemiology in Mental Health, 9*(1), 221–237.
Scheffer, M., Bascompte, J., Brock, W. A., Brovkin, V., Carpenter, S. R., Dakos, V., Held, H., van Nes, E. H., Rietkerk, M., & Sugihara, G. (2009). Early-warning signals for critical transitions. *Nature, 461*(7260), 53–59.
Schipper, K. (1993). *The Taoist body.* (K. C. Duval, Trans.). University of California Press.
Schultz, W. (1998). Predictive reward signal of dopamine neurons. *Journal of Neurophysiology, 80*(1), 1–27.
Schultz, W. (2007). Behavioral dopamine signals. *Trends in Neurosciences, 30*(5), 203–210.
Schultz, W. (2010). Dopamine signals for reward value and risk: Basic and recent data. *Behavioral and Brain Functions, 6*(1), 24.
Schultz, W. (2016). Reward functions of the basal ganglia. *Journal of Neural Transmission, 123*(7), 679–693.
Schwartz, S. (2008). Cultural value orientations: Nature & implications of national differences. *Psychology. Journal of Higher School of Economics, 5*(2), 37–67.
Schwartz, S. (2013). Value priorities and behavior: Applying a theory of integrated value systems. In *The psychology of values* (pp. 1–24). Psychology Press.
Schwartz, S. H. (2003). A proposal for measuring value orientations across nations. *Questionnaire Development Report of the European Social Survey, 259*(290), 261.
Schwartz, S. H. (2006). A theory of cultural value orientations: Explication and applications. *Comparative Sociology, 5*(2–3), 137–182.
Seligman, M. E. P. (2002). *Authentic happiness: Using the new positive psychology to realize your potential for lasting fulfillment.* Free Press.
Seligman, M. E. P. (2011). *Flourish: A visionary new understanding of happiness and well-being.* Free Press.
Seligman, M. E. P., Railton, P., Baumeister, R. F., & Sripada, C. (2013). Navigating into the future or driven by the past. *Perspectives on Psychological Science, 8*(2), 119–141.
Shamay-Tsoory, S. G., Aharon-Peretz, J., & Perry, D. (2007). Two systems for empathy: A double dissociation between emotional and cognitive empathy in inferior frontal gyrus versus ventromedial prefrontal lesions. *Brain, 132*(3), 617–627.

Sharf, R. H. (2015). Buddhist modernism and the rhetoric of meditative experience. In *Buddhism and modernity: Sources from nineteenth-century Japan* (pp. 267–297). University of Hawaii Press.
Sheldon, K. M. (2019). Putting eudaimonia in its place: On the predictor, not the outcome, side of the equation. In J. Vitterso (Ed.), *The handbook of eudaimonic well-being* (pp. 447–462). Springer.
Sheldon, K. M., & Gunz, A. (2009). Psychological needs as basic motives, not just experiential requirements. *Journal of Personality, 77*(5), 1467–1492.
Sheldon, K. M., & Kasser, T. (1998). Pursuing personal goals: Skills enable progress, but not all progress is beneficial. *Personality and Social Psychology Bulletin, 24*(12), 1319–1331.
Sheldon, K. M., & Lyubomirsky, S. (2021). Revisiting the sustainable happiness model and pie chart: Can happiness be successfully pursued? *Journal of Positive Psychology, 16*(2), 145–154.
Shipton, C. (2020). The unity of Acheulean culture. In *Culture history and convergent evolution: Can we detect populations in prehistory?* (pp. 13–27). Springer International Publishing.
Shun, K.-L. (1997). *Mencius and early Chinese thought*. Stanford University Press.
Shun, K.-L. (2004). Conception of the person in early Confucian thought. In K.-L. Shun & D. B. Wong (Eds.), *Confucian ethics: A comparative study of self, autonomy, and community* (pp. 183–199). Cambridge University Press.
Shun, K.-L., & Wong, D. B. (Eds.). (2004). *Confucian ethics: A comparative study of self, autonomy, and community*. Cambridge University Press.
Slingerland, E. (2007). *Effortless action: Wu-wei as conceptual metaphor and spiritual ideal in early China*. Oxford University Press.
Slingerland, E. (2015). *Trying not to try: Ancient China, modern science, and the power of spontaneity*. Crown.
Slingerland, E. (2019). *Mind and body in early China: Beyond orientalism and the myth of harmony*. Oxford University Press.
Smith, H. J., Pettigrew, T. F., Pippin, G. M., & Bialosiewicz, S. (2012). Relative deprivation: A theoretical and meta-analytic review. *Personality and Social Psychology Review, 16*(3), 203–232.
Smith, P. B. (2004). Acquiescent response bias as an aspect of cultural communication style. *Journal of Cross-Cultural Psychology, 35*(1), 50–61.
Sonuga-Barke, E. J. S., & Castellanos, F. X. (2007). Spontaneous attentional fluctuations in impaired states and pathological conditions: A neurobiological hypothesis. *Neuroscience & Biobehavioral Reviews, 31*(7), 977–986.
Spencer-Rodgers, J., Williams, M. J., & Peng, K. (2010). Cultural differences in expectations of change and tolerance for contradiction: A decade of empirical research. *Personality and Social Psychology Review, 14*(3), 296–312.
Srivastava, A., Locke, E. A., & Bartol, K. M. (2001). Money and subjective well-being: It's not the money, it's the motives. *Journal of Personality and Social Psychology, 80*(6), 959–971.
Steel, P., Schmidt, J., & Shultz, J. (2008). Refining the relationship between personality and subjective well-being. *Psychological Bulletin, 134*(1), 138–161.
Stevenson, B., & Wolfers, J. (2013). Subjective well-being and income: Is there any evidence of satiation? *American Economic Review, 103*(3), 598–604.
Sundquist, K., Frank, G., & Sundquist, J. A. N. (2004). Urbanisation and incidence of psychosis and depression: Follow-up study of 4.4 million women and men in Sweden. *The British Journal of Psychiatry, 184*(4), 293–298.
Tan, J. J. X., Kraus, M. W., Carpenter, N. C., & Adler, N. E. (2020). The association between objective and subjective socioeconomic status and subjective well-being: A meta-analytic review. *Psychological Bulletin, 146*(11), 970–1020.
Taras, V., Kirkman, B. L., & Steel, P. (2010). Examining the impact of culture's consequences: A three-decade, multilevel, meta-analytic review of Hofstede's cultural value dimensions. *Journal of Applied Psychology, 95*(3), 405–439.
Tatarkiewicz, W. (1976). *Analysis of happiness*. Martinus Nijhoff.

Tella, R. D., MacCulloch, R. J., & Oswald, A. J. (2003). The macroeconomics of happiness. *Review of Economics and Statistics, 85*(4), 809–827.

Thompson, E. R. (2007). Development and validation of an internationally reliable short-form of the Positive and Negative Affect Schedule (PANAS). *Journal of Cross-Cultural Psychology, 38*(2), 227–242.

Timmermans, T., Van Mechelen, I., & Kuppens, P. (2010). The relationship between individual differences in adult attachment and the experience of negative and positive affect. *Emotion, 10*(1), 10–21.

Tov, W., & Diener, E. (2015). Subjective well-being. In *International encyclopedia of the social & behavioral sciences* (Vol. 23, 2nd ed., pp. 654–660). Elsevier.

Tu, W. (1979). Humanity and self-cultivation: Essays in Confucian thought. Asian Humanities Press.

Tu, W. (1985). *Confucian thought: Selfhood as creative transformation*. State University of New York Press.

Tu, W. (1989). *Centrality and commonality: An essay on Confucian religiousness*. State University of New York Press.

Tugade, M. M., & Fredrickson, B. L. (2007). Regulation of positive emotions: Emotion regulation strategies that promote resilience. *Journal of Happiness Studies, 8*(3), 311–333.

Uchida, Y., & Ogihara, Y. (2012). Personal or interpersonal construal of happiness: A cultural psychological perspective. *International Journal of Wellbeing, 2*(4), 354–369.

Uchida, Y., Norasakkunkit, V., & Kitayama, S. (2004). Cultural constructions of happiness: Theory and empirical evidence. *Journal of Happiness Studies, 5*(3), 223–239.

Valkenburg, P. M. (2022). Social media use and well-being: What we know and what we need to know. *Current Opinion in Psychology, 45*, 101294.

Van de Leemput, I. A., Wichers, M., Cramer, A. O. J., Borsboom, D., Tuerlinckx, F., Kuppens, P., van Nes, E. H., Viechtbauer, W., Giltay, E. J., Aggen, S. H., Derom, C., Jacobs, N., Kendler, K. S., van der Maas, H. L. J., Neale, M. C., Peeters, F., Thiery, E., Zachar, P., & Scheffer, M. (2014). Critical slowing down as early warning for the onset and termination of depression. *Proceedings of the National Academy of Sciences, 111*(1), 87–92.

van de Vijver, F. J. R., & Leung, K. (1997). *Methods and data analysis for cross-cultural research*. Sage.

Van Norden, B. W. (2007). *Virtue ethics and consequentialism in early Chinese philosophy*. Cambridge University Press.

Van Norden, B. W. (2008). *Mengzi: With selections from traditional commentaries*. Hackett Publishing.

Vansteenkiste, M., Ryan, R. M., & Soenens, B. (2020). Basic psychological need theory: Advancements, critical themes, and future directions. *Motivation and Emotion, 44*(1), 1–31.

Veenhoven, R. (2012). Cross-national differences in happiness: Cultural measurement bias or effect of culture? *International Journal of Wellbeing, 2*(4), 333–353.

Verduyn, P., Ybarra, O., Résibois, M., Jonides, J., & Kross, E. (2017). Do social network sites enhance or undermine subjective well-being? A critical review. *Social Issues and Policy Review, 11*(1), 274–302.

Vignoles, V. L., Owe, E., Becker, M., Smith, P. B., Easterbrook, M. J., Brown, R., González, R., Didier, N., Carrasco, D., Cadena, M. P., Lay, S., Schwartz, S. J., Des Rosiers, S. E., Villamar, J. A., Gavreliuc, A., Zinkeng, M., Kreuzbauer, R., Baguma, P., Martin, M., … Bond, M. H. (2016). Beyond the 'east–west' dichotomy: Global variation in cultural models of selfhood. *Journal of Experimental Psychology: General, 145*(8), 966–1000.

Waterman, A. S. (1993). Two conceptions of happiness: Contrasts of personal expressiveness (eudaimonia) and hedonic enjoyment. *Journal of Personality and Social Psychology, 64*(4), 678–691.

Waterman, A. S., Schwartz, S. J., & Conti, R. (2008). The implications of two conceptions of happiness (hedonic enjoyment and eudaimonia) for the understanding of intrinsic motivation. *Journal of Happiness Studies, 9*(1), 41–79.

Watson, B. (1968). *The complete works of Chuang Tzu*. Columbia University Press.

Weiss, L. A., Westerhof, G. J., & Bohlmeijer, E. T. (2016). Can we increase psychological well-being? The effects of interventions on psychological well-being: A meta-analysis of randomized controlled trials. *PLoS One, 11*(6), e0158092.

Wichers, M., Wigman, J. T. W., & Myin-Germeys, I. (2016). Micro-level affect dynamics in psychopathology viewed from complex dynamical system theory. *Emotion Review, 8*(4), 362–367.

Wilkinson, R. G., & Pickett, K. E. (2009). *The spirit level: Why more equal societies almost always do better*. Allen Lane.

Wilkinson, R. G., & Pickett, K. E. (2017). *The inner level: How more equal societies reduce stress, restore sanity and improve everyone's well-being*. Penguin Press.

Williams, P. (2009). *Mahayana Buddhism: The doctrinal foundations* (2nd ed.). Routledge.

Wills, T. A. (1981). Downward comparison principles in social psychology. *Psychological Bulletin, 90*(2), 245–271.

Witter, R. A., Okun, M. A., Stock, W. A., & Haring, M. J. (1985). Religion and subjective well-being in adulthood: A quantitative synthesis. *Review of Religious Research, 26*(4), 332–342.

Wong, P. T. P. (2011). Positive psychology 2.0: Towards a balanced interactive model of the good life. *Canadian Psychology/Psychologie Canadienne, 52*(2), 69–81.

Wong, P. T. P. (2016). Meaning-seeking, self-transcendence, and well-being. In A. Batthyany (Ed.), *Logotherapy and existential analysis: Proceedings of the Viktor Frankl Institute Vienna* (Vol. 1, pp. 311–322). Springer.

Yampolsky, P. B. (2012). *The platform sutra of the sixth patriarch*. (Rev. ed.). Columbia University Press.

Ye, D., Ng, Y.-K., & Lian, Y. (2015). Culture and happiness. *Social Indicators Research, 123*(2), 519–547.

Yearley, L. H. (1990). *Mencius and Aquinas: Theories of virtue and conceptions of courage*. State University of New York Press.

Yonker, J. E., Schnabelrauch, C. A., & DeHaan, L. G. (2012). The relationship between spirituality and religiosity on psychological outcomes in adolescents and emerging adults: A meta-analytic review. *Journal of Adolescence, 35*(2), 299–314.

Yu, A. (2007). *The monkey and the monk: An abridgment of the journey to the west*. University of Chicago Press.

Yuki, M., Schug, J., Horikawa, H., Takemura, K., Sato, K., Yokota, K., & Kamaya, K. (2007). Development of a scale to measure perceptions of relational mobility in society. *CERSS Working Paper Series, 75*, 1–25.

Zhouxiang, L. (Ed.). (2020). *Chinese national identity in the age of globalisation*. Palgrave Macmillan.

Ziporyn, B. (2009). *Zhuangzi: The essential writings with selections from traditional commentaries*. Hackett Publishing Company.

Ziporyn, B. (2020). *Emptiness and omnipresence: An essential introduction to Tiantai Buddhism*. Indiana University Press.

Zubieta, J. K., Smith, Y. R., Bueller, J. A., Xu, Y., Kilbourn, M. R., Jewett, D. M., et al. (2001). Regional mu opioid receptor regulation of sensory and affective dimensions of pain. *Science, 293*(5528), 311–315.

# Chapter 3
# Measuring Well-Being: A Multidimensional Framework

**Abstract** Taking the multidimensionality of well-being as its foundational premise, this chapter moves beyond unidimensional conceptualizations to incorporate hedonic, eudaimonic, and social dimensions. Its central aim is to critically review theoretical frameworks, empirical evidence, and methodological advances in measuring well-being across diverse contexts. Section 3.1 examines the theoretical foundations and typologies of well-being, differentiating hedonic, eudaimonic, and social constructs and their integrative models. Section 3.2 expands into novel domains—social-relational well-being, contextualized well-being in work and spatial environments, and cultural folk models—highlighting contextual and cultural variations. Section 3.3 traces the evolution of measurement paradigms from self-report scales to ecological momentary assessment and objective behavioral and physiological indicators. Finally, Sect. 3.4 addresses validity challenges and emerging opportunities, including system-level modeling and GeoAI. The chapter underscores the necessity of culturally informed, technologically enhanced, and ethically grounded frameworks to advance the holistic and equitable measurement of well-being.

**Keywords** Multidimensional well-being · Hedonic well-being · Eudaimonic well-being · Social well-being · Ecological momentary assessment · Cross-cultural measurement · Measurement equivalence · Cultural folk models · Objective indicators · Validity challenges

Expanding upon the interdisciplinary foundations of well-being established in Chap. 2—which synthesized philosophical origins, biological underpinnings, psychological mechanisms, and socio-cultural influences—this chapter transitions from theoretical exploration to the pragmatic challenges of measuring well-being. While Chap. 2 elucidated the multi-dimensional nature of well-being through historical, evolutionary, and cultural lenses, this chapter focuses on the methodologies and instruments developed to capture these complex constructs empirically. We begin by examining traditional and expanded typologies of well-being, including hedonic, eudaimonic, and social dimensions, and proceed to discuss the evolution


H. Li, G. Shen, *The Anatomy of Subjective Well-Being*, Resilient and Friendly Living Environments, https://doi.org/10.1007/978-981-95-4937-5_3

of measurement paradigms—from self-report scales to ecological momentary assessments and objective behavioral indicators. Emphasis is placed on the integration of these approaches within culturally sensitive frameworks, addressing both the strengths and limitations of current tools. This chapter aims to bridge theoretical insights with methodological rigor, ensuring a coherent progression toward a comprehensive understanding of well-being assessment.

## 3.1 Theoretical Foundations and Typology of Well-Being Constructs

### *3.1.1 Traditional Bipartite Models: Hedonic and Eudaimonic Well-Being*

Hedonic Well-being (HWB) is conceptualized as happiness centered on the occurrence of positive affect and the absence of negative affect, with a focus on pleasurable outcomes and "valued but delimited states" of pleasure and positive affect (Ryan et al., 2008). Its core components include Subjective Well-being (SWB), which encompasses three separate elements: positive affect, negative affect, and life satisfaction (Diener et al., 1985). Life satisfaction, a cognitive and judgmental process, involves evaluating one's circumstances against perceived appropriate standards (Diener et al., 1985), while the affective components reflect emotional experiences. This framework is theoretically anchored in affective adaptation functions from an evolutionary psychology perspective,[1] emphasizing the adaptive nature of affective states in subjective well-being (Das & Diener, 1985).[2] Empirically, HWB is often measured through instruments targeting evaluative (EVA) and emotional (EMO) dimensions, such as the Satisfaction with Life Scale (SWLS) for life satisfaction and the Positive and Negative Affect Schedule (PANAS) for affective states (Ryff, 1989).

Eudaimonic Well-being (EWB), originating from Aristotle's philosophy of "living well," focuses on the pursuit and actualization of valued human potentials and virtues, including personal growth, meaning, self-realization, and excellence, emphasizing processes of living well rather than mere positive affect or pleasure. Core components of EWB include Psychological Well-being (PWB), which Ryff operationalizes into six theory-guided dimensions: self-acceptance, positive relations with others, autonomy, environmental mastery, purpose in life, and personal

[1] Frederick, S., & Loewenstein, G. (1999). Hedonic adaptation. In Kahneman, D., Diener, E. & Schwarz, N. (Eds.), *Well-being: The foundations of hedonic psychology* (pp. 302–329). Russell Sage Foundation.

[2] For a foundational theoretical discussion on hedonic adaptation, which is central to this evolutionary perspective, see Frederick, S., & Loewenstein, G. (1999). Hedonic adaptation. In D. Kahneman, E. Diener, & N. Schwarz (Eds.), *Well-being: The foundations of hedonic psychology* (pp. 302–329). Russell Sage Foundation.

growth, integrating clinical, developmental, and lifespan theories (Ryff, 1989). Additionally, the Self-Determination Theory (SDT)-based model of eudaimonia identifies four core motivational concepts: pursuing intrinsic goals (e.g., personal growth, relationships) for their inherent value, behaving autonomously, being mindful, and satisfying basic psychological needs for autonomy, competence, and relatedness. Empirical evidence shows intrinsic aspirations positively relate to PWB and physical health via need satisfaction, whereas extrinsic aspirations do not enhance well-being and may link to ill-being (Ryan et al., 2008). Cultural variation in EWB is notable, particularly regarding relational fulfillment in collectivist contexts, where eudaimonic living aligns more closely with social connectedness; for example, women often score higher on positive relations with others and personal growth, suggesting relational fulfillment is more central in collectivist or female development contexts, and cross-cultural studies indicate that social harmony and role fulfillment are more strongly tied to eudaimonic well-being in East Asian contexts (Uchida & Ogihara, 2012). In contrast, hedonic happiness often reflects individualistic and market-based cultural values, while eudaimonic outcomes encourage prosocial behavior and collective well-being (Joshanloo, 2014).

### *3.1.2 Expanded Typologies: Subjective, Psychological, and Social Well-Being*

The expanded typologies of well-being—encompassing subjective well-being (SWB), psychological well-being (PWB), and social well-being (SoWB)—represent a multidimensional framework that integrates hedonic experience, eudaimonic functioning, and social embeddedness, respectively. These constructs are conceptually distinct yet interconnected, collectively contributing to a holistic understanding of human flourishing across multidisciplinary research.

Subjective well-being (SWB) is primarily rooted in hedonic traditions, focusing on the experience of positive affect and evaluative judgments of life satisfaction. Operationalized through measures capturing emotional states (e.g., "In general, how happy or unhappy do you usually feel?") and global life appraisals (e.g., "Overall, how satisfied are you with life as a whole these days?"), SWB reflects the subjective quality of an individual's emotional and cognitive experiences. This typology emphasizes immediate affective responses and conscious evaluations of life circumstances, serving as a foundational indicator of hedonic well-being.

Psychological well-being (PWB), by contrast, aligns with eudaimonic perspectives, encompassing broader attributes of self-actualization and personal growth. Core dimensions include autonomy, self-acceptance, optimism, personal growth, engagement, mastery, and meaning or purpose in life. Notably, comprehensive conceptualizations of PWB often extend beyond these traditional markers to include character and virtue—rooted in philosophical frameworks of cardinal virtues such as prudence, justice, fortitude, and temperance—and mental/physical health, which are frequently omitted from conventional PWB scales despite their centrality to holistic flourishing (VanderWeele, 2017). This omission highlights the need for more integrative measures that capture the full spectrum of psychological thriving (Fowers et al., 2021).

Social well-being (SoWB) centers on the appraisal of one's social circumstances and functioning, emphasizing the inherently social nature of human well-being. Keyes (1998) delineated five empirically validated dimensions of SoWB: social integration (feeling part of a community), social contribution (perception of one's social value), social coherence (understanding of the social world), social actualization (belief in societal progress), and social acceptance (trust in others). Complementing this, Larson (1993) conceptualized SoWB as comprising social adjustment (satisfaction with relationships and role performance) and social support (quality and availability of trusted social networks), further underscoring its focus on interpersonal dynamics and social environmental fit. Together, these frameworks position SoWB as a distinct construct that captures the public and relational aspects of well-being often overlooked in individual-focused SWB and PWB models (Keyes & Shapiro, 2004).

The interrelations among SWB, PWB, and SoWB are characterized by moderate correlations yet persistent conceptual distinctness. As components of positive mental health, they collectively define a "complete state" that extends beyond the absence of illness, with empirical evidence confirming their status as separate but correlated dimensions (Keyes, 2005). For instance, an individual may exhibit high life satisfaction (SWB) but lack meaningful social ties (SoWB) or struggle with virtue-related behaviors (PWB), illustrating their unique contributions to overall flourishing. This interdependence is further exemplified by research on loneliness, a social-emotional factor that bridges subjective experience and social context: trait loneliness (chronic social disconnection) disrupts physiological stress regulation via flattened diurnal cortisol slopes, while momentary loneliness amplifies next-day cortisol reactivity, linking transient subjective feelings (SWB) to long-term psychological distress (PWB) and social environmental stressors (SoWB) (Doane & Adam, 2010). Such findings underscore the dynamic interplay between these typologies, where deficits in one domain can cascade across others to impact health outcomes.

In summary, subjective, psychological, and social well-being represent complementary lenses through which to examine human flourishing. Their differentiation allows for nuanced assessment of distinct experiential, developmental, and relational aspects of well-being, while their interconnections highlight the need for integrated measurement approaches in multidisciplinary research. By acknowledging their unique contributions and interdependencies, scholars can better capture the complexity of flourishing and inform interventions that promote holistic well-being. The distinct yet complementary nature of these three typologies is summarized in Table 3.1.

**Table 3.1** Key dimensions of subjective, psychological, and social well-being

| Aspect | Subjective well-being (SWB) | Psychological well-being (PWB) | Social well-being (SoWB) |
|---|---|---|---|
| Core Focus | Hedonic experience, Life evaluation | Eudaimonic functioning, Self-realization | Social functioning, Societal evaluation |
| Theoretical Roots | Affective adaptation, Emotion research | Developmental, Clinical, Humanistic psychology | Sociology, Community psychology |
| Key Dimensions | Positive Affect, Negative Affect, Life Satisfaction | Autonomy, Environmental Mastery, Personal Growth, Positive Relations, Purpose in Life, Self-Acceptance | Social Integration, Social Acceptance, Social Contribution, Social Actualization, Social Coherence |
| Primary Measures | SWLS, PANAS | Ryff's PWB Scales | Keyes's SoWB Scales |

Source: Created by the authors

### *3.1.3 Integrative Frameworks: Multidimensional and Interaction Models*

The emergence of integrative well-being frameworks reflects a paradigm shift toward capturing the complexity of human flourishing by synthesizing hedonic, eudaimonic, and social dimensions. A pivotal example is the dual-continua model of mental health, which posits that mental illness and positive mental health (flourishing) are related but distinct dimensions rather than opposite ends of a single spectrum (Greenspoon & Saklofske, 2001; Keyes, 2002, 2005). This model, supported by confirmatory factor analysis across multiple studies demonstrating optimal fit with a two oblique factor structure, provides a foundational framework for integrating diverse well-being components.

Empirical validation of the dual-continua model includes the identification of distinct subgroups—such as "Complete Mental Health" (high positive mental health, no mental illness), "Vulnerable" (low positive mental health, no mental illness), "Symptomatic but content" (high positive mental health, mental illness present), and "Struggling" (low positive mental health, mental illness present)—each exhibiting differential psychological, behavioral, and educational outcomes, thereby underscoring the multidimensional nature of well-being. Further support for interactional dynamics comes from evidence that positive mental health and mental illness are associated with distinct predictors (e.g., social support, personality traits, coping styles) and outcomes, highlighting separate but interconnected mechanisms that shape holistic flourishing.

Integrative frameworks also emphasize the utility of multidimensional assessment and intervention. Measurement tools such as the Mental Health Continuum-Short Form (MHC-SF), Satisfaction with Life Scale (SWLS), and Positive and

Negative Affect Schedule (PANAS) have demonstrated robust validity… (Diener et al., 1985; Keyes et al., 2008; Watson et al., 1988) have demonstrated robust validity across diverse populations, cultural contexts, age ranges, and mental illness severities, facilitating the operationalization of these multifaceted components. Interventions like Acceptance and Commitment Therapy (ACT) (Hayes et al., 2006) and Positive Psychotherapy (Seligman et al., 2005) further illustrate interactional approaches, with differential yet complementary effects on positive mental health and symptom reduction, emphasizing the need to target multiple dimensions for comprehensive well-being promotion.

Incorporating social-contextual dimensions, technological advances have enabled the integration of objective behavioral data into multidimensional models. For instance, smartphone-based mobile sensors capture multidimensional behavioral categories—including physical movement (e.g., activity patterns), social interactions (e.g., face-to-face encounters, computer-mediated communication), and daily activities—providing ecological validity to traditional self-report measures. By combining such sensor-derived data with subjective assessments, researchers can develop dynamic, interactional models that reflect the interplay between individual experiences (e.g., HWB, SWB) and social-contextual factors (e.g., EWB-related engagement), enhancing the holistic understanding of human flourishing (Harari et al., 2017). This approach aligns with the socio-ecological perspective on well-being, which emphasizes the embeddedness of individuals within broader environmental systems (Bronfenbrenner, 1979).

Together, these advancements underscore the value of integrative frameworks that synthesize HWB, EWB, and SWB through multidimensional and multidimensional and interactional lenses, offering a comprehensive foundation for research, assessment, and intervention in well-being science. This integrative perspective can be visualized as a three-dimensional framework, as illustrated in Fig. 3.1.

## 3.2 Novel Domains and Contextual Expansions of Well-Being

### *3.2.1 Social and Relational Well-Being*

The quality of people's relationships with others and interactions among them are major influences on their feelings of well-being and their evaluations of life satisfaction. Social well-being, as a critical domain of well-being, encompasses social adjustment, support, and community integration, all of which significantly impact psychological well-being but present unique measurement challenges (Maddux, 2017). A foundational framework for understanding social well-being is proposed by Keyes, who identified five distinct yet interrelated dimensions: social integration, social contribution, social coherence, social actualization, and social acceptance. These five dimensions, along with their conceptual definitions, are detailed in

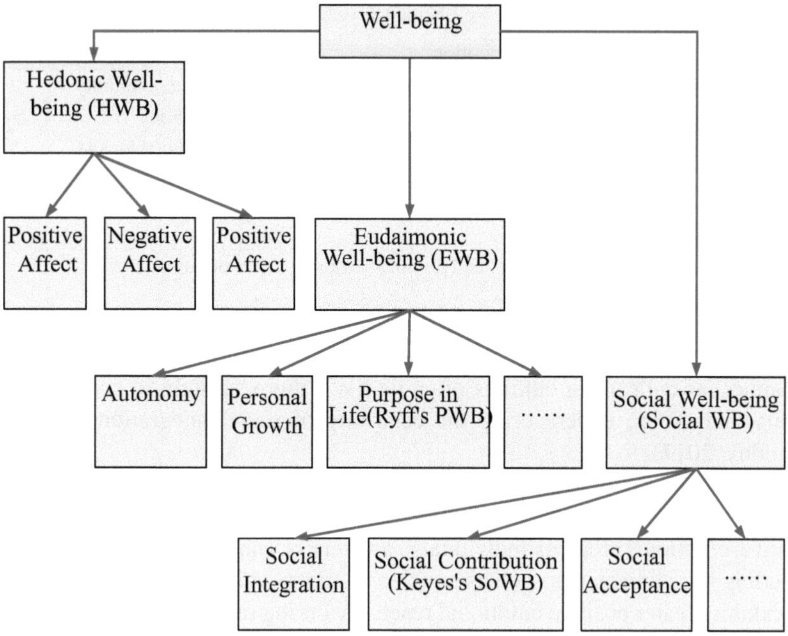


**Fig. 3.1** Three-dimensional framework of well-being. Source: Created by the authors

Table 3.2. These dimensions collectively capture the positive social health of individuals in society, validated through confirmatory factor analysis as latent constructs. Social integration reflects feelings of belonging and shared fate with one's community, drawing on classic sociological concepts like Durkheim's social cohesion; social contribution involves the belief in being a valued, productive member of society, linked to efficacy and generativity; social coherence refers to perceiving the social world as meaningful and predictable; social actualization entails hopefulness about societal progress and potential; and social acceptance involves trusting others and holding favorable views of human nature (Keyes, 1998).

The psychological impact of social and relational well-being is multifaceted. Social support, encompassing tangible assistance, emotional companionship, and informational guidance, is a strong predictor of overall happiness, even after controlling for baseline subjective well-being (SWB) and personality traits. Measurement of social support has demonstrated high reliability (Cronbach's $\alpha = .92$) (Lu, 1999), and frequency and quality of social interactions, along with perceptions of social support, are robust predictors of SWB and life satisfaction. Personality traits further modulate these relationships: extraversion indirectly influences happiness via social support, while neuroticism exerts direct effects on happiness (Suls & Rothman, 2004). Beyond immediate hedonic benefits, social connectedness is linked to both happiness and meaningfulness, with happiness more closely tied to benefits received from others and meaningfulness to benefits bestowed on others (Baumeister & Vohs, 2002). Cultural contexts shape these

**Table 3.2** Five dimensions of social well-being (Keyes, 1998)

| Dimension | Conceptual definition |
|---|---|
| Social Integration | Feeling of belonging to and shared fate with one's community |
| Social Contribution | Belief in being a valued, productive member of society |
| Social Coherence | Perception of the social world as meaningful, predictable, and understandable |
| Social Actualization | Hopefulness about the condition and future of society |
| Social Acceptance | Trusting others and holding favorable views of human nature |

Source: Created by the authors

dynamics, as collectivist cultures often tie SWB more strongly to contributions to family and group, underscoring the centrality of social integration and support (Maddux, 2017).

Additionally, positive emotions play a pivotal role in fostering social well-being: Fredrickson's (2001) broaden-and-build theory of positive emotions highlights that positive emotions broaden individuals' awareness and build personal resources, including social resources such as positive relations with others. People who flourish exhibit greater positive emotional reactivity during routine social activities (e.g., helping others), which predicts growth in complex social understanding and self–other overlap over time. Critical positivity ratio thresholds (e.g., 2.9:1) influence whether such relational resources develop, with sub-threshold ratios failing to support growth (Fredrickson, 2013).

Despite its importance, measuring social and relational well-being presents significant challenges. Keyes notes that developing reliable scales for the five dimensions of social well-being is non-trivial, though empirically validated scales have been derived from large, diverse samples. Broader challenges include difficulties in conceptual and operational definitions, over-reliance on self-report measures, and the need to integrate hedonic (e.g., positive/negative affect) and eudaimonic (e.g., personal growth, positive relations) components of well-being. The dynamic and nonlinear nature of social well-being further complicates measurement, as affective experiences and relational growth are influenced by contextual variability and individual differences, necessitating advanced systems science modeling to capture multicomponent dynamics. These challenges underscore the complexity of operationalizing social well-being but also highlight its critical role in understanding the full spectrum of human flourishing.

### *3.2.2 Contextualized Well-Being: Work, Space, and Environment*

Contextualized well-being in work settings is prominently framed by the Job Demands-Resources (JD-R) theory, which categorizes job characteristics into two core constructs: job demands and job resources. Job demands are physical, psychological, social, or organizational aspects requiring sustained effort, potentially leading to exhaustion and burnout, while job resources are motivating, facilitating goal achievement, and fostering personal growth and engagement (Bakker et al., 2023). Empirically, the JD-R model identifies two distinct processes: a health impairment process, where high job demands drive burnout (characterized by exhaustion and cynicism), and a motivational process, where job resources promote work engagement—a state of vigor, dedication, and absorption. Importantly, job resources can buffer demand-induced strain; for example, autonomy and social support mitigate burnout risk even under high demands (Bakker & Demerouti, 2017).

Employees actively shape their work context through job crafting, a proactive behavior involving modification of job demands and resources to enhance motivation and well-being. Empirical studies show job crafting fosters "gain spirals" of resources and engagement, linking to higher job satisfaction and performance (Bakker & Demerouti, 2014). The theory also acknowledges spatial and environmental influences, with "unfavorable physical working environments" recognized as significant job demands that diminish engagement, highlighting the interplay between physical context and work-related well-being.

Spatial (environmental) well-being centers on the interplay between physical settings and subjective experiences, with public space emerging as a critical domain. Defined by accessibility and use (encompassing streets, plazas, parks), public space differs from privatized areas and is evaluated via five quality dimensions: inclusiveness, meaningful activities, safety, comfort, and pleasurability. The Public Space Index (PSI), developed to measure these dimensions through objective observations and subjective ratings, was applied in Tampa, revealing higher scores for comfort, safety, and inclusiveness but lower scores for meaningfulness and pleasurability, illustrating how spatial qualities shape perceived well-being (Mehta, 2014).

Conceptually, environmental well-being intersects with "livability"—external preconditions like pollution and economic welfare—and "domain satisfaction," where specific contexts (e.g., neighborhood) are enduringly appraised (Veenhoven, 2012). A key construct is sense of place (SOP), capturing emotional connection to environments through place attachment, identity, and dependence (Jorgensen & Stedman, 2001). Stronger SOP correlates with enhanced verbalization of egocentric (self-to-object) and allocentric (object-to-object) navigational strategies, while place imageability—rooted in Lynch's elements (paths, landmarks)—reinforces SOP via cognitive mapping (McCunn & Gifford, 2018).

Methodologically, measuring spatial context faces challenges like the uncertain geographic context problem (UGCoP), where spatial delineation (e.g., census tracts vs. buffers) and temporal lags affect environmental influence analyses. Advances include high-resolution spatiotemporal data (GPS, GIScience) and mixed methods to capture mobility and exposure (Kwan & Schwanen, 2018). Innovations like GIS-based emotional computing integrate georeferenced data and AI to quantify collective emotions, linking naturalness, green spaces, and pollution to spatiotemporal emotional patterns (Huang et al., 2020). Exposure to natural environments, in particular, has been robustly linked to improved well-being through mechanisms such as attention restoration and stress reduction (Kaplan, 1995; Ulrich et al., 1991).

Theoretically, urban form and well-being are explored through Lynch's "performance dimensions" (vitality, sense, fit, access, control) and "learning ecology," framing cities as evolving settlements integrating natural, built, and cultural elements (Banai & Rapino, 2009). Amin further conceptualizes urban well-being via four solidarity registers: repair (infrastructural maintenance), relatedness (inclusion across diversity), rights (participation in urban shaping), and re-enchantment (communal pleasures in public spaces) (Amin, 2006). Environmental psychology integrates these by examining how social design (e.g., lighting, natural materials), home/neighborhood quality (trees, safety), and virtual environments mediate well-being, emphasizing multi-scale interactions from rooms to regions (Gifford, 2014). Material conditions, influenced by household income (rather than GDP alone), also shape livability by improving access to resources like shelter and internet, linking economic context to environmental well-being preconditions (Diener et al., 2013). Furthermore, economic inequality itself has been identified as a significant contextual determinant of population well-being, independent of average income levels (Wilkinson & Pickett, 2009).

### *3.2.3 Cultural Perspectives and Folk Models of Happiness*

Cultural perspectives profoundly shape the conceptualization of well-being, with distinct variations between Western individualistic and Eastern collectivistic traditions. These variations are rooted in differing cultural models that influence self-construal, emotional experiences, and coping strategies. Lu and Gilmour (2004) adopt a cultural psychological approach to highlight this divergence, contrasting Asian socially oriented subjective well-being (SWB) with Euro-American individually oriented SWB. For Asians, SWB is deeply tied to role obligation, interpersonal harmony, and dialectical balance—foundations in Confucian, Taoist, and Buddhist traditions—reflecting an interdependent self-construal focused on fulfilling social roles and maintaining extended relationship harmony. In contrast, Euro-American SWB emphasizes personal accountability, explicit pursuit of happiness, self-autonomy, and agentic mastery of the environment, aligned with an independent self-construal that prioritizes individual rights, achievement, and freedom from social constraints (Lu & Gilmour, 2004). This fundamental distinction between independent and interdependent self-construal was foundational to cultural psychology (Markus & Kitayama, 1991).

The distinction between independent and interdependent self-construal, a core dimension in cultural psychology, arises from collective-level factors such as ecology, economic development, and socialization practices. Western societies prioritize independence, fostering values like autonomy and self-expression, while Eastern societies emphasize interdependence, rooted in relational harmony and social embeddedness. These self-construals are not merely cognitive constructs but are reflected in neural processes; cross-cultural neuroscience reveals systematic East-West differences in brain activations, such as medial prefrontal cortex responses during self-referential processing, facilitating culturally scripted behaviors through neuroplasticity (Kitayama & Uskul, 2011). Earlier behavioral work by Kitayama et al. (2000) laid important groundwork by demonstrating cultural differences in the frequency and social correlates of pleasant emotional experiences.

Conceptualizations of happiness further diverge along these lines. Western individualistic models frame happiness as a primarily positive, hedonic experience centered on personal achievement and positive affect. This is evident in American conceptions, which link happiness to uplifting emotions, personal accomplishment, and the "American Dream" narrative of infinite personal pursuit. In contrast, Eastern collectivistic models adopt a more holistic and dialectical view, where happiness includes ambivalent or negative components. For example, Japanese folk models integrate social harmony, transcendental reappraisal, and even social disruption, allowing for simultaneous positive and negative emotional experiences (Uchida & Kitayama, 2009). Eastern traditions, including Confucianism, Buddhism, and Taoism, reject the stable, bounded self of Western individualism, instead emphasizing self-transcendence, harmony with others and the cosmos, and virtues oriented toward collectivity (Heine et al., 1999). Buddhist teachings, for instance, view the self as an illusion, while Confucianism prioritizes relational harmony over self-enhancement (Joshanloo, 2014).

These cultural models directly influence pathways to happiness. Luo et al. demonstrate that independent self-construal in Western cultures predicts happiness through control beliefs—active mastery of the environment—whereas interdependent self-construal in Eastern cultures predicts happiness through harmony beliefs—maintaining relational balance. Empirical data from Taiwanese (collectivist) and British (individualist) samples support this: Taiwanese participants showed stronger associations between interdependent self-construal/harmony beliefs and happiness, while British participants showed stronger links with independent self-construal/control beliefs (Lu et al., 2001). Cultural values further moderate these relationships: in Taiwanese samples, social integration and human-heartedness (benevolence/compassion) strongly predict happiness, whereas these values are less central for British individuals, reflecting collectivistic versus individualistic priorities (Fulmer et al., 2010). Notably, Confucian work dynamism (diligence, responsibility) emerges as a culture-general predictor of happiness across both contexts (Lu et al., 2001). These culturally divergent pathways to happiness, mediated by control versus harmony beliefs, are illustrated in Fig. 3.2.

Coping strategies also reflect cultural differences in self-construal. Americans cope with unhappiness through externalizing behaviors (e.g., anger, blaming others) to maintain self-esteem, aligning with individualistic emphasis on independence. In contrast, Japanese coping involves transcendental reappraisal and self-improvement, reflecting a self-critical, socially engaging approach. Chinese folk psychology similarly emphasizes dialectical coping, integrating "wisdoms" such as discovery (inward happiness search), contentment (reducing desires), giving (altruism), and self-cultivation (mindfulness, moral development) (Lu, 2001). These strategies reflect the Eastern acceptance of suffering as integral to spiritual growth, as seen in Buddhist practices that embrace others' suffering to overcome one's own.

Cross-cultural variations extend to well-being measurement, challenging the universality of Western instruments. The Satisfaction With Life Scale (SWLS) exhibits structural variation across nations, with items like "In most ways my life is close to my ideal" carrying greater significance in wealthier, individualistic contexts. Economic wealth correlates with life satisfaction cross-nationally, but this relationship weakens when accounting for measurement reliability and cultural differences in item interpretation (Vittersø et al., 2002). Tov and Diener (2009) further note that correlates of well-being differ culturally: self-esteem is more strongly linked to life satisfaction in individualistic societies, while collectivists rely more on social appraisals. These issues of measurement non-equivalence are a central concern in cross-cultural psychology, requiring rigorous methodological approaches to ensure valid comparisons (van de Vijver & Leung, 1997). These findings underscore the need for culturally balanced methodologies that reflect indigenous conceptions—such as the Chinese Happiness Inventory, which integrates Eastern (e.g., relationship harmony) and Western (e.g., self-satisfaction) subscales.

Indigenous psychology movements highlight additional cultural nuances, such as holistic well-being conceptualizations in Indian contexts that integrate physical, mental, and spiritual dimensions, emphasizing harmony with the environment over material wealth. Such perspectives demand methodological pluralism, combining qualitative inquiry and narrative approaches to capture culturally embedded meanings of happiness (Valsiner, 2012). Ultimately, cultural folk models of happiness are dynamic, reflecting not only East-West divides but also context-bound adaptations, as seen in Taiwanese culture's coexistence of traditional collectivistic values and

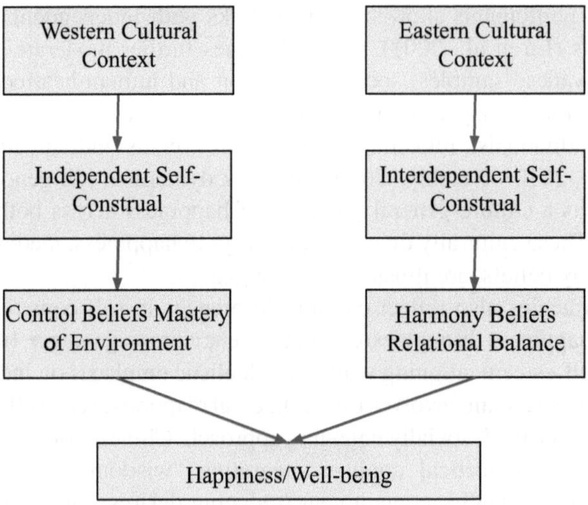


**Fig. 3.2** Cultural pathways to well-being. Source: Created by the authors

modern individualistic influences. Recognizing these variations is critical for advancing a truly global understanding of well-being.

### 3.2.4 The Consensus on the Integration Model

The consensus integration model in well-being research, as articulated by Diener, Ryan, and Keyes, centers on the tripartite interaction framework of hedonic, eudaimonic, and social well-being, which posits well-being as a dynamic integration of Hedonic Well-being (HWB), Eudaimonic Well-being (EWB), and Social Well-being (Social WB). Hedonic Well-being (HWB), often operationalized as Subjective Well-being (SWB), encompasses an individual's cognitive and affective evaluations of life, including emotional reactions to events (e.g., high pleasant emotions, low negative emotions) and cognitive judgments of satisfaction and fulfillment (Diener et al., 2009). Eudaimonic Well-being (EWB), in contrast, focuses on living a meaningful and self-actualizing life aligned with one's inherent potentials, operationalized through dimensions such as autonomy, personal growth, purpose in life, environmental mastery, positive relations, and self-acceptance (i.e., psychological well-being, PWB). Social Well-being (SWB), the third dimension, refers to positive societal functioning and the experience of social value, complementing emotional and psychological facets of well-being.

A key premise of the model is the dynamic interaction among these dimensions. Keyes conceptualizes "flourishing" as the presence of high levels across all three domains, whereas "languishing" denotes low levels in all, underscoring their interdependence (Keyes, 2002). Empirical support for this interaction includes lifespan data revealing differential trajectories: emotional well-being (HWB) increases with age, psychological well-being (EWB) decreases, and social well-being (SWB) remains stable, highlighting their distinct developmental patterns. Person-centered analyses further validate this framework by identifying subgroups that do not fit a unidimensional continuum, such as individuals languishing without mental illness or those with mental illness but moderate well-being.

The integration model is also supported by the "Two Continua Model," which distinguishes mental health (encompassing HWB, EWB, and SWB) from mental illness as related but separate dimensions, supported by factor analyses and empirical data showing older adults report lower mental illness but not necessarily higher positive mental health compared to younger adults (Keyes, 2005). Methodologically, the Mental Health Continuum—Short Form (MHC-SF), developed and validated by Keyes and colleagues, operationalizes these three dimensions with high reliability, providing a tool to measure the integrated construct (Keyes, 2002; Westerhof & Keyes, 2010).

Self-Determination Theory (SDT), proposed by Ryan & Deci, offers a foundational framework for understanding the mechanisms underlying this integration, emphasizing universal basic psychological needs (competence, autonomy, relatedness) whose satisfaction fosters intrinsic motivation and autonomous regulation, thereby enhancing HWB, EWB, and SWB in an integrated manner (Deci & Ryan, 2008). This theoretical and empirical synthesis underscores the importance of measuring and integrating all three dimensions to capture the complexity of well-being.

## 3.3 Evolution of Measurement Paradigms in Well-Being Research

### *3.3.1 Self-Report Scales: Development, Strengths, and Limitations*

Self-report scales have been foundational in well-being research, evolving to capture diverse conceptualizations of hedonic, eudaimonic, and emerging well-being domains. To provide a clear overview of the key instruments discussed in this section, their primary characteristics are compared in Table 3.3.

Hedonic well-being, defined by positive affect and absence of negative affect, has historically been measured with greater precision due to its clear, quantifiable criteria, whereas eudaimonic well-being—centered on intrinsic goal pursuit (e.g., personal growth, relationships) and living in accordance with one's values—has faced challenges of fragmented measurement and theoretical coordination. Emerging domains, such as social well-being, further expand the scope, necessitating critical evaluation of psychometric rigor, cultural sensitivity, and construct validity across instruments.

**1. Hedonic Well-Being Scales: Precision and Narrow Focus**

Hedonic measures primarily assess subjective well-being (SWB), encompassing life satisfaction (cognitive), positive affect, and negative affect (affective components) (Ryan & Deci, 2001). The Satisfaction With Life Scale (SWLS) exemplifies a widely used cognitive hedonic instrument, designed to capture global life satisfaction through self-endorsed weighting of life domains (Pavot & Diener, 2013). Its psychometric strengths include high internal consistency ($\alpha = .87$) and temporal reliability (2-month test-retest $r = .82$), with scores correlating moderately to highly with self-esteem (.54) and inversely with neuroticism (–.48) (Diener et al., 1985). However, the SWLS intentionally excludes affective components, limiting its scope to cognitive judgments, and its reliance on personal standards complicates cross-cultural comparability despite consistent factor structure in adaptations (e.g., French, Mandarin Chinese) (Pavot & Diener, 2013).

Multi-item hedonic scales generally outperform single-item measures (e.g., Cantril's ladder) by reducing response biases (e.g., acquiescence) and improving internal consistency (Kammann & Flett, 1983). For instance, multi-item affective scales differentiate positive and negative affect as independent constructs, enhancing discriminant validity. Yet, even refined hedonic instruments face limitations: single-item scales suffer from skewed scores and limited reliability (Diener, 1984), while multi-item scales may still be influenced by momentary mood, potentially

**Table 3.3** Comparison of selected well-being measurement instruments

| Scale (acronym) | Well-being type | Core construct(s) measured |
|---|---|---|
| Satisfaction with Life Scale (SWLS) | Hedonic | Global cognitive judgment of life satisfaction |
| Positive and Negative Affect Schedule (PANAS) | Hedonic | Frequency of positive and negative affect |
| Ryff's Psychological Well-Being Scales (PWB) | Eudaimonic | 6 dimensions of eudaimonic functioning |
| Scale of Positive and Negative Experience (SPANE) | Hedonic | Balance of positive vs. negative experiences |
| Mental Health Continuum-Short Form (MHC-SF) | Integrated | Emotional, psychological, and social well-being |

Source: Created by the authors

distorting long-term well-being assessments (Diener et al., 2009). Kahneman and Krueger (2006) provide a detailed discussion on the evolution and limitations of subjective well-being measures from an economic perspective.

**2. Eudaimonic Well-Being Scales: Theoretical Grounding and Complexity** Eudaimonic measures address gaps in hedonic paradigms by focusing on self-realization, meaning, and psychological functioning. Ryff's Psychological Well-Being (PWB) scales operationalize six theory-driven dimensions—self-acceptance, positive relations with others, autonomy, environmental mastery, purpose in life, and personal growth—derived from developmental and clinical theories. These scales demonstrate strong internal consistency (Cronbach's $\alpha$ = .86–.93) and test-retest reliability (.81–.88 over 6 weeks), with modest correlations (.25–.45) with traditional hedonic measures (Keyes et al., 2002), confirming their distinct focus on positive functioning. However, the factor structure and cultural applicability of Ryff's model have been subjects of ongoing debate and refinement (Springer & Hauser, 2006).[3]

Self-Determination Theory (SDT) further advances eudaimonic assessment by measuring basic psychological need satisfaction (autonomy, relatedness, competence) and frustration, critical for distinguishing well-being from ill-being. A 24-item SDT-based scale showed good fit for a six-factor model (satisfaction/frustration across three needs) in cross-cultural samples (USA, China, Belgium, Peru), with invariant factor structure supporting universal applicability (Chen et al., 2015). Importantly, SDT measures capture both the "what" (goal content) and "why" (motivational regulation) of behavior, as autonomous motivation toward intrinsic goals correlates more strongly with vitality than controlled motivation (Ryan et al., 2008).

[3] The debate surrounding the factor structure of Ryff's Psychological Well-Being scales is extensive. For a key critique and discussion of methodological effects, see Springer, K. W., & Hauser, R. M. (2006). An assessment of the construct validity of Ryff's scales of psychological well-being: Method, mode, and measurement effects. *Social Science Research*, 35(4), 1080–1102.

Emerging eudaimonic instruments, such as the Hedonic and Eudaimonic Motives for Activities (HEMA) scale and Orientations to Happiness Questionnaire (OTHQ), differentiate hedonic (pleasure-seeking) and eudaimonic (meaning, growth) orientations. These scales exhibit convergent validity with well-being indicators (e.g., life satisfaction, authenticity) and discriminant validity from unrelated constructs, though frequent self-ratings may induce reactivity or alter naturalistic experiences (Huta, 2015). The complexity of measuring eudaimonic well-being, including debates on its distinction from hedonic well-being, is thoroughly discussed in contemporary reviews (Huta & Waterman, 2014).

**3. Emerging Well-Being Types: Expanding the Multidimensional Landscape**
Social well-being represents a key emerging domain, operationalized via five dimensions: social integration (sense of community belonging), social acceptance (positive view of others), social contribution (perceived societal value), social actualization (belief in societal progress), and social coherence (understanding social norms). Confirmatory factor analyses in U.S. samples support this five-factor model, with good internal consistency ($\alpha$ = .70–.86) and construct validity (e.g., inverse correlations with anomie, positive associations with generativity). However, variability in reliability across samples highlights the need for further refinement, particularly regarding item balance and sampling diversity (Keyes, 1998).

**4. Strengths and Limitations of Self-Report Scales**
Strengths of self-report instruments include their practical utility for large-scale surveys, sensitivity to change (e.g., SWLS detecting therapy-induced improvements in life satisfaction; Pavot & Diener, 2013), and alignment with subjective experiences central to well-being. Multi-item scales, in particular, offer a nuanced assessment of distinct components (e.g., affective vs. cognitive SWB) and robust psychometrics (Chen et al., 2015; Ryff, 1989).

Limitations are equally pronounced. Cultural biases persist: hedonic goal importance varies cross-culturally (e.g., pleasure/fun falls midway between intrinsic/extrinsic goals), and SWB self-reports show positivity bias differences (stronger in Puerto Rico, weaker in Japan (Diener et al., 2009). Systematic cross-cultural differences in response styles, such as extreme responding and acquiescence, pose significant challenges to measurement equivalence (van de Vijver & Leung, 1997). Smith (2004) specifically examines acquiescence as a culturally influenced communication style that can bias cross-cultural comparisons. Validity concerns include response styles (e.g., social desirability, though SWLS shows minimal contamination (Diener et al., 1985)), and the narrowness of purely hedonic measures, which overlook vital eudaimonic facets like personal growth. Additionally, self-report reliance on retrospective judgment may disconnect from momentary experiences, as global evaluations are prone to memory distortions (Diener et al., 2009). This retrospective bias is a well-documented limitation of global self-reports (Robinson & Clore, 2002).

**5. Validity Considerations: Beyond Surface-Level Assessment**
Construct validity requires integration of motivational and contextual factors. For example, eudaimonic well-being scales must distinguish autonomous from controlled goal pursuit, as the former uniquely predicts vitality (Ryan et al., 2008). Similarly, social well-being scales demonstrate discriminant validity by correlating minimally with optimism, reflecting distinct social-psychological processes (Keyes, 1998). For a contemporary overview of challenges and future directions in the field, including measurement issues, see Diener et al. (2018) Future advances may lie in hybrid models that combine hedonic, eudaimonic, and social dimensions, as advocated by researchers critical of unidimensional paradigms.

In summary, self-report scales have evolved from a narrow hedonic focus to multidimensional tools capturing eudaimonic and social well-being. While their psychometric strengths (reliability, validity) support widespread use, limitations, including cultural biases, response artifacts, and construct narrowness necessitate cautious interpretation and integration with complementary methods.

### *3.3.2 Ecological Momentary Assessment and Experience Sampling Methods*

The methodological evolution toward real-time experience sampling techniques has been driven by the need to overcome limitations of traditional retrospective and lab-based assessments, leading to the development of approaches such as the Experience-Sampling Method (ESM) and Ecological Momentary Assessment (EMA). ESM, introduced by Csikszentmihalyi and Larson (1987), was designed to capture variations in self-reports of mental processes by prompting individuals to provide immediate self-reports of psychological states, thoughts, and activities in natural environments using signaling devices at random times. Similarly, EMA, as defined by Shiffman et al. (2008), involves repeated sampling of current behaviors and experiences in real time and natural settings, aiming to minimize recall bias, maximize ecological validity, and enable the study of microprocesses influencing behavior in real-world contexts (Shiffman et al., 2008).

This shift was motivated by dissatisfaction with the inability of retrospective self-reports to provide accurate accounts of daily experiences, as these reports are prone to memory bias due to heuristics, mood-dependent memory effects, and the reconstructive nature of autobiographical memory. EMA and ESM address this by focusing on momentary states rather than long-term recall, thereby reducing systematic recall error (Stone & Shiffman, 2002). For example, Maes et al. (2018) found that retrospective global measures of health-related quality of life (HRQOL) failed to predict momentary HRQOL in a study of 139 participants with various health conditions, highlighting the bias in recall-based assessments (Maes et al., 2018).

A key strength of these methods is their ecological validity, derived from collecting data within individuals' real-life contexts. ESM ensures this by assessing subjects "in real time and context", while EMA's focus on real-world environments allows generalization to participants' actual lives. This enables observation of natural fluctuations in psychological and physiological states and their situational influences, which are often missed in laboratory settings (Trull & Ebner-Priemer, 2020).

Methodologically, EMA and ESM employ event-based and time-based sampling strategies. Event-based sampling triggers assessments when predefined events occur, while time-based sampling involves assessments at fixed or random intervals, capturing fluctuating states not tied to discrete events. These strategies provide a "movie" rather than a "snapshot" depiction of experience and behavior dynamics over time (Shiffman, 2009).

Technological advancements have been critical to their evolution. Early approaches used written diaries or telephones, but modern methods leverage electronic diaries, smartphones, and Web-GIS systems (Lwin & Murayama, 2011) with time-stamping and compliance monitoring, reducing falsification and backfilling issues common in paper diaries. Integration of sensor technologies (e.g., GPS, heart rate monitors) further enables passive, continuous collection of objective physiological and behavioral data, enhancing dynamic monitoring through "personal sensing" that links momentary experiences with environmental interactions.

Despite a higher participant burden, compliance with EMA/ESM is high with electronic prompting. Maes et al. (2018) reported a 72% response rate, with participants perceiving the data collection period as representative of a typical week. This feasibility supports application across diverse populations, including clinical groups, for monitoring dynamic changes in mood, symptoms, behaviors, and physiological parameters like heart rate and galvanic skin response.

These methods capture temporal sequences and microprocesses obscured by recall-based assessments, such as mood fluctuations before a smoking lapse or within-person variability in momentary HRQOL. Such insights have facilitated just-in-time adaptive interventions (JITAIs), which deliver context-aware support based on real-time fluctuations, underscoring their methodological significance in advancing well-being research (Nahum-Shani et al., 2018).

### 3.3.3 *Objective Physiological and Behavioral Indicators*

Advances in objective physiological and behavioral indicators have significantly expanded the measurement toolkit for well-being research, complementing traditional subjective self-reports with ecologically valid and scalable markers. Physiological indices, captured via diverse sensing technologies, offer insights into affective styles by reflecting autonomic nervous system activity, neural correlates, and bodily responses to emotional states. For instance, electroencephalography (EEG) measures electrical brain activity across frequency ranges (delta, theta, alpha, beta, gamma), which correlate with distinct emotional states and serve as neural markers of affective processing. Electrocardiography (ECG) and derived heart rate variability (HRV) provide indicators of autonomic regulation, with features like QRS complex dynamics and HRV spectral analysis differentiating emotions such as sadness, joy, and anxiety. Galvanic skin response (GSR), which assesses sweat gland activity via tonic (slow-varying) and phasic (short-term peak) components, is particularly sensitive to arousal states like stress and excitement, while facial electromyography (EMG) captures muscle activity linked to facial expressions associated with strong emotions (Cacioppo et al., 2007). Multimodal integration of these physiological signals—often combined with machine learning algorithms (e.g., SVM, neural networks)—enhances reliability and precision, with approaches like multimodal sensor fusion (e.g., EEG paired with ECG or GSR) emerging as best practices for robust emotion recognition (Dzedzickis et al., 2020).

Behavioral indicators, enabled by smartphone sensing and location-tracking technologies, offer objective insights into naturalistic patterns tied to well-being. Smartphone sensors (e.g., accelerometer, GPS, microphone, Bluetooth) unobtrusively capture social-interaction behaviors (e.g., call logs, conversation turn-taking, in-person group size via Bluetooth scans), daily activities (e.g., physical activity, sleep duration inferred from light, accelerometer, and usage logs with ±42-minute accuracy), and mobility patterns (e.g., time spent at locations, travel distance, routine regularity via GPS/WiFi). These behavioral data can be combined with ecological momentary assessments (EMAs) to partition behaviors by psychological context or trigger context-contingent EMAs (e.g., post-phone call), facilitating fine-grained analysis of behavioral antecedents and consequences of well-being states (Harari et al., 2016). Beyond individual-level tracking, large-scale GPS trajectory data have enabled the development of urban mobility models (e.g., Lévy Walks, Gravity Model, Radiation Model) that mathematically characterize movement regularities, transition probabilities between locations, and temporal routines—objective metrics linked to affective styles and daily well-being. For example, algorithms constructing road network geometries from taxi GPS data (e.g., Karagiorgou & Pfoser, 2012) and deep learning models (e.g., LSTM) predicting mobility with >80% accuracy provide scalable behavioral indices reflecting environmental interactions and routine stability (Crooks et al., 2015; Zhao et al., 2016).

Together, these physiological and behavioral indicators address key limitations of retrospective self-reports and laboratory-based studies by offering objective, ecologically valid, and scalable measures. Psychometric evaluation of these indicators—including reliability, validity, and sampling rate considerations—ensures their utility as complementary markers, advancing well-being measurement paradigms toward a more integrated understanding of affective styles and well-being (Gabriel et al., 2019).[4]

[4] While Gabriel et al. (2019) focus specifically on Experience Sampling Methods, their comprehensive discussion of psychometric considerations for novel data sources provides a valuable evaluative framework that can be aptly applied to the objective physiological and behavioral indicators discussed here.

### *3.3.4 Cultural Adaptation and Cross-Cultural Measurement Equivalence*

Advancing the cross-cultural measurement of well-being requires rigorous methodological innovations to ensure cultural validity and equivalence, particularly in adapting tools to diverse populations and addressing potential biases in item interpretation. A key area of progress lies in the application of advanced analytical techniques to test measurement equivalence, including multigroup structural equation modeling (SEM), multiple indicator multiple cause (MIMIC) models, and item response theory (IRT). These methods collectively contribute to evaluating whether measurement tools function comparably across cultural groups by examining construct equivalence and differential item functioning (DIF) (Byrne & van de Vijver, 2017).

Empirical studies have demonstrated the utility of integrating these techniques. For instance, a study examining the measurement equivalence of the Satisfaction with Life Scale between American and Chinese samples employed multigroup SEM, MIMIC, and IRT to compare item performance across cultural contexts. Whereas multigroup SEM and MIMIC analyses identified only one biased item each (with items 5 and 4, respectively, in the Chinese sample being less robust indicators of life satisfaction than in the American sample), IRT analysis revealed significant DIF in four of the five items. This discrepancy highlights the enhanced rigor of IRT in testing measurement equivalence: unlike SEM and MIMIC approaches, which primarily focus on item discrimination, IRT examines both item discrimination and item difficulty parameters. DIF under the IRT framework is defined as group differences in the conditional probability of endorsing an item given the same level of latent score, with large DIF indicating that cultural groups may interpret or respond to items differently despite equivalent underlying well-being levels.

Notably, the items with the largest DIF in this cross-cultural comparison assessed satisfaction with past accomplishments, whereas the remaining items focused on external living conditions or present satisfaction. This pattern suggests that cultural differences in the conceptualization of life satisfaction—such as varying emphasis on past achievements versus current circumstances—and potential social desirability biases may influence item responses, underscoring the importance of culturally sensitive adaptation and rigorous equivalence testing. Collectively, these findings illustrate how integrating diverse analytical techniques (e.g., IRT alongside SEM/MIMIC) strengthens the validity of cross-cultural well-being measurements by uncovering nuanced cultural variations in item functioning that might otherwise remain undetected (Chen, 2008).

## 3.4 Measurement Validity Challenges and Future Directions in Well-Being Research

### *3.4.1 Controversies in Core Methodologies*

Cross-cultural equivalence in well-being measurement is fundamentally challenged by the embedding of biological and psychological processes within cultural contexts, which provide shared meaning systems shaping experience interpretation. This cultural embedding complicates the assumption that well-being constructs or their correlates are universally comparable. For instance, cultural differences in conceptions of well-being extend to emotion norms, where pride, valued positively in some cultures, may cluster with negative emotions in others, highlighting that even "positive emotions" are not culturally universal in valence (Tov & Diener, 2009). Such differences are reflected in measurement, as seen in the Satisfaction with Life Scale (SWLS), where Item Response Theory (IRT) analysis revealed differential item functioning (DIF) between American and Chinese samples: Chinese respondents with high latent life satisfaction were less likely to endorse items evaluating past accomplishments, aligning with cultural tendencies toward self-criticism rather than self-enhancement (Oishi, 2006).

Structural challenges further emerge in the composition of well-being constructs. The subjective well-being (SWB) framework, typically comprising life satisfaction, positive affect, and negative affect, may exhibit cross-cultural variability in how these components interrelate, and domain satisfaction structures (e.g., organizational domains) vary with cultural and age contexts, undermining universal measurement models (Diener, 1984). Retrospective self-reports exacerbate these issues, as cultural biases in recall—such as mood-dependent memory effects—distort aggregated summaries; Ecological Momentary Assessment (EMA) offers a potential solution by capturing real-time, context-rich data, though it requires careful attention to ecological and context equivalence to ensure comparability across cultures (Shiffman, 2013).

Biological and neural measures introduce additional layers of complexity. Gene-culture interaction models demonstrate that genetic predispositions (e.g., serotonin receptor polymorphisms, oxytocin receptor genes) manifest differently across cultural environments, leading to divergent cognitive, emotional, and well-being outcomes (Kim & Sasaki, 2014). Neural correlates also show culture-specific patterns: similar behavioral outcomes may arise from distinct neural mechanisms, necessitating culturally appropriate experimental tasks and brain templates to avoid misattributing anatomical or scanner variability to cultural differences (Chiao et al., 2010).

Assumptions of universality without empirical testing further threaten validity. The claimed "critical positivity ratio" (Fredrickson & Losada, 2005) exemplifies this pitfall, as no justification was provided for its cultural invariance, with mathematical and empirical flaws revealing the ratio as an artifact of model construction rather than a cross-culturally robust construct (Brown et al., 2013).[5] In contrast, rigorous efforts to establish equivalence have shown promise. For example, Basic Psychological Needs Theory (BPNT) scales assessing autonomy (defined as volition, not independence), relatedness, and competence demonstrated measurement equivalence across Belgium, China, USA, and Peru via multi-group Confirmatory Factor Analysis (CFA), supporting universal links between need satisfaction and well-being (Chen et al., 2015). Similarly, the Maslach Burnout Inventory-General Survey (MBI-GS) showed cross-national invariance across Sweden, Finland, and the Netherlands, though item adaptation (e.g., replacing "work" with "study" for students) requires careful validation to preserve meaning (Schaufeli & Bakker, 2004).

Addressing these challenges demands culturally sensitive methodologies: prioritizing cultural value measurement over stereotypes, developing process-oriented models, and validating scales with non-self-report criteria (e.g., peer ratings, facial coding) to mitigate response biases. Ultimately, cross-cultural well-being research must balance recognition of pancultural elements with sensitivity to culture-specific patterns, ensuring measures capture equivalent meaning rather than assuming universality (Diener & Suh, 2003). This balance is central to the comprehensive volume edited by Diener and Suh (2003), which explores the intricate relationships between culture and subjective well-being. Spatial factors, such as urban green spaces, further complicate equivalence by influencing well-being through culturally specific behavioral and environmental interactions, though their role in measurement remains understudied (Kourtit et al., 2023).

### 3.4.2 *Emerging Challenges and Opportunities*

Current advances in well-being research are accompanied by emerging challenges and opportunities that demand attention to measurement validity. A primary gap lies in the need for robust system-level modeling, which addresses the dynamic and contextual nature of well-being determinants. Traditional approaches often rely on static, administrative areal units to define geographic contexts, but the "uncertain geographic context problem (UGCoP)" highlights the limitations of such methods due to spatial and temporal uncertainties affecting exposure to contextual influences. This calls for dynamic, person-specific conceptualizations that capture complex spatial-temporal configurations of individuals' everyday environments, as enabled by advances in GIS, GPS, and location-aware mobile devices (Kwan, 2012). Similarly, in affect detection, multimodal systems—integrating data from multiple sources like vision, audio, and physiology—have shown promise, with meta-analyses indicating they are consistently more accurate than unimodal

[5] This critique by Brown et al. (2013) was a significant event in the field. In response, Fredrickson (2013) published an update acknowledging the mathematical criticisms and reframing her argument around the non-linear dynamics of positive emotions, while abandoning the specific critical ratio. This exchange highlights the process of scientific self-correction.

counterparts (average improvement of 9.83%), yet challenges remain in modeling interdependencies between modalities at the system level, particularly for naturalistic affective states (D'Mello & Kory, 2015). System-level modeling also extends to movement data analysis, where integrating geospatial and temporal dimensions with uncertainty modeling (e.g., using space-time prisms for uncertain markers) is critical for capturing accurate human mobility patterns relevant to well-being (Senaratne et al., 2017).

Ethical considerations in data collection have become increasingly salient with the proliferation of sensitive well-being data. The collection of detailed geographic and temporal data, for instance, poses significant risks to personal privacy and confidentiality, necessitating rigorous safeguards. Similarly, emotion data—often involving physiological signals, facial expressions, or voice recordings—raises ethical concerns that require responsible use frameworks, including privacy-preserving technologies such as generative adversarial networks (GANs), federated learning, and homomorphic encryption (Rokhsaritalemi et al., 2023). Even aggregated data sources, such as mobile communication records, require attention to biases from sample size and representativeness to ensure ethical integrity in analysis (Lazer et al., 2014).

Leveraging big data and GeoAI technologies presents transformative opportunities for advancing well-being measurement. Large-scale spatiotemporal data, enabled by GIS, GPS, and mobile devices, support the construction of individualized activity spaces, enhancing the accuracy of environmental and social contextual influence measurements. GSM mobile network big data, combined with visual analytics techniques (e.g., matrix views, map views), allows for exploring spatio-temporal patterns and detecting anomalies relevant to urban dynamics and well-being. GeoAI further enables spatial analysis of emotional data, such as sentiment-aware walking route recommendations and geo-visualization of stress for public health and urban planning. Additionally, hybrid models combining CNN and LSTM for multimodal sensor data (on-body, environmental, location data) address system-level modeling needs in emotion recognition, demonstrating the potential of big data analytics (Zou et al., 2024). These advancements collectively push the field toward a more holistic, dynamic, and contextually rich understanding of well-being, while simultaneously demanding renewed focus on ethical frameworks and methodological rigor..

## 3.5 Summary

The synthesis of extant research underscores a multifaceted conceptualization of well-being comprising hedonic, eudaimonic, and social dimensions, each contributing uniquely and interactively to the construct of human flourishing. Traditional bipartite models delineate hedonic well-being as the experience of pleasant affect and life satisfaction, while eudaimonic well-being emphasizes self-actualization, meaning, and intrinsic goal pursuit. Expanded typologies situate subjective, psychological, and social well-being as distinct yet interrelated domains that collectively capture affective experience, personal growth, and social embeddedness. Integrative frameworks, including the dual-continua model (Keyes, 2002) and multidimensional interaction approaches, advance a holistic understanding by recognizing the separability and co-dependence of positive mental health and mental illness, facilitated by robust and culturally sensitive measurement instruments.

Novel domains such as social-relational well-being and contextualized well-being in occupational and environmental settings reveal the critical influence of interpersonal dynamics, spatial factors, and culturally embedded values on well-being outcomes. The incorporation of urban form, place attachment, and work environment characteristics expands the scope beyond individual experience to encompass ecological and sociocultural determinants. Cultural perspectives elucidate profound variations in the self-construal, emotional valence, happiness conceptualizations, and coping strategies, revealing that well-being is deeply rooted in culturally specific meaning systems and social practices. These cultural models challenge universalist assumptions (Markus & Kitayama, 1991) and necessitate culturally nuanced conceptualizations and measurements.

Measurement paradigms have evolved from reliance on retrospective self-reports toward dynamic, multimodal approaches integrating ecological momentary assessment, physiological and behavioral indicators, and advanced psychometric techniques. Self-report scales demonstrate strengths in reliability and construct validity but are limited by cultural biases, retrospective distortions, and narrow operationalizations, particularly regarding eudaimonic and social well-being. Experience sampling methodologies enhance ecological validity by capturing in-the-moment states and contextual contingencies, while objective physiological and behavioral measures afford scalable and unobtrusive markers of affective and social processes. Cross-cultural measurement equivalence remains a significant methodological challenge, emphasizing the need for rigorous statistical approaches (e.g., multigroup CFA, IRT) and culturally sensitive adaptations to ensure construct invariance and minimize differential item functioning.

Emerging challenges in validity pertain to the cultural embedding of psychological processes, biological and neural variability across contexts, and spatial-temporal uncertainties in exposure assessment. Advances in system-level modeling, geospatial technologies, and ethical frameworks for data privacy present opportunities to refine well-being measurement and theory. The dynamic interplay among hedonic, eudaimonic, and social components, moderated by cultural and contextual factors, calls for integrated, multidimensional models that can accommodate complexity and heterogeneity in well-being phenomena.

Overall, the accumulated evidence advocates for a comprehensive and interactional measurement framework of well-being that reconciles diverse theoretical traditions, incorporates cultural specificity, and leverages methodological innovations (Layard, 2011). This framework facilitates nuanced assessment, cross-cultural comparability, and the elucidation of mechanisms underpinning flourishing across diverse populations and contexts. Future research must continue to bridge the gap between sophisticated theoretical models and equally sophisticated, ethically grounded measurement practices.

**Funding** This research was funded by

1. Guangdong Provincial Office of Philosophy and Social Sciences [Grant No. GD24XSH06]
2. Projects of PhDs' Start-up Research of GDUPT [Grant No. 2023bsqd1008]

## References

Amin, A. (2006). The good city. *Urban Studies, 43*(5–6), 1009–1023.

Bakker, A. B., & Demerouti, E. (2014). Job demands–resources theory. In P. Y. Chen & C. L. Cooper (Eds.), *Well-being: A complete reference guide, Volume III: Work and well-being* (pp. 1–28). Wiley-Blackwell.

Bakker, A. B., & Demerouti, E. (2017). Job demands-resources theory: Taking stock and looking forward. *Journal of Occupational Health Psychology, 22*(3), 273–285.

Bakker, A. B., Demerouti, E., & Sanz-Vergel, A. I. (2023). Job demands–resources theory: Ten years later. *Annual Review of Organizational Psychology and Organizational Behavior, 10*, 25–53.

Banai, R., & Rapino, M. A. (2009). Urban theory since A Theory of Good City Form (1981)—A progress review. *Journal of Urbanism: International Research on Placemaking and Urban Sustainability, 2*(3), 259–276.

Baumeister, R. F., & Vohs, K. D. (2002). The pursuit of meaningfulness in life. In C. R. Snyder & S. J. Lopez (Eds.), *Handbook of positive psychology* (pp. 608–618). Oxford University Press.

Bronfenbrenner, U. (1979). *The ecology of human development: Experiments by nature and design*. Harvard University Press.

Brown, N. J. L., Sokal, A. D., & Friedman, H. L. (2013). The complex dynamics of wishful thinking: The critical positivity ratio. *American Psychologist, 68*(9), 801–813.

Byrne, B. M., & van de Vijver, F. J. R. (2017). The maximum likelihood alignment approach to testing for approximate measurement invariance: A paradigmatic cross-cultural application. *Psicothema, 29*(4), 539–551.

Cacioppo, J. T., Tassinary, L. G., & Berntson, G. G. (Eds.). (2007). *Handbook of psychophysiology* (3rd ed.). Cambridge University Press.

Chen, B., Vansteenkiste, M., Beyers, W., Boone, L., Deci, E. L., Van der Kaap-Deeder, J., Duriez, B., Lens, W., Matos, L., Mouratidis, A., Ryan, R. M., Sheldon, K. M., Soenens, B., Petegem, S. V., & Verstuyf, J. (2015). Basic psychological need satisfaction, need frustration, and need strength across four cultures. *Motivation and Emotion, 39*(2), 216–236.

Chen, F. F. (2008). What happens if we compare chopsticks with forks? The impact of making inappropriate comparisons in cross-cultural research. *Journal of Personality and Social Psychology, 95*(5), 1005–1018.

Chiao, J. Y., Hariri, A. R., Harada, T., Mano, Y., Sadato, N., Parrish, T. B., & Iidaka, T. (2010). Theory and methods in cultural neuroscience. *Social Cognitive and Affective Neuroscience, 5*(2–3), 356–361.

Crooks, A., Pfoser, D., Jenkins, A., Croitoru, A., Stefanidis, A., Smith, D. A., Karagiorgou, S., Efentakis, A., & Lamprianidis, G. (2015). Crowdsourcing urban form and function. *International Journal of Geographical Information Science, 29*(5), 720–741.

Csikszentmihalyi, M., & Larson, R. (1987). Validity and reliability of the Experience-Sampling Method. *The Journal of Nervous and Mental Disease, 175*(9), 526–536.

Das, A., & Diener, E. (1985). The subjective well-being of people in India. *Social Indicators Research, 17*(3), 223–242.
Deci, E. L., & Ryan, R. M. (2008). Hedonia, eudaimonia, and well-being: An introduction. *Journal of Happiness Studies, 9*(1), 1–11.
Diener, E. (1984). Subjective well-being. *Psychological Bulletin, 95*(3), 542–575.
Diener, E., Emmons, R. A., Larsen, R. J., & Griffin, S. (1985). The satisfaction with life scale. *Journal of Personality Assessment, 49*(1), 71–75.
Diener, E., Lucas, R. E., & Oishi, S. (2018). Advances and open questions in the science of subjective well-being. *Collabra: Psychology, 4*(1), 15.
Diener, E., Oishi, S., & Lucas, R. E. (2009). Subjective well-being: The science of happiness and life satisfaction. In S. J. Lopez & C. R. Snyder (Eds.), *The Oxford handbook of positive psychology* (2nd ed., pp. 187–194). Oxford University Press.
Diener, E., & Suh, E. M. (Eds.). (2003). *Culture and subjective well-being*. MIT Press.
Diener, E., Tay, L., & Oishi, S. (2013). Rising income and the subjective well-being of nations. *Journal of Personality and Social Psychology, 104*(2), 267–276.
D'Mello, S. K., & Kory, J. (2015). A review and meta-analysis of multimodal affect detection systems. *ACM Computing Surveys, 47*(3), 1–36.
Doane, L. D., & Adam, E. K. (2010). Loneliness and cortisol: Momentary, day-to-day, and trait associations. *Psychoneuroendocrinology, 35*(3), 430–441.
Dzedzickis, A., Kaklauskas, A., & Bucinskas, V. (2020). Human emotion recognition: Review of sensors and methods. *Sensors, 20*(3), 592.
Fowers, B. J., Novak, L. F., & Calder, A. J. (2021). The emergence of the character strengths and virtues in human development: A review of the literature. *Journal of Positive Psychology, 16*(2), 155–170.
Frederick, S., & Loewenstein, G. (1999). Hedonic adaptation. In D. Kahneman, E. Diener, & N. Schwarz (Eds.), *Well-being: The foundations of hedonic psychology* (pp. 302–329). Russell Sage.
Fredrickson, B. L. (2001). The role of positive emotions in positive psychology: The broaden-and-build theory of positive emotions. *American Psychologist, 56*(3), 218–226.
Fredrickson, B. L. (2013). Updated thinking on positivity ratios. *American Psychologist, 68*(9), 814–822.
Fredrickson, B. L., & Losada, M. F. (2005). Positive affect and the complex dynamics of human flourishing. *American Psychologist, 60*(7), 678–686.
Fulmer, C. A., Gelfand, M. J., Kruglanski, A. W., Kim-Prieto, C., Diener, E., Pierro, A., & Higgins, E. T. (2010). On "feeling right" in cultural contexts: How person-culture match affects self-esteem and subjective well-being. *Psychological Science, 21*(11), 1563–1569.
Gabriel, A. S., Podsakoff, N. P., Beal, D. J., Scott, B. A., Sonnentag, S., Trougakos, J. P., & Butts, M. M. (2019). Experience sampling methods: A discussion of critical trends and considerations for scholarly advancement. *Organizational Research Methods, 22*(4), 969–1006.
Gifford, R. (2014). Environmental psychology matters. *Annual Review of Psychology, 65*, 541–579.
Greenspoon, P. J., & Saklofske, D. H. (2001). Toward an integration of subjective well-being and psychopathology. *Social Indicators Research, 54*(1), 81–105.
Harari, G. M., Lane, N. D., Wang, R., Crosier, B. S., Campbell, A. T., & Gosling, S. D. (2016). Using smartphones to collect behavioral data in psychological science: Opportunities, practical considerations, and challenges. *Perspectives on Psychological Science, 11*(6), 838–854.
Harari, G. M., Müller, S. R., Aung, M. S., & Rentfrow, P. J. (2017). Smartphone sensing methods for studying behavior in everyday life. *Current Opinion in Behavioral Sciences, 18*, 83–90.
Hayes, S. C., Luoma, J. B., Bond, F. W., Masuda, A., & Lillis, J. (2006). Acceptance and Commitment Therapy: Model, processes and outcomes. *Behaviour Research and Therapy, 44*(1), 1–25.
Heine, S. J., Lehman, D. R., Markus, H. R., & Kitayama, S. (1999). Is there a universal need for positive self-regard? *Psychological Review, 106*(4), 766–794.

Huang, Y., Fei, T., Kwan, M.-P., Kang, Y., Li, J., Li, Y., Li, X., & Bian, M. (2020). GIS-based emotional computing: A review of quantitative approaches to measure the emotion layer of human–environment relationships. *ISPRS International Journal of Geo-Information, 9*(9), 551.

Huta, V. (2015). The complementary roles of eudaimonia and hedonia and how they can be pursued in practice. In S. Joseph (Ed.), *Positive psychology in practice: Promoting human flourishing in work, health, education, and everyday life* (2nd ed., pp. 159–182). Wiley.

Huta, V., & Waterman, A. S. (2014). Eudaimonia and its distinction from hedonia: Developing a classification and terminology for understanding conceptual and operational definitions. *Journal of Happiness Studies, 15*(6), 1425–1456.

Jorgensen, B. S., & Stedman, R. C. (2001). Sense of place as an attitude: Lakeshore owners attitudes toward their properties. *Journal of Environmental Psychology, 21*(3), 233–248.

Joshanloo, M. (2014). Eastern conceptualizations of happiness: Fundamental differences with Western views. *Journal of Happiness Studies, 15*(2), 475–493.

Kahneman, D., & Krueger, A. B. (2006). Developments in the measurement of subjective well-being. *Journal of Economic Perspectives, 20*(1), 3–24.

Kammann, R., & Flett, R. (1983). Affectometer 2: A scale to measure current level of general happiness. *Australian Journal of Psychology, 35*(2), 259–265.

Kaplan, S. (1995). The restorative benefits of nature: Toward an integrative framework. *Journal of Environmental Psychology, 15*(3), 169–182.

Karagiorgou, S., & Pfoser, D. (2012). On vehicle tracking data-based road network generation. In *Proceedings of the 20th international conference on advances in geographic information systems* (pp. 89–98). ACM.

Keyes, C. L. M. (1998). Social well-being. *Social Psychology Quarterly, 61*(2), 121–140.

Keyes, C. L. M. (2002). The mental health continuum: From languishing to flourishing in life. *Journal of Health and Social Behavior, 43*(2), 207–222.

Keyes, C. L. M. (2005). Mental illness and/or mental health? Investigating axioms of the complete state model of health. *Journal of Consulting and Clinical Psychology, 73*(3), 539–548.

Keyes, C. L. M., & Shapiro, A. D. (2004). Social well-being in the United States: A descriptive epidemiology. In O. G. Brim, C. D. Ryff, & R. C. Kessler (Eds.), *How healthy are we? A national study of well-being at midlife* (pp. 350–372). University of Chicago Press.

Keyes, C. L. M., Shmotkin, D., & Ryff, C. D. (2002). Optimizing well-being: The empirical encounter of two traditions. *Journal of Personality and Social Psychology, 82*(6), 1007–1022.

Keyes, C. L. M., Wissing, M., Potgieter, J. P., Temane, M., Kruger, A., & van Rooy, S. (2008). Evaluation of the mental health continuum-short form (MHC-SF) in Setswana-speaking South Africans. *Clinical Psychology & Psychotherapy, 15*(3), 181–192.

Kim, H. S., & Sasaki, J. Y. (2014). Cultural neuroscience: Biology of the mind in cultural contexts. *Annual Review of Psychology, 65*, 487–514.

Kitayama, S., Markus, H. R., & Kurokawa, M. (2000). Culture, emotion, and well-being: Good feelings in Japan and the United States. *Cognition and Emotion, 14*(1), 93–124.

Kitayama, S., & Uskul, A. K. (2011). Culture, mind, and the brain: Current evidence and future directions. *Annual Review of Psychology, 62*, 419–449.

Kourtit, K., Nijkamp, P., & Toger, M. (2023). Sustainable cities, quality of life, and mobility-related happiness. In *Urban and regional studies* (pp. 103–120). Springer.

Kwan, M.-P. (2012). How GIS can help address the uncertain geographic context problem in social science research. *Annals of GIS, 18*(4), 245–255.

Kwan, M.-P., & Schwanen, T. (2018). Context and uncertainty in geography and GIScience: Advances in theory, method, and practice. *Annals of the American Association of Geographers, 108*(6), 1473–1475.

Larson, J. S. (1993). The measurement of social well-being. *Social Indicators Research, 28*(3), 285–296.

Layard, R. (2011). *Happiness: Lessons from a new science* (2nd ed.). Penguin.

Lazer, D., Kennedy, R., King, G., & Vespignani, A. (2014). The parable of Google Flu: Traps in big data analysis. *Science, 343*(6176), 1203–1205.

Lu, L. (1999). Personal or environmental causes of happiness: A longitudinal analysis. *The Journal of Social Psychology, 139*(1), 79–90.
Lu, L. (2001). Understanding happiness: A look into the Chinese folk psychology. *Journal of Happiness Studies, 2*(4), 407–432.
Lu, L., & Gilmour, R. (2004). Culture and conceptions of happiness: Individual oriented and social oriented SWB. *Journal of Happiness Studies, 5*(3), 269–291.
Lu, L., Gilmour, R., & Kao, S.-F. (2001). Cultural values and happiness: An East-West dialogue. *The Journal of Social Psychology, 141*(4), 477–493.
Lwin, K. K., & Murayama, Y. (2011). Web-based GIS system for real-time field data collection using a personal mobile phone. *Journal of Geographic Information System, 3*(4), 382–389.
Maddux, J. E. (2017). Subjective well-being and life satisfaction: An introduction to the special issue. *Journal of Positive Psychology, 12*(6), 531–532.
Maes, I. H. L., Delespaul, P. A. E. G., Peters, M. L., White, M. P., van Horn, Y., Schruers, K., Anteunis, L., & Joore, M. (2018). Measuring health-related quality of life by experiences: The experience sampling method. *Value in Health, 21*(1), 44–51.
Markus, H. R., & Kitayama, S. (1991). Culture and the self: Implications for cognition, emotion, and motivation. *Psychological Review, 98*(2), 224–253.
McCunn, L. J., & Gifford, R. (2018). Spatial navigation and place imageability in sense of place. *Cities, 74*, 208–218.
Mehta, V. (2014). Evaluating public space. *Journal of Urban Design, 19*(1), 53–88.
Nahum-Shani, I., Smith, S. N., Spring, B. J., Collins, L. M., Witkiewitz, K., Tewari, A., & Murphy, S. A. (2018). Just-in-time adaptive interventions (JITAIs) in mobile health: Key components and design principles for ongoing health behavior support. *Annals of Behavioral Medicine, 52*(6), 446–462.
Oishi, S. (2006). The concept of life satisfaction across cultures: An IRT analysis. *Journal of Research in Personality, 40*(4), 411–423.
Pavot, W., & Diener, E. (2013). Review of the satisfaction with life scale. In *Social indicators research series* (Vol. 54, pp. 101–117). Springer.
Robinson, M. D., & Clore, G. L. (2002). Belief and feeling: Evidence for an accessibility model of emotional self-report. *Psychological Bulletin, 128*(6), 934–960.
Rokhsaritalemi, S., Sadeghi-Niaraki, A., & Choi, S.-M. (2023). Exploring emotion analysis using artificial intelligence, geospatial information systems, and extended reality for urban services. *IEEE Access, 11*, 5056–5075.
Ryan, R. M., & Deci, E. L. (2001). On happiness and human potentials: A review of research on hedonic and eudaimonic well-being. *Annual Review of Psychology, 52*, 141–166.
Ryan, R. M., Huta, V., & Deci, E. L. (2008). Living well: A self-determination theory perspective on eudaimonia. *Journal of Happiness Studies, 9*(1), 139–170.
Ryff, C. D. (1989). Happiness is everything, or is it? Explorations on the meaning of psychological well-being. *Journal of Personality and Social Psychology, 57*(6), 1069–1081.
Schaufeli, W. B., & Bakker, A. B. (2004). Job demands, job resources, and their relationship with burnout and engagement: A multi-sample study. *Journal of Organizational Behavior: The International Journal of Industrial, Occupational and Organizational Psychology and Behavior, 25*(3), 293–315.
Seligman, M. E., Steen, T. A., Park, N., & Peterson, C. (2005). Positive psychology progress: empirical validation of interventions. *American Psychologist, 60*(5), 410.
Senaratne, H., Mueller, M., Behrisch, M., Lalanne, F., Bustos-Jiménez, J., Schneidewind, J., Keim, D., & Schreck, T. (2017). Urban mobility analysis with mobile network data: A visual analytics approach. *IEEE Transactions on Intelligent Transportation Systems, 19*(5), 1537–1546.
Shiffman, S. (2009). Ecological momentary assessment (EMA) in studies of substance use. *Psychological Assessment, 21*(4), 486–497.
Shiffman, S. (2013). Conceptualizing analyses of ecological momentary assessment data. *Nicotine & Tobacco Research, 16*(Suppl_2), S76–S87.

Shiffman, S., Stone, A. A., & Hufford, M. R. (2008). Ecological momentary assessment. *Annual Review of Clinical Psychology, 4*, 1–32.

Smith, P. B. (2004). Acquiescent response bias as an aspect of cultural communication style. *Journal of Cross-Cultural Psychology, 35*(1), 50–61.

Springer, K. W., & Hauser, R. M. (2006). An assessment of the construct validity of Ryff's scales of psychological well-being: Method, mode, and measurement effects. *Social Science Research, 35*(4), 1080–1102.

Stone, A. A., & Shiffman, S. (2002). Capturing momentary, self-report data: A proposal for reporting guidelines. *Annals of Behavioral Medicine, 24*(3), 236–243.

Suls, J., & Rothman, A. (2004). Evolution of the biopsychosocial model: Prospects and challenges for health psychology. *Health Psychology, 23*(2), 119–125.

Tov, W., & Diener, E. (2009). Culture and subjective well-being. In E. Diener (Ed.), *Culture and well-being: The collected works of Ed Diener* (pp. 9–41). Springer.

Trull, T. J., & Ebner-Priemer, U. W. (2020). Ambulatory assessment. In J. S. Comer & P. C. Kendall (Eds.), *The Oxford handbook of research strategies for clinical psychology* (pp. 301–311). Oxford University Press.

Uchida, Y., & Kitayama, S. (2009). Happiness and unhappiness in east and west: Themes and variations. *Emotion, 9*(4), 441–456.

Uchida, Y., & Ogihara, Y. (2012). Personal or interpersonal construal of happiness: A cultural psychological perspective. *International Journal of well-being, 2*(4), 354–369.

Ulrich, R. S., Simons, R. F., Losito, B. D., Fiorito, E., Miles, M. A., & Zelson, M. (1991). Stress recovery during exposure to natural and urban environments. *Journal of Environmental Psychology, 11*(3), 201–230.

Valsiner, J. (2012). *The Oxford handbook of culture and psychology*. Oxford University Press.

VanderWeele, T. J. (2017). On the promotion of human flourishing. *Proceedings of the National Academy of Sciences, 114*(31), 8148–8156.

van de Vijver, F. J. R., & Leung, K. (1997). *Methods and data analysis for cross-cultural research*. Sage.

Veenhoven, R. (2012). Cross-national differences in happiness: Cultural measurement bias or effect of culture? *International Journal of well-being, 2*(4), 333–353.

Vittersø, J., Røysamb, E., & Diener, E. (2002). The concept of life satisfaction across cultures: Exploring its diverse meaning and relation to economic wealth. In E. Gullone & R. A. Cummins (Eds.), *The universality of subjective well-being indicators: A multi-disciplinary and multi-national perspective* (pp. 81–103). Springer.

Watson, D., Clark, L. A., & Tellegen, A. (1988). Development and validation of brief measures of positive and negative affect: The PANAS scales. *Journal of Personality and Social Psychology, 54*(6), 1063–1070.

Westerhof, G. J., & Keyes, C. L. M. (2010). Mental illness and mental health: The two continua model across the lifespan. *Journal of Adult Development, 17*(2), 110–119.

Wilkinson, R. G., & Pickett, K. E. (2009). *The spirit level: Why more equal societies almost always do better*. Allen Lane.

Zhao, K., Tarkoma, S., Liu, S., & Vo, H. (2016). Urban human mobility data mining: An overview. In *2016 IEEE international conference on big data (big data)* (pp. 1911–1920).

Zou, L., Song, Y., & Cervone, G. (2024). Geospatial big data: Theory, methods, and applications. *Annals of GIS, 30*(4), 411–415.

# Chapter 4
# Factors Influencing Subjective Well-Being: Key Dimensions and Research Evolution

**Abstract** This chapter argues that research on subjective well-being (SWB) has evolved from reductionist, single-factor models toward more integrative, multi-level frameworks accounting for biological, psychological, social, situational, and cultural determinants. Its primary aim is to synthesize evidence across these dimensions, delineating a layered conceptual architecture that incorporates endogenous personal attributes, interpersonal socio-demographic factors, and situational environmental influences. Section 4.1 traces the theoretical shift from static classifications to dynamic interactional models, exemplified by the biopsychosocial-ecological framework. Section 4.2 examines endogenous determinants, including genetic and epigenetic mechanisms and psychological traits. Section 4.3 analyzes demographic and socioeconomic variables, while Sect. 4.4 explores interpersonal and situational factors such as social networks and built environment characteristics. Section 4.5 discusses macro-level influences from cultural, institutional, and urban planning contexts. The chapter concludes with Sect. 4.6, which proposes integrative frameworks and highlights implications for policy and interdisciplinary research. By contextualizing SWB within a multi-dimensional ecosystem, this chapter provides a foundational lens for understanding well-being determinants and sets the stage for subsequent empirical analysis of built environment effects in high-density urban settings.

**Keywords** Subjective well-being · Theoretical framework · Biopsychosocial-ecological model · Determinants · Multi-dimensional analysis · Endogenous factors · Interpersonal factors · Situational context · Cultural influences · Policy implications

Building upon the multidimensional conceptualization and measurement frameworks articulated in Chap. 3, this chapter shifts focus toward the determinants of subjective well-being (SWB). While the previous chapter addressed how well-being is assessed, this chapter examines the factors that explain why it varies across individuals and contexts. To this end, it synthesizes evidence across multiple levels of analysis, charting the field's evolution from reductionist, single-factor models to



H. Li, G. Shen, *The Anatomy of Subjective Well-Being*, Resilient and Friendly Living Environments, https://doi.org/10.1007/978-981-95-4937-5_4

sophisticated, dynamic frameworks that acknowledge complex interactions among biological, psychological, social, and environmental determinants. It begins by systematically categorizing these key determinants into endogenous (e.g., genetic and epigenetic mechanisms, personality traits), interpersonal (e.g., demographics, socioeconomic factors, social relationships), and situational dimensions (e.g., physical environment, cultural and institutional contexts), thereby establishing a conceptual framework for analyzing their dynamic interdependence in shaping well-being outcomes.

The review further highlights methodological advances—such as longitudinal designs, multi-method assessments, and machine learning—that facilitate deeper insight into these complex causal pathways and nonlinearities. It also examines how cross-cultural variations, moderated by cultural values, social policies, and urban planning strategies (including the emerging "15-minute city" concept), reveal meaningful differences in the valuation and predictors of well-being. Building on this synthesis, the chapter advocates for operationalizable, testable theoretical frameworks that integrate the experiential and evaluative components of SWB, and emphasizes the necessity of interdisciplinary approaches bridging molecular biology, psychology, social sciences, and urban studies. Finally, it discusses implications for policy and intervention, emphasizing mental health services, social welfare, and sustainable urban development as critical levers for enhancing population well-being.

It is important to note that while this chapter reviews these broad multidimensional determinants of SWB, a focused and detailed examination of one critical situational determinant—the built environment—will be reserved for the subsequent empirical section of this monograph. This deliberate separation allows for a more thorough exploration of its mechanisms within the specific context of high-density urban settings.

## 4.1 Theoretical Framework Transformation in Subjective Well-Being Research

### *4.1.1 Transition from Static Classification to Interactional Modeling of SWB Determinants*

The historical trajectory of subjective well-being (SWB) research has witnessed a profound theoretical shift, moving from fragmented investigations of isolated determinants toward integrative frameworks that emphasize dynamic, multi-level interactions. Early inquiries into SWB primarily adopted a static classification approach, focusing on identifying discrete external life circumstances—such as income, health, or demographic variables—as primary correlates of happiness. This paradigm was rooted in assumptions of single-factor determinism, where phenomena like early economic determinism models posited income as a central explanatory variable.[1] However, empirical evidence consistently revealed that correlations between such objective circumstances and SWB were small to moderate, with demographic factors accounting for limited variance in well-being outcomes (Diener, 1984). For instance, a meta-analysis examining health as a determinant found that it explained only 8.4–14.4% of the variance in SWB, underscoring the inadequacy of single-factor explanations (Okun et al., 1984). Similarly, research on income inequality and SWB failed to establish a universal linear relationship, with associations varying systematically by country economic development—negative in developed nations and positive in developing contexts—further challenging reductionist models (Ngamaba et al., 2018).

This recognition of single-factor limitations catalyzed a theoretical evolution, marked by growing attention to psychological processes and individual differences. Studies began to highlight personality traits—particularly extraversion and neuroticism—as robust predictors of SWB, often exhibiting stronger correlations than situational or demographic variables (Steel et al., 2008). This shift was paralleled by debates between "bottom-up" models, which viewed SWB as an aggregate of life circumstances, and "top-down" models, which emphasized personality-based processes shaping evaluations (Lucas & Diener, 2009). However, neither perspective alone captured the complexity of SWB, leading to the integration of these frameworks into interactional models. For example, the dynamic equilibrium model proposed that SWB reflects both stable personality traits (e.g., extraversion predicting long-term well-being) and transient life events (Headey & Wearing, 1989), though the latter typically lose their impact within 3–6 months as individuals adapt to changes (Suh et al., 1996). This interplay between stable dispositions and situational dynamics revealed SWB as a product of reciprocal person-environment interactions rather than a unidimensional outcome (Magnusson & Stattin, 2006).

The most comprehensive articulation of this transition is embodied in the biopsychosocial-ecological framework, which integrates biological, psychological, and environmental influences across nested levels of analysis. Bronfenbrenner and Morris's bioecological model exemplifies this shift, emphasizing "proximal processes"—reciprocal interactions between an active organism and its immediate environment—as the engines of development, shaped by person characteristics (e.g., dispositions, resources), context (micro-, meso-, exo-, macrosystems), and time (microtime, mesotime, macrotime). This Process-Person-Context-Time (PPCT) model rejects additive single-factor causality, instead positing synergistic, multiplicative interactions; for instance, maternal responsiveness (process) may interact with infant birthweight (person) and social class (context) to influence outcomes (Bronfenbrenner & Morris, 2007). Complementing this, behavioral epigenetics research has demonstrated how early environmental signals (e.g., maternal care) can structurally modify gene expression (e.g., glucocorticoid receptor regulation) (Weaver et al., 2004), providing mechanistic evidence for the biological

[1] The term "economic determinism" in this context refers to early simplistic models in well-being research that over-emphasized the role of income, often neglecting psychological and social dimensions. For a critique of such reductionist approaches in social science, see Tatarkiewicz (1976).

embedding of environmental influences on individual differences in stress reactivity and, by extension, SWB (Meaney, 2010).[2]

Empirical advances in methodological rigor have further solidified this interactional paradigm. Multivariate analyses now routinely control for confounding variables across levels, such as in studies of education's impact on SWB, which demonstrate independent, direct effects beyond income, mediated by health and moderated by socioeconomic context (Striessnig, 2015). These approaches reflect a broader recognition that SWB is determined by dynamic interplay across genetic, psychological (e.g., aspirations, personality), social (e.g., employment, relationships), economic (e.g., income, inequality), and environmental (e.g., culture, living conditions) domains. Collectively, this theoretical transformation has moved SWB research from static classifications of isolated factors to holistic models that prioritize context, temporality, and bidirectional processes, guiding more nuanced empirical inquiry and intervention development. This evolution from simplistic to complex frameworks can be conceptualized as a series of overlapping phases, each refining and incorporating insights from the last. Table 4.1 synthesizes this theoretical progression, outlining the core tenets, key determinants, and methodological signatures of the major paradigms that have shaped the field of SWB research. It serves as an organizing heuristic for understanding how the focus of inquiry has expanded from external circumstances to internal dispositions, and finally to their dynamic interaction within multi-level systems.

### *4.1.2 Conceptual Architecture: Endogenous, Interpersonal, and Situational Layers*

The conceptual architecture of subjective well-being (SWB) has evolved into a multidimensional framework that coherently integrates three interconnected layers: endogenous personal factors, interpersonal socio-demographic characteristics, and situational context factors. This integrated model acknowledges that individual differences and cross-cultural variations in SWB arise from the dynamic interplay of these layers, rather than isolated determinants (Diener et al., 2003). This layered architecture is visualized in Fig. 4.1, which illustrates the hierarchical and interactive nature of SWB determinants. The model posits that endogenous personal factors form a foundational core, which is continuously shaped by, and interacts with, surrounding interpersonal and situational contexts. These contexts are themselves embedded within and moderated by broader macro-level cultural and institutional systems. It is the dynamic, reciprocal transactions across these levels—rather than

[2] The Weaver et al. (2004) study on rats was foundational, showing that high levels of maternal licking and grooming led to decreased methylation of the glucocorticoid receptor gene promoter in the hippocampus of offspring, resulting in a more adaptive stress response throughout life. This provided one of the first direct links between maternal behavior and epigenetic programming of behavior.

**Table 4.1** Evolution of theoretical frameworks in SWB research

| Theoretical framework | Era/key proponents | Core tenet/ mechanism | Key determinants | Typical methods |
|---|---|---|---|---|
| Single-Factor Determinism | 1960s–1970s, Early economic models | Direct, linear influence of objective circumstances | Income, health, demographics | Cross-sectional surveys |
| Top-Down Models | 1980s–1990s, Lucas and Diener (2009) | Stable personality traits shape perception and evaluation of life events | Extraversion, Neuroticism, Optimism | Longitudinal studies, Personality inventories |
| Bottom-Up Models | 1980s, Diener (1984) | SWB is the sum of satisfactions across important life domains | Domain satisfactions (e.g., work, family) | Self-report questionnaires |
| Interactional Models | 1990s–2000s, Suh et al. (1996) | Dynamic interplay between stable dispositions and situational factors | Personality × Life Events, Adaptation | Experience Sampling Method (ESM), Longitudinal designs |
| Biopsychosocial Ecological Framework | 2000s–Present, Bronfenbrenner and Morris (2007) | Multi-level, reciprocal interactions across biological, psychological, and contextual systems | Gene-Environment interplay, Proximal Processes, Epigenetics | Twin studies, Neuroimaging, multi-level modeling |

Source: Created by the authors

the contribution of any single layer in isolation—that ultimately produce SWB outcomes. The following sections will elaborate on the constituents and mechanisms of each layer, beginning with the endogenous determinants.

Endogenous personal factors refer to stable individual attributes rooted in genetics, personality, and psychological dispositions. Genetic influences account for approximately 30–40% of the variance in SWB, indicating a substantial biological basis (Bartels, 2015; Nes & Røysamb, 2017), while environmental factors explain the remaining 60–70%. Personality traits such as extraversion and neuroticism are key endogenous predictors, with direct effects on emotional experiences and indirect effects by moderating responses to situational contexts. Additionally, basic and psychological needs—such as food, shelter, social connection, and respect—serve as inherent drivers, where their fulfillment is positively associated with SWB across diverse cultural contexts (Oishi et al., 2018). Positive emotional traits like hope and curiosity further contribute to resilience, enabling individuals to cope with stress and maintain well-being by broadening cognitive and behavioral repertoires (Tugade et al., 2004). Longitudinal studies have also demonstrated that early adult

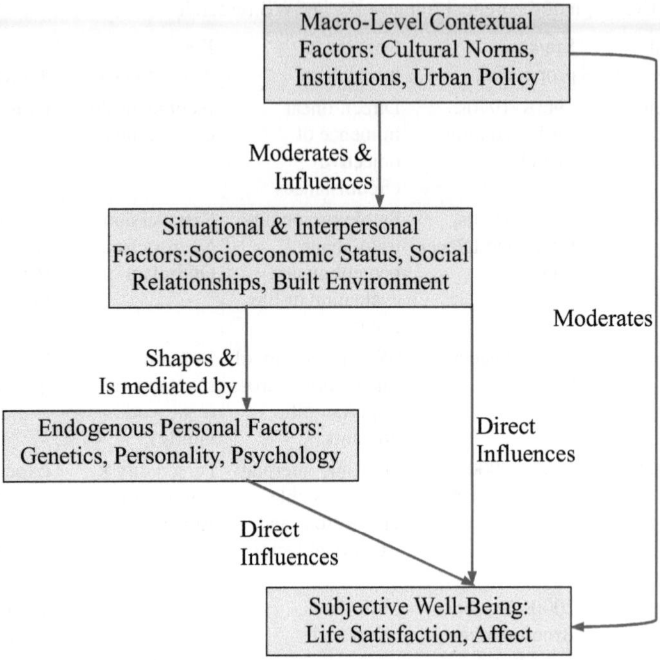


**Fig. 4.1** Conceptual architecture of SWB determinants. Source: Created by the authors

personality traits, including conscientiousness and agreeableness, predict later-life SWB, underscoring the enduring influence of endogenous factors (Friedman et al., 2010).

Interpersonal socio-demographic characteristics encompass social structural positions, relational networks, and comparative processes that shape SWB. Socio-demographic variables such as income, marital status, and education operate through pathways that intersect with endogenous and situational factors. For instance, education enhances SWB indirectly by increasing access to non-alienated work, economic resources, and a sense of personal control, while also fostering supportive social relationships (Ross & Willigen, 1997). Social relationships themselves are multifaceted, involving the quantity of ties (social integration), structural properties (density, reciprocity), and qualitative content (social support), all of which contribute to SWB by buffering stress and regulating emotions (Cohen, 2004). Notably, weak social ties (e.g., acquaintances) play a distinct role, contributing to feelings of belonging and positive affect independently of strong ties, especially when strong tie interactions are limited (Sandstrom & Dunn, 2014), a phenomenon first systematically explored by Granovetter (1973). Social comparison processes further mediate interpersonal influences: subjective socioeconomic status (SES)—rooted in perceived relative social position—has a stronger association with SWB than objective SES (Adler et al., 2000), highlighting the importance of interpersonal evaluation (Tan et al., 2020). Religious involvement also operates through interpersonal mechanisms, with existential certainty (derived from religious beliefs) emerging as a powerful predictor of life satisfaction, particularly among older individuals and those with lower formal education (Ellison, 1991).

Situational context factors encompass the broader cultural, societal, and environmental settings in which individuals are embedded. Cultural norms and values moderate SWB correlates, such as the emphasis on self-enhancement versus self-criticism, which influences average SWB levels across societies. Life events, including marriage, unemployment, and bereavement, trigger initial SWB reactions followed by adaptation, with cognitive well-being (life satisfaction) showing more persistent changes than affective well-being (Luhmann et al., 2012). Economic and political contexts significantly shape SWB: higher levels of socialist welfare state policies, labor decommodification, and leftist party dominance are associated with greater aggregate life satisfaction, independent of individual economic status (Radcliff, 2001). Neighborhood environments, characterized by physical features (e.g., green spaces, pollution) and social dynamics (e.g., collective efficacy), influence SWB through mediators like personal relationships, leisure opportunities, and emotional responses to the built environment (Mouratidis, 2018). The specific attributes of the built environment and their measurement constitute a major area of research in urban studies and environmental psychology. A comprehensive review of these factors, including urban density, diversity, and design access to amenities, will be provided in the empirical chapters of this book, which focus specifically on the mechanisms linking the built environment to well-being. Residential relocation, as a situational change, further modulates SWB by altering exposure to neighborhood contexts, though endogenous preferences (residential self-selection) complicate causal relationships (Boone-Heinonen et al., 2010). Normative contexts, such as peer marriage prevalence, moderate the SWB benefits of marriage, with stronger effects in environments where marriage is more socially normative (Wadsworth, 2016). Cross-nationally, gender differences in SWB vary with situational factors like economic development and cultural traditions, with younger women reporting higher SWB in economically developed societies and older women facing disadvantages in contexts with cultural devaluation of aging (Zuckerman et al., 2017).

The interactions between these layers are integral to the conceptual architecture. For example, income inequality (situational) strengthens the negative association between relative income (endogenous) and life satisfaction, particularly for low-income individuals (Cheung & Lucas, 2016). Neighborhood crime rates (situational) increase fear, but this effect is attenuated by socioeconomic disadvantage and varies with individual characteristics like age and gender (Roman & Chalfin, 2008). Situational social norms in public transit settings can override endogenous personality traits, as even introverted individuals experience increased SWB from social connection with strangers when situational barriers to interaction are reduced (Epley & Schroeder, 2014). These interactions highlight that SWB is not merely the sum of individual, social, and contextual factors but a dynamic outcome of their complex interplay (Das et al., 2020).

This hierarchical and interactive architecture signifies a shift in SWB research from merely cataloging associated factors toward understanding an organic and dynamic ecosystem.

## 4.2 Endogenous Determinants of Subjective Well-Being

### *4.2.1 Biological Attributes and Genetic Influences*

Genetic and biological factors contribute significantly to individual differences in subjective well-being (SWB). Twin and family studies, employing classical designs comparing monozygotic (MZ) and dizygotic (DZ) twins—often extended to include siblings and non-twin relatives—have been pivotal in demonstrating genetic influences, with meta-analyses reporting a weighted average heritability of approximately 36% for overall well-being (confidence interval 34–38%) and 32% for satisfaction with life (confidence interval 29–35%) based on large samples (Bartels, 2015; Bartels & Boomsma, 2009; Nes & Røysamb, 2017). These estimates, however, are not static and vary across age, sex (with inconclusive evidence for differences), measurement methods, and populations, predominantly Western samples. Molecular genetic support further validates these findings, with SNP-based heritability estimates ranging from 12 to 18% after correcting for measurement error, though candidate gene association studies (e.g., 5-HTTLPR, MAOA) remain inconclusive (Bartels, 2015).[3]

Beyond direct genetic effects, epigenetic mechanisms provide a critical link between environmental experiences and stable phenotypic variations in SWB. DNA methylation, an enzymatically-catalyzed modification, enables environmental signals—particularly early life social and physical environments—to become embedded in the genome, affecting long-term phenotypes; early life adversity, for instance, alters DNA methylation across multiple systems (e.g., brain and peripheral tissues), supporting system-wide epigenetic programming. Molecular pathways mediating this include signaling cascades (e.g., serotonin-induced cAMP and PKA activation) that recruit DNA demethylation and histone acetylation activities, modulating expression of stress response genes like the glucocorticoid receptor (GR) and FKBP5 (Szyf, 2013). Rodent studies further demonstrate that variations in maternal care during the early postnatal period alter GR gene promoter methylation in the

[3]The 5-HTTLPR (serotonin-transporter-linked polymorphic region) is a well-studied polymorphism in the *SLC6A4* gene. The short (s) allele has been associated with higher scores on anxiety-related personality traits and an increased susceptibility to depression. However, a large body of research, most famously the study by Caspi et al. (2003), demonstrates that this association is primarily dependent on its interaction with stressful life events (Gene-Environment interaction). The inconsistent replication of a direct main effect of the s allele on its own highlights the complexity of linking single genetic variants to multifaceted psychological constructs like subjective well-being (SWB).

hippocampus, modifying stress responsivity and behavior throughout life, with cross-fostering confirming maternal behavior, not genetics, drives this programming (Meaney, 2010).

Gene-environment interactions are central to understanding these biological contributions, as the notion of independent "main effects" of genes or environment on complex traits like SWB is biologically fallacious; instead, development emerges from their constant interplay. This interplay is realized at the molecular level through transcription factor binding to DNA, altering gene transcription rates. A well-studied example is the serotonin transporter gene (SLC6A4) promoter polymorphism (5-HTTLPR), where effects on emotional behavior depend on environmental exposures like childhood maltreatment, with nonhuman primate studies showing similar genotype-rearing condition interactions modulating emotional reactivity (Caspi et al., 2003). Such interactions highlight that genetic influences on SWB are context-dependent, shaped by the dynamic interplay of genome and environment.

These biological and genetic factors do not directly dictate well-being; rather, they exert their influence by shaping stable psychological tendencies and behavioral patterns, thus highlighting the role of psychological traits.

### 4.2.2 Psychological Traits and Individual Dispositions

Psychological traits and individual dispositions are critical correlates of subjective well-being (SWB), with research indicating that personality is more strongly associated with SWB than life circumstances in many instances. Among personality dimensions, extraversion and neuroticism have been the most frequently studied: extraversion is linked to positive emotions, neuroticism to negative feelings, with meta-analytic correlations showing extraversion-positive affect associations of 0.37–0.44 and neuroticism-negative affect correlations of 0.46–0.54 (Steel et al., 2008). Temperament theories explain these links through behavioral activation system (BAS) sensitivity in extraverts (enhancing reward processing and positive emotions) and behavioral inhibition system (BIS) sensitivity in neurotics (increasing sensitivity to punishment and negative emotions). Personality traits account for greater variance in SWB than demographics, contributing to its stability over time, though SWB can change via life events and processes like emotion regulation (Lucas & Diener, 2009).

Wilson's (1967) review, as highlighted in subsequent research, identified extroversion and optimism as key correlates of SWB, describing the "happy person" as extroverted, optimistic, and worry-free. Psychological traits such as self-esteem and internal locus of control (belief in personal control over outcomes) also correlate positively with SWB, often more strongly than demographic factors. Extraversion correlates with positive affect partly through sociability, while neuroticism correlates with negative affect, supporting distinct affective pathways. Additionally, social contact—linked to traits like extraversion—serves a stress-buffering role, enhancing SWB through interpersonal relationships (Diener, 1984).

An optimistic personality, defined as a dispositional expectancy of favorable outcomes, correlates with SWB by fostering goal setting, effective coping, and persistence amid adversities (Scheier & Carver, 1985). Hope, a related but distinct cognitive-motivational construct involving goal-directed thinking (goals, pathway thinking, and agency), similarly contributes to SWB through resilience and persistence (Snyder, 2002). Both optimism and hope enhance positive emotions, motivation, and stress buffering: optimistic individuals use proactive coping, while hope facilitates finding alternative pathways during challenges, predicting life satisfaction and reducing depressive symptoms across contexts (Dursun, 2021).

Individual differences in cognitive styles and emotion regulation strategies constitute another essential dimension of psychological dispositions influencing SWB. Emotional regulation and cognitive flexibility, associated with personality traits like openness to experience, support the constructive use of internal resources (e.g., creativity, self-transformation) by enabling adaptive processing of emotional experiences, thereby contributing to well-being (Long & Averill, 2003).

In summary, endogenous biological and psychological traits establish a stable baseline from which individuals experience and evaluate their subjective well-being, and engage in complex interactions with interpersonal and situational factors that will be discussed subsequently.

## 4.3 Individual Differences Shaping SWB

### *4.3.1 Demographic Characteristics*

As distal yet influential factors, demographic characteristics, including age, gender, and other variables such as marital status and income, are recognized as key predictors of subjective well-being (SWB), alongside personality and process-oriented factors. SWB, defined as an individual's subjective evaluation of life quality, is shaped by these demographic factors through their association with cultural contexts, life circumstances, and social roles (Diener et al., 2018).

Age exhibits complex and multidimensional associations with SWB, varying across its components: life satisfaction, positive affect, and negative affect. Cross-sectional and longitudinal studies have revealed distinct lifespan trajectories. For life satisfaction, meta-analytic evidence shows a decline during adolescence (ages 9–16, $d = -0.56$), followed by an increase until approximately age 70 ($d = 0.16$), and a subsequent decrease in very late life (after 70, $d = -0.24$) (Buecker et al., 2023). This contrasts with earlier suggestions of a U-shaped trajectory; instead, life satisfaction generally rises from adolescence to midlife before declining in advanced age. Positive affect, by contrast, shows a continuous decline from childhood (age 9) through late adulthood (until age 94, $d = -1.71$). Negative affect follows a curvilinear pattern: after small fluctuations in childhood and adolescence, it decreases from age 22 until around 60 ($d = -0.92$), then increases again until age 87 ($d = 0.58$).

These age-related patterns are robust across gender, country, and ethnicity, with no significant moderation by demographic factors (Buecker et al., 2023). However, large-scale cross-country studies confirm a U-shaped relationship between age and overall SWB, with a global nadir occurring around age 48, observed in both developing and advanced nations (Blanchflower, 2021). Longitudinal data further support that life satisfaction increases from age 40 onward, with a slight drop after 75, though remaining higher than in individuals under 40 (Horley & Lavery, 1995). The "satisfaction paradox" in aging—stable or increasing SWB despite health declines—is often explained by socio-emotional selectivity theory (Carstensen, 1992, 2006), where older adults prioritize emotionally meaningful goals, enhancing life satisfaction (López Ulloa et al., 2013).

Gender differences in SWB are characterized by nuanced patterns in affective experiences rather than overall happiness levels. Women report both more positive and more negative affect than men, with greater emotional intensity, though this does not translate to significant differences in global happiness (Diener, 1984). Specifically, women exhibit higher levels of unpleasant affect, particularly internalizing emotions like sadness and fear, while differences in pleasant affect and life satisfaction are smaller and inconsistent across studies (Lucas & Gohm, 2000). For example, gender accounts for less than 1% of variance in SWB but over 13% of variance in emotional intensity (Diener & Ryan, 2009; Fujita et al., 1991). An age-by-gender interaction further complicates these relationships: younger women (under 45) tend to report higher SWB than younger men (28% vs. 24% "very happy"), whereas this pattern reverses in older age groups (over 54: 20% women vs. 25% men "very happy") (Inglehart, 2002). This reversal may reflect cultural devaluation of older women's social worth in industrialized societies and changing gender roles across the lifespan. Cross-culturally, gender differences in SWB vary with societal gender inequality: smaller differences are observed in countries with lower inequality (e.g., Norway), while larger differences emerge in contexts with higher inequality, particularly in domain-specific SWB like job satisfaction (Tesch-Römer et al., 2008). For instance, national gender inequality predicts greater gender differences in job satisfaction, with women reporting lower satisfaction in more unequal contexts, though global life satisfaction shows minimal gender differences (Batz-Barbarich et al., 2018).

Other demographic variables, including marital status, income, employment, and education, also influence SWB, albeit with modest overall effect sizes. Marital status is a strong predictor: married individuals generally report higher SWB, especially with high marital quality, though adaptation to baseline levels occurs within years post-marriage (Oishi et al., 2018). A meta-analysis found a small positive association between marital status and SWB (mean $r = .14$), which is stronger for men ($r = .17$) than women ($r = 0.12$) and has decreased over time, possibly due to changing social values (Batz-Barbarich et al., 2018). Unemployment is consistently linked to lower SWB, even when controlling for income (Luhmann et al., 2012). Income correlates positively with SWB within countries, particularly at poverty levels, but with diminishing returns at higher incomes (Kahneman & Deaton, 2010). Education shows a weak positive correlation with SWB after controlling for covariates. Collectively, demographic factors explain less than 10–15% of variance in SWB, indicating their influence is subordinate to other individual differences like personality.

These findings suggest that demographic variables primarily serve as proxy indicators for resources, constraints, and social roles, indirectly shaping subjective well-being by influencing individuals' daily experiences and psychological processes.

### *4.3.2 Socioeconomic Status and Resource Access*

Beyond general demographic characteristics, more structurally defined socioeconomic status (SES) factors constitute the key contextual environment influencing well-being. Socioeconomic status (SES) and resource access, encompassing income, education, employment, and perceived social status, play significant roles in shaping subjective well-being (SWB), though their effects are complex and moderated by factors such as social comparison, inequality, and cultural context. These dimensions interact to influence both cognitive (life evaluation) and emotional (daily emotional experience) components of SWB, with distinct pathways and saturation points observed across different socioeconomic contexts.

Income, a core objective SES indicator, exhibits nuanced relationships with SWB. At very low income levels, income strongly predicts SWB by fulfilling basic needs (e.g., food, shelter), but this correlation diminishes in wealthier nations. Kahneman and Deaton further differentiate this relationship: income positively affects life evaluation continuously across the income spectrum, with no satiation, whereas emotional well-being (e.g., positive affect, stress) rises with income only up to approximately $75,000 annual income, after which gains plateau (Kahneman & Deaton, 2010). Cross-nationally, national wealth correlates strongly with mean SWB (correlations of .60 to .70; Diener et al., 2003), yet within countries, relative income effects dominate. For instance, higher neighborhood income negatively predicts individual life satisfaction when controlling for household income, a pattern consistent with the relative income hypothesis. In counties with greater inequality, the negative association between neighborhood income and life satisfaction is stronger, particularly for low-income individuals (Cheung & Lucas, 2016). Such findings align with the Easterlin Paradox (Easterlin, 1974), which posits that rising national income does not increase average SWB over time due to social comparison and adaptation—individuals evaluate their income relative to reference groups (e.g., peers, past income), and adaptation to new income levels erodes long-term gains (Clark et al., 2008).[4] Meta-analytic evidence confirms that subjective SES (e.g.,

[4] The Easterlin Paradox has been debated, with some studies suggesting that long-term economic growth does correlate with increased life satisfaction, especially if it leads to improvements in

perceived social status measured via the MacArthur ladder) more strongly predicts SWB ($r \approx .22$) than objective SES ($r \approx .16$), as it explicitly captures social comparison processes (Adler et al., 2000; Tan et al., 2020). In developing countries, the income-SWB correlation is stronger ($r = .28$) than in high-income economies ($r = .10$), supporting need theory, which emphasizes basic need fulfillment as a primary mechanism (Howell & Howell, 2008).

Education influences SWB indirectly through occupational status rather than income. Witter et al. found education correlates modestly with SWB (weighted mean effect size = .11–.18), with stronger associations for women and older adults (Witter et al., 1984). College graduation enhances life evaluation but has limited or counterintuitive effects on emotional states (e.g., stress). Education may reduce reliance on income for need satisfaction, as highly educated samples show weaker income-SWB correlations, suggesting education provides non-pecuniary resources (e.g., skills, social capital) that buffer SWB.

Employment status and quality are critical for SWB, with unemployment exerting a medium-sized negative effect on mental health (effect size $d = 0.51$) (Paul & Moser, 2009). Job insecurity, rather than past unemployment, is a stronger predictor of reduced SWB: fear of future unemployment ("scaring") outweighs the direct effects of past job loss ("scarring") (Knabe & Rätzel, 2011). Temporary workers report lower well-being primarily due to heightened job insecurity, but controlling for security satisfaction, they may exhibit higher psychological well-being (Dawson et al., 2017). Unemployment's impact varies by gender (men more affected, aligning with the male breadwinner model), age (less severe with age), and labor market context—worse in regions with high unemployment rates (Picchio & Ubaldi, 2024). Paradoxically, individuals unemployed in high-unemployment regions report less mental distress than those in low-unemployment areas, likely due to reduced social stigma via social comparison (Clark & Oswald, 1994). Duration effects are curvilinear: distress rises sharply in the first 9 months, stabilizes, and may worsen for the very long-term unemployed.

Perceived social status, distinct from objective SES, mediates the relationship between material resources and SWB. Subjective perceptions of social standing reflect cultural norms and social comparison processes, which amplify or diminish the impact of objective metrics (Haring et al., 1984). Cultural orientation further moderates these links: individualistic societies prioritize self-satisfaction and freedom for SWB, while collectivist societies emphasize social acceptance.

Moderating factors such as adaptation and homeostatic processes further shape SES-SWB dynamics. Income gains lead to short-term SWB increases, but adaptation to new reference points reduces long-term effects (Clark et al., 2008). Homeostatic theory suggests income acts as an internal buffer (e.g., enhancing perceived control, self-esteem) and external buffer (e.g., purchasing resources to mitigate stressors), protecting SWB homeostasis for the wealthy but leaving the poor

---

health and governance. However, the core insight—that within-country comparisons and adaptation weaken the income-happiness link over time—remains highly influential.

vulnerable to environmental shocks (Cummins, 2000). These complexities highlight that SES influences SWB through interconnected material, psychological, and social pathways, necessitating nuanced consideration of both objective resources and subjective perceptions.

In summary, socioeconomic status enhances well-being not through simple material accumulation, but rather through a complex set of psychological comparison mechanisms, cultural scripts, and adaptation processes, which interact with an individual's inherent traits to collectively shape the ultimate landscape of well-being.

## 4.4 Situational and Interpersonal Factors Affecting SWB

### *4.4.1 Quality of Social Networks and Relationships*

The quality of social networks and relationships, encompassing social integration, social support, community engagement, and social capital, is a critical determinant of subjective well-being (SWB). Subjective evaluations of relationships and social integration correlate more strongly with SWB than external objective conditions, highlighting that the perceived quality of social networks and social support are key factors rather than mere quantity.

Social integration, defined by the degree of involvement in social networks and interactions, exhibits robust associations with SWB. Numerous studies have found correlations between satisfaction with friends, loneliness, and SWB, as well as positive links between objective measures of social activity (e.g., formal and informal social participation) and SWB outcomes. Longitudinal studies further demonstrate that changes in social contact are accompanied by concurrent shifts in SWB, with social participation predicting 2–4% of SWB variance even after controlling for health and socioeconomic status (Diener, 1984). Notably, experimental evidence shows that brief, casual social interactions with strangers—such as spontaneous conversations during commutes—yield more positive experiences than solitude, challenging common misconceptions about the discomfort of such interactions (Epley & Schroeder, 2014). A meta-analysis across 148 studies revealed that complex measures of social integration (e.g., multifaceted network assessments) are particularly impactful, with an odds ratio (OR) of 1.91 for increased survival likelihood, underscoring the depth of integration's influence on well-being-related outcomes (Holt-Lunstad et al., 2010). Prospective studies, such as the seminal Alameda County Study by Berkman and Syme's analysis of over 4700 adults, further confirm that a composite "social network index" predicts mortality with a relative risk ratio of approximately 2.0, indicating that low social integration doubles mortality risk independent of baseline health and behavioral factors (Berkman & Syme, 1979; Cohen, 2004). Additionally, the quantity of daily social interactions shows dose-dependent positive effects on momentary happiness and social connectedness, while conversational depth and relational closeness (e.g., knowing and liking interaction partners) enhance these associations, particularly in self-reported within-person assessments (Sun et al., 2020).

Social support, encompassing both structural (network integration) and functional (received or perceived support) dimensions, also plays a vital role. These components are moderately intercorrelated (r = 0.20–0.30), with functional support operating through mechanisms like the stress buffering model (mitigating stressors) and the main effects model (promoting healthy behaviors and self-esteem). Perceived social support, in particular, emerges as more strongly linked to SWB than objective measures of received support, emphasizing the subjective experience of support availability. Supportive, positive social relationships are deemed necessary for well-being, with experimental evidence indicating harm from ostracism and longitudinal data showing that providing social support predicts greater psychological well-being, sometimes more so than receiving support (Diener & Seligman, 2004).

Community engagement is fostered by environmental and structural factors that facilitate social interaction. Neighborhoods characterized by walkability and mixed land use promote social capital—the social networks and interactions inspiring trust and reciprocity—by enabling both intentional and accidental interactions (e.g., casual conversations at local shops). Multivariate models show that walkable neighborhoods predict higher social capital measures, including knowing neighbors, political participation, and trust, with odds ratios ranging from 1.8 to 2.0 (Leyden, 2003). Such engagement, manifested through activities like volunteering, club memberships, and church participation, correlates with higher community-level SWB.

Social capital itself, at individual and societal levels, strongly influences SWB. Defined by trust, reciprocity, and civic participation, social capital predicts national emotional well-being more robustly than monetary capital (Diener & Ryan, 2009). Cross-nationally, indicators like organizational memberships and societal trust correlate with higher SWB, with communities high in volunteer activity and social entertaining exhibiting greater well-being (Helliwell & Putnam, 2004; Putnam, 2000). However, the relationship between social factors and SWB is complex: effects of social participation depend on moderators such as education, setting constraints, and personality (e.g., extraverts derive more happiness from social settings), and bidirectional causality exists—whereas social contact enhances SWB, happier individuals also tend to be more sociable.

### 4.4.2 *Physical Environment and Built Environment Characteristics*

Beyond direct interpersonal interactions and community atmosphere, the physical and built environments we inhabit, serving as both the container and backdrop for human connection, also profoundly shape our subjective well-being. A substantial body of research confirms that geography, infrastructure, and the design of our everyday spaces operate across multiple scales—from the national level down to the household—to shape access to critical resources and influence well-being through a variety of pathways (Das et al., 2020).

Contact with nature, for instance, is a key component of the physical environment that confers significant mental and physical health benefits through mechanisms such as improved air quality, promotion of physical activity, enhanced social cohesion, and psychological restoration (Hartig et al., 2014; Keniger et al., 2013).[5] Urban form and design characteristics, including density, land use patterns, and the provision of public spaces, present complex relationships with SWB, influencing social opportunities, mobility, and emotional responses to one's neighborhood.

In summary, the physical environment interacts with individual characteristics through complex social, psychological, and behavioral mechanisms, collectively forming an indispensable contextual foundation for well-being. The empirical section of this book will delve deeply into the specific metrics, theoretical models, and evidence surrounding built environment determinants, with a particular focus on the challenges and opportunities presented by high-density urban living. Figure 4.2 presents a conceptual model linking built environment features to SWB through mediating pathways such as social interaction, physical activity, and psychological restoration.

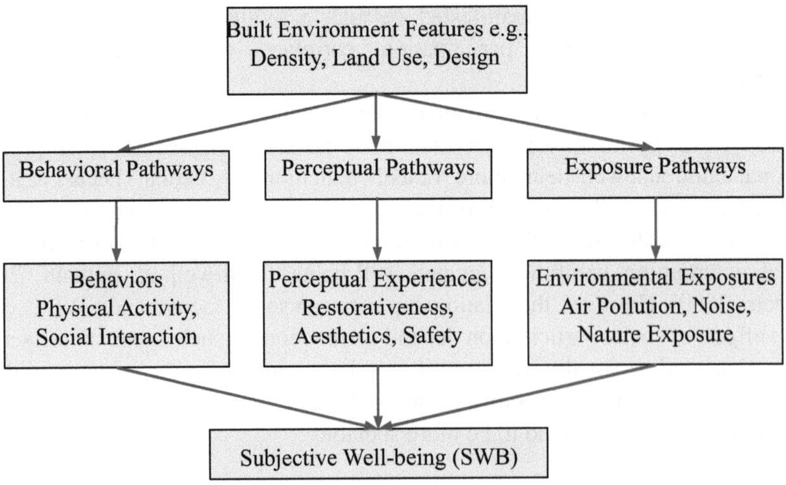


**Fig. 4.2** A conceptual model linking built environment to subjective well-being. Source: Created by the authors

[5] The psychological restoration mechanism is often explained by Attention Restoration Theory (ART), which posits that natural environments engage involuntary "soft fascination," allowing directed attention mechanisms to recover from fatigue. This is a key pathway through which nature exposure reduces stress and improves cognitive function, thereby contributing to SWB.

## 4.5 Macro-Level Influences on Subjective Well-Being

Beyond intangible cultural and institutional forces, the tangible, policy-shaped built environment represents another critical pathway through which macro-level factors influence well-being.

### *4.5.1 Cultural and Institutional Contexts*

Cultural and institutional contexts exert profound influences on subjective well-being (SWB) by shaping the experience, expression, and determinants of well-being, encompassing both universal components (e.g., life satisfaction, pleasant/unpleasant affect) and culture-specific patterns that must be analyzed within each cultural framework. A key dimension along which cultural variations manifest is individualism-collectivism (Hofstede, 1980, 2001; Triandis, 1995), which profoundly impacts the values placed on emotions and their relevance to SWB. For instance, pride—linked to personal achievement in individualist cultures—is often de-emphasized or associated with negative affect in collectivist cultures, where humility and sympathy may be prioritized (Kitayama et al., 1995; Tov & Diener, 2009).

In individualistic cultures (e.g., North American and European contexts), the self is construed as independent and autonomous, leading to happiness being defined as personal achievement and motivated by affirming positive internal attributes. This is reflected in stronger correlations between positive and negative emotions as bipolar opposites, with a focus on maximizing positive affect. In contrast, collectivist cultures (e.g., East Asian contexts) emphasize interdependence and social embeddedness, constructing happiness as the realization of social harmony and maintaining a balance between positive and negative emotions (e.g., positive correlations between positive and negative emotions in Japan). These cultural differences extend to predictors of SWB: self-esteem strongly predicts happiness in individualistic cultures like the US, whereas in collectivist cultures such as Japan and the Philippines, both self-esteem and perceived emotional support are significant, with emotional support in the US influencing happiness primarily through self-esteem (Uchida et al., 2004).

At the societal level, individualism correlates with higher average SWB in wealthy nations, partly due to cultural norms of self-enhancement (e.g., European Americans overestimating past performance) and approach tendencies focused on positive consequences, whereas collectivist societies often exhibit avoidance tendencies and greater self-criticism. However, this association is accompanied by trade-offs, including higher rates of suicide and divorce in individualistic nations, potentially due to reduced social restraints (Diener et al., 2003). Nation-level analyses further indicate that individualism predicts SWB even after adjusting for income, with strong correlations to human rights and gender equality, though its independent effect remains debated. As summarized in Table 4.2, these cultural

**Table 4.2** Cultural variations in the experience and determinants of Subjective Well-being: a comparison of individualistic and collectivistic contexts

| Aspect | Individualistic cultures | Collectivist cultures |
|---|---|---|
| Self-Construal | Independent | Interdependent |
| Key Predictors | Self-esteem, Personal Achievement | Social Harmony, Emotional Support |
| Emotional Focus | High-Arousal Positive Affect | Low-Arousal Positive Affect |
| SWB Expression | Overt Happiness, Life Satisfaction | Balance, Social Acceptance |

Source: Created by the authors

frameworks—individualism and collectivism—systematically shape how well-being is experienced, expressed, and pursued. The table provides a comparative overview of key differences, highlighting how self-construal, emotional priorities, and the very predictors of happiness vary across cultural contexts. This organized comparison underscores that cultural values are not merely background variables but active forces structuring the ecosystem of well-being.

Cultural differences also extend to emotional valuation, such as European-Americans prioritizing high-arousal positive affect (excitement) versus Asian-Americans and Chinese valuing low-arousal positive affect (calmness) (Diener & Ryan, 2009). The buffering effect of institutional and social policies is critical for mitigating well-being shocks like unemployment. Unemployment significantly reduces SWB beyond income loss alone, with unemployed individuals reporting lower happiness even after controlling for income differences (Diener, 1984). However, welfare policies can ameliorate these impacts: time series analyses of 12 European nations found unemployment benefits reduced the negative effect of unemployment on SWB, and similar research across 18 industrial democracies linked welfare state generosity (e.g., pensions, disability support) to higher SWB (Di Tella et al., 2003). Additionally, multi-level analyses across 41 nations associated welfare expenditure with greater life satisfaction (Radcliff, 2001). Social security policies that ensure stability in income, work, and health are identified as essential for happiness, providing a foundation of security that supports SWB (Layard, 2011). Notably, the buffering effect may also operate through social comparison: unemployed individuals in high-unemployment areas report relatively higher SWB than those in low-unemployment environments, potentially due to reduced stigma. Conversely, some studies suggest welfare benefits do not buffer unemployment's impact, attributing the negative effect to psychological factors like loss of self-respect rather than material deprivation, particularly in wealthier countries.

Religion and religious participation are significant cultural forces shaping SWB through normative frameworks, emotional norms, and social support. Religious participation, such as church attendance and engagement in religious groups, generally correlates positively with SWB, though findings are mixed. Mechanisms include the provision of meaning, purpose, and social capital—factors that enhance coping with stressors like unemployment and low income (Das et al., 2020). For example, religious belief buffers individuals against stressors, and across nations, belief in a god correlates with higher life satisfaction and lower suicide rates (Diener & Seligman, 2004). Doctrinal differences further shape emotional experiences: Christians report higher happiness and less shame than Muslims, linked to religious emphasis on emotions like joy (Christianity) versus shame and guilt (Islam). At the individual level, intrinsic religious motivation (e.g., personal faith) is more strongly associated with SWB than extrinsic motivation, with benefits observed across gender, age, and racial groups (Smith et al., 2003). Nationally, Christian majority populations correlate positively with SWB, while Buddhist-majority nations have shown negative associations in some analyses.

### *4.5.2 Built Environment Policy and Urban Planning*

Urban planning strategies and policies represent a powerful macro-level influence on subjective well-being (SWB) by structurally shaping the environments in which people live, work, and interact. Concepts such as the "15-minute city" (Moreno et al., 2021) emphasize the importance of proximity and accessibility to daily needs, aiming to foster livability, reduce environmental burdens, and enhance community connections.[6] Empirical research provides support for the SWB benefits of such planning paradigms, particularly through enhanced social capital and satisfaction with neighborhood characteristics.

The design of neighborhoods—encompassing factors like walkability, mixed land use, and access to public transport and green spaces—can facilitate or hinder the social interactions and daily experiences that contribute to well-being. While specific findings on urban density, spatial configuration, and sustainability assessment tools will be explored in detail later in the empirical chapter dedicated to planning implications, it is clear that strategic urban planning prioritizes dimensions that overlap significantly with known contributors to quality of life. Integrating objective indicators with subjective stakeholder preferences can inform context-specific interventions, enhancing SWB through improved built environments (Khalil, 2012).

In summary, while culture and institutions set the basic 'climate' for a population's well-being, urban planning directly shapes the 'habitat' of people's daily lives. The detailed analysis of how specific planning policies and built environment features translate into well-being outcomes will be a central focus of the forthcoming empirical analysis.

[6]The "15-minute city" model, prominently advocated by Carlos Moreno (2021), proposes that urban neighborhoods should be designed or retrofitted so that residents can meet most of their daily needs (work, shopping, education, healthcare, leisure) within a 15-minute walk or bike ride from their homes. This model is seen as a pathway to sustainability, equity, and well-being.

## 4.6 Integrative Perspectives and Future Directions in SWB Research

### *4.6.1 Toward Operationalizable Theoretical and Empirical Frameworks*

To advance the scientific understanding of subjective well-being (SWB), integrating its multi-dimensional determinants into coherent, testable frameworks remains a central challenge. SWB is influenced by a multiplicity of interrelated factors, including basic demographics, socioeconomic status (SES), health and functioning, personality, social support, religion and culture, and geography and infrastructure, requiring theoretical models that accommodate this high dimensionality (Das et al., 2020). A critical first step involves addressing the complexity of operationalizing SWB itself, which encompasses distinct yet interrelated components: life satisfaction judgments (evaluative), positive affect, and negative affect (experiential), with positive and negative affect showing near independence (Diener, 1984). This distinction between experiential (moment-to-moment emotional states) and evaluative (reflective life satisfaction) measures highlights the need for frameworks that account for these complexities (Diener et al., 2018).

This section begins by proposing an integrative, multi-level framework (Fig. 4.3) that visually synthesizes the diverse determinants of SWB discussed throughout this chapter. This framework serves as a foundational model for understanding the complex interplay of factors and sets the stage for discussing the challenges and innovations in operationalizing such models.

While frameworks like this are essential for conceptual clarity, their development and testing face significant challenges, not least of which are the measurement limitations and causal inference problems outlined below.

**1. Challenges in Integration**

Key challenges hinder the development of such frameworks. Measurement limitations persist: single-item scales cannot capture the full breadth of SWB, relying instead on respondents' subjective integration, while multi-item scales, despite demonstrating substantial convergent validity, require ongoing psychometric refinement to ensure they adequately operationalize multidimensional constructs. Additionally, SWB assessments are susceptible to bias from current mood states, though large-sample studies indicate these effects are generally small. Time frame variance further complicates integration, as determinants may operate differently across varying temporal contexts, necessitating models that explicitly account for state versus trait components. Methodologically, the overreliance on correlational and cross-sectional designs limits causal inference, as few studies employ longitudinal, experimental, or quasi-experimental approaches to disentangle the direction of influence between variables (Cohen, 2004). Finally, interactions among determinants—such as the interplay of gender, age, and cultural context—introduce layers of complexity that traditional linear models struggle to capture (Diener et al., 2020).

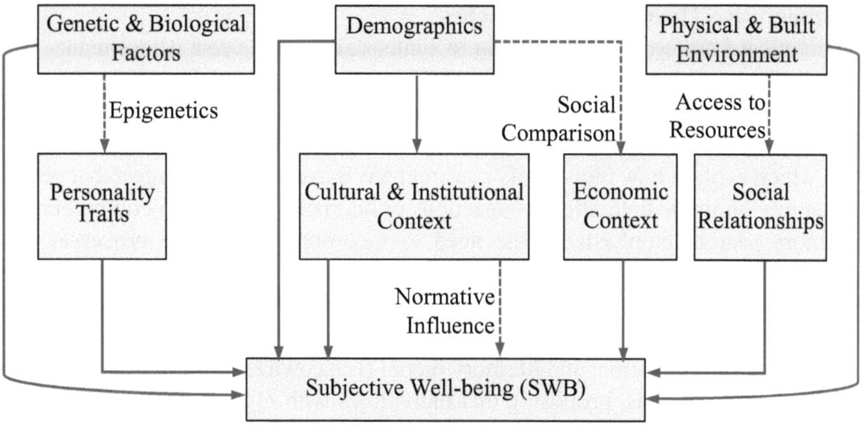


**Fig. 4.3** Integrative framework of SWB determinants. Source: Created by the authors

**2. Methodological Innovations**

To address these challenges, methodological innovations have emerged. In measurement, multi-item scales (e.g., the Satisfaction with Life Scale, Positive and Negative Affect Scale) have been developed to capture specific components of SWB, while combining self-report measures with non-self-report methods (e.g., observer reports, physiological measures, facial expression analysis) enhances objectivity and reduces bias. Longitudinal and experimental designs have become increasingly prevalent, allowing researchers to clarify temporal dynamics and causal relationships between predictors and SWB outcomes. Advanced analytical techniques now enable integration of multi-dimensional data: meta-analytic approaches, for instance, have distinguished subjective SES (relative rank) from objective SES (income, education), revealing stronger associations between subjective SES and SWB ($r = .22$ vs. $r = .16$ for objective SES) and identifying moderators like income inequality and cultural orientation (Tan et al., 2020). Machine learning methods, such as fully convolutional networks (FCN) for quantifying street space quality and natural language processing (NLP) for sentiment analysis of social media data, have facilitated large-scale, fine-grained measurement of environmental and affective determinants.[7] Nonlinear relationships are increasingly modeled using tools like restrictive cubic spline (RCS) regression, which identifies critical thresholds in associations between built environment variables (e.g., green space agglomeration, land mixed use) and SWB outcomes (Ma et al., 2024).[8]

[7] For example, FCN models can analyze street-level imagery (e.g., from Google Street View) to automatically assess urban design qualities like "greenness," "enclosure," or "imageability." NLP applied to social media posts (e.g., Twitter) can generate large-scale, real-time maps of collective mood or well-being, offering a complement to traditional survey methods.

[8] RCS regression is a flexible statistical technique that does not assume a linear relationship between the predictor and outcome variables. It fits a smooth curve defined by a series of polynomial segments, allowing researchers to visually and statistically identify points where the relation-

**3. Integrative Theoretical Approaches**
Theoretical frameworks are evolving to synthesize these diverse determinants. The top-down versus bottom-up heuristic provides a useful organizing structure, positing that SWB is influenced by both dispositional factors (e.g., personality traits) and evaluative reactions to external life circumstances (Diener, 1984). Judgment models, which explore how individuals construct SWB assessments, suggest that people often rely on immediate affective reactions or heuristics rather than comprehensive memory search, emphasizing the need to incorporate cognitive processes into empirical models. Parducci's range-frequency theory offers testable predictions about how distributions of life events shape SWB (Parducci, 1995), demonstrating the potential for integrating contextual and cognitive determinants. The AIM (Attention, Interpretation, and Memory model (Isaacowitz, 2006) further highlights cognitive mechanisms, proposing that individuals with higher SWB exhibit biases toward attending to positive stimuli, interpreting events positively, and recalling past experiences with a positive memory bias (Diener & Ryan, 2009).

Recent integrative frameworks explicitly incorporate mediating pathways to link multi-level determinants. For example, Mouratidis's model connects neighborhood characteristics (objective physical, sociodemographic, and perceived) to SWB dimensions (hedonic well-being, life satisfaction, eudaimonia) through mediators like personal relationships, leisure activities, and health, using structural equation modeling to test these complex relationships (Mouratidis, 2018). Such approaches move beyond correlations to unpack indirect mechanisms, facilitating more precise theoretical predictions.

**4. Future Directions**
Progress toward operationalizable frameworks requires stronger connections between theory and empirical data, with constructs rigorously defined and falsifiable propositions developed. Interdisciplinary collaboration is essential, as variables from weather to social networks to gene-environment interactions collectively influence SWB. Standardizing measurement across studies—consistently assessing experiential and evaluative components—and combining quantitative and qualitative methods will enhance comparability and depth of understanding. Ultimately, integrating these advances will enable the development of models that not only explain SWB's complexity but also inform interventions to promote well-being across diverse contexts.

ship changes direction or intensity, thus revealing potential "tipping points" or optimal ranges for environmental features.

### 4.6.2 Implications for Policy, Intervention, and Well-Being Enhancement

A comprehensive understanding of subjective well-being (SWB) provides critical insights for shaping practical approaches across public policy, mental health interventions, and urban development, moving beyond narrow economic metrics to holistic quality of life enhancement. In public policy, national accounts of subjective well-being are increasingly adopted to inform decisions at community and societal levels, recognizing that economic factors such as GDP per capita (cross-sectional $r = 0.83$, $p < 0.001$) and sociopolitical factors including lower inequality ($r = 0.47$, $p < 0.001$), political freedoms ($r = 0.55$, $p < 0.001$), and reduced corruption (accounting for 13% of variance in national life satisfaction over time) are key determinants of societal well-being (Helliwell et al. 2020; Oishi et al., 2018). Layard extends this by advocating the Greatest Happiness Principle as a foundation for policy, emphasizing security across income, work, family, community, and health domains, with mental health treatment and social safety nets as essential intervention pillars. Diener and Seligman argue that policy should transcend economic indicators, as well-being in wealthy societies is more strongly linked to social relationships, work enjoyment, social capital, and democratic governance; they propose a national well-being index measuring positive/negative emotions, engagement, purpose, and life satisfaction to guide evidence-based policy (Diener & Seligman, 2004). Globally, cross-national analyses confirm that while economic growth boosts SWB in lower-income nations, sustainable development and social policies—such as unemployment benefits and progressive taxation—are critical to mitigate adverse economic impacts and enhance satisfaction with public services (Helliwell et al., 2020).

For mental health interventions, SWB research highlights mental health as a cornerstone of well-being. Layard underscores that untreated mental illness undermines happiness, advocating expanded mental health treatment and social safety nets to alleviate insecurity. He further suggests integrating emotional intelligence education—covering emotion management, family, work, and values—into curricula, alongside promoting goals beyond self-interest and service to others, to inform programmatic interventions (Layard, 2011). Diener and Seligman reinforce this, noting mental disorders cause widespread suffering and call for policies to expand treatment access and support caregivers, as mental health is integral to quality of life.

In urban development, the built and natural environments profoundly influence SWB, guiding planning for social and environmental sustainability. As will be discussed in greater depth in the context of our empirical findings, compact urban forms, when addressing urban problems like crime fear, noise, and litter, can correlate positively with life satisfaction by enhancing personal relationships and health while reducing anxiety. Natural environments contribute via multiple pathways: improving air quality, enhancing physical activity and social cohesion, and reducing stress through restoration, supporting policies for accessible, well-maintained green spaces (Hartig et al., 2014).

### *4.6.3 Future Direction*

A critical future direction in subjective well-being (SWB) research lies in the integration of interdisciplinary approaches, as understanding the complex interplay of factors influencing SWB demands a synthesis of insights from diverse fields. Central to this integration is the recognition that genetic influences on complex traits like SWB cannot be fully comprehended in isolation but must be considered within the environmental context in which they operate. Epigenetic processes, such as DNA methylation and histone modification, provide a molecular mechanism through which environmental signals can stably program gene expression, mediating enduring effects of early experiences on behavioral and emotional phenotypes over the lifespan (Bird, 2007). For instance, parental care in early life has been shown to epigenetically program gene expression via changes in DNA methylation patterns in offspring brain regions, leading to long-lasting influences on stress reactivity and behavior—highlighting the need to bridge developmental psychology with molecular biology in SWB research (Meaney & Szyf, 2005).

Moreover, gene–environment interactions, exemplified by polymorphisms in the serotonin transporter gene promoter and their modulation by early environmental factors like maternal care, further illustrate the dynamic interplay between genomic and environmental factors that shape emotional and neural phenotypes relevant to SWB (Caspi et al., 2003). Understanding such bidirectional biological embedding between environmental signals and genome function requires moving beyond traditional disciplinary boundaries. Given that epigenetic marks can be dynamically altered by environmental stimuli even in mature cells, an interdisciplinary framework combining genetics, molecular biology, developmental psychology, and social science perspectives is essential to fully elucidate the mechanisms underpinning SWB (Meaney, 2010). This integrated approach not only enhances our grasp of how early life experiences influence SWB across the lifespan but also paves the way for more nuanced and holistic models of SWB that capture its multifaceted nature.

## 4.7 Summary

The comprehensive examination of subjective well-being (SWB) research demonstrates a paradigmatic shift from reductionist, static classifications of isolated determinants toward integrative, interactional models that encapsulate complex, multi-level influences. Empirical and theoretical advancements have converged on frameworks that emphasize the dynamic interplay among genetic predispositions, psychological traits, social relationships, situational contexts, and broader cultural-institutional environments. This multidimensional perspective underscores SWB as a reciprocal outcome of intertwined endogenous dispositions, interpersonal factors, and situational characteristics, embedded within temporal and ecological systems.

At the endogenous level, genetic and biological substrates substantially contribute to individual differences in SWB, with heritability estimates indicating moderate genetic influence modulated by epigenetic mechanisms that integrate environmental exposures, particularly during sensitive developmental periods. Psychological traits such as extraversion, neuroticism, optimism, and emotion regulation capacities are pivotal in explaining affective patterns and stable well-being dispositions, operating both independently and interactively with environmental cues.

Interpersonal and demographic variables—including socioeconomic status, education, income, age, gender, and social roles—exert nuanced effects contingent upon subjective perceptions, comparative social contexts, and cultural norms. These factors manifest through both direct and mediated pathways, reflecting the influence of social integration, social support quality, and normative structures. Importantly, subjective social status frequently exhibits stronger associations with SWB than objective indicators, thereby highlighting the salience of social comparison and perception processes in shaping well-being.

The situational and physical environments are critical domains shaping SWB through access to natural spaces, built environment characteristics, neighborhood design, and urban planning policy. As this chapter has outlined, built environment features contribute to SWB. The specific mechanisms and empirical evidence detailing how these built environment attributes function as determinants will be extensively reviewed and tested in the subsequent empirical section of this monograph.

Cultural and institutional contexts provide overarching frameworks that modulate the experience and evaluation of well-being. The individualism-collectivism continuum elucidates variations in emotional valuation, self-construal, and social embeddedness affecting SWB patterns across societies. Welfare provisions and religious participation further buffer adverse life events, underscoring the relevance of normative and institutional structures in well-being dynamics.

Methodologically, the integration of multi-component SWB measures alongside longitudinal, experimental, and advanced analytic techniques enables more precise modeling of causal pathways and interactive effects. Interdisciplinary approaches incorporating genetic, molecular, psychological, social, environmental, and cultural dimensions are imperative to unravel the complexity inherent in SWB phenomena.

Future research trajectories necessitate operationalizable frameworks that articulate testable hypotheses accounting for temporal dynamics, multi-level interactions, and cross-cultural applicability. Bridging molecular biology insights—such as epigenetic programming—with social and developmental sciences offers promising avenues for elucidating mechanisms of resilience and vulnerability in SWB. Furthermore, translation of empirical findings into targeted socio-political and environmental interventions requires rigorous evaluation to optimize strategies fostering sustainable well-being enhancements at individual and societal scales.

**In sum, this chapter has provided a comprehensive overview of the multidimensional determinants of SWB, establishing a layered conceptual architecture that situates built environment factors within the broader situational context. This sets the stage for the following part of the book, which will narrow its focus to a rigorous examination of the built environment-well-being nexus, thereby extending the general framework presented here into a specific, empirically-grounded investigation.****FundingThis research was funded by**

**1**. Guangdong Provincial Office of Philosophy and Social Sciences [Grant No. GD**24**XSH**06**]

**2**. Projects of PhDs' Start-up Research of GDUPT [Grant No. **2023**bsqd**1008**]

## References

Adler, N. E., Epel, E. S., Castellazzo, G., & Ickovics, J. R. (2000). Relationship of subjective and objective social status with psychological and physiological functioning: Preliminary data in healthy White women. *Health Psychology, 19*(6), 586–592.

Bartels, M. (2015). Genetics of well-being and its components satisfaction with life, happiness, and quality of life: A review and meta-analysis of heritability studies. *Behavior Genetics, 45*(2), 137–156.

Bartels, M., & Boomsma, D. I. (2009). Born to be happy? The etiology of subjective well-being. *Behavior genetics, 39*(6), 605–615.

Batz-Barbarich, C., Tay, L., Kuykendall, L., & Cheung, H. K. (2018). A meta-analysis of gender differences in subjective well-being: Estimating effect sizes and associations with gender inequality. *Psychological Science, 29*(9), 1491–1503.

Berkman, L. F., & Syme, S. L. (1979). Social networks, host resistance, and mortality: A nine-year follow-up study of Alameda County residents. *American Journal of Epidemiology, 109*(2), 186–204.

Bird, A. (2007). Perceptions of epigenetics. *Nature, 447*(7143), 396–398.

Blanchflower, D. G. (2021). Is happiness U-shaped everywhere? Age and subjective well-being in 145 countries. *Journal of Population Economics, 34*(2), 575–624.

Boone-Heinonen, J., Guilkey, D. K., Evenson, K. R., & Gordon-Larsen, P. (2010). Residential self-selection bias in the estimation of built environment effects on physical activity between adolescence and young adulthood. *International Journal of Behavioral Nutrition and Physical Activity, 7*, 70.

Bronfenbrenner, U., & Morris, P. A. (2007). The bioecological model of human development. In W. Damon & R. M. Lerner (Eds.), *Handbook of child psychology* (Vol. 1, 6th ed., pp. 793–828). Wiley.

Buecker, S., Luhmann, M., Haehner, P., Bühler, J. L., Dapp, L. C., Luciano, E. C., & Orth, U. (2023). The development of subjective well-being across the life span: A meta-analytic review of longitudinal studies. *Psychological Bulletin, 149*(7–8), 418–447.

Carstensen, L. L. (1992). Social and emotional patterns in adulthood: Support for socioemotional selectivity theory. *Psychology and Aging, 7*(3), 331–338.

Carstensen, L. L. (2006). The influence of a sense of time on human development. *Science, 312*(5782), 1913–1915.

Caspi, A., Sugden, K., Moffitt, T. E., Taylor, A., Craig, I. W., Harrington, H., McClay, J., Mill, J., Martin, J., Braithwaite, A., & Poulton, R. (2003). Influence of life stress on depression: Moderation by a polymorphism in the 5-HTT gene. *Science, 301*(5631), 386–389.

Cheung, F., & Lucas, R. E. (2016). Income inequality is associated with stronger social comparison effects: The effect of relative income on life satisfaction. *Journal of Personality and Social Psychology, 110*(2), 332–341.

Clark, A. E., Frijters, P., & Shields, M. A. (2008). Relative income, happiness, and utility: An explanation for the Easterlin paradox and other puzzles. *Journal of Economic Literature, 46*(1), 95–144.

Clark, A. E., & Oswald, A. J. (1994). Unhappiness and unemployment. *The Economic Journal, 104*(424), 648–659.

Cohen, S. (2004). Social relationships and health. *American Psychologist, 59*(8), 676–684.

Cummins, R. A. (2000). Personal income and subjective well-being: A review. *Journal of Happiness Studies, 1*(2), 133–158.

Das, K. V., Jones-Harrell, C., Fan, Y., Ramaswami, A., Orlove, B., & Botchwey, N. (2020). Understanding subjective well-being: Perspectives from psychology and public health. *Public Health Reviews, 41*(1), 25.

Dawson, C., Veliziotis, M., & Hopkins, B. (2017). Temporary employment, job satisfaction and subjective well-being. *Economic and Industrial Democracy, 38*(1), 69–98.

Diener, E. (1984). Subjective well-being. *Psychological Bulletin, 95*(3), 542–575.

Diener, E., Lucas, R. E., & Oishi, S. (2018). Advances and open questions in the science of subjective well-being. *Collabra: Psychology, 4*(1), 15.

Diener, E., Oishi, S., & Lucas, R. E. (2003). Personality, culture, and subjective well-being: Emotional and cognitive evaluations of life. *Annual Review of Psychology, 54*, 403–425.

Diener, E., & Ryan, K. (2009). Subjective well-being: A general overview. *South African Journal of Psychology, 39*(4), 391–406.

Diener, E., & Seligman, M. E. (2004). Beyond money: Toward an economy of well-being. *Psychological Science in the Public Interest, 5*(1), 1–31.

Diener, E., Thapa, S., & Tay, L. (2020). Positive emotions at work. *Annual Review of Organizational Psychology and Organizational Behavior, 7*(1), 451–477.

Di Tella, R., MacCulloch, R. J., & Oswald, A. J. (2003). The macroeconomics of happiness. *Review of Economics and Statistics, 85*(4), 809–827.

Dursun, P. (2021). Optimism, hope and subjective well-being: A literature overview. *Çatalhöyük Uluslararası Turizm ve Sosyal Araştırmalar Dergisi, 6*, 61–74.

Easterlin, R. A. (1974). Does economic growth improve the human lot? Some empirical evidence. In P. A. David & M. W. Reder (Eds.), *Nations and households in economic growth* (pp. 89–125). Academic Press.

Ellison, C. G. (1991). Religious involvement and subjective well-being. Journal of health and social behavior, 80–99.

Epley, N., & Schroeder, J. (2014). Mistakenly seeking solitude. *Journal of Experimental Psychology: General, 143*(5), 1980–1999.

Friedman, H. S., Kern, M. L., & Reynolds, C. A. (2010). Personality and health, subjective well-being, and longevity. *Journal of Personality, 78*(1), 179–216.

Fujita, F., Diener, E., & Sandvik, E. (1991). Gender differences in negative affect and well-being: The case for emotional intensity. *Journal of Personality and Social Psychology, 61*(3), 427–434.

Granovetter, M. S. (1973). The strength of weak ties. *American Journal of Sociology, 78*(6), 1360–1380.

Haring, M. J., Stock, W. A., & Okun, M. A. (1984). A research synthesis of gender and social class as correlates of subjective well-being. *Human Relations, 37*(8), 645–657.

Hartig, T., Mitchell, R., De Vries, S., & Frumkin, H. (2014). Nature and health. *Annual Review of Public Health, 35*, 207–228.

Headey, B., & Wearing, A. (1989). Personality, life events, and subjective well-being: Toward a dynamic equilibrium model. *Journal of Personality and Social Psychology, 57*(4), 731–739.

Helliwell, J. F., Layard, R., Sachs, J. D., & De Neve, J. E. (Eds.). (2020). *World happiness report 2020*. Sustainable Development Solutions Network.

Helliwell, J. F., & Putnam, R. D. (2004). The social context of well-being. *Philosophical Transactions of the Royal Society B: Biological Sciences, 359*(1449), 1435–1446.

Hofstede, G. (1980). *Culture's consequences: International differences in work-related values*. Sage.

Hofstede, G. (2001). *Culture's consequences: Comparing values, behaviors, institutions, and organizations across nations* (2nd ed.). Sage.

Holt-Lunstad, J., Smith, T. B., & Layton, J. B. (2010). Social relationships and mortality risk: A meta-analytic review. *PLoS Medicine, 7*(7), e1000316.

Horley, J., & Lavery, J. J. (1995). Subjective well-being and age. *Social Indicators Research, 34*(2), 275–282.

Howell, R. T., & Howell, C. J. (2008). The relation of economic status to subjective well-being in developing countries: A meta-analysis. *Psychological Bulletin, 134*(4), 536–560.

Inglehart, R. (2002). Gender, aging, and subjective well-being. *International Journal of Comparative Sociology, 43*(3–5), 391–408.

Isaacowitz, D. M. (2006). Motivated gaze: The view from the gazer. *Current Directions in Psychological Science, 15*(2), 68–72.

Kahneman, D., & Deaton, A. (2010). High income improves evaluation of life but not emotional well-being. *Proceedings of the National Academy of Sciences, 107*(38), 16489–16493.

Keniger, L. E., Gaston, K. J., Irvine, K. N., & Fuller, R. A. (2013). What are the benefits of interacting with nature? *International Journal of Environmental Research and Public Health, 10*(3), 913–935.

Khalil, H. A. E. E. (2012). Enhancing quality of life through strategic urban planning. *Sustainable Cities and Society, 5*, 77–86.

Kitayama, S., Markus, H. R., & Matsumoto, H. (1995). Culture, self, and emotion: A cultural perspective on "self-conscious" emotions. In J. P. Tangney & K. W. Fischer (Eds.), *Self-conscious emotions: The psychology of shame, guilt, embarrassment, and pride* (pp. 439–464). Guilford Press.

Knabe, A., & Rätzel, S. (2011). Scarring or scaring? The psychological impact of past unemployment and future unemployment risk. *Economica, 78*(310), 283–293.

Layard, R. (2011). *Happiness: Lessons from a new science* (2nd ed.). Penguin.

Leyden, K. M. (2003). Social capital and the built environment: The importance of walkable neighborhoods. *American Journal of Public Health, 93*(9), 1546–1551.

Long, C. R., & Averill, J. R. (2003). Solitude: An exploration of benefits of being alone. *Journal for the Theory of Social Behaviour, 33*(1), 21–44.

López Ulloa, B. F., Møller, V., & Sousa-Poza, A. (2013). How does subjective well-being evolve with age? A literature review. *Journal of Population Ageing, 6*(3), 227–246.

Lucas, R. E., & Diener, E. (2009). Personality and subjective well-being. In E. Diener (Ed.), *The science of well-being: The collected works of Ed Diener* (pp. 75–102). Springer.

Lucas, R. E., & Gohm, C. L. (2000). Age and sex differences in subjective well-being across cultures. In E. Diener & E. M. Suh (Eds.), *Culture and subjective well-being* (pp. 291–317). MIT Press.

Luhmann, M., Hofmann, W., Eid, M., & Lucas, R. E. (2012). Subjective well-being and adaptation to life events: A meta-analysis. *Journal of Personality and Social Psychology, 102*(3), 592–615.

Ma, S., Wang, B., Liu, W., Zhou, H., Wang, Y., & Li, S. (2024). Assessment of street space quality and subjective well-being mismatch and its impact, using multi-source big data. *Cities, 147*, 104797.

Magnusson, D., & Stattin, H. (2006). The person in context: A holistic-interactionistic approach. In W. Damon & R. M. Lerner (Eds.), *Handbook of child psychology: Vol. 1. Theoretical models of human development* (6th ed., pp. 400–464). Wiley.

Meaney, M. J. (2010). Epigenetics and the biological definition of gene × environment interactions. *Child Development, 81*(1), 41–79.

Meaney, M. J., & Szyf, M. (2005). Environmental programming of stress responses through DNA methylation: Life at the interface between a dynamic environment and a fixed genome. *Dialogues in Clinical Neuroscience, 7*(2), 103–123.

Moreno, C., Allam, Z., Chabaud, D., Gall, C., & Pratlong, F. (2021). Introducing the "15-Minute City": Sustainability, resilience and place identity in future post-pandemic cities. *Smart Cities, 4*(1), 93–111.

Mouratidis, K. (2018). Rethinking how built environments influence subjective well-being: A new conceptual framework. *Journal of Urbanism, 11*(1), 24–40.
Nes, R. B., & Røysamb, E. (2017). The heritability of subjective well-being: Review and meta-analysis. In M. Pluess (Ed.), *Genetics of psychological well-being* (pp. 75–96). Oxford University Press.
Ngamaba, K. H., Panagioti, M., & Armitage, C. J. (2018). Income inequality and subjective well-being: A systematic review and meta-analysis. *Quality of Life Research, 27*(3), 577–596.
Oishi, S., Diener, E., & Lucas, R. E. (2018). *Subjective well-being: The science of happiness and life satisfaction*. Oxford University Press.
Okun, M. A., Stock, W. A., Haring, M. J., & Witter, R. A. (1984). Health and subjective well-being: A meta-analysis. *International Journal of Aging & Human Development, 19*(2), 111–132.
Parducci, A. (1995). *Happiness, pleasure, and judgment: The contextual theory and its applications*. Lawrence Erlbaum.
Paul, K. I., & Moser, K. (2009). Unemployment impairs mental health: Meta-analyses. *Journal of Vocational Behavior, 74*(3), 264–282.
Picchio, M., & Ubaldi, M. (2024). Unemployment and health: A meta-analysis. *Journal of Economic Surveys, 38*(4), 1437–1472.
Putnam, R. D. (2000). *Bowling alone: The collapse and revival of American community*. Simon & Schuster.
Radcliff, B. (2001). Politics, markets, and life satisfaction: The political economy of human happiness. *American Political Science Review, 95*(4), 939–952.
Roman, C. G., & Chalfin, A. (2008). Fear of walking outdoors: A multi-level ecologic analysis of crime and disorder. *American Journal of Preventive Medicine, 34*(4), 306–312.
Ross, C. E., & Willigen, M. V. (1997). Education and the subjective quality of life. *Journal of Health and Social Behavior, 38*(3), 275–297.
Sandstrom, G. M., & Dunn, E. W. (2014). Social interactions and well-being: The surprising power of weak ties. *Personality and Social Psychology Bulletin, 40*(7), 910–922.
Scheier, M. F., & Carver, C. S. (1985). Optimism, coping, and health: Assessment and implications of generalized outcome expectancies. *Health Psychology, 4*(3), 219–247.
Smith, T. B., McCullough, M. E., & Poll, J. (2003). Religiousness and depression: Evidence for a main effect and the moderating influence of stressful life events. *Psychological Bulletin, 129*(4), 614–636.
Snyder, C. R. (2002). Hope theory: Rainbows in the mind. *Psychological Inquiry, 13*(4), 249–275.
Steel, P., Schmidt, J., & Shultz, J. (2008). Refining the relationship between personality and subjective well-being. *Psychological Bulletin, 134*(1), 138–161.
Striessnig, E. (2015). Too educated to be happy? An investigation into the relationship between education and subjective well-being. *IIASA Interim Report*. International Institute for Applied Systems Analysis.
Suh, E., Diener, E., & Fujita, F. (1996). Events and subjective well-being: Only recent events matter. *Journal of Personality and Social Psychology, 70*(5), 1091–1102.
Sun, J., Harris, K., & Vazire, S. (2020). Is well-being associated with the quantity and quality of social interactions? *Journal of Personality and Social Psychology, 119*(6), 1478–1496.
Szyf, M. (2013). DNA methylation, behavior and early life adversity. *Journal of Genetics and Genomics, 40*(7), 331–338.
Tan, J. J. X., Kraus, M. W., Carpenter, N. C., & Adler, N. E. (2020). The association between objective and subjective socioeconomic status and subjective well-being: A meta-analytic review. *Psychological Bulletin, 146*(11), 970–1020.
Tatarkiewicz, W. (1976). *Analysis of happiness*. Martinus Nijhoff.
Tesch-Römer, C., Motel-Klingebiel, A., & Tomasik, M. J. (2008). Gender differences in subjective well-being: Comparing societies with respect to gender equality. *Social Indicators Research, 85*(2), 329–349.
Tov, W., & Diener, E. (2009). Culture and subjective well-being. In E. Diener (Ed.), *Culture and well-being* (pp. 9–41). Springer.
Triandis, H. C. (1995). *Individualism & collectivism*. Westview Press.

Tugade, M. M., Fredrickson, B. L., & Barrett, L. F. (2004). Psychological resilience and positive emotional granularity: Examining the benefits of positive emotions on coping and health. *Journal of Personality, 72*(6), 1161–1190.

Uchida, Y., Norasakkunkit, V., & Kitayama, S. (2004). Cultural constructions of happiness: Theory and empirical evidence. *Journal of Happiness Studies, 5*(3), 223–239.

Wadsworth, T. (2016). Marriage and subjective well-being: How and why context matters. *Social Indicators Research, 126*(3), 1025–1048.

Weaver, I. C., Cervoni, N., Champagne, F. A., D'Alessio, A. C., Sharma, S., Seckl, J. R., Dymov, S., Szyf, M., & Meaney, M. J. (2004). Epigenetic programming by maternal behavior. *Nature Neuroscience, 7*(8), 847–854.

Wilson, W. R. (1967). Correlates of avowed happiness. *Psychological Bulletin, 67*(4), 294–306.

Witter, R. A., Okun, M. A., Stock, W. A., & Haring, M. J. (1984). Education and subjective well-being: A meta-analysis. *Educational Evaluation and Policy Analysis, 6*(2), 165–173.

Zuckerman, M., Li, C., & Diener, E. (2017). Societal conditions and the gender difference in well-being: Testing a three-stage model. *Personality and Social Psychology Bulletin, 43*(3), 329–336.

# Part II
# Empirical Investigations: The Built Environment and Well-Being in High-Density Cities

# Chapter 5
# Multidimensional Environmental Influences on Subjective Well-Being

**Abstract** This chapter demonstrates that the study of environmental influences on subjective well-being (SWB) has progressed toward multidimensional and integrative frameworks, incorporating natural, sociocultural, and built environmental dimensions. It aims to systematically examine how these environmental layers interact to shape SWB through behavioral, perceptual, and exposure mechanisms, with particular emphasis on the built environment as a modifiable and proximal determinant. Section 5.1 outlines the chapter's objectives and theoretical positioning. Section 5.2 reviews foundational theories—including Attention Restoration Theory, Stress Reduction Theory, and sociocultural frameworks—and presents conceptual models linking environmental perception to SWB. Section 5.3 analyzes macro-level sociocultural, economic, and institutional factors that moderate environmental effects. Section 5.4 details proximal built environment elements (e.g., density, walkability, land use mix) and their mediating pathways. Section 5.5 synthesizes empirical evidence and methodological challenges, while Sect. 5.6 integrates trade-offs and synergistic effects across environmental dimensions. The chapter concludes by positioning the built environment as a critical intervention point for enhancing SWB in high-density urban settings, providing a theoretical foundation for subsequent empirical investigations.

**Keywords** Subjective well-being · Built environment · Compact city paradox · Restorative environments · Mediating pathways · Sociocultural context · Environmental perception · Urban density · Theoretical frameworks · Policy intervention

Having established in Chap. 4 a comprehensive framework of the factors influencing subjective well-being (SWB)—spanning genetic, psychological, social, and macroeconomic dimensions—this chapter narrows its focus to the role of environmental contexts, particularly the built environment, in shaping SWB. While Chap. 4 emphasized the multi-layered and interactive nature of SWB determinants, this review synthesizes multidisciplinary evidence encompassing natural, sociocultural, and built environmental dimensions to elucidate their complex effects on subjective



H. Li, G. Shen, *The Anatomy of Subjective Well-Being*, Resilient and Friendly Living Environments, https://doi.org/10.1007/978-981-95-4937-5_5

well-being (SWB). This chapter moves from a broad understanding of well-being influencers to a specific examination of key physical elements—such as urban design, walkability (Gehl, 2010; Jacobs, 1962), density, land use mix (Cervero & Kockelman, 1997), infrastructure, and environmental noise—and their pathways impacting SWB through behavioral (e.g., physical activity, social interaction), perceptual (e.g., restorative experiences, aesthetic evaluations), and exposure mechanisms (e.g., pollution mitigation, noise attenuation), thereby articulating a coherent conceptual model that links environmental features to SWB. The review further integrates psychological evolutionary theories (Appleton, 1975; Ulrich, 1983), explaining natural environment benefits alongside socio-structural factors such as income inequality, institutional trust, and cultural values that moderate environmental effects.

It highlights critical trade-offs, particularly the "Compact City Paradox" (Mouratidis, 2021; Nielsen & Hansen, 2007), illustrating tensions between densification benefits and adverse social or environmental consequences, while also underscoring synergistic and compensatory interactions between environmental dimensions. Methodological challenges, including self-selection bias (Cao, 2016), exposure measurement heterogeneity (Kwan, 2012), and causal inference limitations, are discussed, motivating the advancement of longitudinal, experimental, and multi-scale research designs. Concluding with an integrative multi-level conceptual framework, the review positions the built environment as a pivotal intervention point for urban planning aimed at enhancing SWB, especially within high-density urban contexts, and provides a foundation for subsequent empirical investigations that detail mediating mechanisms and boundary conditions.

## 5.1 Introduction: From a Multi-Level Framework to Deepening Environmental Dimensions

### *5.1.1 Bridging the Preceding and Following Contents*

As noted, the discussion of the environment in the previous chapter was summary in nature. In contrast, this chapter (Sect. 5.1) aims to provide a panoramic and institutional review by synthesizing earlier frameworks and recent empirical data, which not only extends and refines previous literature but also serves as theoretical and methodological guidance for further empirical research. Specifically, this review focuses on the evidence regarding the range of pathways linking the built environment to subjective well-being (SWB), identifying and reviewing seven potential pathways: travel, leisure, work, social relationships, residential well-being, emotional responses, and health. Additionally, it presents a conceptual model that organizes these pathways, with the review structured into sections including the conceptual model outlining the seven pathways, empirical evidence for these pathways, urban planning strategies based on the review, and a summary of findings and conclusions. Through such a comprehensive institutional review that integrates multiple environmental dimensions and life domains, this chapter links built environment features to subjective well-being, thereby anchoring the ensuing empirical research (Mouratidis, 2021).

### *5.1.2 Chapter Objectives and Roadmap*

The chapter objectives are to systematically review how three types of environments influence subjective well-being (SWB) through distinct mechanisms, examine their interactions, and ultimately argue for the prioritization of the built environment in empirical research. To achieve this, the first focus is on reviewing mechanisms: one type of environment, encompassing diverse settings such as neighborhoods and transit systems, affects SWB through mechanisms like walkability, transit access, and green spaces, while also involving interactions between transportation modes and social engagement, and is highlighted as a modifiable determinant warranting research attention (Dannenberg et al., 2011). Another type, residential greenspace, operates through three biopsychosocial pathways—reducing harm (e.g., mitigating pollutants), restoring capacities (e.g., stress recovery), and building capacities (e.g., promoting physical activity)—with complex interactions among these pathways contributing to its overall impact on SWB (Markevych et al., 2017). The third type, built environment at neighborhood and city scales, influences SWB through mechanisms related to transit-oriented development (TOD), including transport modes, walkability, and urban vibrancy, and is advocated for focused research due to its role in fostering sustainable travel behavior and enhancing livability (Loo & du Verle, 2016).

## 5.2 Theoretical Foundations and Conceptual Models of Environmental Influences on SWB

### *5.2.1 Theories in Urban Sociology*

The focus on well-being in urban sociology can be traced back to the period of rapid urbanization brought about by Western industrialization in the nineteenth century. The swift development of cities led to various urban issues, prompting scholars to reflect on urban development (Durkheim, 2019). Relevant studies in urban sociology suggest that fast-paced living and high-density urban environments not only create psychological burdens but also increase interactions among individuals. However, rather than fostering interpersonal relationships, this often leads to a sense of loneliness (Simmel, 2016). Urban life theory further posits that urbanization causes negative issues such as the alienation of interpersonal relationships, behavioral deviance, and structural anomie, thereby reducing residents' well-being (Wirth, 1938). Increases in population density and size can lead to anonymity and superficial interactions among residents, accelerating the dilution of interpersonal relationships. It may also cause sensory overload, leading individuals to withdraw from noisy social lives. The heterogeneity of residents and organizational life can result in negative behaviors such as interpersonal alienation, deviance, and social anomie (Simmel, 2016). This provides theoretical insights for research on the built environment and well-being.

### *5.2.2 Social Capital Theory*

Research on social capital originated in the 1970s. Since the 1990s, social capital theory has gradually become a frontier and focal issue in academia, studied from the perspectives of sociology, political science, and other disciplines. Over the past two decades, it has remained an important concept in public health and epidemiology. After reviewing definitions of social capital, it can be categorized into four main types: as a social network, a general connection or social relationship, resources embedded in social structures, and trust, networks, norms, and institutions. It is used to explain various prosocial behaviors, including information and knowledge exchange. Although social capital has many dimensions, the most common classification includes structural, relational, and cognitive capital. Structural capital refers to the connections between individuals, relational capital comprises assets rooted in sustained interpersonal relationships (with an inherent relationship between structural and relational capital), and cognitive capital denotes resources that provide shared interpretations and collective vision within a group. Existing literature confirms that built environment elements influence residents' social capital (Mazumdar et al., 2018), and social capital is a significant factor affecting subjective well-being (Cramm et al., 2013).

### *5.2.3 Conceptual Models of Environmental Perception and Satisfaction*

Several conceptual models have been developed to explain how objective environments influence SWB through perceptual and evaluative pathways:

**1. Campbell's Model**
With the improvement of human settlement quality, research on residential satisfaction has gained increasing attention. Residential satisfaction refers to residents' overall subjective evaluation of their community across various aspects. It serves as a multi-level indicator system shaped by accumulated daily experiences, and is an

important measure of community construction and management quality, closely linked to a sense of belonging and social participation.

Campbell et al. (1976) proposed a theoretical model to explain the formation of residential satisfaction. According to this model, satisfaction arises from an individual's contact with and cognition of the objective environment, leading to domain-specific evaluations and ultimately influencing overall life satisfaction. The model incorporates attributes at different scales—community, neighborhood, and housing—which shape perception and evaluation, thereby affecting the corresponding levels of satisfaction. As shown in Fig. 5.1, perception and evaluation act as mediating variables between objective attributes and satisfaction.

Moreover, due to differences in personal attributes, residents may perceive the same environmental features differently, resulting in varied evaluations (Amerigo & Aragones, 1990). Thus, the perceived environment plays a mediating role between the objective environment and satisfaction. Consequently, focusing only on the objective or subjective environment alone may lead to biased or incomplete conclusions.

Although Campbell's model distinguishes environmental scales, it does not clearly define the objective versus perceived environments. This lack of clarity, combined with the broad and multidimensional nature of both environments, has contributed to the absence of a consistent research paradigm in later studies.

Subsequent theoretical work has extended Campbell's model. Galster and Hesser suggested that residents form neighborhood satisfaction by constructing reference standards based on their needs. Amérigo et al. emphasized that subjective evaluations are influenced by both the objective environment and personal attributes or residential standards. Canter (1977) further highlighted that the objective environment affects satisfaction not only through perception, but also through emotional responses and urban activities—such as cultural, leisure, and social interactions—which reflect how people engage with their living environment.

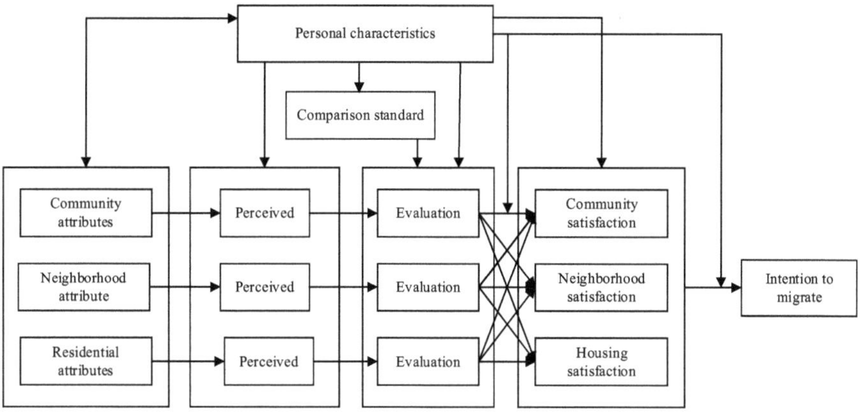


**Fig. 5.1** Campbell's model. Source: Created by the authors from the references (Campbell et al., 1976)

Overall, the core of Campbell's model framework lies in emphasizing that the objective environment influences subjective perception, which in turn affects satisfaction, highlighting the mediating role of subjective perception (Yuan et al., 2018).

**2. Sirgy's Model of Neighborhood Characteristics and Life Satisfaction**
Sirgy et al. proposed a model framework exploring how neighborhood and community satisfaction influence life satisfaction, examining the relationship between community characteristics and life satisfaction. In this framework, neighborhood economic characteristics affect housing satisfaction and family satisfaction; neighborhood physical characteristics affect housing satisfaction and neighborhood satisfaction; and neighborhood social characteristics affect neighborhood satisfaction and community satisfaction. Additionally, neighborhood satisfaction influences life satisfaction by affecting housing satisfaction and then family satisfaction, as well as affecting community satisfaction. This framework connects neighborhood, housing, family, and community with life satisfaction, forming a model of how neighborhood characteristics influence life satisfaction (Fig. 5.2) (Sirgy & Cornwell, 2002).

**3. Sugiyama's Framework on Neighborhood Environment and Quality of Life**
In recent years, many scholars have attempted to build on previous research to explore theories linking the built environment and subjective well-being. Based on research in health, psychology, urban design, and related fields, Sugiyama et al. analyzed the relationship between the residential environment and the well-being of older adults. The study suggested that the impact of the residential environment on the quality of life of older adults primarily lies in its promotion of outdoor activities (Sugiyama & Thompson, 2007). The research was grounded in the self-efficacy aspect of social cognitive theory. Studies were conducted from perspectives such as physical activity (Strawbridge et al., 2002), exposure to natural environments (Kaplan, 2001), and social interaction (Mouratidis, 2018; Sugiyama & Thompson, 2007). Accordingly, Sugiyama proposed a conceptual framework linking neighborhood environment and quality of life. As shown in Fig. 5.3, the framework includes personal factors (e.g., socioeconomic demographics, physiological and cognitive conditions), environmental support (the support of outdoor environments for individual participation in outdoor activities), and outdoor activities (including physical activity, nature exposure, and social interaction).

**4. Refined Campbell's Model**
Relevant studies have shown that, after controlling for key personal characteristic variables, residential satisfaction is positively correlated with life satisfaction (Cao, 2016) and associated with higher levels of well-being (Amerigo & Aragones, 1997). Based on Campbell's model, Cao (2016) linked the objective built environment to life satisfaction through perception and residential satisfaction, exploring the relationship between the built environment and subjective well-being. In the refined model shown in Fig. 5.4, objective built environment characteristics of the community, such as density, diversity, and design, affect residents' perceptions of accessibility and aversion, which in turn influence residential satisfaction and subsequently life satisfaction.

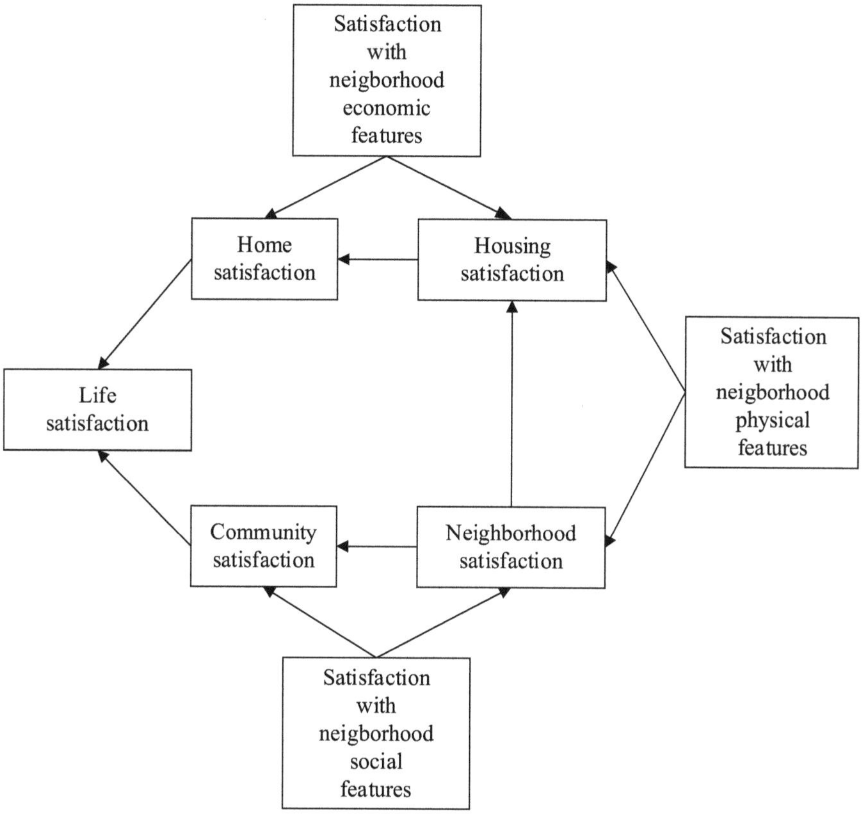


**Fig. 5.2** Sirgy's model of neighborhood characteristics and life satisfaction. Source: Created by the authors from the references (Sirgy & Cornwell, 2002)

**5. Built Environment Elements Directly Linked to Resident Well-Being Proposed by Pfeiffer**

Existing literature indicates that although theoretical research on the built environment and subjective well-being has progressed, a unified framework has not yet been established, and theoretical development remains insufficient. Subsequently, Pfeiffer et al. summarized three categories of built environment elements that may be directly linked to resident well-being through a literature review (as shown in Table 5.1): first, elements that improve health, such as natural or open green spaces; second, elements that facilitate social interaction; and third, elements that enhance safety. Additionally, potential links between elements such as housing conditions, housing types, and transportation facilities and well-being were mentioned. However, due to insufficient theoretical support, the relationships between these built environment elements and well-being require further research.

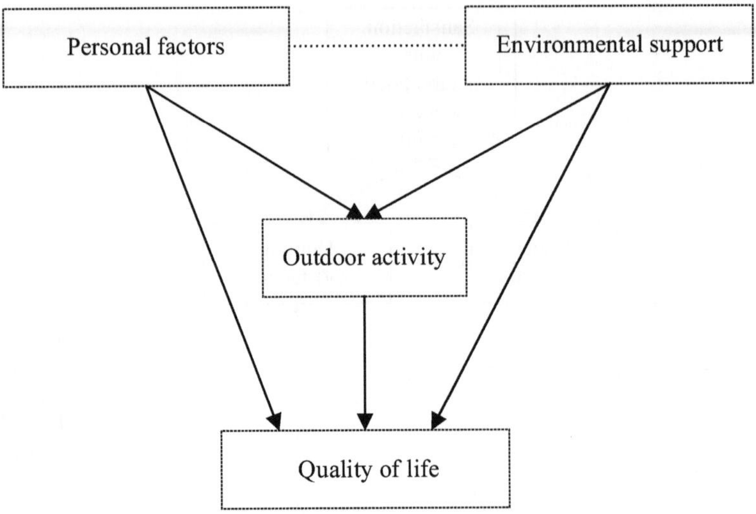


**Fig. 5.3** Sugiyama's framework on neighborhood environment and quality of life. Source: Created by the authors from the references (Sugiyama & Thompson, 2007)

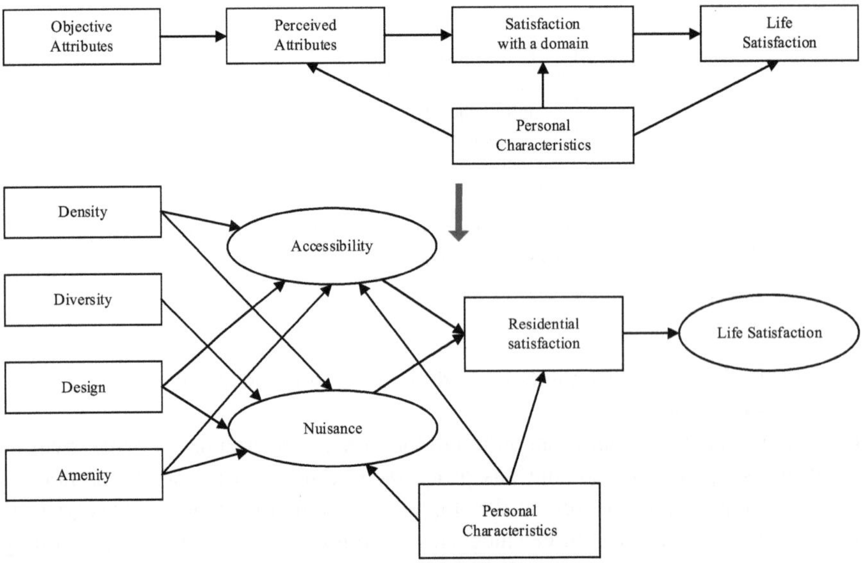


**Fig. 5.4** Cao's revised model. Source: Created by the authors from the references (Cao, 2016)

## 6. New Framework on Resident Well-Being Proposed by Mouratidis

In light of the continuous development of subjective well-being theories, Mouratidis systematically explored a research framework linking neighborhood built environment and subjective well-being. Figure 5.5 presents the new conceptual framework proposed by Mouratidis. This framework integrates different philosophical perspectives on subjective well-being—Hedonic Well-being, Life Satisfaction, and Eudaimonia—as three elements of SWB. Meanwhile, environmental characteristics

**Table 5.1** Built environment pathways to resident's well-being

| Pathway | Key components of the built environment | Proposed mechanism of impact |
|---|---|---|
| The Natural Environment | Access to open space, natural areas, parks, and green spaces | Improves physical and mental health; reduces stress; encourages exploration and curiosity |
| The Social Environment | High density, mixed land use, street connectivity, walkable design, public spaces, building design (e.g., porches) | Facilitates and encourages social interaction and community engagement |
| The Safe Environment | "Eyes on the street" (e.g., windows facing sidewalks), adequate lighting, maintained buildings, mixed use | Promotes perceived and actual safety; reduces fear of crime and physical hazards |

Source: Created by the authors (Pfeiffer & Cloutier, 2016)

affecting subjective well-being are categorized into the objective environment and the perceived environment. The objective environment affects subjective well-being both directly and indirectly through the perceived environment. Additionally, Mouratidis identified factors through which the neighborhood environment influences subjective well-being (Mouratidis, 2018):

First, interpersonal relationships, which may be the most important determinant of subjective well-being (Diener & Seligman, 2002; Robinson & Martin, 2008). Studies have shown that walkable and livable communities may enhance social interaction and thereby improve residents' well-being (Mouratidis, 2018).

Second, leisure activities, mainly referring to activities influenced by the environment, which can affect residents' happiness and personal growth, thereby influencing well-being.

Third, emotions, as the built environment can elicit either positive or negative emotions, affecting well-being.

Fourth, health, as extensive literature indicates that the built environment may directly impact health (e.g., air pollution, noise) or influence physical activity, interpersonal communication, dietary behavior, and other factors that affect resident health, thereby interfering with subjective well-being.

These four factors, consistent with most existing literature, determine subjective well-being. They may not be independent but rather interconnected and jointly influential, though more empirical research is needed to verify this.

## 5.3 The Restorative Pathway of Natural Environments: Theories, Evidence, and Debates

### *5.3.1 Core Psychological Evolutionary Theoretical Frameworks*

Core psychological evolutionary theoretical frameworks explaining how natural environments influence subjective well-being include the Attention Restoration Theory (ART), Stress Reduction Theory (SRT), and Biophilia Hypothesis.

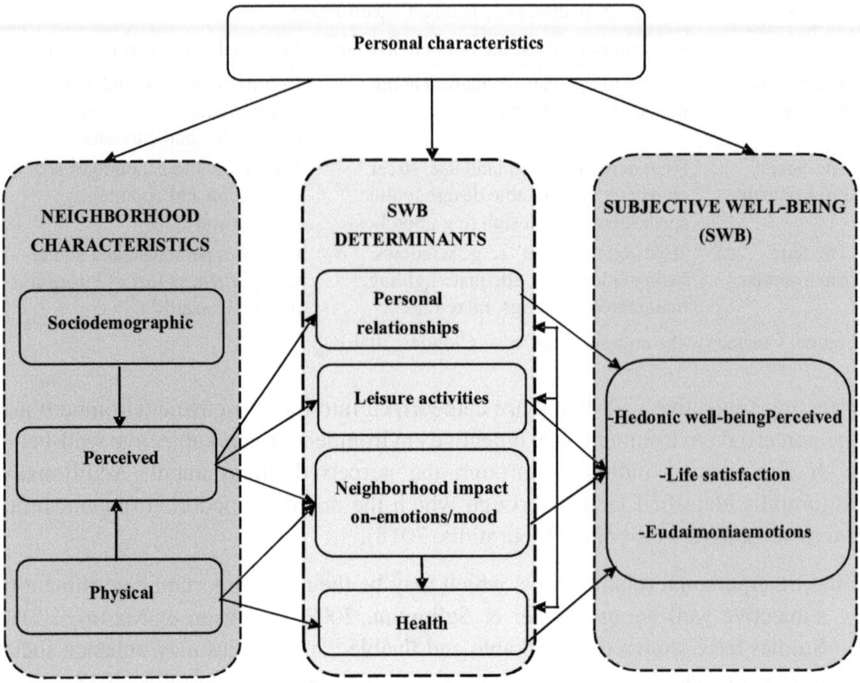


**Fig. 5.5** New framework on resident well-being proposed by Mouratidis. Source: Created by the authors from the references (Mouratidis, K. (2018). Built environment and social well-being: How does urban form affect social life and personal relationships?. *Cities*, *74*, 7–20)

ART posits that natural environments enhance subjective well-being by restoring directed-attention capacities (Kaplan, 1995). It distinguishes between involuntary attention (bottom-up, triggered by inherently intriguing stimuli in nature) and voluntary/directed attention (top-down controlled), arguing that natural environments modestly engage involuntary attention, allowing directed attention mechanisms critical to cognitive control to replenish, unlike urban environments that impose additional demands on directed attention. Experimental evidence supporting ART includes two studies demonstrating that walking in nature or viewing natural scenes significantly improved cognitive performance on tasks requiring directed attention, specifically the backwards digit-span task and the executive component of the Attention Network Task (ANT); these improvements were selective to executive attention and independent of changes in mood states or environmental peacefulness (Berman et al., 2008).

The Biophilia Hypothesis, defined as "the innate tendency to focus on life and lifelike processes", suggests that humans have an evolutionarily shaped propensity for connection with life and lifelike processes, which underlies their attraction to natural environments and subsequent support for subjective well-being. This tendency is thought to have been shaped by natural selection over millions of years, as human existence and psychological spirit depend on this affinity (Wilson, 1984).

Additionally, SRT is recognized as another key framework, positing that natural environments reduce stress to enhance well-being, though it is not explicitly tested in the provided literature.[1] Collectively, these theories highlight evolutionary and psychological mechanisms through which natural environments contribute to subjective well-being (Fig. 5.6).

### *5.3.2 Key Environmental Dimensions and Impact Mechanisms*

Critical natural environmental dimensions that impact subjective well-being include recreational time spent in natural environments, biodiversity, and accessible urban park settings, each operating through distinct yet interrelated impact mechanisms. Recreational time in natural environments has been identified as a pivotal dimension, with studies indicating that spending at least 120 min/week in natural settings is associated with significantly higher odds of reporting good health and high subjective well-being,[2] suggesting a potential threshold effect in nature exposure benefits. This dimension emphasizes the primacy of direct recreational nature contact, as it is a more precise predictor of well-being improvements compared to indirect exposure measures such as residential proximity to greenspace alone. The mechanism underlying this relationship is partially independent of physical activity, with evidence that passive exposure (e.g., sitting calmly in nature) can confer psychophysiological benefits, highlighting distinct restorative pathways triggered by natural environmental exposure (White et al., 2019).

Biodiversity represents another critical environmental dimension, characterized by multiple components including species richness, abundance, identity, functional traits, and genetic diversity, which can be assessed through objective measures (actual biodiversity) or subjective assessments (perceived biodiversity).[3] The impact of biodiversity on well-being operates through four key domains: reducing harm (e.g., mitigating air and noise pollution, providing medicinal resources), restoring capacities (e.g., stress recovery, attention restoration), building capacities (e.g., promoting physical activity, social cohesion), and causing harm (e.g., exposure to allergens, infectious diseases). Notably, the "restoring capacities" domain aligns with core restorative mechanisms, where complex and biodiverse environments facilitate

---

[1] Ulrich, R. S. (1983). Aesthetic and affective response to natural environment. In I. Altman & J. F. Wohlwill (Eds.), *Behavior and the natural environment* (pp. 85–125). Springer US.

[2] White, M. P., Alcock, I., Grellier, J., Wheeler, B. W., Hartig, T., Warber, S. L., Bone, A., Depledge, M. H., & Fleming, L. E. (2019). Spending at least 120 minutes a week in nature is associated with good health and well-being. *Scientific Reports*, *9*(1), 7730.

[3] Marselle, M. R., Hartig, T., Cox, D. T., De Bell, S., Knapp, S., Lindley, S., Triguero-Mas, M., Böhning-Gaese, K., Braubach, M., Cook, P. A., de Vries, S., Heintz-Buschart, A., Hofmann, M., Irvine, K. N., Kabisch, N., Kolek, F., Kraemer, R., Markevych, I., Martens, D., ... Bonn, A. (2021). Pathways linking biodiversity to human health: A conceptual framework. *Environment International*, *150*, 106420.

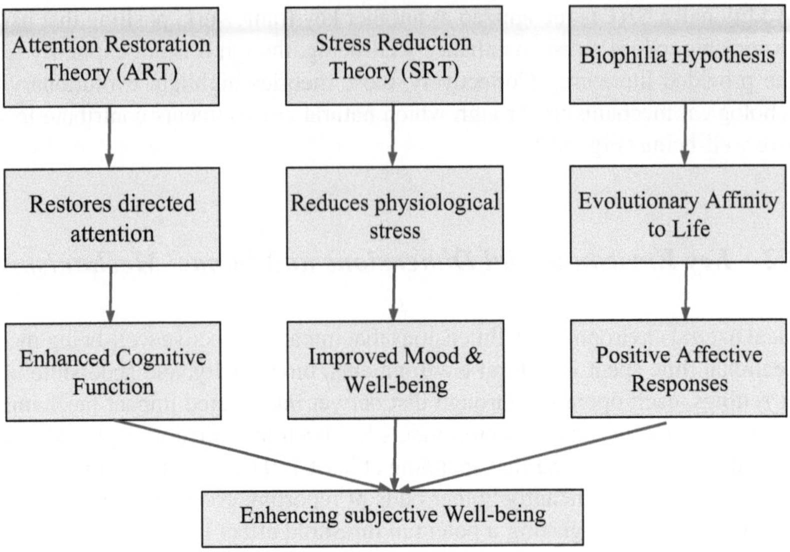


**Fig. 5.6** Theoretical frameworks of natural environment restoration. Source: Created by the authors

psycho-physiological and cognitive restoration, supported by empirical evidence linking greater plant and bird species richness to reduced stress and improved mood. Contact with biodiversity involves two critical mechanisms—exposure (frequency, duration, physical proximity) and experience (sensory modalities, intentionality of interaction)—both of which influence health outcomes through these mediating pathways (Marselle et al., 2021).

Urban parks, as accessible natural environments in urban contexts, are crucial for population-level health promotion by providing opportunities for nature contact through viewing natural scenes, physical presence, recreational activities, and nature-based therapy programs (Maller et al., 2009). These environments exhibit restorative properties by satisfying Kaplan and Kaplan's four elements of restorative environments: fascination (involuntary attention), a sense of being away (escape from usual settings), extent (feeling part of a larger whole), and compatibility with individual inclinations, which collectively contribute to mental fatigue recovery. Empirical evidence further supports that contact with natural environments in parks reduces physiological stress markers (e.g., heart rate, blood pressure, muscle tension) and enhances positive psychological responses such as increased positive mood and improved attention, underscoring their role in promoting subjective well-being (Maller et al., 2009).

### 5.3.3 Frontier Discussions and Critiques

Contemporary discussions around the restorative pathway of natural environments highlight several critical debates and limitations. One prominent issue is environmental justice, as urban green space distribution often disproportionately benefits predominantly White and more affluent populations, raising concerns about equitable access to the psychological and physical health benefits of green environments. Compounding this, the creation of new urban green spaces can paradoxically trigger gentrification, displacing the very communities they aim to assist through increased housing costs and property values—a phenomenon termed the "green space paradox"[4]—as observed in cases like New York's High Line and greening efforts in Chinese cities such as Hangzhou. Additionally, methods for measuring access to urban green spaces, such as GIS-based spatial analyses, have significant limitations: they fail to capture heterogeneity in park quality, available facilities, perceptions of safety, and congestion, all of which influence actual usage and subsequent health outcomes, necessitating more nuanced assessment approaches (Wolch et al., 2014).

Another key controversy lies in the relationship between biodiversity and psychological well-being, particularly regarding the discrepancy between perceived and actual species richness. Research indicates no consistent association between objective measures of plant, butterfly, or bird species richness and the psychological well-being of urban greenspace visitors; instead, well-being shows a positive correlation with the richness users perceive to be present.[5] Notably, well-being even declined with greater actual plant species richness, further complicating the link between ecological metrics and human outcomes. This mismatch stems partly from poor biodiversity identification skills among the public: only 0.002% of participants could identify all presented species, 27.3% could not identify any, and bird identification skills were superior to those for plants or butterflies. The disconnect between perception and reality, which is negatively correlated with wildlife knowledge, poses challenges for aligning conservation goals (focused on actual biodiversity) with human well-being agendas (reliant on perceived biodiversity), highlighting the need for strategies to enhance public ecological knowledge and engagement (Dallimer et al., 2012).

[4]Gould, K., & Lewis, T. (2016). *Green gentrification: Urban sustainability and the struggle for environmental justice*. Routledge.

[5]Dallimer, M., Irvine, K. N., Skinner, A. M., Davies, Z. G., Rouquette, J. R., Maltby, L. L., Warren, P. H., Armsworth, P. R., & Gaston, K. J. (2012). Biodiversity and the feel-good factor: understanding associations between self-reported human well-being and species richness. *BioScience, 62*(1), 47–55.

## 5.4 Role of Socio-Cultural Environment

### 5.4.1 *Beyond Social Networks: Macro Sociocultural Factors*

The understanding of factors shaping health behaviors and, by extension, subjective well-being has undergone a fundamental paradigm shift, moving from an exclusive focus on individual and interpersonal determinants to recognizing the critical role of macro sociocultural and structural contexts (Owen & Fisher, 2008). Within this macro framework, the erosion of social capital, as highlighted by Putnam's research on declining participation in civic, social, and religious organizations, represents a significant macro social trend affecting societal connectedness and well-being (Putnam, 1995/2000). Social trust, extending beyond close social networks to include trust in strangers and institutions, emerges as a powerful macro sociocultural factor, with effects on subjective well-being comparable to substantial income gains; combined trust in domains such as social relations, police, and legal system can produce even greater impacts (Helliwell et al., 2016). Social ecological theory further underscores that macro social background factors, including institutional structures, community contexts, and policy environments, dynamically interact to shape well-being, highlighting the interconnectedness of individual experiences with broader sociocultural systems (Stokols, 1996).

### 5.4.2 *Economic and Institutional Environment: The Bedrock of Security and Equity*

Income inequality, a core dimension of the economic environment, operates through mechanisms aligned with the relative deprivation theory, intensifying social comparison and status anxiety to undermine subjective well-being.[6] Using a large US dataset of over 1.7 million participants across 2425 counties, Cheung and Lucas (2015) found that higher income inequality (measured by the Gini coefficient) strengthened relative income effects: individuals with the same household income reported lower life satisfaction when living in richer counties with higher inequality, particularly for those with lower household incomes.[7] This aligns with cross-national evidence showing that income inequality correlates with a range of social problems—including mental illness, violence, and poor educational performance—through psychosocial pathways involving status competition and material inequality (Wilkinson & Pickett, 2009). In the US context, Oishi et al. (2011) further

[6] Stouffer, S. A., Suchman, E. A., DeVinney, L. C., Star, S. A., & Williams, R. M. (1949). *The American soldier: Adjustment during army life* (Vol. 1). Princeton University Press.

[7] Cheung, F., & Lucas, R. E. (2016). Income inequality is associated with stronger social comparison effects: The effect of relative income on life satisfaction. *Journal of Personality and Social Psychology*, *110*(2), 332.

demonstrated an inverse association between national income inequality and happiness, mediated by reduced perceived fairness and general trust, with disproportionately negative impacts on lower-income respondents.[8] Additionally, income inequality moderates the relationship between economic growth and subjective well-being: economic growth correlates with increased happiness only when inequality is low, whereas this positive effect disappears or reverses under high inequality, confirming the role of relative deprivation in shaping well-being outcomes (Oishi & Kesebir, 2015). Cross-national studies summarized by Tony Veal reinforce these findings, showing that more equal societies exhibit better outcomes across indicators like life expectancy, infant mortality, and crime rates, with benefits extending beyond disadvantaged groups to society at large (Veal, 2011; Wilkinson & Pickett, 2009).

Institutional quality and corruption significantly shape subjective well-being by influencing trust in institutions and perceptions of fairness. Helliwell and Huang (2006) found that life satisfaction is more strongly linked to World Bank measures of government quality (e.g., trustworthy and efficient service delivery) than to real per capita incomes, with workplace trust showing particularly large compensating differentials—equivalent to a 48% income increase for moving from the 75th to the top percentile of workplace trust.[9] Corruption, conversely, undermines well-being through multiple pathways: it erodes institutional trust, reduces public resources available for social welfare programs, and creates political barriers to universal welfare policies, leading to an "inequality trap" where low trust and persistent inequality reinforce each other (Rothstein & Uslaner, 2006). At both individual and national levels, corruption negatively impacts subjective well-being by reducing income, institutional trust, and perceptions of fairness, with stronger effects observed in Western countries due to established anti-corruption norms (Tay et al., 2014).

Access to public services, though not always the primary focus of empirical studies, indirectly influences subjective well-being by reducing life uncertainty and enhancing equity. Cheung and Lucas (2016) noted that richer counties often correlate with better public goods (e.g., green space, school systems), hinting at institutional mechanisms linking economic context to well-being through service provision. Universal social welfare programs—such as healthcare and education—are particularly impactful, as they promote economic equality and opportunity, reduce status anxiety by avoiding stigma associated with means-tested benefits, and foster social solidarity and optimism for the future. These programs exemplify how institutional environments can mitigate the negative psychological effects of inequality by ensuring basic security and equitable access to resources.

[8] Oishi, S., Kesebir, S., & Diener, E. (2011). Income inequality and happiness. *Psychological Science*, *22*(9), 1095–1100.

[9] Helliwell, J. F., & Huang, H. (2006). How's your government? International evidence linking good government and well-being. *British Journal of Political Science, 38*(4), 595–619.

### 5.4.3 Cultural Values: Defining Well-Being and Regulating Paths

Cultural values significantly shape the definition of subjective well-being (SWB) and regulate the pathways through which environmental factors influence it, with individualism/collectivism and cultural tightness-looseness being two critical dimensions.

In terms of individualism/collectivism, this dimension determines the relative priority of social relations versus personal achievement in SWB. Individualistic cultures, such as many European-American societies, prioritize personal achievement and autonomy as key predictors of happiness, while collectivistic cultures, like East Asian contexts, emphasize interpersonal connectedness, harmony, and social support (Uchida & Ogihara, 2012). Hofstede's cultural dimensions framework,[10] which includes individualism-collectivism, provides a theoretical foundation for understanding these differences, highlighting how cultural values shape social regulation mechanisms and influence the balance between individual goals and communal bonds in defining well-being (Triandis, 1993). For example, in collectivistic settings, an overemphasis on achievement orientation may undermine social relationships, thereby lowering well-being, illustrating the moderating role of cultural context in SWB pathways.

Cultural tightness-looseness,[11] distinct from individualism-collectivism, reflects the strength of social norms and tolerance for deviance, regulating the intensity of negative impacts from norm deviation on SWB. Tight cultures are characterized by strictly enforced rules, low tolerance for deviance, and often arise from higher ecological or historical threats (e.g., natural disasters, disease prevalence), leading to greater social stability but potentially lower happiness due to excessive behavioral constraint. In contrast, loose cultures have fewer rigid norms, tolerate greater deviance, and exhibit higher creativity, openness, and happiness, though they may experience more social disorder (Harrington & Gelfand, 2014). This dimension operates by enforcing conformity in tight cultures, where norm violation incurs stronger negative social consequences that reduce happiness intensity, whereas loose cultures mitigate such impacts through greater tolerance for deviance, thereby shaping the regulatory function of cultural values on SWB pathways.

### 5.4.4 Critical Summary: Foundational and Regulatory Roles

The sociocultural environment functions as a contextual moderator rather than a direct determinant of subjective well-being (SWB). For example, when operationalized as income inequality, it intensifies the negative association between regional

[10] Hofstede, G. (1980). *Culture's consequences: International differences in work-related values*. Sage.

[11] Gelfand, M. J., Nishii, L. H., & Raver, J. L. (2006). On the nature and importance of cultural tightness-looseness. *Journal of Applied Psychology*, 91(6), 1225–1244.

(county-level) income and individual life satisfaction, thereby magnifying social comparison effects on SWB. Controlling for individual household income, living in richer counties is associated with lower life satisfaction, replicating the relative income hypothesis that relative income influences SWB through social comparison mechanisms. Additionally, income inequality increases the salience and frequency of upward social comparisons, resulting in stronger negative effects of relative income among residents in regions with high income inequality, a regulatory role that is especially pronounced for low-income individuals, who report greater decreases in life satisfaction when exposed to higher neighborhood incomes under conditions of elevated income inequality. Furthermore, the interaction between income inequality and county income suggests that the sociocultural environment moderates the impact of the built environment's socioeconomic characteristics on SWB by shaping psychological processes (e.g., social comparison) through which local environmental factors influence individual outcomes, illustrating its foundational and regulatory function as a backdrop that influences how individual-level and built environment factors exert their effects on SWB (Cheung & Lucas, 2015).

## 5.5 Proximal Mechanisms of Built Environment

### *5.5.1 Elements of Built Environment*

Key built environment elements directly influencing subjective well-being encompass urban design qualities, walkability, density, land use mix, infrastructure, and environmental noise, among others. Urban design qualities play a pivotal role, with studies identifying specific attributes such as imageability (distinct, memorable places with visible people, historic buildings, and parks), enclosure (street definition by buildings, walls, or trees), human scale (elements matching human proportions like street furniture and first-floor windows), transparency (visibility of activity via ground-floor windows), and complexity (visual richness through diverse colors, public art, and human activity) as critical for enhancing walkability and pedestrian experience, thereby contributing to subjective well-being (Ewing & Handy, 2009). Additionally, urban design elements related to safety, such as building layout, site plans, and orientation facilitating natural surveillance, along with infrastructure like lighting, signage, and closed-circuit television (CCTV), are emphasized in defensible space and Crime Prevention Through Environmental Design (CPTED) frameworks, which influence perceptions of security and territoriality, reducing crime and fear of crime to improve quality of life (Cozens & Love, 2015).

Walkability, a multifaceted element, is characterized by components including net residential density, retail floor area ratio (indicating smaller setbacks and pedestrian access), land use mix (diversity of uses), and intersection density (street connectivity), as operationalized in walkability indices (Frank et al., 2009). Supporting features include pedestrian-friendly sidewalks, street trees, crossings, traffic calming measures, and proximity to services, which collectively enhance nonmotorized accessibility and encourage walking, linked to improved well-being (Ewing & Dumbaugh, 2009). Jane Jacobs' advocacy for mixed uses, small blocks, and intricate land use patterns further underscores the importance of walkability and transportation integration in fostering social fabric and functionality (Jacobs, 1962).

Density, a core element of compact urban form, is measured by population, jobs, or housing units per area and is associated with supporting accessible, mixed-use environments and reducing vehicle travel (Burton et al., 1996; Litman, 2023). In high-density contexts like Hong Kong and Singapore, tall buildings (including "super tall" structures of 50+ storeys) and density management through plot ratios are used to enhance liveability, though concerns about visual impact, ventilation, and heat islands necessitate urban design guidelines to mitigate negative effects (Yuen & Yeh, 2011).

Land use mix, integrating residential, commercial, institutional, and recreational uses, is a key attribute of sustainable urban form, with measures such as the ratio of nonresidential to residential land and entropy scores quantifying diversity (Burton et al., 1996; Frank et al., 2009). Mixed-use development supports accessibility, reduces trip lengths, and fosters social interaction, contributing to subjective well-being by enhancing convenience and community vibrancy (Litman, 2023; Yuen & Yeh, 2011).

Infrastructure, particularly transportation infrastructure, includes roads, parking facilities, sidewalks, and public transit, which shape development patterns, travel behavior, and accessibility (Litman, 2023). Services such as hospitals, parks, and schools, along with traffic-calming devices, access management, and intersection control measures, further influence liveability and safety (Ewing & Dumbaugh, 2009; Yuen & Yeh, 2011).

Environmental noise from transportation infrastructure, including road traffic, aircraft, and railway noise, emerges as a critical built environment element impacting subjective well-being through physiological stress and psychological annoyance, with adverse effects on mental health, cognition, and sleep disturbance (Clark et al., 2018; Clark & Paunovic, 2018). To summarize the key elements discussed, Table 5.2 provides a synthesized overview of the primary built environment dimensions, their defining characteristics, and their general impacts on subjective well-being.

### 5.5.2 *Mediating Pathways Influencing Subjective Well-Being (SWB)*

The mediating pathways linking the built environment features to subjective well-being (SWB) encompass interconnected behavioral, perceptual, and exposure mechanisms, as organized within frameworks distinguishing three general functional domains: reducing harm, restoring capacities, and building capacities (Markevych et al., 2017).

Behavioral pathways primarily involve patterns of physical activity and social interaction. Built environment attributes such as residential density, public transport density, and park availability collectively explain 45–59% of recommended weekly physical activity, indicating substantial behavioral mediation (Sallis et al., 2016). For neighborhood open spaces (NOSs), attractiveness and size are critical: an attractive nearby NOS encourages recreational walking, while a large attractive NOS supports sufficient physical activity for health benefits, directly linking environmental quality to behavioral engagement. Walkable, mixed-use neighborhoods further enhance social capital through frequent social interactions, political participation, and social trust, with each unit increase in walkability associated with higher odds of social engagement—a behavioral mechanism tied to SWB (Leyden, 2003). Compact urban forms also promote active travel (walking, cycling) by reducing travel times and improving access to activities, reinforcing physical activity as a key behavioral mediator.

Perceptual pathways center on psychological restoration and perception formation. Restorative capacities, grounded in Stress Reduction Theory and Attention Restoration Theory, explain how natural environment exposure reduces physiological stress and restores directed attention, with studies supporting partial mediation of health benefits by these mechanisms (Markevych et al., 2017). Perceived greenness and aesthetic quality of built environments, rather than objective measures alone, strongly predict physical activity and restorative experiences, highlighting subjective perception as a critical mediator (Sugiyama et al., 2008). Additionally, emotional responses to built environments—such as positive affect from well-maintained green spaces, aesthetically pleasing design, and perceptions of safety—mediate the relationship between environmental features and SWB (Mouratidis, 2021).

Exposure pathways involve mitigating adverse environmental factors and enhancing beneficial exposures. Greenspace reduces air pollution via particulate deposition and emission reduction, though evidence for partial mediation of greenness-health associations (e.g., mortality, cognitive development) is limited. Conversely, air pollution itself mediates urban residency and adolescent mental health: higher $NO_2$ and $NO_x$ exposure explains ~60% of the association between urban living and adolescent psychotic experiences, linking environmental exposure to SWB-relevant outcomes (Newbury et al., 2019). Noise reduction is another key exposure pathway, with greenspace providing physical acoustic attenuation (5–10 dB reduction) and psychological buffering (reduced annoyance), independently influencing well-being. Exposure to natural environments through large, attractive green spaces also supports restorative experiences, directly contributing to mental health via environmental exposure.

Notably, these pathways interact: physical activity in greenspace may foster social cohesion and further restoration, while perceived greenness enhances both behavioral engagement and restorative exposure, underscoring the complexity of mediating mechanisms linking built environments to SWB.

Synthesizing the theoretical foundations and mediating pathways discussed above, this study proposes a comprehensive conceptual framework (Fig. 5.7) to

**Table 5.2** Key factors of the built environment and their link to SWB

| Element | Description | Link to SWB |
|---|---|---|
| Urban Design Qualities | Imageability, enclosure, human scale, transparency, complexity | Enhances walkability, safety, aesthetic pleasure |
| Walkability | Density, land use mix, intersection density, supporting features | Promotes physical activity, social interaction |
| Density | Population/jobs per unit area | Supports accessibility, may reduce social cohesion |
| Land Use Mix | Integration of residential, commercial, institutional uses | Increases convenience, fosters social vibrancy |
| Infrastructure | Transportation networks, public services, traffic calming | Shapes accessibility, safety, and mobility |
| Environmental Noise | Noise from traffic, aircraft, railways | Increases stress, sleep disturbance, annoyance |

Source: Created by the authors

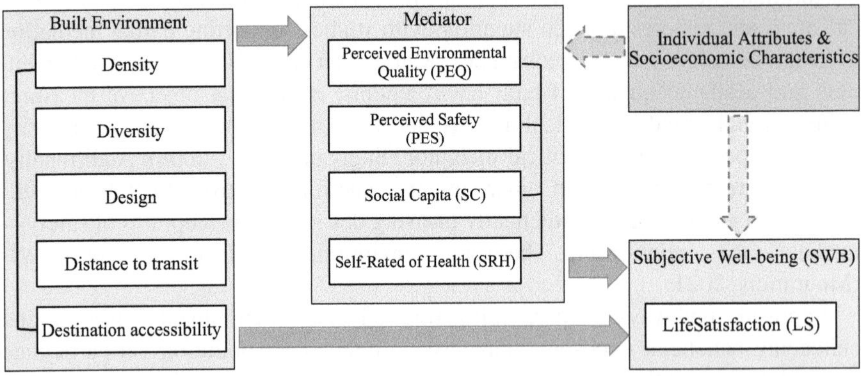


**Fig. 5.7** The proposed conceptual framework of built environment's impact on life satisfaction. Source: Created by the authors

investigate the impact of the built environment on residents' life satisfaction in high-density urban contexts.

This framework posits that the objective built environment, characterized by the five core dimensions (Density, Diversity, Design, Distance to transit, and Destination accessibility), influences residents' Subjective Well-being (SWB) and its cognitive component, Life Satisfaction (LS), primarily through four key mediating pathways: Perceived Environmental Quality (PEQ), Perceived Safety (PES), Social Capital (SC), and Self-Rated Health (SRH). Furthermore, individual socioeconomic attributes are incorporated as control variables to isolate the net effect of the built environment.

### 5.5.3 Empirical Evolution

Empirical research on the proximal mechanisms linking built environment to subjective well-being has evolved through diverse methodological approaches, including cross-sectional, longitudinal, and experimental designs, across varied geographical and cultural contexts. Cross-sectional studies have provided foundational insights into associations between specific built environment elements and mediating pathways. For instance, a study in Seoul, Korea, examined residential built environment features (e.g., land use mix, density, block size, building age, accessibility) and their relationship with walking activity—a proxy for urban vitality and potential well-being. Using multi-level regression models within 500 m buffers, it identified positive associations between walking activity and mixed land use, small block sizes, older building age, higher development densities, and proximity to natural features (rivers, streams), while border vacuums (e.g., large single-use residential complexes) showed negative associations, highlighting the value of micro-level environmental analysis (Sung & Lee, 2015). Internationally, a cross-sectional study across 14 cities in 10 countries found that objectively measured proximal elements—net residential density, intersection density, public transport density, and number of parks within 0.5–1.0 km buffers—were significantly and linearly associated with moderate-to-vigorous physical activity (MVPA), explaining 11–12% of MVPA variance. These associations were generalizable across diverse socioeconomic and cultural contexts, with differences in MVPA between the least and most supportive neighborhoods ranging from 68 to 89 min/week, underscoring population-level public health impacts (Sallis et al., 2016).

Longitudinal and experimental studies have advanced the understanding of potential causal relationships. A systematic review incorporating longitudinal designs found limited evidence for a causal link between surrounding greenness and adult mental health, with some studies showing improved mental health after moving to greener areas (Gascon et al., 2015). For children, however, evidence remained inadequate, with mixed findings on emotional and behavioural problems. Focusing on rigorous experimental and longitudinal designs (excluding cross-sectional studies), another review identified consistent associations between urban green space exposure and improved attention, mood, physical activity, and reduced mortality and violence, with greening interventions (e.g., vacant lot cleaning, tree planting) linked to crime reduction and enhanced safety perceptions (Kondo et al., 2018). A meta-analysis of 143 studies (observational and interventional) further validated greenspace benefits, reporting significant associations with reduced salivary cortisol (stress biomarker), improved cardiovascular function, and lower odds of type II diabetes, all-cause mortality, and preterm birth, with stronger effects among socioeconomically deprived groups (Twohig-Bennett & Jones, 2018). Conversely, research on environmental stressors like particulate matter (PM) has shown negative impacts: a meta-analysis found long-term PM2.5 exposure associated with higher depression prevalence (OR = 1.102 per 10 $\mu g/m^3$) and short-term PM10 exposure linked to increased suicide risk, with proposed biological mechanisms including neuroinflammation and hypothalamic–pituitary–adrenal axis dysregulation (Braithwaite et al., 2019).

Across these studies, heterogeneity in measurement approaches—such as buffer distances (100 m–3 km), exposure metrics (e.g., NDVI vs. land-cover maps), and outcome assessments—has complicated comparability. Context-specific findings, such as the positive association between older building age and walking activity in Seoul versus generalizable density-MVPA links, highlight the need for nuanced, standardized methodologies. Future research priorities include longitudinal designs to establish causality, improved exposure quantification (e.g., activity spaces over residential proxies), and integration of mediators like physical activity, stress reduction, and social interaction, alongside moderators such as socioeconomic status and green space quality (Braithwaite et al., 2019; Gascon et al., 2015; Kondo et al., 2018).

## 5.6 Summary of Framework Development

Synthesizing the reviewed literature reveals an integrated multi-level framework that organizes pathways linking the built environment to subjective well-being (SWB) across multiple domains, such as travel, leisure, work, social relationships, and health, thereby illustrating complex multi-level connections between environmental features and SWB outcomes. A central tension within this framework is the "Compact City Paradox," which encapsulates the inherent trade-offs of high-density urban forms: while compact designs enable shorter distances to destinations, promote walkability, and reduce travel times, they may also correlate with lower neighborhood social cohesion, increased noise, overcrowding, and impersonal interactions (Mouratidis, 2021). This paradox extends beyond urban form to other environmental dimensions, such as biodiversity, where benefits like access to restorative natural elements and ecosystem services coexist with potential harms including allergen exposure and zoonotic disease risks (Marselle et al., 2021); similarly, mixed-use, pedestrian-friendly neighborhood designs (density, diversity, design) can reduce personal vehicle miles traveled but may face complexities where certain street elements (e.g., quadrilateral blocks) inadvertently facilitate vehicle travel (Cervero & Kockelman, 1997). Urban greenspace further exemplifies this trade-off, as its benefits (e.g., pollution mitigation, stress recovery) vary with spatial scale, quality, and accessibility, leading to unequal distribution of advantages (Markevych et al., 2017), while vibrant urban soundscapes—characterized by human activity and moderate loudness variability—enhance engagement but may introduce noise-related disamenities due to acoustic roughness (Aletta & Kang, 2018).

Contextual dependence strongly modulates these relationships, with environmental and socio-cultural contexts shaping the impact of built environment features on subjective experiences. Cultural values, urban vs. rural settings, and individual sensitivities act as modifiers, for instance, influencing how biodiversity exposure translates to health outcomes. Regional disparities are evident in urban models: Western low-density, car-oriented environments often correlate with longer commutes and increased car dependency, whereas Asian high-density, transit-oriented models promote active travel but may induce stress from density-related factors. Empirical findings also reflect geographic biases, as most evidence on environmental-health links originates from high-income countries, with limited data from Asia, Africa, and developing regions. Cross-cultural studies further highlight context-specific moderators, such as the critical role of visual context and social cues (e.g., presence of people) in shaping vibrancy perception, which shows consistency between UK and Chinese urban sites despite potential differences in urban density models.

Core methodological challenges hinder causal inference in this field, with self-selection bias remaining the primary obstacle: individuals often choose neighborhoods aligned with their preferences (e.g., travel behavior or lifestyle), complicating the isolation of built environment effects from inherent predispositions (Mokhtarian & Cao, 2008). This is compounded by difficulties in controlling for confounding variables (e.g., socio-demographic factors, multicollinearity among built environment metrics) and measuring complex exposures (e.g., actual vs. perceived biodiversity, multisensory inputs in soundscapes); Addressing these challenges requires rigorous approaches, including longitudinal designs, intervention studies, and mediation analyses to untangle pathways, as well as improved measurement tools (e.g., GPS data for actual exposure, integrated audio-visual metrics) to capture the multidimensionality of environmental influences (Markevych et al., 2017; Mokhtarian & Cao, 2008).

## 5.7 Integration Perspective: Environmental Dimensions of Trade-Offs, Trade-Offs and Policy Implications

Interacting effects between environmental dimensions manifest through synergistic and compensatory/buffer mechanisms, shaping subjective well-being (SWB) through complex interplays. Synergistic effects occur when multiple environmental dimensions amplify each other's benefits. For instance, communities supporting transit, walking, and bicycling (built environment) are associated with improved air quality (natural environment), reduced greenhouse gas emissions, and increased physical activity, collectively lowering chronic disease risk (Dannenberg et al., 2011). Natural elements like water and vegetation act as "preferenda" that quickly elicit positive affective reactions, and features such as water enhance perceived depth, synergistically boosting positive environmental responses. Urban green spaces (built environment) promote physical activity and psychological well-being, with accessibility and design amplifying these health benefits (Wolch et al., 2014). Land use, transport systems, and urban design synergistically facilitate travel, while green communal spaces strengthen neighborhood social ties (Mouratidis, 2021).

Social contact in natural settings (e.g., parks) aids community cohesion by reducing prejudices, synergistically driving personal and neighborhood transformation (Maller et al., 2009). Trees and landscaping in built environments provide dual benefits: contributing positive factors like phytoncides and mitigating negatives such as air pollution, urban heat, and violence (Kuo, 2015). Cultural tightness (sociocultural environment) correlates with urbanization (built environment) and economic growth, interacting synergistically to support health and tolerance (Chua et al., 2019). Well-maintained built environments (e.g., walkable areas, park accessibility) enhance motivation to use green spaces, amplifying natural health benefits (Lachowycz & Jones, 2013).

Compensatory or buffer effects mitigate negative environmental impacts via moderating factors. Prior affective states and cognitive history moderate responses: stressed individuals experience greater restorative effects from natural settings due to positive affective modulation. High social capital buffers built environment deficiencies, such as a lack of public space; community activism, for example, is critical to designing "just green enough" spaces that prevent green gentrification and displacement. In the bioecological model, responsive mother-child interaction (proximal process) reduces behavioral disturbance and diminishes social disadvantage impacts, buffering contextual deficits. Individual factors like socioeconomic status moderate nature-health associations, with social capital potentially buffering deficient built environment effects (Markevych et al., 2017); similarly, tighter provinces with strong social regulation show higher life satisfaction, indicating social capital buffers restrictive environment impacts. High social capital may also buffer the adverse effects of built environment deficiencies, as highlighted in integrated nature-health pathway frameworks (Markevych et al., 2017).

Conceptual models integrate these dynamics, illustrating how nature, sociocultural, and built environments influence SWB via mediators (behavior, perception, exposure) and moderating individual factors. Ulrich's model distinguishes interrelated affective and cognitive systems: initial affective reactions (e.g., liking) arise from environmental "preferenda" (complexity, focality, natural content) with minimal cognition, shaping cognitive appraisal; individual factors (prior affect, cognitive history, culture) moderate both, guiding adaptive behavior (approach, restoration) (Ulrich, 1983). The biodiversity-health framework shows environmental dimensions affecting well-being through mediators like behavior and perception, moderated by socioeconomic status and personality (Marselle et al., 2021). Bratman et al.'s model outlines steps from natural features (size, biodiversity) to exposure (proximity, duration), experience (sensory modalities), and effects (mental health via stress reduction, social cohesion), moderated by individual traits (Bratman et al., 2019). The bioecological model emphasizes interactions between Process (proximal processes), Person, Context, and Time, with proximal processes as developmental engines shaped by person and context (Bronfenbrenner & Morris, 2006). Shankardass et al.'s ecological framework positions individuals within overlapping digital and physical place interactions, where static/dynamic traits (demographics, personality) moderate place effects on well-being (Shankardass et al., 2018). These models highlight environmental integration via mediators like perception and behavior, with individual and sociocultural factors moderating outcomes, informing policy to balance trade-offs.

## 5.8 Conclusion: To Establish the Theory for the Empirical Study of the Built Environment

Environmental impact on subjective well-being (SWB) is multi-dimensional, involving multiple mechanisms and complex interactions across different environmental dimensions. This complexity is manifested in overlapping and interacting pathways across life domains such as travel, leisure, work, and social relationships (Mouratidis, 2021), with hypothetical causal pathways involving mediators like perceptions of the environment and socialization, and acknowledging multiple interacting mediating and moderating factors (Lachowycz & Jones, 2013). Such multi-dimensionality is further underscored by the intricate interactions between spatial scales and actors in environmental governance, as observed in regional governance transformations.

Among the three major environmental categories (natural, social, and built), the built environment, characterized by proximality and high malleability, emerges as the most direct and effective lever for urban planning interventions aimed at enhancing well-being, particularly in the context of global high-density urbanization (Sung & Lee, 2015). Its critical role stems from being a proximate and modifiable domain (Dannenberg et al., 2011), with controllability and urban relevance making it a key intervention point for public health and urban policy (Marselle et al., 2021). This is supported by evidence that the built environment influences travel demand through its relative malleability and direct linkage to daily activity patterns (Cervero & Kockelman, 1997), and serves as an immediate and modifiable lever in promoting health and well-being, including through urban green spaces (Markevych et al., 2017). Empirical studies in high-density Asian cities, such as Shanghai, further demonstrate that dense, mixed, and pedestrian-oriented built environments can reduce travel distance and encourage sustainable travel modes, reinforcing the built environment's role as an intervention lever (Sun et al., 2016). Additionally, governance mechanisms like the recentralization of land management highlight the built environment's manipulability as a policy tool (Li & Wu, 2012a, 2012b).

This is corroborated by our empirical findings in Guangzhou, which demonstrated that built environment factors at housing, neighborhood, and community scales have distinct and significant impacts on residents' life satisfaction, with the strength of influence diminishing from proximal to distal scales (Li et al., 2025).

Despite progress, current studies face significant gaps. First, the unique characteristics of high-density Asian cities, such as Seoul and Shanghai, are insufficiently addressed, with limited attention to local contexts like housing market dynamics and governance structures. Second, empirical examinations of mediating mechanisms (e.g., how built environment factors affect SWB through walking or stress reduction) and boundary conditions (e.g., which population subgroups are most affected) remain limited. For instance, the influence of compact urban form on social equity shows complex, multi-dimensional effects, with benefits like improved public transport use counterbalanced by issues such as reduced affordable housing, indicating the need to explore boundary conditions (Burton, 2000). Methodologically, challenges persist, including self-selection bias in residential location choice, reliance on self-reported data, and the need for improved exposure measurement and longitudinal designs. Furthermore, governance gaps in regional coordination, such as fragmented institutional mechanisms in China, hinder effective implementation of built environment interventions, emphasizing the need for rigorous empirical examination of operational mechanisms (Li & Wu, 2012).

## 5.9 Summary and Synthesis

The comprehensive review elucidates a multidimensional and multi-level framework in which built environment factors interact with natural and sociocultural environmental dimensions to influence subjective well-being (SWB) through diverse pathways. This integrative perspective identifies key environmental elements—ranging from urban design attributes, walkability, density, and land use mix to exposure to natural settings and socioeconomic institutional contexts—that jointly shape behavioral patterns, perceptual experiences, and physiological exposure mechanisms mediating well-being outcomes. Behavioral mediators predominantly include physical activity and social engagement, facilitated or hindered by proximate environmental characteristics. Perceptual pathways emphasize psychological restoration and affective responses regulated by environmental aesthetics, safety perceptions, and cultural values. Exposure mechanisms capture both beneficial agents, such as greenspace-induced air quality improvements and noise attenuation, and adverse factors, such as pollution and noise stressors.

The review highlights inherent trade-offs within the compact urban form paradigm, where densification fosters transit accessibility and walkability but may concomitantly increase social stressors, noise, and perceived overcrowding, reflecting the complexity of optimizing built environment interventions. Parallel trade-offs emerge in biodiversity management and urban green space provision, underscoring the necessity for nuanced, context-sensitive strategies that balance ecological and social objectives. Sociocultural environments modulate these effects by shaping social comparisons, institutional trust, economic equity, and cultural norms, which collectively regulate individual susceptibility and resilience to environmental influences.

Empirical evidence presents consistent associations between built environment attributes and SWB mediators, although methodological challenges—including residential self-selection, heterogeneous measurement approaches, and limited longitudinal data—constrain causal inference and generalizability, especially within high-density Asian urban contexts. Additionally, discrepancies between perceived and actual environmental qualities, such as biodiversity, reveal critical gaps in public ecological literacy and accessibility equity, complicating the alignment of environmental and well-being goals.

**Funding** This research was funded by

1. Guangdong Provincial Office of Philosophy and Social Sciences [Grant No. GD24XSH06]
2. Projects of PhDs' Start-up Research of GDUPT [Grant No. 2023bsqd1008]

## References

Aletta, F., & Kang, J. (2018). Towards an urban vibrancy model: A soundscape approach. *International Journal of Environmental Research and Public Health, 15*(8), 1712.

Amerigo, M., & Aragones, J. I. (1990). Residential satisfaction in council housing. *Journal of Environmental Psychology, 10*(4), 313–325.

Amerigo, M., & Aragones, J. I. (1997). A theoretical and methodological approach to the study of residential satisfaction. *Journal of Environmental Psychology, 17*(1), 47–57.

Appleton, J. (1975). *The experience of landscape*. Wiley.

Berman, M. G., Jonides, J., & Kaplan, S. (2008). The cognitive benefits of interacting with nature. *Psychological Science, 19*(12), 1207–1212.

Braithwaite, I., Zhang, S., Kirkbride, J. B., Osborn, D. P., & Hayes, J. F. (2019). Air pollution (particulate matter) exposure and associations with depression, anxiety, bipolar, psychosis and suicide risk: A systematic review and meta-analysis. *Environmental Health Perspectives, 127*(12), 126002.

Bratman, G. N., Anderson, C. B., Berman, M. G., Cochran, B., de Vries, S., Flanders, J., Folke, C., Frumkin, H., Gross, J. J., Hartig, T., Kahn Jr., P. H., Kuo, M., Lawler, J. J., Levin, P. S., Lindahl, T., Meyer-Lindenberg, A., Mitchell, R., Ouyang, Z., Roe, J., … Daily, G. C. (2019). Nature and mental health: An ecosystem service perspective. *Science Advances, 5*(7), eaax0903.

Bronfenbrenner, U., & Morris, P. A. (2006). The bioecological model of human development. In W. Damon & R. M. Lerner (Eds.), *Handbook of child psychology: Vol. 1. Theoretical models of human development* (6th ed., pp. 793–828). Wiley.

Burton, E. (2000). The compact city: Just or just compact? A preliminary analysis. *Urban Studies, 37*(11), 1969–2006.

Burton, E., Jenks, M., & Williams, K. (Eds.). (1996). *The compact city: A sustainable urban form?* E & FN Spon.

Campbell, A., Converse, P. E., & Rodgers, W. L. (1976). *The quality of American life: Perceptions, evaluations, and satisfactions*. Russell Sage Foundation.

Canter, D. (1977). *The psychology of place*. St Martin's Press.

Cao, X. J. (2016). How does neighborhood design affect life satisfaction? Evidence from twin cities. *Travel Behaviour and Society, 5*, 68–76.

Cervero, R., & Kockelman, K. (1997). Travel demand and the 3Ds: Density, diversity, and design. *Transportation Research Part D: Transport and Environment, 2*(3), 199–219.

Cheung, F., & Lucas, R. E. (2015). When does money matter most? Examining the association between income and life satisfaction over the life course. *Psychology and Aging, 30*(1), 120–135.

Cheung, F., & Lucas, R. E. (2016). Income inequality is associated with stronger social comparison effects: The effect of relative income on life satisfaction. *Journal of Personality and Social Psychology, 110*(2), 332–341.

Chua, R. Y., Huang, K. G., & Jin, M. (2019). Mapping cultural tightness and its links to innovation, urbanization, and happiness across 31 provinces in China. *Proceedings of the National Academy of Sciences, 116*(14), 6720–6725.
Clark, C., & Paunovic, K. (2018). WHO Environmental Noise Guidelines for the European Region: A systematic review on environmental noise and cognition. *International Journal of Environmental Research and Public Health, 15*(2), 285.
Clark, C., Paunovic, K., & WHO Noise Guidelines Development Group. (2018). WHO Environmental Noise Guidelines for the European Region: A systematic review on environmental noise and quality of life, well-being and mental health. *International Journal of Environmental Research and Public Health, 15*(11), 2400.
Cozens, P., & Love, T. (2015). A review and current status of Crime Prevention Through Environmental Design (CPTED). *Journal of Planning Literature, 30*(4), 393–412.
Cramm, J. M., Van Dijk, H. M., & Nieboer, A. P. (2013). The importance of neighborhood social cohesion and social capital for the well being of older adults in the community. *The Gerontologist, 53*(1), 142–152.
Dallimer, M., Irvine, K. N., Skinner, A. M., Davies, Z. G., Rouquette, J. R., Maltby, L. L., Warren, P. H., Armsworth, P. R., & Gaston, K. J. (2012). Biodiversity and the feel-good factor: Understanding associations between self-reported human well-being and species richness. *BioScience, 62*(1), 47–55.
Dannenberg, A. L., Frumkin, H., & Jackson, R. J. (Eds.). (2011). *Making healthy places: Designing and building for health, well-being, and sustainability*. Island Press.
Diener, E., & Seligman, M. E. (2002). Very happy people. *Psychological science, 13*(1), 81–84.
Durkheim, E. (2019). The division of labor in society. In D. Grusky (Ed.), *Social stratification, class, race, and gender in sociological perspective* (2nd ed., pp. 178–183). Routledge.
Ewing, R., & Dumbaugh, E. (2009). The built environment and traffic safety: A review of empirical evidence. *Journal of Planning Literature, 23*(4), 347–367.
Ewing, R., & Handy, S. (2009). Measuring the unmeasurable: Urban design qualities related to walkability. *Journal of Urban Design, 14*(1), 65–84.
Frank, L. D., Sallis, J. F., Saelens, B. E., Leary, L., Cain, K. L., Conway, T. L., & Hess, P. M. (2009). The development of a walkability index: Application to the Neighborhood Quality of Life Study. *British Journal of Sports Medicine, 44*(13), 924–933.
Galster, G. C., & Hesser, G. W. (1981). Residential satisfaction: Compositional and contextual correlates. *Environment and Behavior, 13*(6), 735–758.
Gascon, M., Triguero-Mas, M., Martínez, D., Dadvand, P., Forns, J., Plasència, A., & Nieuwenhuijsen, M. J. (2015). Mental health benefits of long-term exposure to residential green and blue spaces: A systematic review. *International Journal of Environmental Research and Public Health, 12*(4), 4354–4379.
Gehl, J. (2010). *Cities for people*. Island Press.
Gelfand, M. J., Nishii, L. H., & Raver, J. L. (2006). On the nature and importance of cultural tightness-looseness. *Journal of Applied Psychology, 91*(6), 1225–1244.
Gould, K., & Lewis, T. (2016). *Green gentrification: Urban sustainability and the struggle for environmental justice*. Routledge.
Harrington, J. R., & Gelfand, M. J. (2014). Tightness–looseness across the 50 united states. *Proceedings of the National Academy of Sciences, 111*(22), 7990–7995.
Helliwell, J. F., & Huang, H. (2006). How's your government? International evidence linking good government and well-being. *British Journal of Political Science, 38*(4), 595–619.
Helliwell, J. F., Huang, H., & Wang, S. (2016). *New evidence on trust and well-being* (NBER Working Paper No. 22450). National Bureau of Economic Research.
Hofstede, G. (1980). *Culture's consequences: International differences in work-related values*. Sage.
Jacobs, J. (1962). *The death and life of great American cities*. Vintage Books.
Kaplan, R. (2001). The nature of the view from home: Psychological benefits. *Environment and Behavior, 33*(4), 507–542.

Kaplan, S. (1995). The restorative benefits of nature: Toward an integrative framework. *Journal of Environmental Psychology, 15*(3), 169–182.
Kondo, M. C., Fluehr, J. M., McKeon, T., & Branas, C. C. (2018). Urban green space and its impact on human health. *International Journal of Environmental Research and Public Health, 15*(3), 445.
Kuo, M. (2015). How might contact with nature promote human health? Promising mechanisms and a possible central pathway. *Frontiers in Psychology, 6*, 1093.
Kwan, M.-P. (2012). The uncertain geographic context problem. *Annals of the Association of American Geographers, 102*(5), 958–968.
Lachowycz, K., & Jones, A. P. (2013). Towards a better understanding of the relationship between greenspace and health: Development of a theoretical framework. *Landscape and Urban Planning, 118*, 62–69.
Leyden, K. M. (2003). Social capital and the built environment: The importance of walkable neighborhoods. *American Journal of Public Health, 93*(9), 1546–1551.
Li, H., Pan, C., Qiu, N., Guo, J., & Wu, J. (2025). Does the multi-scale built environment impact on residents' subjective well-being? *Buildings, 15*(18), 3311.
Li, Z., & Wu, F. (2012). The transformation of regional governance in China: The rescaling of statehood. *Progress in Planning, 78*(2), 55–99.
Litman, T. (2023). *Evaluating transportation land use impacts*. Victoria Transport Policy Institute.
Loo, B. P., & du Verle, F. (2016). Transit-oriented development in future cities: Towards a two-level sustainable mobility strategy. *International Journal of Urban Sciences, 20*(1), 1–16.
Maller, C., Townsend, M., St Leger, L., Henderson-Wilson, C., Pryor, A., Prosser, L., & Moore, M. (2009). Healthy parks, healthy people: The health benefits of contact with nature in a park context. In *The George Wright Forum* (Vol. 26, No. 2, pp. 51–83). George Wright Society.
Markevych, I., Fuertes, E., Tiesler, C. M., Birk, M., Bauer, C. P., Koletzko, S., von Berg, A., Berdel, D., & Heinrich, J. (2017). Surrounding greenness and birth weight: Results from the GINIplus and LISAplus birth cohorts in Munich. *Health & Place, 46*, 39–46.
Markevych, I., Schoierer, J., Hartig, T., Chudnovsky, A., Hystad, P., Dzhambov, A. M., de Vries, S., Triguero-Mas, M., Brauer, M., Nieuwenhuijsen, M. J., Lupp, G., Richardson, E. A., Astell-Burt, T., Dimitrova, D., Feng, X., Sadeh, M., Standl, M., Heinrich, J., & Fuertes, E. (2017). Exploring pathways linking greenspace to health: Theoretical and methodological guidance. *Environmental Research, 158*, 301–317.
Marselle, M. R., Hartig, T., Cox, D. T., de Bell, S., Knapp, S., Lindley, S., Triguero-Mas, M., Böhning-Gaese, K., Braubach, M., Cook, P. A., de Vries, S., Heintz-Buschart, A., Hofmann, M., Irvine, K. N., Kabisch, N., Kolek, F., Kraemer, R., Markevych, I., Martens, D., … Bonn, A. (2021). Pathways linking biodiversity to human health: A conceptual framework. *Environment International, 150*, 106420.
Mazumdar, S., Learnihan, V., Cochrane, T., & Davey, R. (2018). The built environment and social capital: A systematic review. *Environment and Behavior, 50*(2), 119–158.
Mokhtarian, P. L., & Cao, X. (2008). Examining the impacts of residential self-selection on travel behavior: A focus on methodologies. *Transportation Research Part B: Methodological, 42*(3), 204–228.
Mouratidis, K. (2018). Built environment and social well-being: How does urban form affect social life and personal relationships? *Cities, 74*, 7–20.
Mouratidis, K. (2021). Urban planning and quality of life: A review of pathways linking the built environment to subjective well-being. *Cities, 115*, 103229.
Newbury, J. B., Arseneault, L., Beevers, S., Kitwiroon, N., Roberts, S., Pariante, C. M., Kelly, F. J., & Fisher, H. L. (2019). Association of air pollution exposure with psychotic experiences during adolescence. *JAMA Psychiatry, 76*(6), 614–623.
Nielsen, T. A. S., & Hansen, K. B. (2007). Do green areas affect health? Results from a Danish survey on the use of green areas and health indicators. *Health & Place, 13*(4), 839–850.
Oishi, S., & Kesebir, S. (2015). Income inequality explains why economic growth does not always translate to an increase in happiness. *Psychological Science, 26*(10), 1630–1638.

Oishi, S., Kesebir, S., & Diener, E. (2011). Income inequality and happiness. *Psychological Science, 22*(9), 1095–1100.

Owen, N., & Fisher, E. B. (2008). Ecological models of health behavior. In K. Glanz, B. K. Rimer, & K. Viswanath (Eds.), *Health behavior and health education: Theory, research, and practice* (4th ed., pp. 465–485). Jossey-Bass.

Pfeiffer, D., & Cloutier, S. (2016). Planning for happy neighborhoods. *Journal of the American Planning Association, 82*(3), 267–279.

Putnam, R. D. (2000). *Bowling alone: The collapse and revival of American community*. Simon & Schuster.

Robinson, J. P., & Martin, S. (2008). What do happy people do? *Social Indicators Research, 89*(3), 565–571.

Rothstein, B., & Uslaner, E. M. (2006). All for all: Equality, corruption, and social trust. *World Politics, 58*(1), 41–72.

Sallis, J. F., Bull, F., Burdett, R., Frank, L. D., Griffiths, P., Giles-Corti, B., & Stevenson, M. (2016). Use of science to guide city planning policy and practice: How to achieve healthy and sustainable future cities. *The Lancet, 388*(10062), 2936–2947.

Sallis, J. F., Cerin, E., Conway, T. L., Adams, M. A., Frank, L. D., Pratt, M., Salvo, D., Schipperijn, J., Smith, G., Cain, K. L., Davey, R., Kerr, J., Lai, P. C., Mitáš, J., Reis, R., Sarmiento, O. L., Schofield, G., Troelsen, J., Van Dyck, D., … Owen, N. (2016). Physical activity in relation to urban environments in 14 cities worldwide: A cross-sectional study. *The Lancet, 387*(10034), 2207–2217.

Shankardass, K., Muntaner, C., Kokkinen, L., Shahidi, F. V., Freiler, A., Oneka, G., Bayoumi, A. M., & O'Campo, P. (2018). The implementation of health in all policies initiatives: A systems framework for government action. *Health Research Policy and Systems, 16*(1), 26.

Simmel, G. (2016). The metropolis and mental life. In W. Longhofer & D. Winchester (Eds.), *Social theory re-wired* (pp. 469–477). Routledge.

Sirgy, M. J., & Cornwell, T. (2002). How neighborhood features affect quality of life. *Social indicators research, 59*(1), 79–114.

Stokols, D. (1996). Translating social ecological theory into guidelines for community health promotion. *American Journal of Health Promotion, 10*(4), 282–298.

Stouffer, S. A., Suchman, E. A., DeVinney, L. C., Star, S. A., & Williams, R. M. (1949). *The American soldier: Adjustment during army life* (Vol. 1). Princeton University Press.

Strawbridge, W. J., Deleger, S., Roberts, R. E., & Kaplan, G. A. (2002). Physical activity reduces the risk of subsequent depression for older adults. *American Journal of Epidemiology, 156*(4), 328–334.

Sugiyama, T., Leslie, E., Giles-Corti, B., & Owen, N. (2008). Associations of neighbourhood greenness with physical and mental health: do walking, social coherence and local social interaction explain the relationships? *Journal of Epidemiology & Community Health, 62*(5), e9.

Sugiyama, T., & Thompson, C. W. (2007). Outdoor environments, activity and the well-being of older people: conceptualising environmental support. *Environment and Planning A, 39*(8), 1943–1960.

Sun, B., He, Z., Zhang, T., & Wang, R. (2016). Urban spatial structure and commute duration: An empirical study of China. *International Journal of Sustainable Transportation, 10*(7), 638–644.

Sung, H., & Lee, S. (2015). Residential built environment and walking activity: Empirical evidence of Jane Jacobs' urban vitality. *Transportation Research Part D: Transport and Environment, 41*, 318–329.

Tay, L., Herian, M. N., & Diener, E. (2014). Detrimental effects of corruption on subjective well-being: Whether, how, and when. *Social Psychological and Personality Science, 5*(7), 751–759.

Triandis, H. C. (1993). Collectivism and individualism as cultural syndromes. *Cross-Cultural Research, 27*(3–4), 155–180.

Twohig-Bennett, C., & Jones, A. (2018). The health benefits of the great outdoors: A systematic review and meta-analysis of greenspace exposure and health outcomes. *Environmental Research, 166*, 628–637.

Uchida, Y., & Ogihara, Y. (2012). Personal or interpersonal construal of happiness: A cultural psychological perspective. *International Journal of Well-Being, 2*(4), 354–369.

Ulrich, R. S. (1983). Aesthetic and affective response to natural environment. In I. Altman & J. F. Wohlwill (Eds.), *Behavior and the natural environment* (pp. 85–125). Springer.

Veal, A. J. (2011). *The leisure society*. Palgrave Macmillan.

White, M. P., Alcock, I., Grellier, J., Wheeler, B. W., Hartig, T., Warber, S. L., Bone, A., Depledge, M. H., & Fleming, L. E. (2019). Spending at least 120 minutes a week in nature is associated with good health and well-being. *Scientific Reports, 9*(1), 7730.

Wilkinson, R. G., & Pickett, K. E. (2009). *The spirit level: Why more equal societies almost always do better*. Allen Lane.

Wilson, E. O. (1984). *Biophilia*. Harvard University Press.

Wirth, L. (1938). Urbanism as a way of life. *American Journal of Sociology, 44*(1), 1–24.

Wolch, J. R., Byrne, J., & Newell, J. P. (2014). Urban green space, public health, and environmental justice: The challenge of making cities 'just green enough'. *Landscape and Urban Planning, 125*, 234–244.

Yuan, Y., Ding, K., Cao, X., & Wu, X. (2018). A review of neighborhood satisfaction. *Urban Development Studies, 25*(10), 105–111.

Yuen, B., & Yeh, A. G. (Eds.). (2011). *High-rise living in Asian cities*. Springer.

# Chapter 6
# Research Design and Methodology

**Abstract** This chapter outlines the research design and methodology for empirically examining the relationship between the built environment and residents' subjective well-being (SWB) in high-density urban areas, with Guangzhou's central urban district as the case study. It aims to provide a rigorous, multi-method framework for measuring both subjective perceptions and objective built environment features, and to delineate the pathways through which environmental factors influence SWB. Section 6.1 introduces the study area—characterized by high population and building density—and details the multi-stage stratified sampling procedure. Section 6.2 elaborates on the measurement of core variables, including Life Satisfaction (SWLS), Perceived Environmental Quality, Social Capital, and optimism (LOT-R), using validated scales and Likert-type instruments. Section 6.3 reports pilot testing and item discrimination analysis. Section 6.4 assesses the reliability and validity of all instruments. Section 6.5 describes formal data collection—including geocoding and the use of 1000 m network buffers to address geographic context uncertainty—and integration of multi-source geographic data. Finally, Section 6.6 outlines analytical strategies, including correlation analysis, regression, structural equation modeling (SEM), and Bootstrap mediation tests. This chapter establishes a robust empirical foundation for identifying built environment interventions that enhance well-being in high-density urban contexts.

**Keywords** Research methodology · Subjective Well-being · Built environment · High-density city · Structural equation modeling · Reliability and validity · Geographic contextual uncertainty · Mediation analysis · Guangzhou · Sampling design

Grounding the multidimensional theoretical framework established in Chap. 5, which synthesized natural, sociocultural, and built environmental influences on subjective well-being (SWB), this chapter transitions to the empirical phase of the



H. Li, G. Shen, *The Anatomy of Subjective Well-Being*, Resilient and Friendly Living Environments, https://doi.org/10.1007/978-981-95-4937-5_6

research. It details the methodological approach adopted to examine these relationships within the high-density urban context of Guangzhou, China. Specifically, this chapter outlines the study area selection, questionnaire design, sampling strategy, and measurement tools used to capture both subjective perceptions and objective built environment features. By grounding the research in a robust methodological foundation, this chapter aims to operationalize the conceptual models discussed earlier and provide a clear pathway for testing the hypothesized pathways linking built environment elements to residents' life satisfaction.

## 6.1 Study Area and Data Collection and Sampling

### *6.1.1 Study Area*

The central urban area of Guangzhou City was selected as the study area for this research based on several key considerations that make it an ideal and representative context for investigating the relationship between multi-scale built environments and subjective well-being.

First, as a premier first-tier Chinese city and a core economic hub within the Guangdong-Hong Kong-Macao Greater Bay Area, Guangzhou exemplifies the rapid urbanization, high population density, and complex urban fabric that characterize modern Chinese metropolitan development. These features provide a critical context for studying how intense urban living conditions impact residents' quality of life.

Second, Guangzhou exhibits significant socio-spatial diversity, encompassing a wide variety of residential built environments. These range from historic urban villages and traditional neighborhoods to modern high-rise commercial housing estates and large-scale newly built districts. This heterogeneity allows for a robust examination of how different housing, neighborhood, and community-scale environmental factors operate across varied settings.

Third, the city has been an active pioneer in implementing urban renewal projects and policies aimed at creating sustainable and livable communities. Consequently, understanding the determinants of life satisfaction in Guangzhou yields valuable insights that can inform urban planning and policy interventions, not only locally but also in other major Chinese cities facing similar developmental challenges.

According to the definition of a high-density city and the analysis of characteristic attributes in Chap. 1, the main characteristics of a high-density city are high population density and high building density. Guangzhou City is a national central city and a South China economic, military, cultural and other multi-center city. According to the data of the seventh census, the resident population is about 18.6766 million,[1] with an average annual growth rate of about 3.93%. The resident

[1] *Bulletin of the Seventh National Population Census of Guangdong Province (No. 2)*, Guangdong Provincial Bureau of Statistics, May 2021.

population of its central Urban area is about 6.3378 million, accounting for 33.93% of the city's population, down 5.8% from the sixth census. The above data show that although the population growth of the central Urban area is smaller than that of the peripheral Urban area, the population density reaches 22,900 people/km$^2$ because the area of the central Urban area is only 276.6 km$^2$.

According to the administrative area of Guangzhou in 2020 reflected by the data of Guangzhou Statistical Yearbook 2021 (Guangzhou City Bureau of Statistics, 2021) and the resident population at the end of the year,[2] calculate its population density. The data analysis in Table 6.1 shows that the population density in the central Urban area of Guangzhou City, namely Yuexiu District, Liwan District, Haizhu District, and Tianhe District, all exceeds 20,000 people/km2. The data in Table 6.2 shows that the building density in the above four districts all exceeds 0.4. In addition, the population density of the other 6 districts and Conghua District in the peripheral Urban area of Guangzhou City is close to 5000 or less.

The four core urban districts of Guangzhou—Yuexiu, Liwan, Haizhu, and Tianhe—all exhibit the characteristic high population density noted by Li and Liu (2019). It is worth emphasizing that the building density referenced in the suggested standards specifically refers to the Floor Area Ratio (FAR), defined as FAR = Total Floor Area of the Building/ Total Area of the Land Parcel. As a result, FAR values are generally greater than the Building Density values used in this study, where Building Density = (Total Building Footprint Area/Total Site Area) × 100%. Due to data availability constraints, this study adopts the latter metric. The building densities across these four districts range between 40% and 50%, which still align with the high building density characteristics described in the reference literature.[3]

In summary, according to the definition standard of high-density city in Table 1.1 and the data results in Tables 6.1 and 6.2, it is proven that the central Urban area of Guangzhou conforms to the characteristics of a high-density city with high population density and high building density. The selection of this area is further justified by its successful use as a case study in our previous investigation into multi-scale built environment effects on well-being, which confirmed the feasibility of collecting high-quality data and the relevance of the context for this line of inquiry (Li et al., 2025).

In order to study the relationship between the community built environment and residents 'Subjective Well-being under the characteristics of a high-density city, the central Urban area of Guangzhou City is taken as the research area, that is, Liwan

[2] *Guangzhou Population Census Yearbook 2020 (Volume I)*, Guangzhou Municipal Bureau of Statistics, https://tjj.gz.gov.cn/stats_newtjyw/tjsj/pcsj/d7crkpc/index.html (accessed December 11, 2022).

[3] The classification of high building density (≥40%) is consistent with the standards adopted in studies such as Li and Liu (2019) and the high-density urban design literature.

**Table 6.1** Administrative district, permanent resident population (at Year-end) and Population Density of Guangzhou

| District | Permanent Resident Population (10,000 people) | Area ($km^2$) | Population Density (10,000 people /$km^2$) |
|---|---|---|---|
| **Yuexiu District** | **103.660** | **33.800** | **3.067** |
| **Liwan District** | **124.160** | **59.100** | **2.101** |
| **Haizhu District** | **181.980** | **90.400** | **2.013** |
| **Tianhe District** | **225.100** | **96.330** | **2.337** |
| Baiyun District | 375.910 | 795.790 | 0.472 |
| Huangpu District | 126.920 | 484.170 | 0.262 |
| Panyu District | 267.260 | 529.940 | 0.504 |
| Nansha District | 84.900 | 783.860 | 0.108 |
| Huadu District | 165.090 | 970.040 | 0.170 |
| Conghua District | 71.980 | 1974.500 | 0.036 |
| Zengcheng District | 147.070 | 1616.470 | 0.091 |

Source: Created by the authors

**Table 6.2** Building density in Guangzhou central urban area

| District | Area ($km^2$) | Building Area ($km^2$) | Building density (%) |
|---|---|---|---|
| Yuexiu District | 33.800 | 15.459 | 45.7 |
| Liwan District | 59.100 | 27.589 | 46.7 |
| Haizhu District | 90.400 | 36.599 | 40.5 |
| Tianhe District | 96.330 | 40.474 | 42.0 |

Source: Created by the authors

District, Yuexiu District, Haizhu District and Tianhe District within the inner ring road are taken as the regional objects.

### 6.1.2 *Questionnaire Design and Tools*

Considering that the research primarily explores subjective well-being within the Chinese context, the questionnaire was developed in Chinese. It consists of three main sections: a cover letter and consent form, the main questionnaire, and an acknowledgments section. The cover letter explains the background and motivation of the study to respondents and assures them that their responses will remain confidential and be used solely for academic purposes (Dillman et al., 2014). The main questionnaire includes the following parts:

Part 1: Basic Residential Information—designed to collect data on the respondent's administrative district, neighborhood, and year of move-in.
Part 2: Social Capital—assesses residents' neighborhood relationships.
Part 3: Perceptions of the Community (within a 1-kilometer radius)—evaluates perceptions of diversity and convenience, transportation accessibility, aesthetic quality, and safety and security.
Part 4: Life Satisfaction—Evaluation measures satisfaction with life conditions, ideals, achievements, changes, and overall life satisfaction.
Part 5: Life Orientation Test-Revised (LOT-R)—assesses individual optimistic personality traits.
Part 6: Personal Health Status—examines self-reported health conditions.
Part 7: Socioeconomic Attributes—collects demographic and socioeconomic information.

Table 6.3 summarizes the measured constructs, measurement tools, and supporting literature references for the designed questionnaire.

Among the seven parts of the questionnaire, the main variables included Life Satisfaction, Perceived Environmental Quality, Perceived Safety, Social Capital, Self-Rated Health, and Individual Socioeconomic Characteristics. An analysis of measurement content and measurement tools is provided below.

## 6.2 Measurement of Primary Variables

### 6.2.1 *Life Satisfaction (LS)*

According to the definition of Subjective Well-being in Chap. 2, this study focuses on the cognitive evaluation of Subjective Well-being, that is, Life Satisfaction. At present, a large number of scholars at home and abroad from philosophy, ethics, sociology, economics and other perspectives to exploring the study, but also developed different measurement methods. According to literature studies, Life Satisfaction evaluation mainly measures individual satisfaction with life as a whole (Diener, 1984; Diener, 2009), and has subjectivity, integrity and relative stability (Diener, 1984; Diener et al., 1985). Evaluation characteristics. In this study, satisfaction with life was measured by the SWLS (Satisfaction With Life Scale) compiled by Diener et al. (1985). The scale has been extensively reviewed and applied in subsequent research (Pavot & Diener, 2008). The scale has been applied in many fields and countries and has proved to have good reliability.[4] In our prior study conducted in the same urban context, the scale exhibited high reliability (Cronbach's $\alpha = 0.877$), supporting its appropriateness for this research (Li et al., 2025). In order to ensure the research effect, the scale was adjusted according to the literature suggestion, and the Likert 7-point scale was used to measure the cognitive evaluation

[4] The SWLS has been validated in cross-cultural contexts, including Chinese populations, demonstrating high reliability and validity (e.g., Bai et al., 2011).

**Table 6.3** Questionnaire design and main measurements

| Measurement | Measurement content | Measurement tools | literature basis |
|---|---|---|---|
| Basic information about residence | a. Administrative district of residence<br>b. Residential communities<br>c. Check-in time | Questionnaire | |
| Social capital | a. Network size<br>b. Frequency of interaction<br>c. Sense of belonging<br>d. Sense of trust | Likert scale | (Tao et al., 2020; Yang et al., 2021; Zhang et al., 2020) |
| Perceive environment | a. Multiple convenience perception<br>b. Traffic convenience perception<br>c. Aesthetic perception<br>d. Perceived safety | Likert scale | (Cao, 2016; Zhang et al., 2019; Das, 2020) |
| Life satisfaction | a. Life ideals<br>b. Life conditions<br>c. Satisfied with life<br>d. Life gains<br>e. Life changes | Satisfaction with life scale (SWLS) | (Diener et al., 1985) |
| Life orientation test-revised (LOT-R) | a. Optimistic expectations<br>b. Pessimistic expectations | LOT-R scale | (Scheier et al., 1994). |
| Personal health status | a. Self-rated health<br>b. Height<br>c. Weight | Likert scale<br>Questionnaire<br>Questionnaire | (Liu et al., 2018) |
| Individual socioeconomic attributes | a. Age<br>b. Gender<br>c. Household status<br>d. Marital status<br>e. Educational attainment<br>f. Employment status<br>g. Personal monthly income<br>h. Type of housing<br>i. Per capita living area | Questionnaire | |

Source: Created by the authors

of the respondents on the following 5 questions from "1 point" indicating "strongly disagree" to "7 points" indicating "strongly agree". The higher the score for each item, the higher the Life Satisfaction.

The title of the section reads as follows:

(a) In most ways, my life is close to my ideal (Life ideals = LS 1).
(b) The conditions of my life are excellent (Life conditions = LS 2).

(c) I am satisfied with my life (Satisfied with life = LS 3).
(d) So far, I have gotten the important things I want in life (Life gains = LS 4).
(e) If I could live my life over, I would change almost nothing (Life changes = LS 5).

### 6.2.2 *Perceived Environmental*

(1) Perceived Environmental Quality (PEQ)
According to the current literature (Das, 2020; Shen & Fu, 2019), divided the perception evaluation of the community into facility perception meeting residents 'basic survival needs, i.e. diversified convenience perception evaluation, perception meeting residents' basic safety needs, i.e. Perceived Safety safety perception evaluation, perception meeting residents 'travel activities and social infrastructure perception, i.e. traffic convenience perception evaluation, and environmental perception meeting residents' aesthetic respect needs, i.e. aesthetic perception evaluation (Das, 2020) Four parts. Using the Likert 7-point scale, respondents were measured from "1-very dissatisfied" to "7-very satisfied" on the following four items:

1. The first aspect mainly includes residents 'perception of various conveniences in their community (within 1000 meters of the community center). The topics are as follows:

   (a) Your rating of the diversity and convenience of community business services facilities (DCP1).
   (b) Your rating of the diversity and convenience of community leisure sports facilities (DCP2).
   (c) Your rating of the diversity and convenience of community educational facilities (primary and secondary)(DCP3).
   (d) Your evaluation of the diversity and convenience of community health facilities (outpatient hospital)(DCP4).

2. The second aspect mainly includes residents 'perception of traffic convenience in their community (within 1000 meters from the community center). The topics are as follows:

   (a) Your rating of community bus stop accessibility (TCP1).
   (b) Your rating of the accessibility of community subway stations (TCP2).
   (c) Your rating of the smooth flow of driving in the community (no traffic congestion)(TCP3).

3. The third aspect mainly includes residents 'aesthetic perception of their community (within 1000 meters of the community center). The topics are as follows:

   (a) Your evaluation of community park green space aesthetics (AEP1).
   (b) Your aesthetic rating of the neighborhood streetscape (AEP2).
   (c) Your evaluation of the aesthetics of community buildings (AEP3).

(2) Perceived Safety (PES)

According to the research of relevant literature, Perceived Safety can be measured from two aspects: one is objective measurement, that is, the number of Perceived Safety and crime cases in a certain period; the other is the concern and evaluation of residents in the community for Perceived Safety. Because the former has the problem of spatial imbalance, this study adopts the method of residents 'perception evaluation to measure it (Foster et al., 2010). Residents 'perception of Perceived Safety and safety in their community (within 1000 meters of the community center). The topics are as follows:

(a) Your rating of lighting equipment, such as community street lights (PES1).
(b) Your rating of community policing equipment (PES2).
(c) Your assessment of community environmental security and crime safety (PES3).

(3) Social Capital (SC)

Due to different definitions of Social Capital, the concept also contains different elements (Putnam, 2000). Referring to the corresponding literature in the urban planning discipline (Zhang et al., 2020), it mainly includes structural capital measured by network scale and interaction frequency and cognitive Social Capital measured by sense of belonging and trust (Moore & Kawachi, 2017). Social Capital is measured in four dimensions: trust, belonging, network and interaction. Using Likert's 7-point scale, respondents were asked to rate the following 9 items on a scale from "1" for "strongly disagree" to "7" for "strongly agree." The Social Capital scale consists of the following topics:

(a) Neighbors are trustworthy (SC1).
(b) Neighbors are generally harmonious within the community (SC2).
(c) Likes the community and has a sense of belonging to the community (SC3).
(d) Recognize most of the neighbors in the cell (SC4).
(e) High number of close neighbors (SC5).
(f) Frequent chatting with neighbors (SC6).
(g) Eating with neighbors (SC7).
(h) In case of emergency, you can get help from neighbors (SC8).
(i) The community regularly organizes colorful activities (SC9).

(4) Life Orientation Test-Revised (LOT-R)

The measurement of the optimistic personality trait uses the revised Life Orientation Test (LOT-R; Scheier et al., 1994), which is a widely accepted measure known for its reliability and validity. The Life Orientation Test-Revised was used to measure dispositional optimism. It consists of 6 items: 3 measuring optimism, 3 measuring pessimism. Responses were recorded on a 5-point Likert scale ranging from 1 (strongly disagree) to 5 (strongly agree). The three optimism items (items a, c, and f) were positively scored. The three pessimism items (items b, d, and e) were reverse-scored.

(a) In uncertain times, I usually expect the best (OPT1).
(b) If something can go wrong for me, it will (PEM1).

(c) I'm always optimistic about my future (OPT2).
(d) I hardly ever expect things to go my way (PEM2).
(e) I rarely count on good things happening to me (PEM3).
(f) Overall, I expect more good things to happen to me than bad (OPT3).

Although the LOT-R scale is widely accepted and renowned for its reliability and validity, the literature shows ongoing debate regarding its dimensionality, with some studies supporting a two-factor structure (Wen, 2012) and others a unidimensional structure (Gu, 2012). This study focuses solely on three items representing the optimistic factor from the aforementioned literature: "Under uncertain circumstances, I often expect the best results (OPT1)", "I am optimistic about my future (OPT2)", "Overall, I expect good things rather than bad things to happen to me (OPT3)".

(5) Self-Rated Health (SRH)

At present, the research on the impact of the built environment on health has become a hot spot. In the corresponding studies, two main methods are used to measure health. One is objective measurement, such as height, weight, blood sugar, blood pressure, blood lipids, etc.; the other is subjective report, such as self-evaluation of physical health, mental health, etc. In the discipline of urban planning, objective measurements are limited by equipment, sample tracking, and respondent confidentiality, and most studies use subjective reports to measure health (Idler & Benyamini, 1997). Considering that the main variable of Life Satisfaction in this study is subjective cognitive evaluation, the study adopts the form of self-reported health to measure. A 10-point scale was used to measure respondents 'Self-rated health from "1″ for "very unhealthy" to "10″ for "very healthy." The title is as follows:

Your self-rating of your physical health (excluding psychological aspects)(SRH).

(6) Measurement of Other Variables

The first part of the questionnaire is the basic residential characteristics, such as residential administrative district, residential community, year of residence, and the sixth part is the basic information of respondents, such as age, gender, marital status, working status, education status, personal monthly income, per capita housing area, etc.

## 6.3 Data Sampling

This study mainly discusses the impact of the built environment on residents 'Subjective Well-being and its mechanism, mainly involving residential conditions, Perceived Environmental Quality evaluation, Social Capital, Subjective Well-being (Life Satisfaction), health, individual social and economic characteristics and other aspects of information. According to the research purpose, the sample targets in the study area are the heads, spouses or main family members aged over 18 in the households living in the built-up areas of the central Urban area.

The survey employed a combined method of multi-stage stratified sampling and probability sampling. In the first stage, the overall study area was determined based on density-based zoning criteria, identifying four administrative districts (namely, Liwan District, Yuexiu District, Haizhu District, and Tianhe District). In the second stage, streets within each district were sorted according to the size of the resident population and selected using random systematic sampling. During the third stage, neighborhood committees within the selected streets were sorted by population size and similarly chosen through random systematic sampling. The fourth stage involved random sampling within the selected neighborhood committees; Fig. 6.1 illustrates the survey scope and the distribution of the sampled communities. In the fifth and final stage, households were randomly selected based on the population proportion within the chosen communities for household questionnaire surveys.

According to the objective of the study and the reality of the subject, the subjective evaluation data were collected by means of a social questionnaire. This method has been widely used in urban planning, architecture, landscape architecture and other disciplines. With the improvement of the Internet and information technology, paperless network investigation can not only improve efficiency but also effectively reduce costs. However, taking into account the age, body, lifestyle or other constraints of the respondents, the study developed both paper and online questionnaires. The investigation process consists of three parts. First, prior to the start of the survey, respondents were asked to read the cover and sign the Informed Consent Form. According to the two types of questionnaire design, there are two ways to complete the questionnaire. One is for the selected interviewees to scan the QR code provided by the research team and enter the link to fill in the questionnaire, with the investigator assisting them on the side. If there are any questions, the investigator will answer them according to the previous training; the other is for the selected interviewees to fill in the paper questionnaire provided by the research team or read the questions one by one by the members of the research team, and the interviewees will answer the questions one by one in writing or orally.

In order to avoid possible bias in formal surveys and to ensure the reliability and validity of measurements, pre-survey tests were conducted between formal surveys. According to the sampling procedure mentioned above, 171 samples were selected from the target population for a pre-test of items, and the total scores of each component item were summed up and grouped into high and low quantiles of 27 and 73 (Gelman & Park, 2009; Kelley, 1939). A mean equivalence T-test was performed. If $T > 1.96$ and $p < 0.05$, which proves that all items have good discrimination and discrimination. Therefore, each item is retained to form a formal questionnaire for formal sample collection.

Based on the pilot testing analysis of the project, Tables 6.4, 6.5, 6.6, 6.7, and 6.8 report the t-values, degrees of freedom, means, and mean scores of the high- and low-scoring groups for each item, respectively. For all items measuring Life Satisfaction (LS), Perceived Environmental Quality (PEQ), Perceived Safety (PES), Social Capital (SC), and Self-Rated Health (SRH). The t-values exceeded 1.96, with $p < 0.05$, indicating adequate discriminatory power of each item. Therefore, all

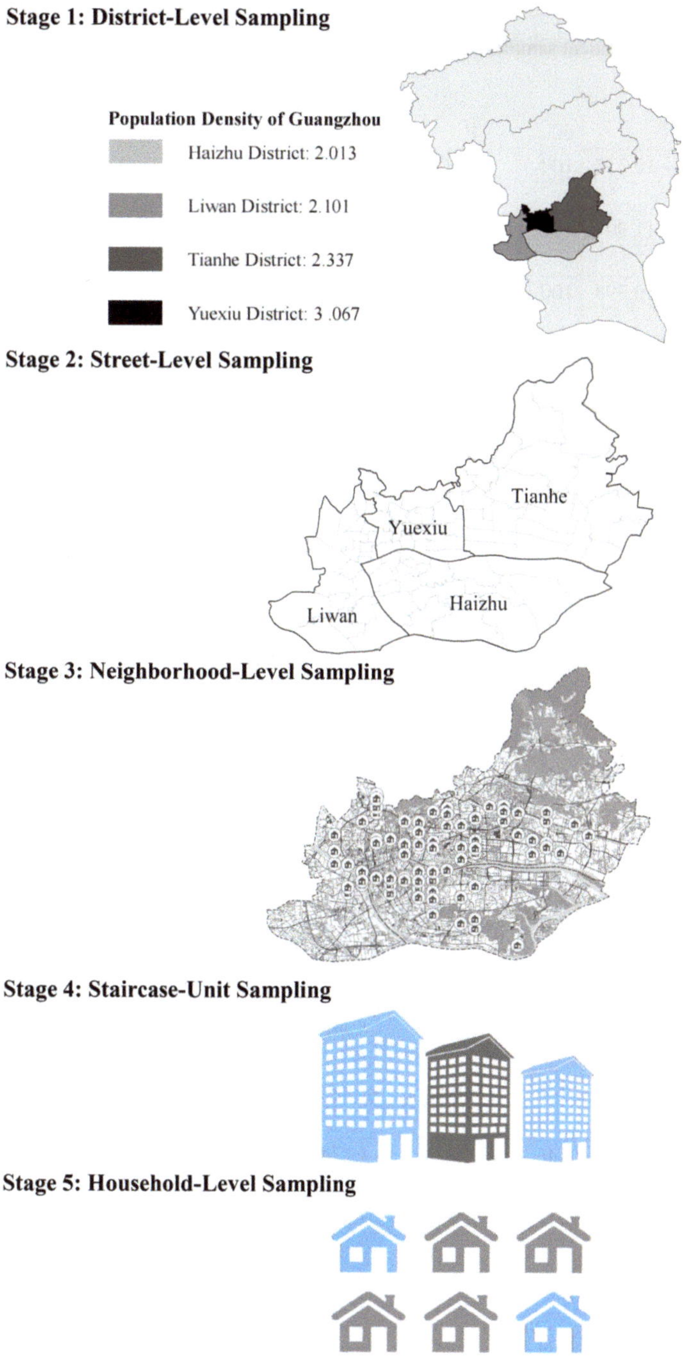


**Fig. 6.1** Schematic diagram of the multi-stage sampling process Source: Created by the authors

**Table 6.4** T-test for high and low groups of Life Satisfaction clusters

| | Independent samples T-test | | | | | | | |
|---|---|---|---|---|---|---|---|---|
| | T | df | P | Mean Difference (MD) | High vs low groups | N | Mean | Standard Deviation |
| LS 1 | −10.737 | 100 | .000 | −1.980 | L | 51 | 4.290 | 1.205 |
| | | | | | H | 51 | 6.270 | .532 |
| LS 2 | −11.907 | 100 | .000 | −2.333 | L | 51 | 3.880 | 1.275 |
| | | | | | H | 51 | 6.220 | .577 |
| LS 3 | −10.598 | 100 | .000 | −2.059 | L | 51 | 4.290 | 1.221 |
| | | | | | H | 51 | 6.350 | .658 |
| LS 4 | −12.862 | 100 | .000 | −2.549 | L | 51 | 3.760 | 1.320 |
| | | | | | H | 51 | 6.310 | .510 |
| LS 5 | −14.197 | 100 | .000 | −2.608 | L | 51 | 3.510 | 1.138 |
| | | | | | H | 51 | 6.120 | .653 |

Source: Created by the authors

**Table 6.5** T-test for high and low groups of community perception

| | Independent samples T-test | | | | | | | |
|---|---|---|---|---|---|---|---|---|
| | T | df | P | Mean Difference (MD) | High vs low groups | N | Mean | Standard Deviation |
| DCP1 | −11.335 | 101 | .000 | −2.063 | L | 46 | 4.670 | 1.283 |
| | | | | | H | 57 | 6.740 | .444 |
| DCP2 | −12.807 | 101 | .000 | −2.105 | L | 46 | 4.460 | .912 |
| | | | | | H | 57 | 6.560 | .756 |
| DCP3 | −16.077 | 101 | .000 | −2.515 | L | 46 | 4.240 | 1.058 |
| | | | | | H | 57 | 6.750 | .474 |
| DCP4 | −13.473 | 101 | .000 | −2.423 | L | 46 | 4.260 | 1.255 |
| | | | | | H | 57 | 6.680 | .469 |
| TCP1 | −10.883 | 146 | .000 | −1.448 | L | 50 | 4.960 | 1.106 |
| | | | | | H | 98 | 6.410 | .514 |
| TCP2 | −12.291 | 146 | .000 | −1.887 | L | 50 | 4.440 | 1.327 |
| | | | | | H | 98 | 6.330 | .533 |
| TCP3 | −13.568 | 146 | .000 | −1.874 | L | 50 | 4.320 | 1.133 |
| | | | | | H | 98 | 6.190 | .550 |
| AEP1 | −15.349 | 111 | .000 | −2.240 | L | 38 | 4.030 | 1.078 |
| | | | | | H | 75 | 6.270 | .475 |
| AEP2 | −17.244 | 111 | .000 | −2.532 | L | 38 | 3.870 | 1.044 |
| | | | | | H | 75 | 6.400 | .520 |
| AEP3 | −19.778 | 111 | .000 | −2.821 | L | 38 | 3.550 | .978 |
| | | | | | H | 75 | 6.370 | .540 |

Source: Created by the authors

**Table 6.6** T-test for high and low groups of perceived safety

| | Independent samples T-test | | | | | | | |
|---|---|---|---|---|---|---|---|---|
| | T | df | P | Mean Difference (MD) | High vs low groups | N | Mean | Standard Deviation |
| PES1 | −11.553 | 100 | .000 | −2.155 | L | 50 | 4.460 | 1.182 |
| | | | | | H | 52 | 6.620 | .631 |
| PES2 | −13.497 | 100 | .000 | −2.430 | L | 50 | 4.320 | 1.203 |
| | | | | | H | 52 | 6.750 | .480 |
| PES3 | −16.255 | 100 | .000 | −2.587 | L | 50 | 4.240 | 1.080 |
| | | | | | H | 52 | 6.830 | .382 |

Source: Created by the authors

**Table 6.7** T-test for high and low clusters of social capital

| | Independent samples T-test | | | | | | | |
|---|---|---|---|---|---|---|---|---|
| | T | df | P | Mean Difference (MD) | High vs low groups | N | Mean | Standard Deviation |
| SC1 | −10.934 | 92 | .000 | −2.159 | L | 46 | 4.200 | 1.167 |
| | | | | | H | 48 | 6.350 | .699 |
| SC2 | −9.438 | 92 | .000 | −2.312 | L | 46 | 4.020 | 1.570 |
| | | | | | H | 48 | 6.330 | .630 |
| SC3 | −13.220 | 92 | .000 | −2.284 | L | 46 | 4.170 | 1.060 |
| | | | | | H | 48 | 6.460 | .544 |
| SC4 | −14.244 | 92 | .000 | −2.941 | L | 46 | 3.410 | 1.309 |
| | | | | | H | 48 | 6.350 | .565 |
| SC5 | −15.462 | 92 | .000 | −3.122 | L | 46 | 3.070 | 1.237 |
| | | | | | H | 48 | 6.190 | .641 |
| SC6 | −16.951 | 92 | .000 | −3.082 | L | 46 | 3.020 | 1.043 |
| | | | | | H | 48 | 6.100 | .692 |
| SC7 | −12.002 | 92 | .000 | −2.934 | L | 46 | 2.590 | 1.127 |
| | | | | | H | 48 | 5.520 | 1.238 |
| SC8 | −11.563 | 92 | .000 | −2.509 | L | 46 | 3.780 | 1.381 |
| | | | | | H | 48 | 6.290 | .582 |
| SC9 | −11.645 | 92 | .000 | −2.870 | L | 46 | 3.130 | 1.276 |
| | | | | | H | 48 | 6.000 | 1.111 |

Source: Created by the authors

**Table 6.8** T-test for high and low groups of self-rated health

| | Independent samples T-test | | | | | | | |
|---|---|---|---|---|---|---|---|---|
| | T | df | P | Mean Difference (MD) | High vs low groups | N | Mean | Standard Deviation |
| SRH | −21.304 | 114 | .000 | −4.681 | L | 47 | 4.640 | 1.737 |
| | | | | | H | 69 | 9.320 | .469 |

Source: Created by the authors

items were retained to form the formal questionnaire for subsequent data collection from the main sample.

## 6.4 Reliability and Validity Analysis of Research Instruments

### *6.4.1 Reliability Analysis*

Before empirical research, the reliability and validity of the research tools need to be tested. Reliability is mainly used to determine the stability and consistency of tool measurement concepts in order to evaluate the goodness of measurement (Nunnally & Bernstein, 1994; Sekaran & Bougie, 2016). According to Hair and Black (2005), the reliability of a measurement tool can be assessed using Cronbach's alpha. A coefficient below 0.6 is generally considered to indicate low reliability, while a value of 0.7 or above is acceptable for further analysis. The reliability of the Detection Tools was evaluated by Cronbach's alpha coefficient and overall reliability.

Tables 6.9, 6.10, 6.11, 6.12, 6.13, 6.14, and 6.15 respectively report the results of reliability analysis on Life Satisfaction, Perceived Convenience of Community Built Environment Diversity, Perceived Traffic Convenience in Community Built Environment, Aesthetic Perception of Community Built Environment, Perceived Safety, Social Capital structure, Optimism Personality. Cronbach's $\alpha > 0.7$, which indicates that the scale had good internal consistency and could be used to collect formal samples.

### *6.4.2 Validity Analysis*

To ensure that measurements reflect established concepts and what is being tested, tools need to be validated (Hair et al., 2007)。To ensure content validity, experts were invited to discuss and review the questions. In addition, factor loads and mean variance extractions were used to verify the construct validity of the construct.

**Table 6.9** Reliability analysis of life satisfaction

| | LS 1 | LS 2 | LS 3 | LS 4 | LS 5 | CITC | Cronbach's α |
|---|---|---|---|---|---|---|---|
| LS 1 | 1.000 | | | | | .755 | .903 |
| LS 2 | .686 | 1.000 | | | | .789 | |
| LS 3 | .638 | .643 | 1.000 | | | .752 | |
| LS 4 | .662 | .720 | .705 | 1.000 | | .800 | |
| LS 5 | .631 | .662 | .623 | .656 | 1.000 | .740 | |

Caption: CITC = Corrected Item-Total Correlation; values <0.30 indicate poor item reliability (Field, 2020)

Source: Created by the authors

**Table 6.10** Reliability analysis of perceived convenience of community built environment diversity

| | DOP1 | DOP2 | DOP3 | DOP4 | CITC | Cronbach's α |
|---|---|---|---|---|---|---|
| DCP1 | 1.000 | | | | .729 | .868 |
| DCP2 | .552 | 1.000 | | | .618 | |
| DCP3 | .631 | .551 | 1.000 | | .749 | |
| DCP4 | .700 | .551 | .739 | 1.000 | .784 | |

Source: Created by the authors

**Table 6.11** Reliability analysis of perceived traffic convenience in community built environment

| | TAP1 | TAP2 | TAP3 | CITC | Cronbach's α |
|---|---|---|---|---|---|
| TCP1 | 1.000 | | | .621 | .786 |
| TCP2 | .532 | 1.000 | | .622 | |
| TCP3 | .570 | .569 | 1.000 | .649 | |

Source: Created by the authors

**Table 6.12** Reliability analysis of aesthetic perception of community built environment

| | AEP1 | AEP2 | AEP3 | CITC | Cronbach's α |
|---|---|---|---|---|---|
| AEP1 | 1.000 | | | .694 | .847 |
| AEP2 | .657 | 1.000 | | .742 | |
| AEP3 | .615 | .677 | 1.000 | .711 | |

Source: Created by the authors

**Table 6.13** Reliability analysis of community security

| | SEP1 | SEP2 | SEP3 | CITC | Cronbach's α |
|---|---|---|---|---|---|
| PES1 | 1.000 | | | .592 | .830 |
| PES2 | .584 | 1.000 | | .764 | |
| PES3 | .523 | .745 | 1.000 | .716 | |

Source: Created by the authors

**Table 6.14** Reliability analysis of social capital

| | SC1 | SC2 | SC3 | SC4 | SC5 | SC6 | SC7 | SC8 | SC9 | CITC | Cronbach's α |
|---|---|---|---|---|---|---|---|---|---|---|---|
| SC1 | 1.000 | | | | | | | | | .789 | .931 |
| SC2 | .822 | 1.000 | | | | | | | | .766 | |
| SC3 | .735 | .721 | 1.000 | | | | | | | .758 | |
| SC4 | .634 | .658 | .627 | 1.000 | | | | | | .797 | |
| SC5 | .636 | .620 | .600 | .764 | 1.000 | | | | | .809 | |
| SC6 | .645 | .563 | .670 | .755 | .805 | 1.000 | | | | .789 | |
| SC7 | .433 | .404 | .401 | .580 | .620 | .594 | 1.000 | | | .628 | |
| SC8 | .739 | .737 | .677 | .627 | .673 | .664 | .504 | 1.000 | | .812 | |
| SC9 | .548 | .561 | .572 | .511 | .500 | .427 | .568 | .671 | 1.000 | .656 | |

Source: Created by the authors

**Table 6.15** Reliability analysis of optimism

| | OPT1 | OPT1 | OPT1 | CITC | Cronbach's α |
|---|---|---|---|---|---|
| OPT1 | 1.000 | | | .653 | .804 |
| OPT2 | .667 | 1.000 | | .717 | |
| OPT3 | .491 | .575 | 1.000 | .586 | |

Source: Created by the authors

Validity studies are used to analyze whether research questions are reasonable and valid. Bartlett's spherical test was approximately chi-square 5094.57 and KMO 0.907, indicating that the test passed and factor analysis could be performed. Exploratory factor analysis (EFA) was performed using principal component analysis (PCA). The results showed that there were 5 common factors explaining the questionnaire. According to the literature, it is suggested to delete the two topics "community often organizes colorful activities" and "your evaluation of community street lamps and other equipment" with cross-factor load>0.5, and then re-examine the whole item and re-examine it. According to previous literature, 6 dimensions were extracted forcibly, and the total explained variance was 78.79, which indicated that the information in the questions could be extracted effectively, and the validity was good.

## 6.5 Data Collection and Processing Methods

### *6.5.1 Collection of Questionnaire Data*

In order to scientifically, reasonably and effectively study the relationship between the built environment and Subjective Well-being, the study collected formal sample data according to the predetermined sampling process in the proposed target population. Before the formal survey, referring to the suggestions in the literature (MacCallum et al., 1996), the sample size was estimated using the R language (Preacher & Coffman, 2006). The results showed that the maximum required sample size was 147–195, indicating that a sample size greater than 195 could better support the study and ensure its smooth progress.

The survey data collection spanned a period of 9 months (from July 2021 to March 2022). The survey started at the end of that wave of the pandemic. Meanwhile, based on the sampling results, communities that were unaffected or less affected by that wave of the pandemic were surveyed first, followed by large-scale surveys after the pandemic was completely under control. Considering the practicality of the survey and the work schedules of residents, the surveys were mainly conducted on weekdays or weekends.

The survey process was divided into three parts. First, before the official start of the survey, respondents were invited to read the cover letter or consent form and confirm "whether they agreed to participate in this survey". If they did not agree, the survey for this respondent would be terminated. Then, another respondent would be randomly selected until a respondent who agreed to participate in the survey was found. If they agreed, they would sign the "Informed Consent Form" and fill out the questionnaire. Second, during the survey, respondents with normal abilities were asked to fill out the questionnaire independently, while the researchers were on standby to answer any questions. For respondents with disabilities such as visual impairment, the researchers would read out the questionnaire, and the respondents would answer orally, with the researchers recording their answers. If respondents encountered sentences that were difficult to read or understand while filling out the questionnaire, the researchers from the research group would provide further explanations. The survey process took approximately 25–40 min. Finally, after the survey, the questionnaires were collected, and small gifts were given to the respondents as a token of gratitude. In total, the research group collected 593 questionnaires from 81 communities (Fig. 6.2), among which 572 were valid, with an effective rate of approximately 96.46%. In the four administrative districts surveyed, 20 communities in Yuexiu District were surveyed with 116 samples, 18 communities in Liwan District with 98 samples, 28 communities in Haizhu District with 185 samples, and 21 communities in Tianhe District with 173 samples. The average number of valid questionnaires per surveyed community was 6.58. The minimum number of valid questionnaires in a surveyed community was 5, and the maximum was 19.

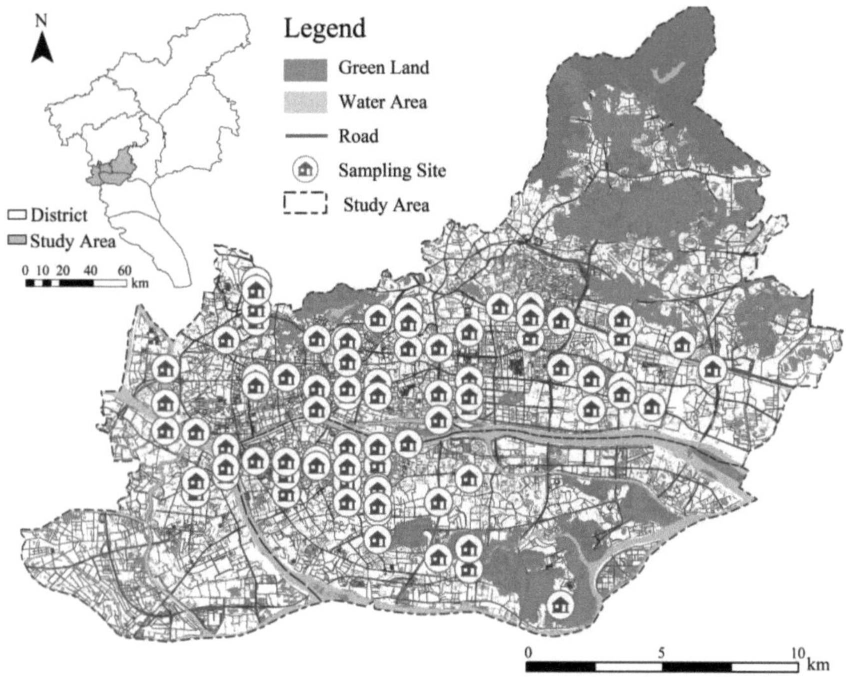


**Fig. 6.2** The distribution of sample communities. Source: Created by the authors

### 6.5.2 Data Collection of the Objective Environment

In order to further explore the relationship between the built environment and Subjective Well-being, the research collected data from the subjective measurement part. The acquisition of geospatial information requires geographical referencing based on the questionnaire information and then geocoding, so that the samples with spatial associations can be better estimated when processing data on the built environment, health, and well-being (Grafström & Tillé, 2013). Although there are still objective differences between the information obtained through geocoding and the actual data, address-based geocoding remains a widely used and practical method in academic research. Therefore, following common practices (e.g., Goldberg et al., 2007), the spatial information was reverse-geocoded, and residential communities were converted into geographical coordinates for the collection of objective data. This approach aligns with our previous methodological application, where a 1000 m network buffer was similarly employed to define neighborhood boundaries and calculate objective built environment metrics such as density, diversity, and design, effectively capturing community-scale influences on well-being (Li et al., 2025).

It is important to note that communities can be defined using various spatial boundaries. In measuring the built environment, most studies adopt units such as Traffic Analysis Zones (TAZs), Zip Codes, census tracts, blocks, jiedao (streets), or buffer-based radii (ranging from 100 m to 2 km) (Ewing & Handy, 2009), while a limited number of studies have considered cognitive map-based spaces. Even when examining the same built environment features, analytical results may differ depending on the choice of geographical unit, giving rise to the "Uncertain Geographic Context Problem (UGCoP)". This concept refers to the phenomenon where "the effect of spatial variables on individual behavior may be influenced by the delineation of geographic contextual units or neighbourhood boundaries and their degree of deviation from the true geographic context of causal processes"(Cheng, 2021). This issue is reflected in studies across multiple contexts: for instance, research on social capital and mental health in deprived areas[5] (Mitchell & LaGory, 2002), spatial regression on youth depressive symptoms (Duncan et al., 2013), and comparisons between objective and subjective measures of the built environment in capturing walking behaviours (Lin & Moudon, 2010).

The issue of geographic unit definition, often conceptualized as the Modifiable Areal Unit Problem (MAUP) (Wong, 2004) or the Uncertain Geographic Context Problem (UGCoP) (Kwan, 2012; Cheng, 2021), is a fundamental methodological challenge in spatial and contextual analysis. Similarly, reviews on the built environment and obesity (Feng et al., 2010), neighbourhood effects on health in the U.S. (Arcaya et al., 2016), and intra-urban disparities in health behaviours such as

[5]The Uncertain Geographic Context Problem (UGCoP) is a well-recognized issue in geographic and health studies, emphasizing the sensitivity of results to the delineation of spatial units (Kwan, 2012).

excessive drinking all highlight the critical importance of geographic unit selection in shaping analytical outcomes and interpretive validity (Sun et al., 2019). This concern is further echoed in empirical studies comparing analytical approaches (Chaix et al., 2005) and discussing the spatial dimensions of neighborhood effects (Spielman & Yoo, 2009), underscoring that the choice of geographic unit is not merely a technical decision but one that shapes the very phenomena we seek to understand. Given this methodological imperative, this study adopts a network buffer of 1000 meters to define neighborhood boundaries,[6] an approach commonly used and justified in prior literature (e.g., Lin & Moudon, 2010), to capture the objective built environment.

Relevant studies have shown that the impact of the administrative boundaries of communities on residents' health, Life Satisfaction, and Subjective Well-being is unclear. Therefore, in current studies on the built environment of communities, the administrative boundary of the community is rarely used as a basis. Instead, in most cases, a buffer zone with a certain radius is drawn centered on the centroid of the surveyed community as the community scope. This study draws on relevant research designs and sets a buffer zone with a radius of 1000 meters as the scope of the surveyed community to examine the Objective Built Environment of the community. The Objective Built Environment data used in the study mainly come from the population data of Guangzhou City released by Worldpop in 2021,[7] point of interest (POI) data and road network data obtained from Open Street Map, Amap, Baidumaps, and a certain map network, as well as data obtained from some real estate websites. Table 6.16 reports the measurement dimensions, measurement variables, measurement tools, and data sources of the Objective Built Environment.

## 6.6 Analysis Methods

After data collection, the study used the Statistical Package for the Social Sciences (SPSS 22.0) and Analysis of Moment Structures (AMOS 24.0) for analysis. SPSS was used for data cleaning, screening, coding, transformation, analysis, and plotting. AMOS 24.0 software was employed to conduct statistical analysis and testing of the measurement model and the structural equation model (Collier, 2020).

To grasp the overall trend of the relationship between the built environment and residents' Subjective Well-being, the micro-survey data were aggregated to the community scale. The Geographic Information System software (ArcGIS 10.1) was used to visualize the spatial attribute data. The Geographic Information

[6] The 1000 m network buffer is a commonly used geographic unit in urban studies to capture walkable neighborhood environments (e.g., Frank et al., 2005).

[7] Worldpop provides high-resolution population distribution data widely used in health and urban studies (Tatem, 2017). POI data were extracted from Amap and Baidu Maps, which are authoritative sources for geospatial data in China.

**Table 6.16** Measurement of objective built environment variables and data sources

| Measurement dimension | Measured variables | Measurement tools | Data sources |
|---|---|---|---|
| Density | a. Population density (PD)<br>b. Building density (BD) | ArcGIS/calculation<br>ArcGIS/calculation | Worldpop (Tatem 2017)<br>Baidumaps (2021) |
| Diversity | Functional mix diversity (FMD) | ArcGIS/calculation | POI; Amap |
| Design | Road intersection density (RID) | ArcGIS/calculation | Openstreetmap |
| Distance to transit | a. Bus line density (BLD)<br>b. Bus stop density (BSD) | ArcGIS/calculation<br>ArcGIS/calculation | Amap (2021)<br>Amap |
| Destination accessibility | POI density (POID) | ArcGIS/calculation | Amap |

Source: Created by the authors

visualization technology (ArcScene 3D visualization) and scatter plot Fitting the Trend lines were employed to conduct a descriptive analysis at the aggregation scale of the correlation between the built environment elements, which are the core concerns of this paper, and the average Life Satisfaction of community residents.

(1) Data Analysis

The T-test was employed to examine the discrimination and discriminatory power of the items in the pretest questionnaire. Descriptive analysis was mainly used to analyze the frequency, proportion, maximum, minimum, mean, standard deviation, skewness, kurtosis, etc., of variables, so as to describe the central tendency and dispersion of the data and test the normality of the data.

Correlation analysis is mainly used to examine the changing trend and linkage direction of the linear relationship between two variables. For continuous variables, the Pearson correlation coefficient is primarily employed in correlation analysis, aiming to explore whether the relationship between variables is positive, negative, or zero correlation. In this study, correlation analysis is used to explore the relationships among objective built-environment elements, subjective built-environment perception elements, and their respective relationships with Subjective Well-being.

(2) Regression Analysis

Regression analysis is an analytical technique used to examine the influence of variables on the dependent variable, aiming to evaluate the statistical significance of the overall model, $R^2$, as well as the direction and strength of the relationships among multiple variables (Hair et al., 2007). To conduct a more rigorous examination of the impact of the built environment on Subjective Well-being, this study employed regression analysis to test the influence of Objective Built Environment elements on residents' Life Satisfaction.

(3) Structural Equation Modeling

The structural equation model consists of two parts: the measurement model and the structural model (Hair et al., 2011; Wu, 2009). The measurement model can be used to extract latent variables from observable indicators, and the estimation method is confirmatory factor analysis (CFA); the structural model is used to analyze the relationships among variables (Byrne, 2009; Kline, 2015). The analysis of the measurement model in SEM is completed through CFA. CFA is an important part of SEM. Before analyzing the structural model, one should start with CFA. After verification and reliability and validity tests of the measurement models for each construct (Brown, 2015; Hu & Bentler, 1999), it can be ensured that the measurement model correctly reflects the constructs of the study (Thompson, 2000), and then SEM is worth conducting (Brown, 2015; Kline, 2015). Structural Equation Modeling (SEM) integrates "factor analysis" and "regression analysis" from traditional multivariate statistical analysis (Kline, 2023). It is a statistical model that can explore the functional relationships between latent exogenous variables and latent endogenous variables, and can conduct model identification, estimation, and verification for various causal models.

To this end, this paper uses a structural equation model to analyze the action path of the built environment on Subjective Well-being. In Chap. 6, the measurement model is first used to measure the latent variables of Perceived Environmental Quality, Social Capital, and Life Satisfaction. Then, a structural path analysis of the structural model is conducted.

(4) Mediating Effect Testing Techniques

① Causal Steps Approach

In structural equation modeling, although there are numerous methods available for testing hypotheses regarding the effects of mediating variables, the most widely used approach is the causal steps approach (Baron & Kenny, 1986; Hayes, 2009). This method requires estimating each path in the model and then determining whether the estimated values meet certain statistical criteria to identify mediating variables. In practical operations, researchers primarily assess whether the paths of a and b are statistically significant, and whether c' is closer to 0. When the total effect of X on Y is also statistically significant, it indicates the existence of a mediating effect (Baron & Kenny, 1986).

However, among the methods for testing the effect of the mediating variable, the causal steps approach also has low efficiency (Fritz & Mackinnon, 2007; Mackinnon et al., 2002; Shrout & Bolger, 2002). Suppose the influence of X on Y is partially indirect through the mediating variable M. In that case, the causal steps approach needs to test the statistical significance of each path, so the efficiency is relatively low in actual testing. In addition, when both a and b are not equal to 0 and are significant, this method identifies that the variable has a mediating effect. However, since the mediating effect is a × b, only testing the significance of a and b without testing the significance of a*b may lead to incorrect conclusions.

② Sobel Test

To address the above issues and quantify the mediating effect, Sobel proposed the product—of—coefficients method (Sobel, 1982, 1986). This method requires estimating the standard errors of a and b (Preacher & Hayes, 2004), and then using the ratio of ab to the standard error as the test statistic to test the existence of the indirect effect. Sobel's method compensates for the shortcomings of the causal steps approach and played an important role for some time. However, a major drawback of the Sobel method is that it assumes the sampling distribution of the indirect effect is normal, while the sampling distribution of ab is usually asymmetric (Bollen & Stine, 1990; Stone & Sobel, 1990). This drawback becomes more prominent when testing multiple mediating effects.

③ Bootstrap Mediation Test

To address this issue, scholars have proposed the Bootstrap and M test methods as alternatives to the Sobel test. The Bootstrap Mediation Test can directly test the product of coefficients and is considered a better option among the two, which has been widely recognized in the academic community (Hayes, 2009; Wen & Ye, 2014). The Bootstrap Mediation Test does not rely on the population distribution and does not require any prior information.[8] Instead, it randomly samples from the population with replacement to increase the sample size, obtaining a stable distribution of the mean, asymmetric confidence intervals, and standard error estimates. Among the confidence intervals estimated by various Bootstrap methods, the bias-corrected non-parametric percentile method (BC) can provide the most accurate confidence interval for the mediating effect (Hayes, 2009; Wen & Ye, 2014).

Drawing on the steps recommended in relevant literature, this study employed the bias-corrected non-parametric percentile estimation method to test the mediating effect of the structural equation model (Fang et al., 2011). According to the suggestions in the literature, the number of samplings was set at 5000 (Hayes, 2009; Preacher & Hayes, 2008). The 95% and 90% confidence intervals of the bias-corrected non-parametric percentile estimation were reported to test the mediating effect.

**Funding** This research was funded by

1. Guangdong Provincial Office of Philosophy and Social Sciences [Grant No. GD24XSH06]
2. Projects of PhDs' Start-up Research of GDUPT [Grant No. 2023bsqd1008].

[8] Bootstrap methods are recommended for mediation analysis due to their robustness to non-normal distributions and higher statistical power (Hayes, 2017; Preacher & Hayes, 2008).

# References


Amap. (2021). *Point of Interest (POI) data for Guangzhou, China* [Data set]. Retrieved from https://www.amap.com

Arcaya, M. C., Tucker-Seeley, R. D., Kim, R., Schnake-Mahl, A., So, M., & Subramanian, S. V. (2016). Research on neighborhood effects on health in the United States: A systematic review of study characteristics. *Social Science & Medicine, 168*, 16–29.

Bai, X., Wu, C., Zheng, R., & Ren, X. (2011). The psychometric evaluation of the satisfaction with life scale using a nationally representative sample of China. *Journal of Happiness Studies, 12*(2), 183–197.

Baidu Maps. (2021). *Point of Interest (POI) data for Guangzhou, China* [Data set]. Retrieved from https://map.baidu.com

Baron, R. M., & Kenny, D. A. (1986). The moderator-mediator variable distinction in social psychological research: Conceptual, strategic, and statistical considerations. *Journal of Personality and Social Psychology, 51*(6), 1173–1182.

Bollen, K. A., & Stine, R. (1990). Direct and indirect effects: Classical and bootstrap estimates of variability. *Sociological Methodology, 20*, 115–140.

Brown, T. A. (2015). *Confirmatory factor analysis for applied research* (2nd ed.). Guilford Press.

Byrne, B. M. (2009). *Structural equation modeling with AMOS: Basic concepts, applications, and programming* (2nd ed.). Routledge.

Cao, X. J. (2016). How does neighborhood design affect life satisfaction? Evidence from twin cities. *Travel Behaviour and Society, 5*, 68–76.

Chaix, B., Merlo, J., & Chauvin, P. (2005). Comparison of a spatial approach with the multi-level approach for investigating place effects on health: The example of healthcare utilisation in France. *Journal of Epidemiology & Community Health, 59*(6), 517–526.

Cheng, Y. (2021). The uncertain geographic context problem (UGCoP) in measuring neighborhood effects: A systematic review. *Urban Science, 5*(2), 45.

Collier, J. (2020). *Applied structural equation modeling using AMOS: Basic to advanced techniques*. Routledge.

Das, K. V. (2020). The influence of neighborhood residential density on satisfaction with life: Evidence from Austin, TX. *Journal of Planning Education and Research, 40*(2), 175–187.

Diener, E. (1984). Subjective Well-being. *Psychological Bulletin, 95*(3), 542–575.

Diener, E. (2009). Assessing subjective Well-being: Progress and opportunities. In E. Diener (Ed.), *The science of Well-being* (pp. 25–65). Springer.

Diener, E., Emmons, R. A., Larsen, R. J., & Griffin, S. (1985). The satisfaction with life scale. *Journal of Personality Assessment, 49*(1), 71–75.

Dillman, D. A., Smyth, J. D., & Christian, L. M. (2014). *Internet, phone, mail, and mixed-mode surveys: The tailored design method*. John Wiley & Sons.

Duncan, D. T., Piras, G., & Aldstadt, J. (2013). The spatial patterning of county-level teen birth rates in the United States. *Demographic Research, 28*, 105–118.

Ewing, R., & Handy, S. (2009). Measuring the unmeasurable: Urban design qualities related to walkability. *Journal of Urban Design, 14*(1), 65–84.

Fang, J., Wen, Z., & Ye, B. (2011). Mediation analysis using structural equation modeling. *Advances in Psychological Science, 19*(5), 765–774.

Feng, J., Glass, T. A., Curriero, F. C., Stewart, W. F., & Schwartz, B. S. (2010). The built environment and obesity: A systematic review of the epidemiologic evidence. *Health & Place, 16*(2), 175–190.

Field, A. (2020). *Discovering statistics using IBM SPSS statistics* (5th ed.). Sage.

Foster, S., Giles-Corti, B., & Knuiman, M. (2010). Neighbourhood design and fear of crime: A social-ecological examination of the correlates of residents' fear in new suburban housing developments. *Health & Place, 16*(6), 1156–1165.

Frank, L. D., Schmid, T. L., Sallis, J. F., Chapman, J., & Saelens, B. E. (2005). Linking objectively measured physical activity with objectively measured urban form: Findings from SMARTRAQ. *American Journal of Preventive Medicine, 28*(Suppl. 2), 117–125.

Fritz, M. S., & Mackinnon, D. P. (2007). Required sample size to detect the mediated effect. *Psychological Science, 18*(3), 233–239.

Gelman, A., & Park, D. K. (2009). Splitting a predictor at the upper quarter or third and the lower quarter or third. *The American Statistician, 63*(1), 1–8.

Grafström, A., & Tillé, Y. (2013). Doubly balanced spatial sampling with spreading and restitution of auxiliary totals. *Environmetrics, 24*(2), 120–131.

Goldberg, D. W., Wilson, J. P., & Knoblock, C. A. (2007). From text to geographic coordinates: The current state of geocoding. *URISA Journal, 19*(1), 33–46.

Gu, H. (2012). The dimensionality of the Chinese version of the life orientation test-revised. *Chinese Journal of Clinical Psychology, 20*(2), 151–154.

Hair, J. F., & Black, W. C. (2005). *Multivariate data analysis* (6th ed.). Prentice Hall.

Hair, J. F., Black, W. C., Babin, B. J., & Anderson, R. E. (2007). *Multivariate data analysis* (7th ed.). Pearson.

Hair, J. F., Ringle, C. M., & Sarstedt, M. (2011). PLS-SEM: Indeed a silver bullet. *Journal of Marketing Theory and Practice, 19*(2), 139–152.

Hayes, A. F. (2009). Beyond Baron and Kenny: Statistical mediation analysis in the new millennium. *Communication Monographs, 76*(4), 408–420.

Hayes, A. F. (2017). *Introduction to mediation, moderation, and conditional process analysis: A regression-based approach* (2nd ed.). Guilford Press.

Hu, L., & Bentler, P. M. (1999). Cutoff criteria for fit indexes in covariance structure analysis: Conventional criteria versus new alternatives. *Structural Equation Modeling: A Multidisciplinary Journal, 6*(1), 1–55.

Idler, E. L., & Benyamini, Y. (1997). Self-rated health and mortality: A review of twenty-seven community studies. *Journal of Health and Social Behavior, 38*(1), 21–37.

Kelley, T. L. (1939). The selection of upper and lower groups for the validation of test items. *Journal of Educational Psychology, 30*(1), 17–24.

Kline, R. B. (2015). *Principles and practice of structural equation modeling* (4th ed.). Guilford Press.

Kline, R. B. (2023). *Principles and practice of structural equation modeling*. Guilford Press.

Kwan, M.-P. (2012). The uncertain geographic context problem. *Annals of the Association of American Geographers, 102*(5), 958–968.

Li, H., Pan, C., Qiu, N., Guo, J., & Wu, J. (2025). Does the multi-scale built environment impact on residents' subjective Well-being? *Buildings, 15*(18), 3311.

Li, Y., & Liu, Y. (2019). Measuring high-density urban form and its evolution in Guangzhou, China. *Environment and Planning B: Urban Analytics and City Science, 46*(5), 871–888.

Lin, L., & Moudon, A. V. (2010). Objective versus subjective measures of the built environment, which are most effective in capturing associations with walking? *Health & Place, 16*(2), 339–348.

Liu, Y., Wang, R., Xiao, Y., Huang, B., & Chen, H. (2018). Exploring the linkage between green space exposure and obesity among Chinese adults. *International Journal of Environmental Research and Public Health, 15*(9), 2058.

MacCallum, R. C., Browne, M. W., & Sugawara, H. M. (1996). Power analysis and determination of sample size for covariance structure modeling. *Psychological Methods, 1*(2), 130–149.

Mackinnon, D. P., Lockwood, C. M., Hoffman, J. M., West, S. G., & Sheets, V. (2002). A comparison of methods to test mediation and other intervening variable effects. *Psychological Methods, 7*(1), 83–104.

Mitchell, C. U., & LaGory, M. (2002). Social capital and mental distress in an impoverished community. *City & Community, 1*(2), 199–222.

Moore, S., & Kawachi, I. (2017). Twenty years of social capital and health research: A glossary. *Journal of Epidemiology and Community Health, 71*(5), 513–517.

Nunnally, J. C., & Bernstein, I. H. (1994). *Psychometric theory* (3rd ed.). McGraw-Hill.

Pavot, W., & Diener, E. (2008). The satisfaction with life scale and the emerging construct of life satisfaction. *The Journal of Positive Psychology, 3*(2), 137–152.

Preacher, K. J., & Coffman, D. L. (2006). *Computing power and minimum sample size for RMSEA* [Computer software]. Available from http://quantpsy.org/.

Preacher, K. J., & Hayes, A. F. (2004). SPSS and SAS procedures for estimating indirect effects in simple mediation models. *Behavior Research Methods, Instruments, & Computers, 36*(4), 717–731.

Preacher, K. J., & Hayes, A. F. (2008). Asymptotic and resampling strategies for assessing and comparing indirect effects in multiple mediator models. *Behavior Research Methods, 40*(3), 879–891.

Putnam, R. D. (2000). *Bowling alone: The collapse and revival of American community.* Simon & Schuster.

Scheier, M. F., Carver, C. S., & Bridges, M. W. (1994). Distinguishing optimism from neuroticism (and trait anxiety, self-mastery, and self-esteem): A reevaluation of the life orientation test. *Journal of Personality and Social Psychology, 67*(6), 1063–1078.

Sekaran, U., & Bougie, R. (2016). *Research methods for business: A skill-building approach* (7th ed.). Wiley.

Shen, Y., & Fu, X. (2019). Impact of the built environment and perceived neighborhood characteristics on residents' satisfaction: Evidence from the Shanghai suburbs. *Progress in Geography, 38*(5), 686–697.

Shrout, P. E., & Bolger, N. (2002). Mediation in experimental and nonexperimental studies: New procedures and recommendations. *Psychological Methods, 7*(4), 422–445.

Sobel, M. E. (1982). Asymptotic confidence intervals for indirect effects in structural equation models. *Sociological Methodology, 13*, 290–312.

Sobel, M. E. (1986). Some new results on indirect effects and their standard errors in covariance structure models. *Sociological Methodology, 16*, 159–186.

Spielman, S. E., & Yoo, E. (2009). The spatial dimensions of neighborhood effects. *Social Science & Medicine, 68*(6), 1098–1105.

Stone, C. A., & Sobel, M. E. (1990). The robustness of estimates of total indirect effects in covariance structure models estimated by maximum likelihood. *Psychometrika, 55*(2), 337–352.

Sun, Y., Du, Y., & Wang, Y. (2019). Spatial disparities of excessive drinking in urban China: A geographical detector study. *Health & Place, 57*, 100–108.

Tao, Y., Yang, J., & Chai, Y. (2020). How does neighborhood social capital affect mental health? Evidence from Beijing. *Social Science & Medicine, 255*, 113000.

Tatem, A. J. (2017). WorldPop, open data for spatial demography. *Scientific Data, 4*(1), 170004.

Thompson, B. (2000). Ten commandments of structural equation modeling. In L. G. Grimm & P. R. Yarnold (Eds.), *Reading and understanding more multivariate statistics* (pp. 261–283). American Psychological Association.

Wen, Z. (2012). The dimensionality of the Chinese version of the life orientation test-revised. *Chinese Journal of Clinical Psychology, 20*(2), 151–154.

Wen, Z., & Ye, B. (2014). Analyses of mediating effects: The development of methods and models. *Advances in Psychological Science, 22*(5), 731–745.

Wong, D. W. (2004). The modifiable areal unit problem (MAUP). In *WorldMinds: Geographical perspectives on 100 problems: Commemorating the 100th anniversary of the Association of American Geographers 1904–2004* (pp. 571–575). Springer.

Wu, M. L. (2009). *Structural equation modeling: Operation and application of AMOS* (2nd ed.). Chongqing University Press.

Yang, J., Wang, H., & Chai, Y. (2021). How does neighborhood social capital influence residents' mental health? A mediation analysis. *Health & Place, 68*, 102533.

Zhang, Y., Li, Y., & Liu, Y. (2019). The relationship between built environment and life satisfaction: Evidence from Beijing. *Journal of Environmental Psychology, 65*, 101335.

Zhang, Y., Li, Y., Liu, Y., & Li, X. (2020). Social capital and life satisfaction: The mediating role of neighborhood cohesion. *Cities, 106*, 102870.

# Chapter 7
# Descriptive Patterns and Initial Insights

**Abstract** This chapter lays a descriptive foundation for analyzing life satisfaction determinants within urban residential contexts, focusing on Guangzhou, China. Its primary objective is to characterize the socioeconomic attributes of the sampled population, evaluate the distribution and normality of key variables—including life satisfaction, perceived built environment quality, objective built environment metrics, social capital, and self-rated health—and preliminarily examine community-level variations in life satisfaction. Section 7.1 analyzes the socioeconomic composition of the valid samples (N = 572), highlighting representativeness in gender, income, and housing conditions relative to official census data. Section 7.2 delves into the measurement and normative properties of core constructs: life satisfaction (using the Satisfaction with Life Scale), subjective built environment perceptions (diverse convenience, traffic convenience, aesthetic perception, and perceived safety), objective built environment indicators (density, diversity, design, transit accessibility, and destination accessibility), social capital, and self-rated health. Rigorous normality checks (skewness and kurtosis thresholds) confirm suitability for parametric analyses. Community-level life satisfaction is further explored, revealing significant inter-community variation despite no notable differences across housing construction eras. By systematically documenting variable distributions and initial patterns, this chapter provides a critical empirical groundwork for subsequent multivariate analyses, underscoring the nuanced roles of both subjective and objective environmental factors in shaping urban residents' well-being.

**Keywords** Life satisfaction · Built environment · Perceived environmental quality · Subjective Well-being · Social capital · Self-rated health · Descriptive analysis · Normality test · Community-level variation · Guangzhou

Following the comprehensive methodological outline presented in Chap. 6, this chapter proceeds to the preliminary analysis of the collected data, offering descriptive insights into the sample characteristics and key variables under study. It begins by examining the socioeconomic profile of respondents and their overall life satisfaction levels, then delves into the objective and perceived built environment, social



H. Li, G. Shen, *The Anatomy of Subjective Well-Being*, Resilient and Friendly Living Environments, https://doi.org/10.1007/978-981-95-4937-5_7

capital, and health status. These initial analyses not only provide a foundational understanding of the data structure and distribution but also prepare the ground for subsequent multivariate and causal analyses.

## 7.1 Characteristic Analysis and Differences of the Life Satisfaction Samples

### *7.1.1 Socioeconomic Characteristics of the Samples*

In the questionnaire, the content of the personal socioeconomic attributes section covers aspects such as the respondents' age, gender, marital status, employment, education level, personal income, per capita housing area, household registration, and social security to examine other characteristics of the respondents. Table 7.1 reports the names, definitions, and other descriptive characteristics of the variables. Among all the valid samples, males account for 51.22% and females account for 48.78%. This proportion is basically consistent with the proportion of males and females in the seventh National Population Census Bulletin, indicating that the gender ratio of the sample is basically in line with the social reality.

Among the respondents participating in the survey, the youngest male was 18 years old, the oldest was 81 years old, and the average age was 39.66 years old; the youngest female was 18 years old, the oldest was 83 years old, and the average age was 41.05 years old. Respondents aged 18–60 accounted for 84.27%, and those over 60 years old accounted for 15.73%. According to the composition of the household-registered population in Guangzhou in the 2020 Statistical Yearbook of Guangzhou, after calculation, the proportion of people aged 18–60 in the central urban area was 71.16%, and the proportion of those over 60 years old was 28.84%. There was a certain difference between the age distribution of the overall sample and this data. The main reason is that the overall sample of respondents was permanent residents over 18 years old, while this data was for the household-registered population. The difference might be caused by the fact that among the permanent residents, a large number of non—Guangzhou household—registered people were aged 18–60.

In terms of marital status, 60.49% of the respondents were married, and 39.51% were in other marital statuses. Regarding educational attainment, 50.87% of the respondents had received a college education (including junior college and undergraduate) or above, while 49.13% had received an education below high school (including secondary vocational school and technical school). In terms of employment, 65.73% of the respondents had jobs. In terms of economic status, 38.11% of the respondents had a personal monthly income of less than 5000 yuan, and 61.89% had a personal monthly income of more than 5001 yuan. According to the Statistical Yearbook of Guangzhou in 2020, the annual per capita disposable income of urban

**Table 7.1** Definitions and descriptive statistics of socioeconomic attribute variables

| Variable | Type | Frequency | Percentage (%) |
|---|---|---|---|
| Age | 18–29 | 193 | 33.74 |
| | 30–39 | 114 | 19.93 |
| | 40–49 | 78 | 13.64 |
| | 50–59 | 97 | 16.96 |
| | ≥ 60 | 90 | 15.73 |
| Gender | Male | 293 | 51.22 |
| | Female | 279 | 48.78 |
| Marriage | Married | 346 | 60.49 |
| | Never married | 178 | 31.12 |
| | Divorced | 12 | 2.10 |
| | Other | 36 | 6.29 |
| Education level | Junior high or below | 130 | 22.73 |
| | Senior high/vocational school | 151 | 26.40 |
| | College (associate/bachelor's) | 247 | 43.18 |
| | Master's or higher | 44 | 7.69 |
| Occupation | Government/public institution | 69 | 12.06 |
| | Enterprise employee | 159 | 27.80 |
| | Freelancer | 148 | 25.87 |
| | Unemployed~ | 31 | 5.42 |
| | Retired | 81 | 14.16 |
| | Other | 84 | 14.69 |
| Monthly income (RMB) | ≤5000 | 218 | 38.11 |
| | 5001 ~ 8000 | 153 | 26.75 |
| | 8001 ~ 10,000 | 73 | 12.76 |
| | 10,001 ~ 13,000 | 66 | 11.54 |
| | ≥13,001 | 62 | 10.84 |
| Hukou status | Local urban hukou | 287 | 50.17 |
| | Non-local urban hukou | 169 | 29.55 |
| | Non-local rural hukou | 95 | 16.61 |
| | Other residency status | 21 | 3.67 |
| Housing tenure | Commercial housing | 418 | 73.08 |
| | Danwei welfare housing | 42 | 7.34 |
| | Resettlement housing | 18 | 3.15 |
| | Subsidized affordable housing | 15 | 2.62 |
| | Urban village housing | 79 | 13.81 |
| Per capita living area ($m^2$) | ≤20m² | 155 | 27.10 |
| | 21 ~ 30m² | 164 | 28.67 |
| | 31 ~ 40m² | 106 | 18.53 |
| | 41 ~ 50m² | 54 | 9.44 |
| | 51 ~ 60m² | 24 | 4.20 |
| | ≥61m² | 69 | 12.06 |

(continued)

**Table 7.1** (continued)

| Variable | Type | Frequency | Percentage (%) |
|---|---|---|---|
| Home ownership status | Privately owned | 419 | 73.25 |
| | Danwei-owned housing | 30 | 5.24 |
| | Informal tenure | 79 | 13.81 |
| | Other tenure types | 44 | 7.69 |

Caption: Marital status categories are defined in accordance with the World Health Organization's International Classification of Diseases (ICD-11) guidelines for personal life circumstances (WHO, J. (2018). International classification of diseases, 11th revision (ICD-11). *Geneva, WHO*). In China, Local urban hukou refers to non-agricultural household registration registered in the target city. Non-local urban hukou refers to household registration registered outside the target city. Non-local rural hukou refers to household registration registered outside the target city. Other household registration status refers to those that cannot be classified, or special household registrations (such as military household registration, and household registrations of Hong Kong, Macao, Taiwan, and foreign nationals). Commercial housing refers to fully-owned housing developed through the market mechanism. Danwei welfare housing refers to the housing allocated by units during the planned economy period. Resettlement housing refers to the housing provided as compensation by the government after land expropriation. Subsidized affordable housing refers to government-subsidized affordable housing (public rental housing/low-rent housing/affordable housing). Urban village housing refers to the housing on collective land remaining from urban expansion. In China, Owner-occupied Housing refers to properties fully owned by individuals. Danwei-owned Housing refers to properties allocated by state-affiliated employers in China's planned economy era. Informal tenure generally refers to the situation where urban migrants live in housing without formal tenure or rental tenure. Other Tenure Types can cover various forms, including Use Rights Housing, Trust Ownership, Co-ownership, etc
Source: Created by the authors

residents was approximately 68,300 yuan[1], indicating that the characteristics of the sample could basically and effectively reflect the actual income situation of residents.

Among the respondents participating in the survey, those with Guangzhou household registration accounted for 50.17%, which was relatively close to the proportion of the registered population in Guangzhou to the permanent population, which was 52.75%. In terms of housing property rights, 73.25% of the respondents had the independent property rights of their current residences. Among the respondents, the average per capita living area was 2.71, falling within the area range of $31 \sim 40\text{m}^2$. According to the Statistical Yearbook of Guangzhou in 2020, the per capita housing area of urban residents was about $34.61\text{m}^2$[2] (Guangzhou Municipal Bureau of Statistics, 2020), indicating that the characteristics of the sample could basically and effectively reflect the actual housing living conditions of residents.

[1] Guangzhou Municipal Bureau of Statistics, *Guangzhou Statistical Yearbook 2020*, China Statistics Press, 2020.

[2] Guangzhou Municipal Bureau of Statistics, *Guangzhou Statistical Yearbook 2020*, China Statistics Press, 2020.

### 7.1.2 *Life Satisfaction Characteristics and Differences of Samples*

In the questionnaire of Life Satisfaction (Fig. 7.1), which is based on the widely used Satisfaction with Life Scale (Diener et al., 1985) to measure the cognitive dimension of subjective well-being (Diener, 1984), five questions mainly reflect "Life ideals", "Life conditions", "Satisfaction with life", "Life gains " and "Life changes", and their scores are added up to obtain the total score. The total score of the five questions should range from 5 to 35 points. The questionnaire results showed that the total scores of respondents ranged from 9 to 35, with a mean of 26.261 and a standard deviation of 4.917. The mean of the total score indicates that the Life Satisfaction of the investigated samples is in the satisfaction interval, which means that most people have a high evaluation of Life Satisfaction.

There are no missing values for all five items. Life ideal (Mean = 5.510, SD = 1.082), life condition (Mean = 5.240, SD = 1.192), Life Satisfaction (Mean = 5.400, SD = 1.135), and life achievement (Mean = 5.200, SD = 1.223), the mean of which was between comparative agreement (5) and agreement (6), indicating that these four aspects were in a higher range. The mean of life change (Mean = 4.910, SD = 1.340) falls between uncertain (4) and comparative agreement (5), indicating closer to comparative agreement in this respect. Table 7.2 reports descriptive statistics on Life Satisfaction. The minimum and maximum values indicate that there is no missing value in this part of the data sample, and the average value is between 4.910 and 5.510, indicating that the identification of Life Satisfaction is between comparative agreement and agreement, that is, Life Satisfaction is between relatively high and high. Before the follow-up analysis, according to the statistical technique requirements adopted in this study, the graphical method and statistical test method were used to test the normality of the dependent variable, Life Satisfaction.

Figure 7.2 presents a visual histogram of the total scores for the five items measuring Life Satisfaction. The graph suggests a distribution that approximately follows a normal curve. To complement this graphical assessment, the skewness and

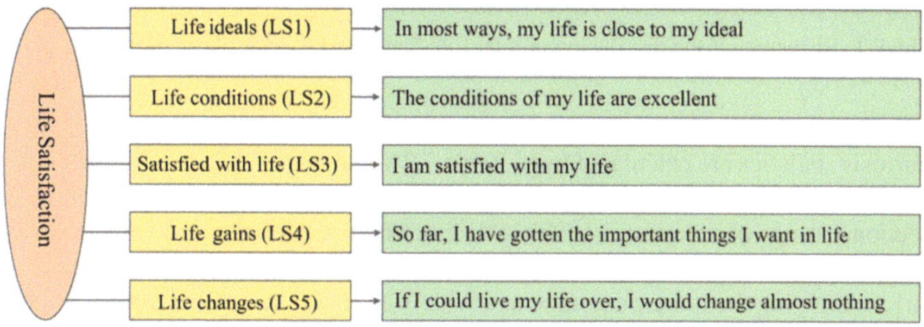


**Fig. 7.1** The five questions of life satisfaction scale. Source: Adapted from Diener et al. (1985)

**Table 7.2** Descriptive statistics of the total score on the satisfaction with life and its subscale items

| | N | Min | Max | Mean | SD | Skewness | Kurtosis |
|---|---|---|---|---|---|---|---|
| LS | 572 | 9.000 | 35.000 | 26.261 | 4.917 | −0.784 | 1.620 |
| LS 1 | 572 | 2.000 | 7.000 | 5.510 | 1.082 | −1.230 | 1.761 |
| LS 2 | 572 | 1.000 | 7.000 | 5.240 | 1.192 | −.937 | .947 |
| LS 3 | 572 | 1.000 | 7.000 | 5.400 | 1.135 | −.917 | 1.525 |
| LS 4 | 572 | 1.000 | 7.000 | 5.200 | 1.223 | −.856 | 1.129 |
| LS 5 | 572 | 1.000 | 7.000 | 4.910 | 1.340 | −.587 | .194 |

Source: Created by the authors

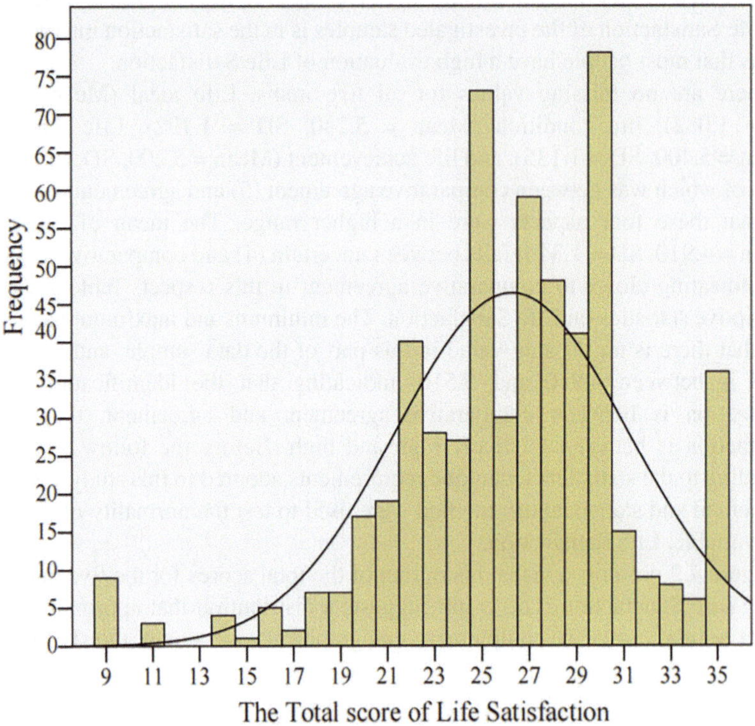


**Fig. 7.2** Histogram of the distribution of total Life Satisfaction scores ($N = 572$). Caption: The curve represents the fitted normal distribution. Source: Created by the authors

kurtosis statistics are calculated (e.g., Table 7.2). The interpretation of these values, however, relies on choosing an appropriate benchmark. As discussed in the previous section, a range of criteria exists. A highly conservative rule of thumb (skewness <1, kurtosis <3) would be met by this dataset, but such a standard is often unnecessarily strict. Applying the more pragmatic and widely accepted thresholds for general linear models (skewness <2, kurtosis <7; Finney & DiStefano, 2006) would also lead to a conclusion of no severe violation. Furthermore, since the analysis plan involves structural equation modeling—a technique robust to minor deviations—the even more liberal thresholds suggested by Kline (2015; skewness < 3, kurtosis < 10) are highly applicable. The computed values for the life satisfaction variable fall well within these lenient limits. This example underscores a key practical recommendation: for applied research, prioritizing absolute benchmarks over statistical significance tests is more informative. While one could calculate z-scores for skewness and kurtosis, this method's well-documented sensitivity in large samples (Curran et al., 1996) makes it less reliable for assessing the practical severity of non-normality. The case of the life satisfaction data demonstrates how aligning the choice of criteria with the research context (e.g., the statistical techniques employed) leads to a sound and reasonable conclusion that the data satisfactorily meet the normality assumption. As shown in Table 7.2, the computed values for the life satisfaction variable fall well within the liberal thresholds, indicating that the data satisfactorily meet the assumption of normality for the planned analyses.

The frequency analysis of the life satisfaction scores (N = 572) revealed a score range from 9 to 35. The distribution was slightly negatively skewed, which is consistent with the overall high levels of reported satisfaction. The mode was 30 (N = 78, 13.6%). The scores for the majority of respondents were clustered in the moderate to high range, with the most frequent scores being 25 (12.8%), 27 (10.3%), 28 (8.4%), 29 (7.2%), and 30 (13.6%). The cumulative percentages showed that only 6.6% of respondents scored below 20, whereas a significant majority, 87.1%, scored 20 or above (please verify this percentage based on your actual data, as 87.1% scoring 20+ seems very high if 87.1% was intended for 30+).

As illustrated in Fig. 7.2 and Table 7.3, the frequency distribution is plotted against a normal curve. This visualization confirms that, despite a slight negative skew, the distribution of life satisfaction scores aligns sufficiently with a normal distribution based on the applied criteria (Byrne, 2016; Finney & DiStefano, 2006; Kline, 2015).

As depicted in Fig. 7.3, the frequency distribution and normal curve are presented for Life Ideals (LS 1) among the sampled respondents. The corresponding frequency analysis indicates that the most frequently selected response was option 6 ("agree"), which was endorsed 279 times, accounting for 48.8% of the total responses. This was followed by option 5 ("somewhat agree"), selected 152 times (26.6%), and option 2 ("somewhat disagree"), recorded 14 times (2.4%). The normal curve reveals that the distribution of life ideal ratings is slightly left-skewed, yet continues to satisfy the assumptions of normality according to standard statistical criteria. Overall, both the frequency distribution and the normal curve demonstrate considerable variation at the individual level in respondents' evaluations of their life ideals, suggesting meaningful diversity in perceived attainment of life aspirations.

As illustrated in Fig. 7.4, the frequency distribution and normal curve are presented for Life Conditions (LS 2) among the sample respondents. The results of the frequency analysis showed that the most frequently selected response was option 6 ("agree"), which was endorsed 207 times, accounting for 36.2% of the total responses. This was followed by option 5 ("somewhat agree"), selected 193 times (33.7%), and option 1 ("strongly disagree"), which received 6 responses (1.0%).

**Table 7.3** Frequency analysis of the total score on the life satisfaction and its subscale items

| Score | Frequency | Percent | Valid Percent | Cumulative Percent |
|---|---|---|---|---|
| 9.000 | 9.000 | 1.600 | 1.600 | 1.600 |
| 11.000 | 3.000 | 0.500 | 0.500 | 2.100 |
| 14.000 | 4.000 | 0.700 | 0.700 | 2.800 |
| 15.000 | 1.000 | 0.200 | 0.200 | 3.000 |
| 16.000 | 5.000 | 0.900 | 0.900 | 3.800 |
| 17.000 | 2.000 | 0.300 | 0.300 | 4.200 |
| 18.000 | 7.000 | 1.200 | 1.200 | 5.400 |
| 19.000 | 7.000 | 1.200 | 1.200 | 6.600 |
| 20.000 | 17.000 | 3.000 | 3.000 | 9.600 |
| 21.000 | 19.000 | 3.300 | 3.300 | 12.900 |
| 22.000 | 40.000 | 7.000 | 7.000 | 19.900 |
| 23.000 | 28.000 | 4.900 | 4.900 | 24.800 |
| 24.000 | 27.000 | 4.700 | 4.700 | 29.500 |
| 25.000 | 73.000 | 12.800 | 12.800 | 42.300 |
| 26.000 | 30.000 | 5.200 | 5.200 | 47.600 |
| 27.000 | 59.000 | 10.300 | 10.300 | 57.900 |
| 28.000 | 48.000 | 8.400 | 8.400 | 66.300 |
| 29.000 | 41.000 | 7.200 | 7.200 | 73.400 |
| 30.000 | 78.000 | 13.600 | 13.600 | 87.100 |
| 31.000 | 15.000 | 2.600 | 2.600 | 89.700 |
| 32.000 | 9.000 | 1.600 | 1.600 | 91.300 |
| 33.000 | 8.000 | 1.400 | 1.400 | 92.700 |
| 34.000 | 6.000 | 1.000 | 1.000 | 93.700 |
| 35.000 | 36.000 | 6.300 | 6.300 | 100.000 |
| Total | 572 | 100.000 | 100.000 | |

Source: Created by the authors

The normal curve indicates that the distribution of Life Conditions is moderately left-skewed, yet still conforms to normality according to standard statistical tests. Taken together, the frequency distribution and normal curve not only reflect a concentration of positive evaluations but also suggest considerable heterogeneity in respondents' evaluations of their life conditions at the individual level.

As shown in Fig. 7.5, the frequency distribution and normal curve are presented for Satisfaction with Life (LS 3) among the sample respondents. The frequency analysis revealed that the most common response was option 6 ("agree"), which was selected 206 times, accounting for 36.0% of the total. This was followed by option 5 ("somewhat agree"), chosen 187 times (32.7%). Additionally, options 1 ("strongly disagree") and 2 ("somewhat disagree") were each selected 8 times, together accounting for 1.4% of responses. The normal curve suggests that the distribution of Satisfaction with Life is moderately negatively skewed; however, it still conforms to the normality assumption based on standard statistical tests.

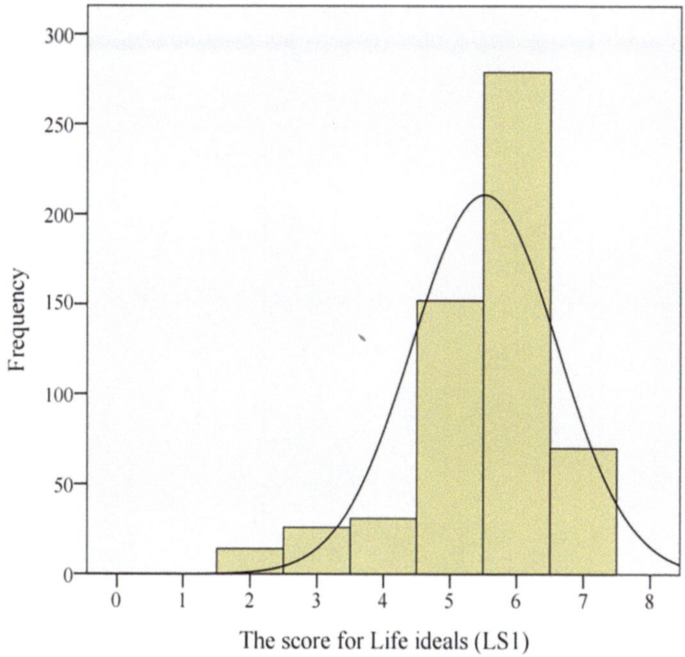


**Fig. 7.3** Histogram of the distribution for the score of life ideals. Caption: The curve represents the fitted normal distribution. Source: Created by the authors

The overall shape of the distribution, along with the skewness observed, indicates that while most responses are concentrated on the higher end of the scale, there remains noticeable individual variability in how respondents assess their life satisfaction.

As depicted in Fig. 7.6, the frequency distribution and normal curve are presented for Life Gains (LS 4) among the sampled respondents. The frequency analysis reveals that the most frequently selected response was option 5 ("somewhat agree"), which was chosen 198 times, accounting for 34.6% of the total responses. This was followed by option 6 ("agree"), selected 177 times (30.9%), and option 1 ("strongly disagree"), recorded 8 times (1.4%). The normal curve suggests a moderately left-skewed distribution for Life Gains, yet the data still conform to normality based on established statistical testing. Therefore, the frequency distribution and normal curve collectively indicate notable individual-level variation in respondents' perceptions of their life gains, reflecting divergent experiences and evaluations among the participants.

As illustrated in Fig. 7.7, the frequency distribution and normal curve are presented for Life Changes (LS 5) among the sampled respondents. Results of the frequency analysis showed that the most frequently selected response was option 5 ("somewhat agree"), which was chosen 162 times, accounting for 28.3% of the total responses. This was followed by option 6 ("agree"), selected 154 times (26.9%),

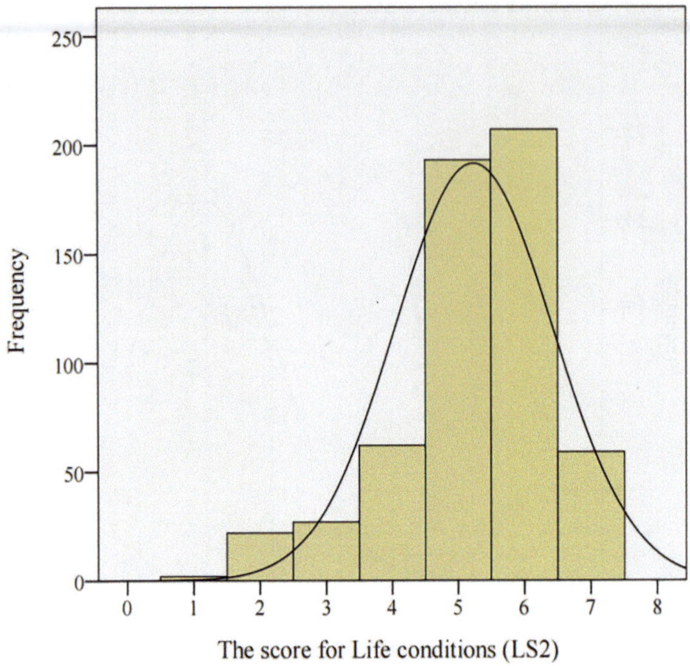


**Fig. 7.4** Histogram of the distribution for the score of life conditions. Caption: The curve represents the fitted normal distribution. Source: Created by the authors

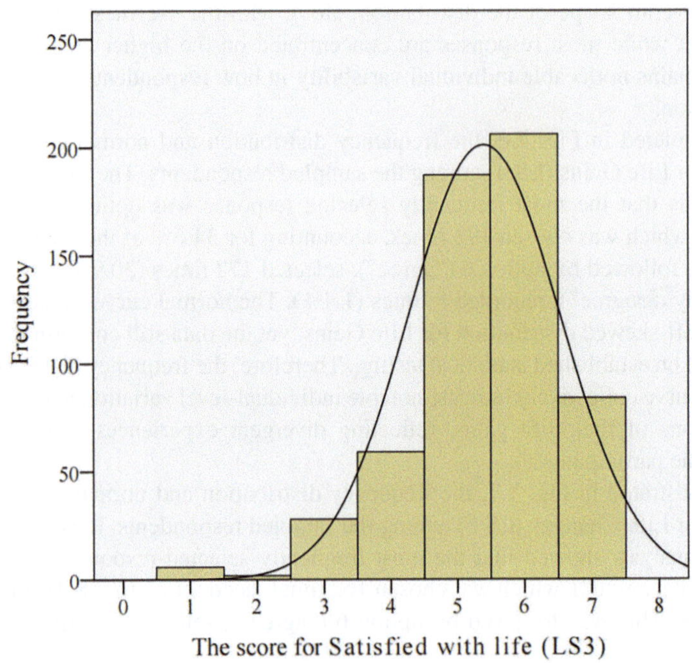


**Fig. 7.5** Histogram of the distribution for the score of satisfied with life. Caption: The curve represents the fitted normal distribution. Source: Created by the authors

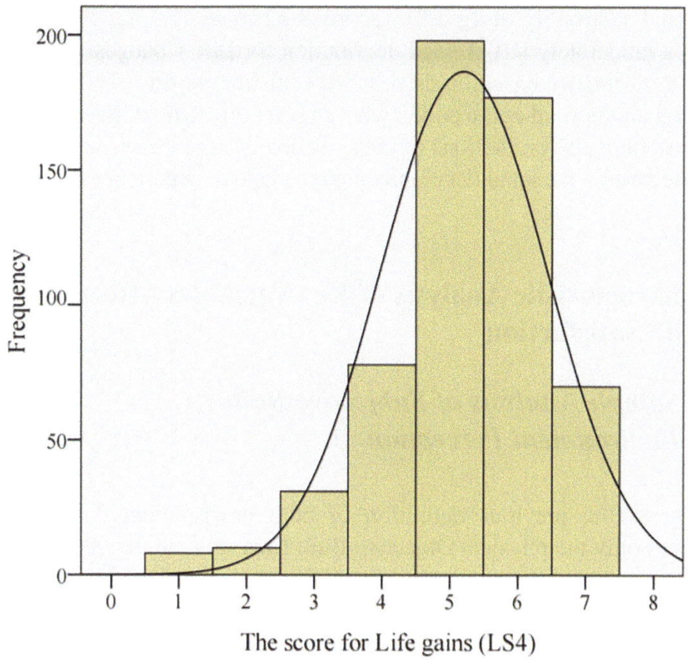


**Fig. 7.6** Histogram of the distribution for the score of life gains. Caption: The curve represents the fitted normal distribution. Source: Created by the authors

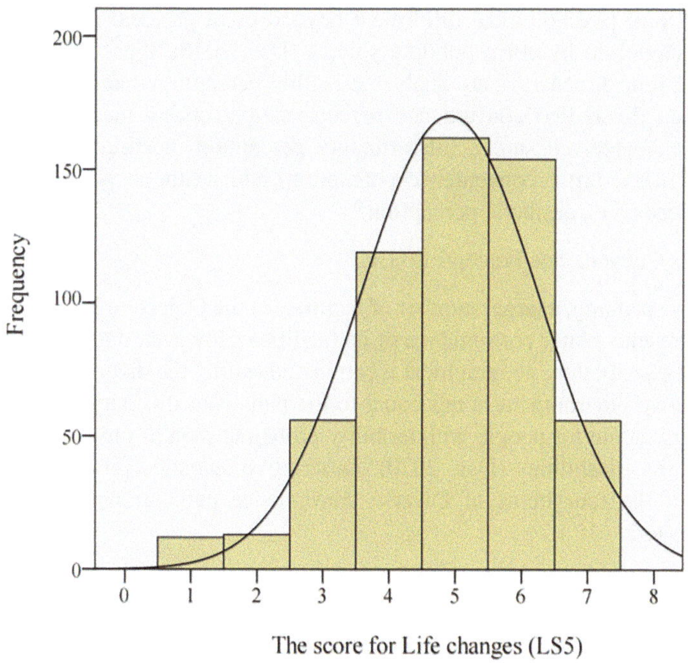


**Fig. 7.7** Histogram of the distribution for the score of life changes scores. Caption: The curve represents the fitted normal distribution. Source: Created by the authors

and option 1 ("strongly disagree"), recorded 12 times (2.1%). The normal curve indicates a moderately left-skewed distribution for Life Changes, yet the data still conform to normality based on established statistical testing. Thus, the frequency distribution and normal curve collectively suggest discernible individual-level variation in respondents' evaluations of their life changes, reflecting a spectrum of perceptions regarding personal development and adaptive experiences.

## 7.2 Characteristic Analysis of Key Variables Affecting Life Satisfaction

### *7.2.1 Sample Analysis of Subjective Built Environment Perception*

According to the previous definition of built environment, built environment includes not only the physical Objective Built Environment, but also the subjective built environment. Although there are errors in subjective characteristics of the built environment, subjective characteristics of the built environment may have more direct and significant effects on Subjective Well-being than Objective Built Environment characteristics (Cao & Zhang, 2016; Hur & Nasar, 2014). Different studies will design targeted indicators based on case and data conditions, so the model cannot be simply copied (Cao & Zhang, 2016). This emphasis on subjective perception aligns with Maslow's (1943) hierarchy of needs, which highlights the importance of psychological fulfillment beyond mere physical conditions, a connection supported by corresponding studies (Das, 2020; Shen & Fu, 2019). The following four dimensions are analyzed: facility perception meeting basic survival needs, i.e., diversified convenience perception, perception meeting basic safety needs, i.e., perceived safety, infrastructure perception meeting travel and social activities, i.e., traffic convenience perception, and aesthetic perception meeting respect needs, i.e., aesthetic perception.[3]

1. Divers Convenience Perception (DCP)

Generally speaking, a larger number of facilities in the Objective Built Environment often represents better convenience of its facilities. However, due to the limitations of spatial distribution, geographical location and traffic, the diversity of facilities in the objective environment is not equal to the perceived diversity. This problem is particularly acute for people with mobility problems such as physical impairments and walking disabilities (Das, 2020). Descriptive statistics for the measurement sample of the four items of Diverse convenience perception are shown in the table below.

[3] Shen, Y., & Fu, X. (2019). Impact of the built environment and perceived neighborhood characteristics on residents' satisfaction: Evidence from the Shanghai suburbs. Progress in Geography, 38(5), 686–697.

Table 7.4 shows descriptive statistics for the four items on Diverse convenience perception. The minimum value in the table is 1.000 and the maximum value is 7.000, indicating no missing values in the data. The average number in the table ranged from 5.630 to 5.810, indicating that residents 'evaluation of diversified convenience was between very satisfactory and satisfactory, and closer to the satisfactory level. The results of Table 7.3 show that the data basically conform to the normal distribution. More stringent normality test skewness and kurtosis are reported in the table and verified to meet the corresponding recommended standards. DCP2 is the lowest in the table, indicating that residents 'evaluation of "leisure sports facilities" is lower than the other three items. DCP1 has the largest average, indicating that residents 'evaluation of "commercial service facilities" is higher than the other three items, and it is closer to the satisfaction level. Grade difference/standard deviation between 1.204 ~ 1.273, indicating that the discrimination of the topic is good.

Figures 7.8, 7.9, 7.10 and 7.11 respectively present the frequency distributions with the normal curve for the scores of the diversity and convenience of business service facilities, leisure sports facilities, educational facilities, and health facilities. Furthermore, the histogram indicated an approximately bell-shaped distribution. It was therefore concluded that the deviation from normality was not substantial enough to undermine the assumptions of subsequent parametric analyses.

2. Traffic Convenience Perception (TCP)

Traffic convenience perception mainly measures the convenience evaluation of public transport travel. As with the Objective Built Environment, perceived traffic convenience is measured in three aspects, namely rail transit, ground public transport and individualized traffic travel. The measurement topics include three topics: "Evaluation of community bus stop convenience (TCP1)", "Evaluation of community subway stop convenience (TCP2)" and "Evaluation of smooth traffic flow (TCP3)". Table 7.5 shows descriptive statistics for the three questions of traffic convenience perception. The minimum value in the table is 1.000 and the maximum value is 7.000, indicating no missing values in the data. The average number in the table ranged from 5.490 to 5.860, indicating that residents 'evaluation of diversified convenience was between satisfactory and very satisfactory. TCP1 has the smallest average in the table, indicating that residents rated "ease of driving (no congestion)" lower than the other two items. TCP1 has the largest average, indicating that residents 'evaluation of "bus stop convenience" is higher than the other two items, which is closer to the satisfaction level. The grade difference/standard deviation is between 1.031 and 1.243, which indicates that the discrimination of the questions is good, the skewness and kurtosis meet the corresponding suggested standards, and the test conforms to the normal distribution.

Figures 7.12, 7.13 and 7.14 respectively, present the frequency distributions with the normal curve for the scores of the accessibility of bus stops, subway stations, and smooth traffic flow. Furthermore, the histogram indicated an approximately bell-shaped distribution. It was therefore concluded that the deviation from

**Table 7.4** Descriptive statistics for multiple convenience perceptions

| | N | Min | Max | Mean | SD | Skewness | Kurtosis |
|---|---|---|---|---|---|---|---|
| DCP1 | 572 | 1.000 | 7.000 | 5.810 | 1.204 | −1.511 | 2.853 |
| DCP2 | 572 | 1.000 | 7.000 | 5.630 | 1.262 | −1.007 | .824 |
| DCP3 | 572 | 1.000 | 7.000 | 5.640 | 1.273 | −1.017 | .725 |
| DCP4 | 572 | 1.000 | 7.000 | 5.800 | 1.247 | −1.523 | 2.643 |

Source: Created by the authors

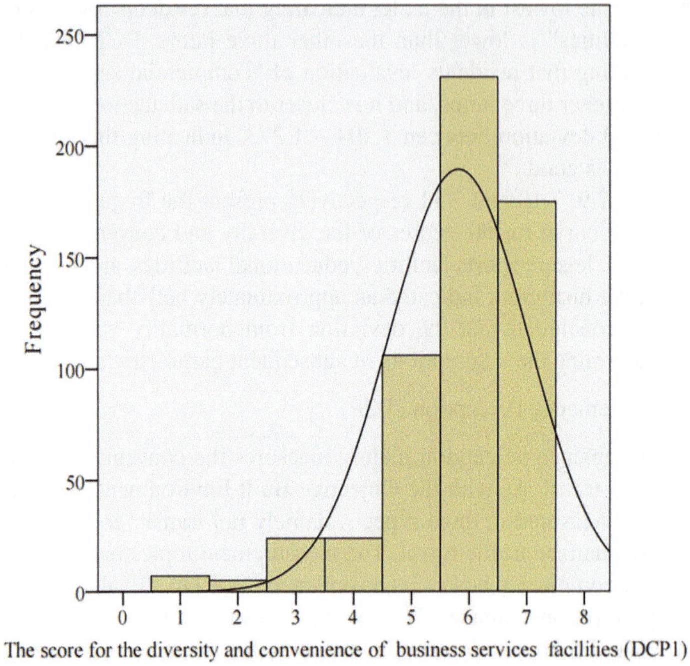


**Fig. 7.8** Histogram of the distribution for the score for the diversity and convenience of business services facilities. Caption: The curve represents the fitted normal distribution. Source: Created by the authors

normality was not substantial enough to undermine the assumptions of subsequent parametric analyses.

3. Aesthetic Perception (AEP)

Aesthetic perception of the built environment mainly measures residents 'aesthetic evaluation of the environment. The measurement topics include "aesthetic evaluation of community park green space (AEP1)", "aesthetic evaluation of street landscape (AEP2)" and "aesthetic evaluation of community building appearance (AEP3)". Table 7.6 shows descriptive statistics for the three aesthetic perception items. The minimum value in the table is 1.000 and the maximum value is 7.000, indicating no missing values in the data. The average number in the table ranged from 5.370 to 5.470, indicating that residents 'evaluation of aesthetics and

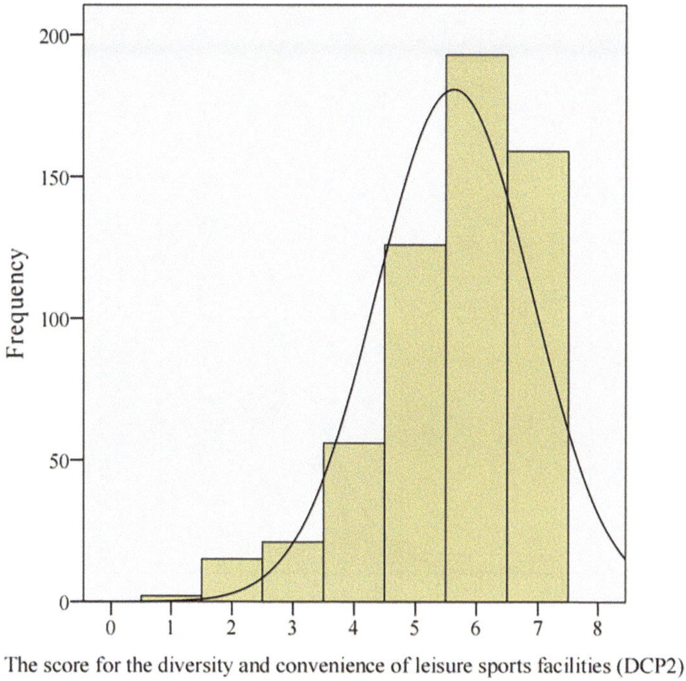


**Fig. 7.9** Histogram of the distribution for the score of the diversity and convenience of leisure sports facilities. Caption: The curve represents the fitted normal distribution. Source: Created by the authors

convenience was concentrated in the middle level of satisfaction and satisfaction. AEP2 has the smallest average in the table, indicating that residents rated "aesthetics of street view" lower than the other two items. AEP1 had the largest average, indicating that residents rated the aesthetics of park green space higher than the other two items. The grade difference/standard deviation is between 1.222 and 1.330, which indicates that the discrimination of the questions is good, the skewness and kurtosis meet the corresponding suggested standards, and the test conforms to the normal distribution.

Figures 7.15, 7.16 and 7.17 respectively present the frequency distributions with the normal curve for the scores of the aesthetics of park and green spaces, and the community streetscapes, the community buildings. Furthermore, the histogram indicated an approximately bell-shaped distribution. It was therefore concluded that the deviation from normality was not substantial enough to undermine the assumptions of subsequent parametric analyses.

4. Perceived Safety (PES)

Perceived Safety (PES) of the built environment mainly measures residents' safety evaluation of the environment, which is influenced by factors such as lighting and surveillance (Foster et al., 2010). The measurement topics include "The

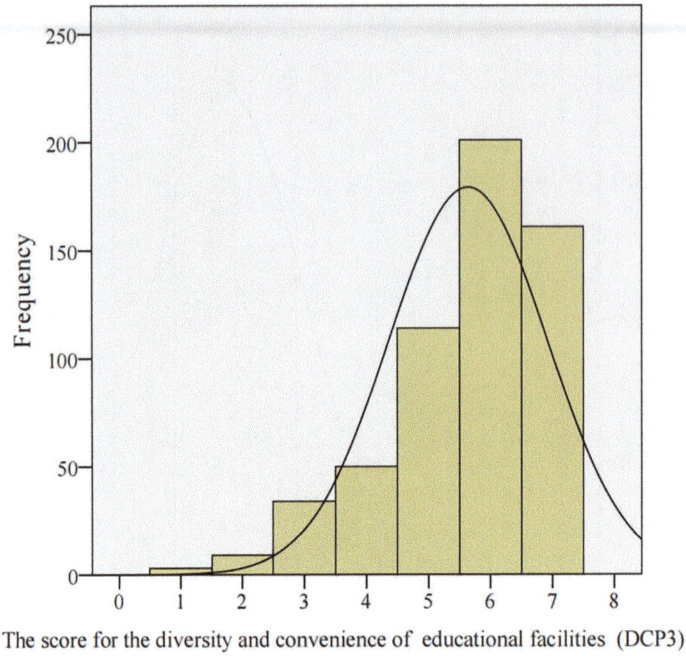


**Fig. 7.10** Histogram of the distribution for the score of the diversity and convenience of educational facilities. Caption: The curve represents the fitted normal distribution. Source: Created by the authors

evaluation of the street lighting equipment (PES1)", "The evaluation of the policing equipment (PES2)", " The evaluation of the environmental security and crime safety (PES3)". Table 7.7 shows descriptive statistics for the three perceived Safety items. The minimum value in the table is 1.000 and the maximum value is 7.000, indicating no missing values in the data. The average number in the table ranged from 5.590 to 5.640, indicating that residents 'evaluation of perceived safety was concentrated in the middle level of satisfaction and satisfaction. PES1 has the smallest average in the table, indicating that residents rated "the evaluation of the street lighting equipment" lower than the other two items. PES3 had the largest average, indicating that residents rated the evaluation of the environmental security and crime safety higher than the other two items. The grade difference/standard deviation is between 1.210 and 1.267, which indicates that the discrimination of the questions is good, the skewness and kurtosis meet the corresponding suggested standards, and the test conforms to the normal distribution.

Figures 7.18, 7.19 and 7.20 respectively present the frequency distributions with the normal curve for the evaluation score of the street lighting equipment, the policing equipment, the environmental security and crime safety. Furthermore, the histogram indicated an approximately bell-shaped distribution. It was therefore concluded that the deviation from normality was not substantial enough to undermine the assumptions of subsequent parametric analyses.

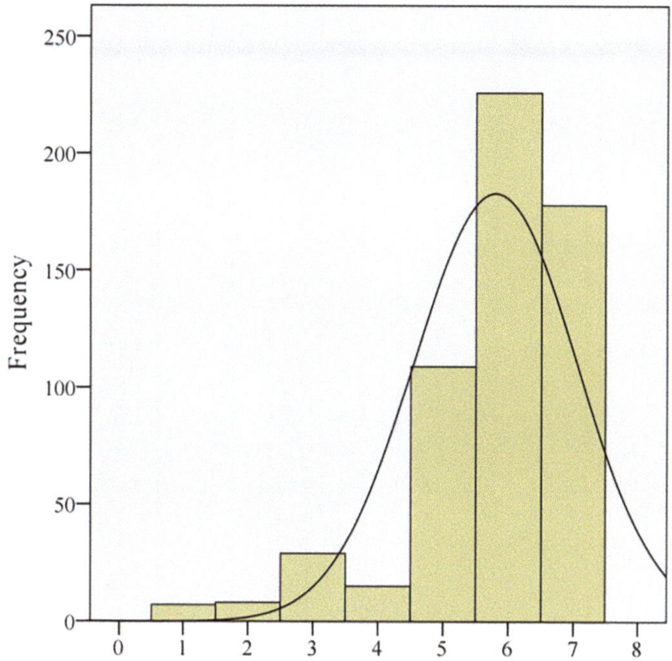


**Fig. 7.11** Histogram of the distribution for the score of the diversity and convenience of health facilities. Caption: The curve represents the fitted normal distribution. Source: Created by the authors

**Table 7.5** Descriptive statistics for traffic convenience perception

| | N | Min | Max | Mean | SD | Skewness | Kurtosis |
|---|---|---|---|---|---|---|---|
| TCP1 | 572 | 2.000 | 7.000 | 5.860 | 1.031 | −1.224 | 1.735 |
| TCP2 | 572 | 1.000 | 7.000 | 5.680 | 1.243 | −1.380 | 2.053 |
| TCP3 | 572 | 1.000 | 7.000 | 5.490 | 1.221 | −1.065 | 1.159 |

Source: Created by the authors

## *7.2.2 Community Objective Built Environment Measurement and Data Characteristics*

1. Density

According to the "5D" model (Density, Diversity, Design, Distance to transit, Destination accessibility) initially proposed by Cervero and Kockelman (1997)[4], density mainly refers to population density, that is, the number of people per unit of

[4] Ewing, R., & Cervero, R. (2010). Travel and the built environment: A meta-analysis. *Journal of the American planning association*, *76*(3), 265-294.

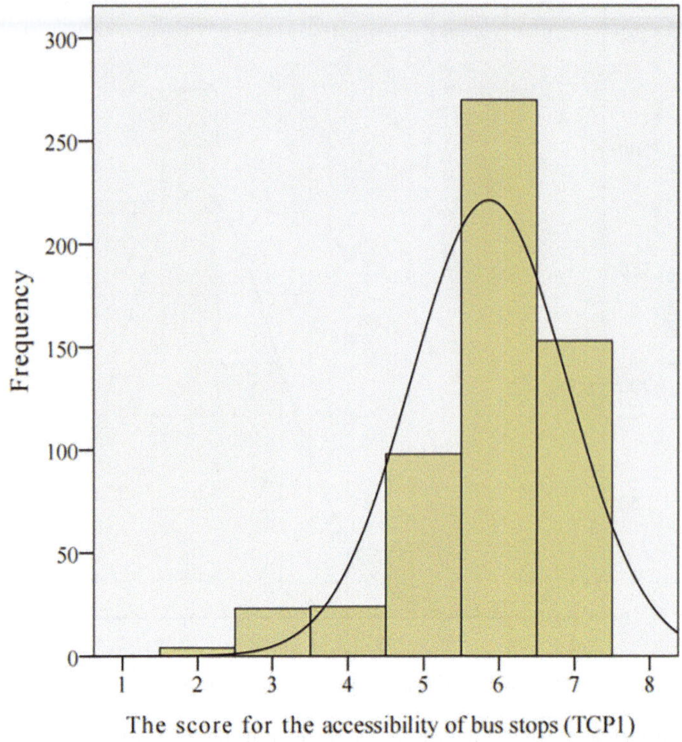


**Fig. 7.12** Histogram of the distribution for the score of the accessibility of bus stops. Caption: The curve represents the fitted normal distribution. Source: Created by the authors

land area, expressed in 10,000 people/km$^2$. The average population density of the surveyed communities is 4262 Persons/km$^2$, and the community with the lowest population density is 3030 persons/km$^2$. It is located at the junction of the Pearl River and Haizhu Wetland Park, and its buffer zone divides the Pearl River and Wetland Park into smaller population density. As important as population density, building density is often considered in the built environment literature. Building density is the ratio of the sum of the building base area (km$^2$) to the geographical unit area (km$^2$). Among the communities surveyed, the average building density was 31%, the lowest community was 13%, and the highest community was 49%.

2. Diversity

Diversity captures the degree to which different land uses are mixed within a given area.—Cervero & Kockelman (1997, Transportation Research Part D). Due to the limited availability of data, this study used Functional Mix Diversity as a proxy variable. POI categories in the study area include accommodation service facilities, scientific research and education service facilities, catering services, financial service facilities, government agencies, tourism (scenic spots), shopping service

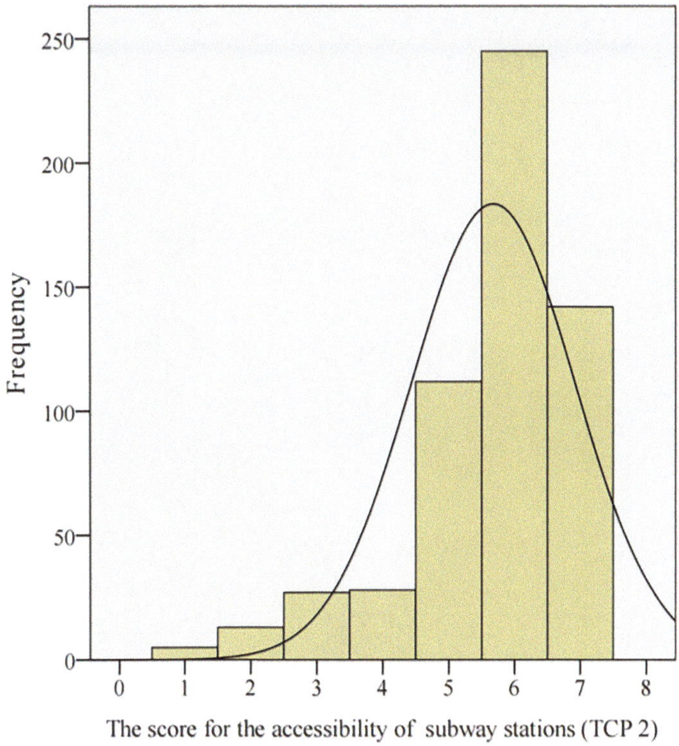


**Fig. 7.13** Histogram of the distribution for the score for the accessibility of subway stations. Caption: The curve represents the fitted normal distribution. Source: Created by the authors

facilities, leisure and entertainment service facilities, large shopping malls, and other facilities. The calculation formula is:

$$MIX = \sum_{i=1}^{n} S_i \ln S_i$$

In this formula, $S_i$ represents the proportion of the number of POI in category $i$ to the total number of POI in the 1000-meter buffer, $n$ represents the number of all POI in the buffer, and the larger the MIX value, the greater the degree of Functional Mix Diversity. Within the communities surveyed, the average Functional Mix Diversity was 0.73, the lowest community was 0.46, and the highest was 0.86.

3. Design

Street design mainly emphasizes the constituent attributes of the street network (such as connectivity and intersection density), often measured through metrics like walkability indices (Ewing & Handy, 2009), but does not typically emphasize aesthetic attributes. Limited by the availability of data, street design mainly includes two aspects: road network density and intersection density of street connections. In

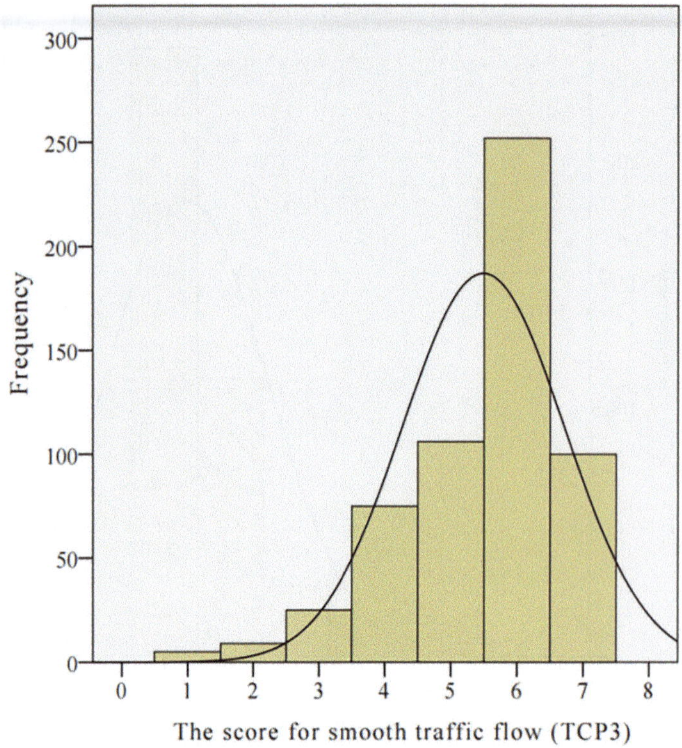


**Fig. 7.14** Histogram of the distribution for the score for smooth traffic flow. Caption: The curve represents the fitted normal distribution. Source: Created by the authors

**Table 7.6** Descriptive statistics for aesthetic perception

| | N | Min | Max | Mean | SD | Skewness | Kurtosis |
|---|---|---|---|---|---|---|---|
| AEP1 | 572 | 2.000 | 7.000 | 5.470 | 1.222 | −1.018 | .594 |
| AEP2 | 572 | 1.000 | 7.000 | 5.370 | 1.325 | −.906 | .430 |
| AEP3 | 572 | 1.000 | 7.000 | 5.390 | 1.330 | −.860 | .170 |

Source: Created by the authors

related studies, road network density is often regarded as a proxy variable for street connectivity, and the logic is that higher road network density usually means better street connectivity, which can increase the route choice of road traffic to the destination, thus improving the destination accessibility. The measure of road network density, a key component in walkability indices (Frank et al., 2010), mainly refers to the ratio of the total length of all roads in the investigated community, such as major roads, minor roads and other roads, to the total area of the community, i.e., expressed in $km/km^2$.

Intersection density is another important aspect of street design, and many studies use intersection density as a proxy variable for street design. Calculated as the

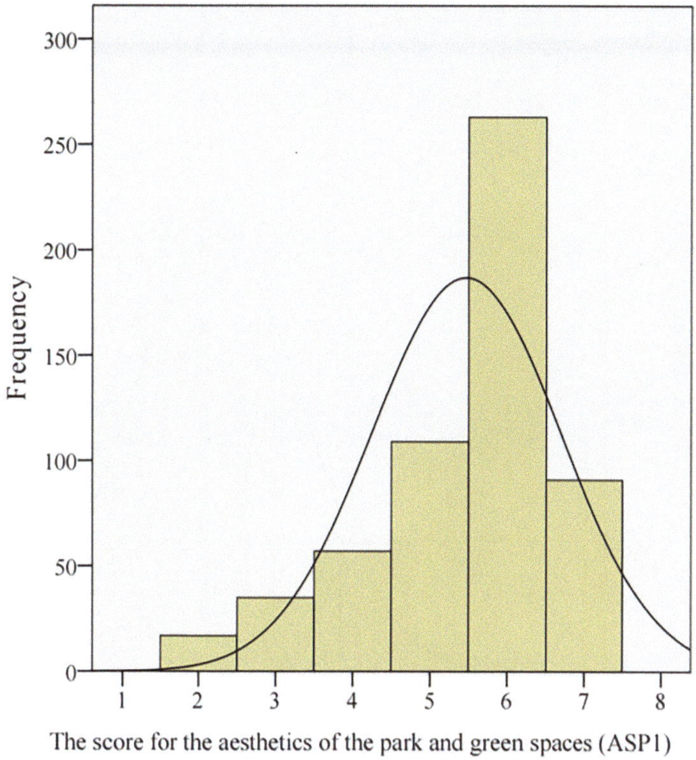


**Fig. 7.15** Histogram of the distribution for the score for the aesthetics of park and green spaces. Caption: The curve represents the fitted normal distribution. Source: Created by the authors

ratio of the number of road intersections to the area of geographic units (units/km$^2$). The road intersections summarized in this study are based on the number of intersections; that is, the intersection statistics are 4 intersections, and the T-intersection statistics are 3 intersections. The more intersections, the better the accessibility of the street. Within the communities surveyed, the average intersection density was 142.61/km$^2$, with a minimum of 14.65/km$^2$ and a maximum of 359.24/km$^2$.

4. Distance to Transit

At present, most of the related researches on the built environment, as summarized in meta-analyses (e.g., Ewing & Cervero, 2010), consider the influence of bus distance or bus line density on health, living satisfaction and Life Satisfaction. Distance to the bus stop is mainly used to measure a public transportation trip. In this study, two kinds of variables are measured: ground transportation (bus) and rail transportation (subway). Ground transportation distance is measured as the distance from the centroid of the research community to the nearest bus stop, while rail transportation distance is measured as the distance from the centroid of the research community to the nearest subway stop. To a large extent, the number of bus stops

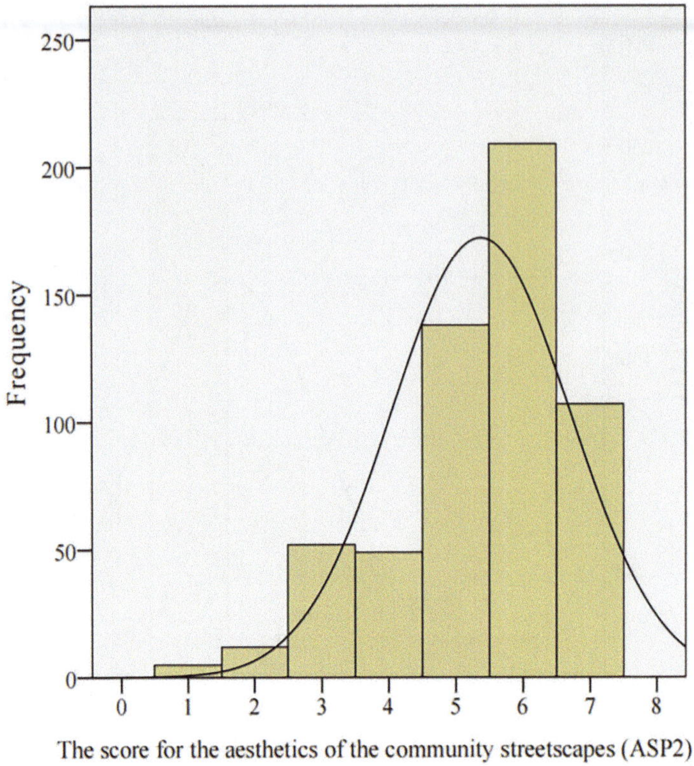


**Fig. 7.16** Histogram of the distribution for the score for the aesthetics of the community streetscapes. Caption: The curve represents the fitted normal distribution. Source: Created by the authors

and subway stops within the buffer zone of 1000 meters of the research community centroid means that the distance from the research community to the bus stop or subway stop is relatively close. Therefore, considering the complex travel choices of individual residents, the density of bus stops and subway stops can be used as proxy variables for distance. Within the sample communities surveyed, the average bus stop density is 8.35/km$^2$, the maximum is 16.88/km$^2$, and the minimum is 1.27/km$^2$. Considering the complexity of residents 'travel choices, bus stop density and subway stop density are combined and called bus stop density.

Strictly speaking, the number of bus stops and subway stops or the high density of bus stops does not necessarily mean that the range of destinations that can be reached is wider, while the density of bus lines and subway lines is are important variable for travel accessibility and convenience. In this study, bus line density was considered as a variable of the built environment to explore Subjective Well-being. The measurement of bus line density mainly refers to the ratio of the total road length of all bus lines in the survey community to the total area of the area, i.e., expressed in km/km$^2$. Within the communities surveyed, the average bus line density is 9.77 km/km$^2$, the maximum is 14.94 km/km$^2$, and the minimum is 2.24 km/km$^2$.

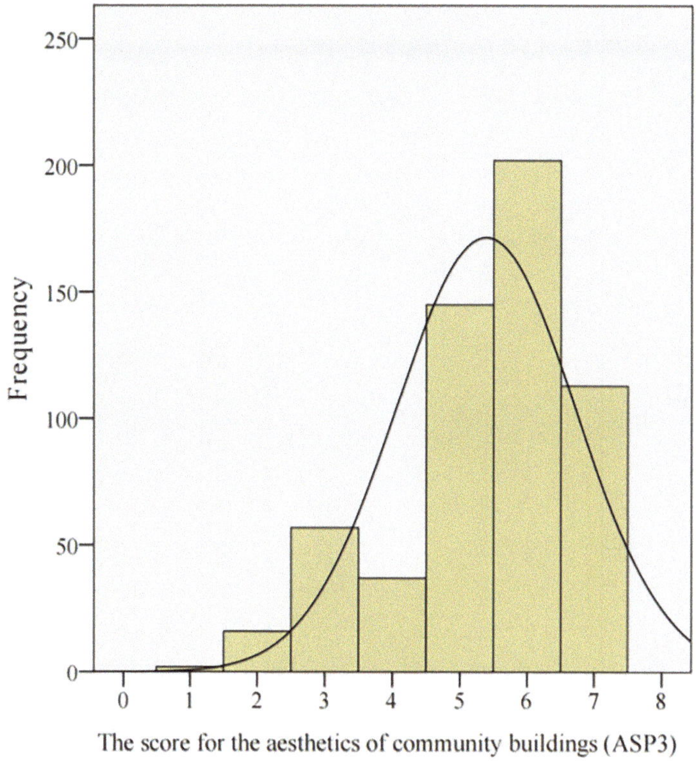


**Fig. 7.17** Histogram of the distribution for the score for the aesthetics of the community buildings. Caption: The curve represents the fitted normal distribution. Source: Created by the authors

**Table 7.7** Descriptive statistics for perceived safety

| | N | Min | Max | Mean | SD | Skewness | Kurtosis |
|---|---|---|---|---|---|---|---|
| PES1 | 572 | 1.000 | 7.000 | 5.620 | 1.210 | −1.256 | 1.753 |
| PES2 | 572 | 1.000 | 7.000 | 5.590 | 1.267 | −1.188 | 1.281 |
| PES3 | 572 | 1.000 | 7.000 | 5.640 | 1.236 | −1.018 | .818 |

Caption: PES1 refers to the evaluation of the street lighting equipment
PES2 refers to the evaluation of the policing equipment. PES3 refers to the evaluation of environmental security and crime safety

5. Destination Accessibility

Destination accessibility refers primarily to the ease with which important destinations are reached. POI density is often used to quantify the spatial distribution of urban facilities. POI refers to a point of interest, meaning a point of interest that is not geographically meaningful on a map. POI density refers to the number of points of interest (POI) per unit area, reflecting the concentration of urban functions and service accessibility. POI density in this study refers to the ratio of the sum of the number of POI points in 10 categories, including accommodation, catering,

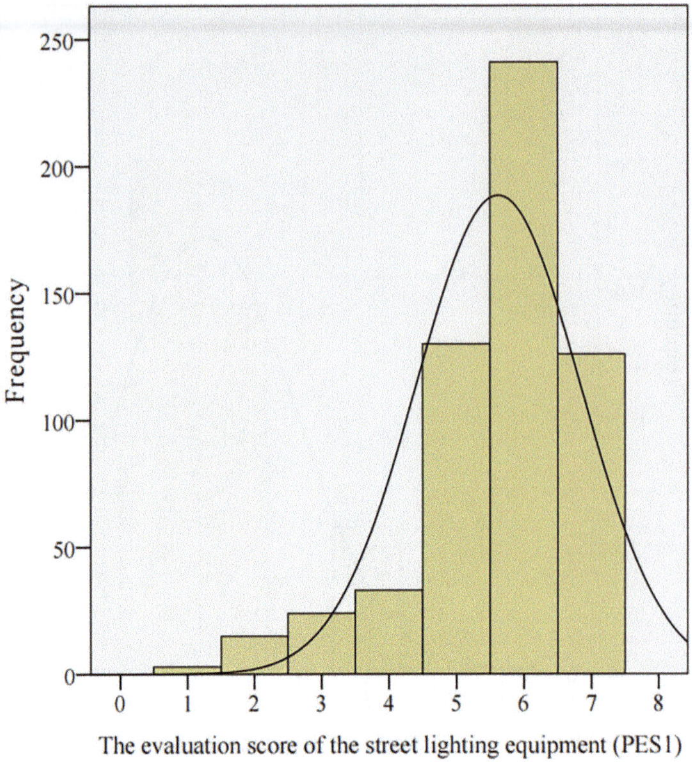


**Fig. 7.18** Histogram of the distribution for the score for the evaluation score of the street lighting equipment. Caption: The curve represents the fitted normal distribution. Source: Created by the authors

shopping, leisure and entertainment, tourism, scientific research and education, government agencies, large-scale shopping, financial services and other facilities, to the area of geographical units (units/km$^2$) in the buffer zone. The calculation formula is:

$$POID = \frac{\sum_{k=1}^{n} N_k}{A}$$

In this formula, $N_k$ represents the number of the kth POI, $n$ represents the number of POI types in the buffer, $A$ represents the area of the geographical unit, i.e., km$^2$, and the larger the POID value, the higher the function density. Within the communities surveyed, the average POI Density was 624.39/km$^2$, with a minimum of 27.07/km$^2$ and a maximum of 1618.16/km$^2$.

Table 7.8 reports descriptive statistical results for the seven elements of the Objective Built Environment described above. According to the literature

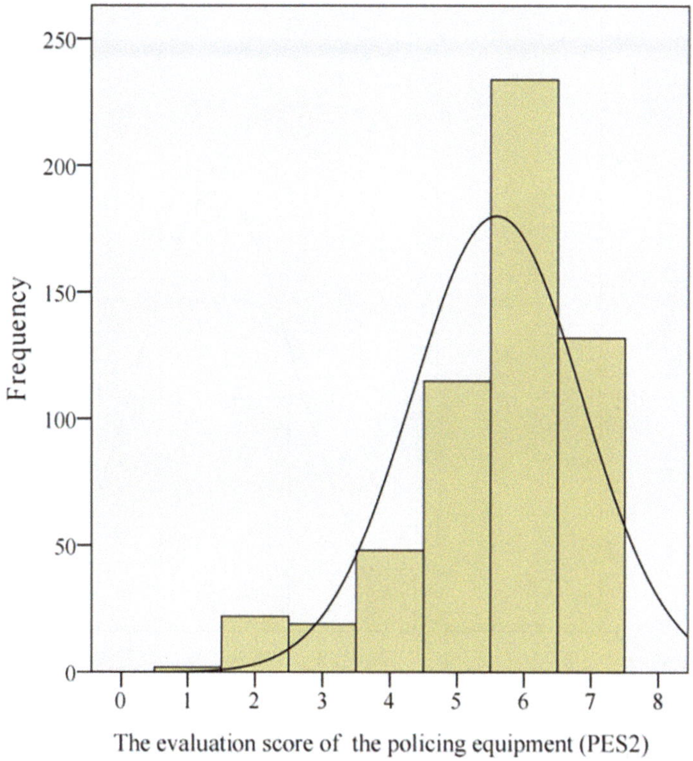


**Fig. 7.19** Histogram of the distribution for the score for the policing equipment. Caption: The curve represents the fitted normal distribution. Source: Created by the authors

recommendations, skewness and kurtosis results show that the seven elements are in accordance with normal distribution.

### *7.2.3 Measurement of Other Important Variables and Sample Characteristics*

1. Measurement of Social Capital and Its Sample Characteristics

Table 7.9 shows descriptive statistics for eight items of positive Social Capital, which have been linked to various community benefits including walkability and well-being (Putnam, 2000; Leyden, 2003). The minimum value in the table is 1.000 and the maximum value is 7.000, indicating no missing values in the data. The average number in the table is between 4.410 and 5.460, indicating that residents 'evaluation of Social Capital is between satisfactory and very satisfactory. SC7 is the lowest in the table, indicating that residents 'evaluation of "eating with neighbors"

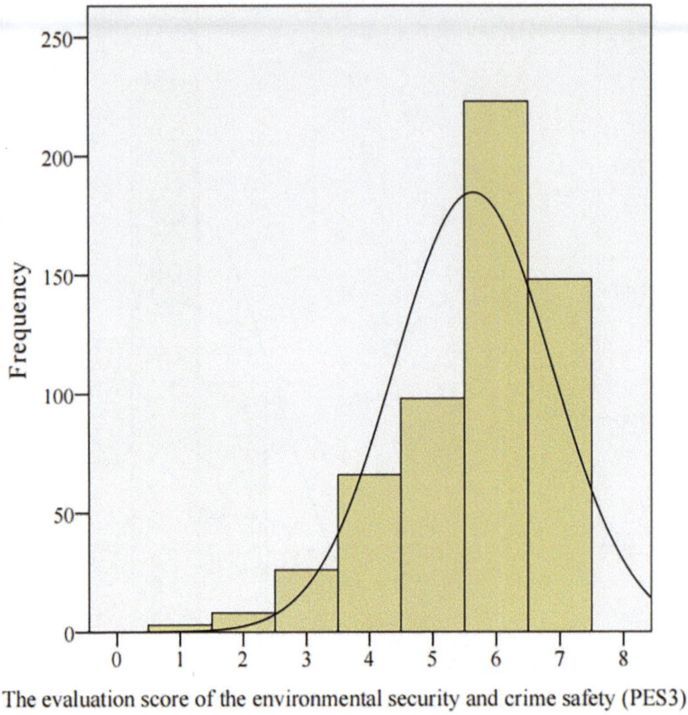


**Fig. 7.20** Histogram of the distribution for the score for the evaluation score of the environmental security and crime safety. Caption: The curve represents the fitted normal distribution. Source: Created by the authors

is in the evaluation range of general and relatively satisfactory, so the evaluation is relatively general. This reflects perhaps a broader trend of declining informal social connections discussed in works like Putnam (2000) 'Bowling Alone'. SC1 has the largest average, which indicates that residents 'evaluation of "neighbors are trustworthy" is in the evaluation interval of agreement and agreement, so the evaluation is higher. The grade difference/standard deviation ranged from 3.503 to 5.435, indicating that the item discrimination was good, the skewness and kurtosis met the corresponding suggested standards, and the test met the normal distribution.

2. Measures of Self-Rated Health (SRH) and its Sample Characteristics

Descriptive statistics for respondents' Self-rated health (SRH), a well-established predictor of mortality and morbidity (Idler & Benyamini, 1997), are reported in Tables 7.10, which show that respondents participating in the survey had a minimum value of 1.000 and a maximum value of 10.00, indicating no missing values. The average score of Self-Rated Health was 7.510, indicating that most people had high Self-Rated Health scores. Skewness and kurtosis are consistent with the corresponding normal standards (Curran et al., 1996; Kline, 2015). It shows that the data do not deviate from normal.

**Table 7.8** Descriptive statistics for the objective built environment (OBE)

| OBE Variables | Min | Max | Mean | SD | Skewness | Kurtosis |
|---|---|---|---|---|---|---|
| PD | .303 | 9.274 | 4.262 | 2.364 | .224 | −.982 |
| BD | .130 | .490 | .310 | .072 | .088 | .251 |
| FMD | .460 | .860 | .730 | .060 | −1.464 | 4.723 |
| RID | 14.650 | 359.240 | 142.608 | 75.994 | .506 | −.157 |
| BLD | 2.240 | 14.940 | 9.772 | 3.034 | −.471 | −.596 |
| BSD | 1.270 | 16.880 | 8.354 | 3.183 | .547 | .196 |
| POID | 27.070 | 1618.160 | 624.393 | 347.915 | .732 | .206 |

Caption: PD refers to Population Density and BD refers to Building Density. Together, they reflect community density. FMD refers to the degree of Functional Mix Diversity, reflecting the degree of community Functional Mix Diversity. RID refers to the density of Road Intersections, reflecting road design characteristics. BLD refers to Bus Line Density, and BSD refers to Bus Stop Density, both reflect the distance to bus stop. POID refers to the density of primary functional points of interest, reflecting the characteristics of the community to the destination
Source: Created by the authors

**Table 7.9** Descriptive statistics for social capital (SC)

| | N | Min | Max | Mean | SD | Skewness | Kurtosis |
|---|---|---|---|---|---|---|---|
| SC1 | 572 | 1.000 | 7.000 | 5.460 | 1.104 | −1.002 | 1.631 |
| SC2 | 572 | 1.000 | 7.000 | 5.350 | 1.196 | −1.075 | 1.680 |
| SC3 | 572 | 1.000 | 7.000 | 5.360 | 1.179 | −.759 | .677 |
| SC4 | 572 | 1.000 | 7.000 | 4.970 | 1.535 | −.731 | −.189 |
| SC5 | 572 | 1.000 | 7.000 | 4.930 | 1.535 | −.646 | −.398 |
| SC6 | 572 | 1.000 | 7.000 | 4.820 | 1.565 | −.481 | −.611 |
| SC7 | 572 | 1.000 | 7.000 | 4.410 | 1.713 | −.275 | −.811 |
| SC8 | 572 | 1.000 | 7.000 | 5.210 | 1.330 | −1.037 | 1.006 |

Source: Created by the authors

**Table 7.10** Descriptive statistics for self-rated health (SRH)

| | N | Min | Max | Mean | SD | Skewness | Kurtosis |
|---|---|---|---|---|---|---|---|
| SRH | 572 | 1.000 | 10.000 | 7.510 | 2.134 | −1.183 | 1.111 |

Source: Created by the authors

### *7.2.4 Characteristics of the Mean Score of Residents' Life Satisfaction at the Community Level*

The preceding description analyzes Life Satisfaction characteristics, Perceived Environmental Quality characteristics, Objective Built Environment characteristics, and other important variable characteristics of the survey sample. This study focuses on the objective environment and Life Satisfaction, so it is necessary to explore whether there are differences in residents 'Life Satisfaction at the community level.

The total life satisfaction scores of residents across the 87 sampled communities ranged from 17 to 33, with a mean score of 26.020. The absolute values of both

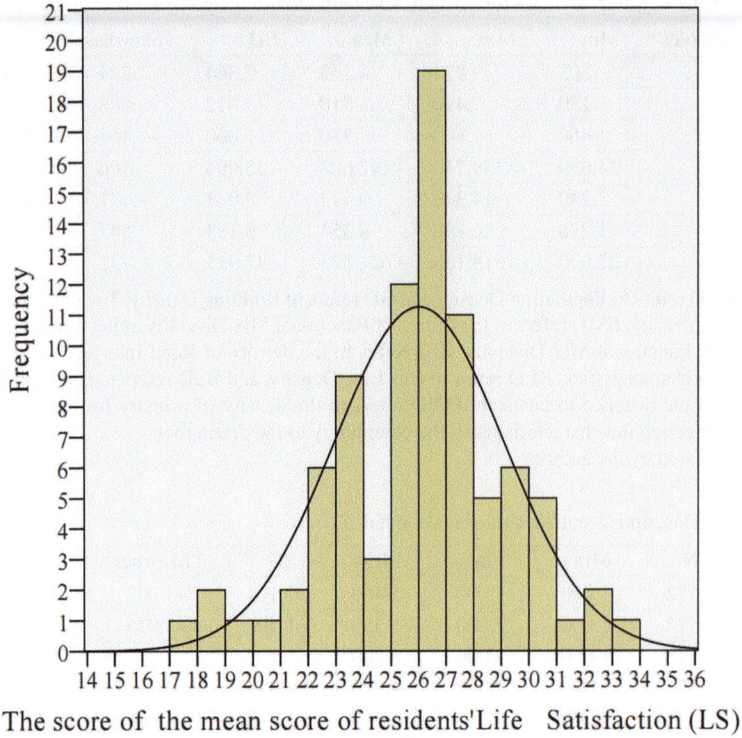


**Fig. 7.21** Histogram of the distribution for the mean score of residents' life satisfaction in the sampled community. Caption: The curve represents the fitted normal distribution. Source: Created by the authors

skewness (−0.367) and kurtosis (0.580) were well below the recommended thresholds (Skewness <2 and Kurtosis <7, respectively) (Kline, 2015). Furthermore, as shown in Fig. 7.21, the frequency distribution and normal curve show that the average Life Satisfaction of community residents is slightly negative skewness, but it accords with an approximately bell-shaped distribution. It was therefore concluded that the deviation from normality was not substantial enough to undermine the assumptions of subsequent parametric analyses.

The preceding descriptive results indicate significant variations in life satisfaction across communities, underscoring the need to identify the underlying environmental determinants. In a related study utilizing the same survey data, Li et al. (2025) employed a multi-scale hierarchical regression framework to disentangle the effects of built environment factors at the housing, neighborhood, and community levels. Their findings confirm that environmental characteristics across these three scales—such as housing construction quality at the housing scale, neighborhood facilities and management at the neighborhood scale, and density and transit accessibility at the community scale—collectively and significantly explain the variance in residents' life satisfaction. This established multi-scale influence provides a

**Table 7.11** Frequency analysis of the construction era of the community

| Group | Construction Era | Frequency | Percentage (%) |
|---|---|---|---|
| 1 | Before 1995 | 15 | 17.24 |
| 2 | 1996–1999 | 20 | 22.99 |
| 3 | 2000–2004 | 22 | 25.29 |
| 4 | 2005–2010 | 14 | 16.09 |
| 5 | 2011–2018 | 16 | 18.39 |
| Total | | 87 | 100 |

Source: Created by the authors

robust empirical foundation and analytical framework for the subsequent multivariate analyses in this work.

### 7.2.5 *Differences in Life Satisfaction among Communities of Different Construction Era*

The foregoing describes the characteristics of average Life Satisfaction of residents in different communities, and finds that there are differences in average Life Satisfaction of residents in different communities. The objective data used in this study are limited by the consideration that it is difficult to measure the age composition of different buildings in the community, and the relationship between community types and residents 'average Life Satisfaction is explored by using the age of community construction as a variable. Table 7.11 reports the type, year and proportion of the distribution of the 87 communities in which residents live. Table 7.12 reports the variance analysis and Scheffe method post-hoc multiple comparison results of the Life Satisfaction of residents in communities with different construction years.[5] It is proven that there is no significant difference in the average Life Satisfaction among residents in communities with different construction years (Hur & Morrow-Jones, 2008). This finding aligns with some studies (e.g., Hur & Morrow-Jones, 2008) that suggest housing age may be less influential on satisfaction than other neighborhood characteristics.

## 7.3 Summary of this Chapter

This chapter analyzes the characteristics of individual Life Satisfaction, Perceived Environmental Quality, built environment, Perceived Safety, Social Capital, Self-Rated Health and other variables of the formal sample. The results show that there

[5] Hur, M., & Morrow-Jones, H. (2008). Factors that influence residents' satisfaction with neighborhoods. *Environment and Behavior*, 40(5), 619–635.

**Table 7.12** Post Hoc comparisons of life satisfaction by construction era

| (I) Group | (J) Group | MD (I-J) | SE | P | 95% CI | |
|---|---|---|---|---|---|---|
| | | | | | Lower | Upper |
| 1 | 2 | .004 | 1.055 | 1.000 | −3.320 | 3.330 |
| | 3 | .207 | 1.034 | 1.000 | −3.050 | 3.470 |
| | 4 | −1.589 | 1.148 | .751 | −5.210 | 2.030 |
| | 5 | .198 | 1.110 | 1.000 | −3.300 | 3.700 |
| 2 | 1 | −.004 | 1.055 | 1.000 | −3.330 | 3.320 |
| | 3 | .203 | .954 | 1.000 | −2.800 | 3.210 |
| | 4 | −1.594 | 1.076 | .701 | −4.990 | 1.800 |
| | 5 | .194 | 1.036 | 1.000 | −3.070 | 3.460 |
| 3 | 1 | −.207 | 1.034 | 1.000 | −3.470 | 3.050 |
| | 2 | −.203 | .954 | 1.000 | −3.210 | 2.800 |
| | 4 | −1.797 | 1.056 | .578 | −5.120 | 1.530 |
| | 5 | −.009 | 1.015 | 1.000 | −3.210 | 3.190 |
| 4 | 1 | 1.589 | 1.148 | .751 | −2.030 | 5.210 |
| | 2 | 1.594 | 1.076 | .701 | −1.800 | 4.990 |
| | 3 | 1.797 | 1.056 | .578 | −1.530 | 5.120 |
| | 5 | 1.787 | 1.130 | .646 | −1.780 | 5.350 |
| 5 | 1 | −.198 | 1.110 | 1.000 | −3.700 | 3.300 |
| | 2 | −.194 | 1.036 | 1.000 | −3.460 | 3.070 |
| | 3 | .009 | 1.015 | 1.000 | −3.190 | 3.210 |
| | 4 | −1.787 | 1.130 | .646 | −5.350 | 1.780 |

Caption: MD refers to Mean difference. SE refers to standard error. CI refers to the confidence interval

Source: Created by the authors

are differences among individual residents 'Life Satisfaction, and there are also significant differences in Perceived Environmental Quality, Perceived Safety, Social Capital, and Self-Rated Health at the individual level. The average Life Satisfaction among residents in different types of communities built in different years. There was no significant difference in Life Satisfaction among different groups of residents.

This study focuses on the impact of built environment factors at the community level, so it analyzes the characteristics of average Life Satisfaction at the community level. The results show that there are significant differences in average Life Satisfaction at the community level. What factors influence the difference in Life Satisfaction, and whether there is any relationship between it and Objective Built Environment, Perceived Environmental Quality, Perceived Safety, Social Capital and self-rated health, need to be further discussed.

**Funding** This research was funded by

1. Guangdong Provincial Office of Philosophy and Social Sciences [Grant No. GD24XSH06]
2. Projects of PhDs' Start-up Research of GDUPT [Grant No. 2023bsqd1008].

# References


Byrne, B. M. (2016). Structural equation modeling with AMOS: Basic concepts, applications, and programming (3rd ed.). Routledge.

Cao, Y., & Zhang, Y. (2016). The impact of the built environment on subjective Well-being: A case study of Guangzhou. *Journal of Urban Planning and Development, 142*(4), 05016007.

Cervero, R., & Kockelman, K. (1997). Travel demand and the 3Ds: Density, diversity, and design. *Transportation Research Part D: Transport and Environment, 2*(3), 199–219.

Curran, P. J., West, S. G., & Finch, J. F. (1996). The robustness of test statistics to nonnormality and specification error in confirmatory factor analysis. *Psychological Methods, 1*(1), 16–29.

Das, A. (2020). Mobility impairment and access to urban facilities: A study of perceived barriers. *Disability & Society, 35*(6), 897–917.

Diener, E. (1984). Subjective Well-being. *Psychological Bulletin, 95*(3), 542–575.

Diener, E., Emmons, R. A., Larsen, R. J., & Griffin, S. (1985). The satisfaction with life scale. *Journal of Personality Assessment, 49*(1), 71–75.

Ewing, R., & Cervero, R. (2010). Travel and the built environment: A meta-analysis. *Journal of the American Planning Association, 76*(3), 265–294.

Ewing, R., & Handy, S. (2009). Measuring the unmeasurable: Urban design qualities related to walkability. *Journal of Urban Design, 14*(1), 65–84.

Finney, S. J., & DiStefano, C. (2006). Nonnormal and categorical data in structural equation modeling. In G. R. Hancock & R. O. Mueller (Eds.), *Structural equation modeling: A second course* (pp. 269–314). Information Age Publishing.

Foster, S., Giles-Corti, B., & Knuiman, M. (2010). Neighbourhood design and fear of crime: A social-ecological examination of the correlates of residents' fear in new suburban housing developments. *Health & Place, 16*(6), 1156–1165.

Frank, L. D., Sallis, J. F., Saelens, B. E., Leary, L., Cain, K., Conway, T. L., & Hess, P. M. (2010). The development of a walkability index: Application to the neighborhood quality of life study. *British Journal of Sports Medicine, 44*(13), 924–933.

Guangzhou Municipal Bureau of Statistics. (2020). *Guangzhou statistical yearbook 2020*. China Statistics Press.

Hur, M., & Morrow-Jones, H. (2008). Factors that influence residents' satisfaction with neighborhoods. *Environment and Behavior, 40*(5), 619–635.

Hur, M., & Nasar, J. L. (2014). Physical upkeep, perceived upkeep, and residential satisfaction: A dynamic perspective. *Environment and Behavior, 46*(6), 643–660.

Idler, E. L., & Benyamini, Y. (1997). Self-rated health and mortality: A review of twenty-seven community studies. *Journal of Health and Social Behavior, 38*(1), 21–37.

Kline, R. B. (2015). *Principles and practice of structural equation modeling* (4th ed.). Guilford Press.

Leyden, K. M. (2003). Social capital and the built environment: The importance of walkable neighborhoods. *American Journal of Public Health, 93*(9), 1546–1551.

Maslow, A. H. (1943). A theory of human motivation. *Psychological Review, 50*(4), 370–396.

Li, H., Pan, C., Qiu, N., Guo, J., & Wu, J. (2025). Does the multi-scale built environment impact on residents' subjective Well-being? *Buildings, 15*(18), 3311.

Putnam, R. D. (2000). *Bowling alone: The collapse and revival of American community*. Simon & Schuster.

Shen, Y., & Fu, X. (2019). Impact of the built environment and perceived neighborhood characteristics on residents' satisfaction: Evidence from the Shanghai suburbs. *Progress in Geography, 38*(5), 686–697.

WHO. (2018). *International classification of diseases, 11th revision (ICD-11)*. WHO.

# Chapter 8
# Key Factors Linking Built Environment and Well-Being

**Abstract** Following the preceding descriptive analyses, this chapter empirically investigates the linkages between the Objective Built Environment (OBE) and residents' Life Satisfaction (LS) within Guangzhou's high-density urban context. Its primary aim is to examine both the geographic and statistical **relationships** between OBE elements—conceptualized through the "5D" framework (Density, Diversity, Design, Distance to transit, and Destination accessibility)—and LS, **testing direct effects via multiple linear regression**. Section 8.1 employs geographic information visualization and correlation matrices to explore bivariate relationships between each OBE factor (e.g., Population Density, Building Density, Functional Mix Diversity, Road Intersection Density, Bus Line Density, Bus Stop Density, POI Density) and community-level average LS. These preliminary analyses reveal mixed spatial patterns and weak-to-moderate correlations. Section 8.2 specifies a multivariate regression model controlling for individual socioeconomic attributes to test hypotheses on OBE effects. **Results show significant negative effects of Building Density ($\beta = -0.106, p < 0.05$), Road Intersection Density ($\beta = -0.164, p < 0.01$), and Bus Stop Density ($\beta = -0.135, p < 0.1$), and a positive effect of Bus Line Density ($\beta = 0.146, p < 0.05$). Functional mix diversity was marginally negative ($\beta = -0.075, p < 0.1$), while population and POI Density were non-significant. By regression analysis, this chapter provides strong evidence to reveal whether specific OBE factors affect LS, highlighting their nuanced and sometimes countervailing roles**. These findings **establish a foundation** for subsequent mechanistic analyses via structural equation modeling.

**Keywords** Objective built environment · Life satisfaction · 5D model · Geographic spatial relationship · Multiple regression · Density · Diversity · Destination accessibility · Perceived environmental quality · Urban Well-being

Building upon the comprehensive descriptive analysis presented in Chap. 7, which revealed significant variations in life satisfaction both at the individual and community levels, underscoring the need to examine how specific elements of the built environment may influence these outcomes. Chap. 8, therefore, transitions to a more



H. Li, G. Shen, *The Anatomy of Subjective Well-Being*, Resilient and Friendly Living Environments, https://doi.org/10.1007/978-981-95-4937-5_8

analytical focus, employing geographic visualization and multiple regression techniques to examine the spatial and statistical relationships between objective built environment factors—structured around the classic "5D" dimensions (density, diversity, design, distance to transit, and destination accessibility)—and residents' life satisfaction, thereby bridging descriptive insights with causal inquiry.

In the preceding part, the study reports and analyzes the measurement of Subjective Well-being and sample characteristics, and measures and briefly describes the main variables of the community built environment, especially the characteristics of residents 'Life Satisfaction, and finds that there are large differences between samples. Based on the five dimensions (density, diversity, design, distance to transit, and destination accessibility) of the classic "5D" built environment framework (Cervero & Kockelman, 1997; Ewing & Cervero, 2010, 2017), this chapter mainly discusses the geographical spatial relationship and correlation between Objective Built Environment elements and Life Satisfaction, and discusses the impact of Objective Built Environment elements on Life Satisfaction by using multiple linear regression model.

## 8.1 Relationship Between Objective Built Environment and Life Satisfaction

Based on the "5D" built environment framework (Ewing & Cervero, 2010, 2017), this study explores the relationship between density (population density), building density, diversity (Functional Mix Diversity), design (intersection density), distance to public transport (Bus Line Density), bus stop density, destination accessibility (POI point density) and Life Satisfaction. Figure 8.1 plots the matrix of relationships between the five dimensions of the Objective Built Environment, or seven elements, and Life Satisfaction. To further explore the strength of these associations, Fig. 8.2 visualizes the correlation matrix as a heatmap, focusing on the links between the built environment elements and community-level average life satisfaction. The figure shows that there is a clear correlation between the seven Objective Built Environment factors reflecting the built environment characteristics of high-density cities, that is, there is a clear positive correlation between population density and building density, road intersection density, bus line density, bus stop density, POI point density, etc. The relationship between building density, bus line density, bus stop density and Life Satisfaction is basically clear, while the relationship between other built environment elements and Life Satisfaction is not clear, which needs to be further discussed.

In order to better grasp the overall trend of the relationship between built environment and Subjective Well-being and study the influencing factors of Life Satisfaction, from an intuitive point of view, the following sets the individual sample data of micro survey to its community scale, and uses the method of combining geographic information visualization technology with scatter point Fitting the Trend

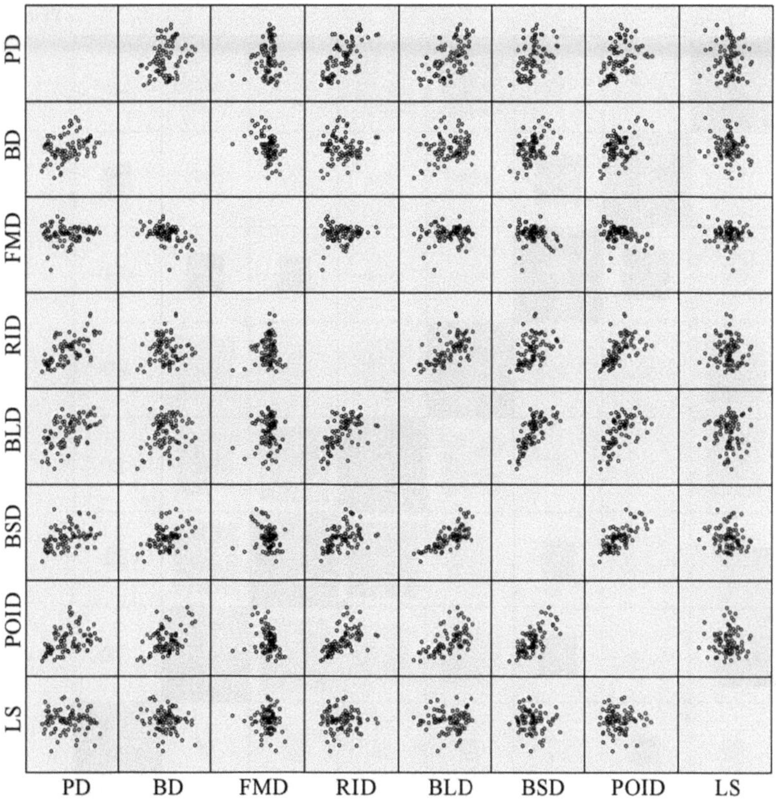


**Fig. 8.1** Matrix diagram of relationship between objective built environment factors and average life satisfaction of residents. Source: Created by the authors

map to fit the geographical spatial relationship and data of the relationship between Objective Built Environment, Perceived Environmental Quality and average Life Satisfaction of residents on the community scale. To more intuitively describe the relationship between two.[1]

### 8.1.1 Density and Life Satisfaction

1. Description of the relationship between Community Population Density and Life Satisfaction

[1] Geospatial visualization and trend fitting were performed using Arc GIS 10.8. The coordinate system was set to WGS_1984_UTM_Zone_49N.

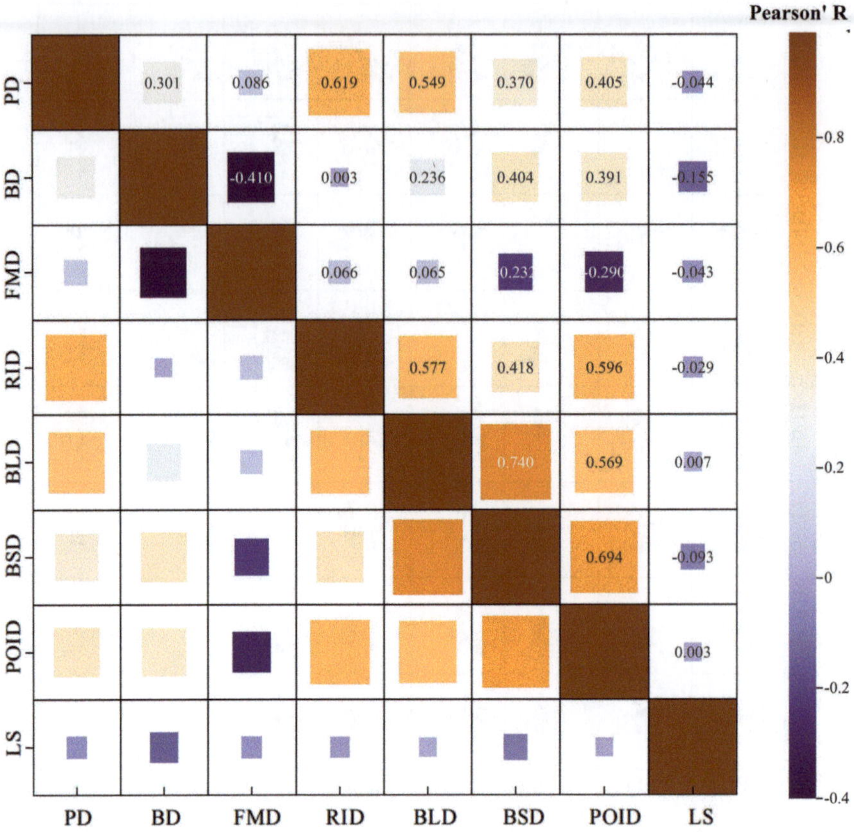


**Fig. 8.2** Heat map of correlation between objective built environment factors and average life satisfaction of residents. Source: Created by the authors

Figures 8.3 and 8.4 show the geospatial relationship between community population density and average Life Satisfaction and the trend of data fitting. The geographical spatial relationship between them shows that the population density of high-density Urban areas in Haizhu District and Yuexiu District is large,[2] and the average Life Satisfaction is high. However, in Tianhe District and Liwan District, the average satisfaction degree of the community is also relatively high. Therefore, the relationship between population density and average Life Satisfaction is unclear. The trend graph of data fitting shows that the population density of the community has a small negative correlation with Life Satisfaction.[3] The Pearson correlation

[2] To obtain higher spatial resolution population distribution data, this study employed a spatial disaggregation method based on the principle of dasymetric mapping. This approach utilizes ancillary data, such as high-resolution remote sensing imagery, to allocate census data aggregated at administrative units into regular grids according to weighting schemes.

[3] Building density was calculated from interpreted high-resolution remote sensing imagery (2020).

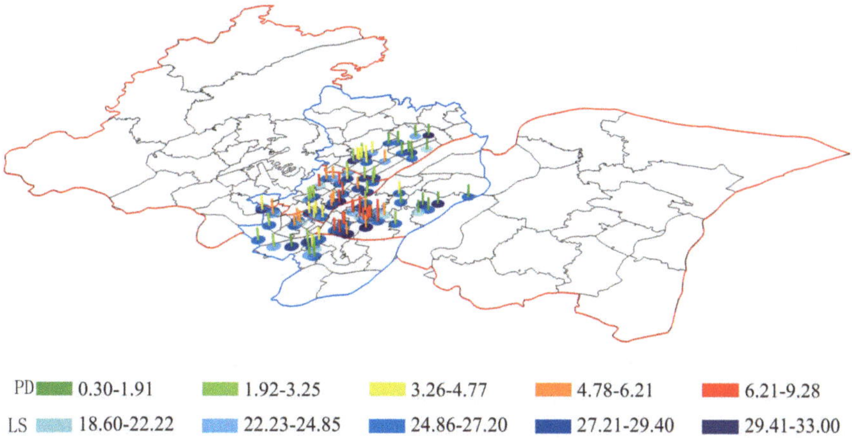


**Fig. 8.3** Geospatial relationship between community population density and average life satisfaction of residents. Source: Created by the authors

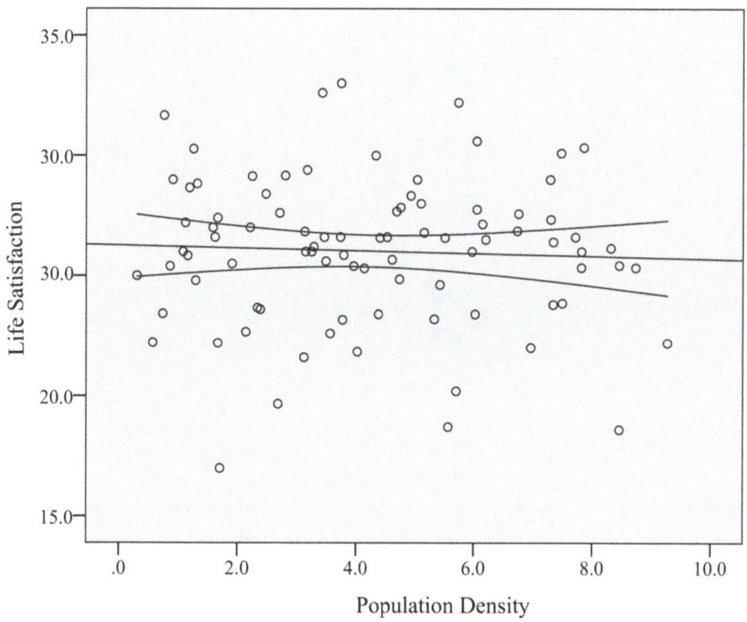


**Fig. 8.4** Fitting the trend of the relationship between community population density and average life satisfaction of residents. ○ Average Life Satisfaction of survey community residents.—Fitting the Trend line and confidence interval. Source: Created by the authors

coefficient was −0.044, which indicated that the relationship between them was not significant.

2. Description of the Relationship between Building Density and Life Satisfaction in the Community

Figures 8.5 and 8.6 show the geospatial relationship between building density and Life Satisfaction and the trend of data fitting. The geographical spatial relationship between them shows that the communities with higher building density in Liwan District, Yuexiu District and Haizhu District have lower average Life Satisfaction. In some communities with relatively low building density, the averag Life Satisfaction of residents is higher. Therefore, the relationship between building density and average Life Satisfaction is relatively certain. The trend graph of data fitting shows that the population density of the community has a small negative correlation with Life Satisfaction. The trend line results of the data fit plot are consistent with those of the geospatial relationships. The Pearson correlation coefficient is −0.155, which indicates that building density may be one of the Objective Built Environment factors that potentially affect residents 'Life Satisfaction. More rigorous models are needed to explore the impact of building density on Life Satisfaction.

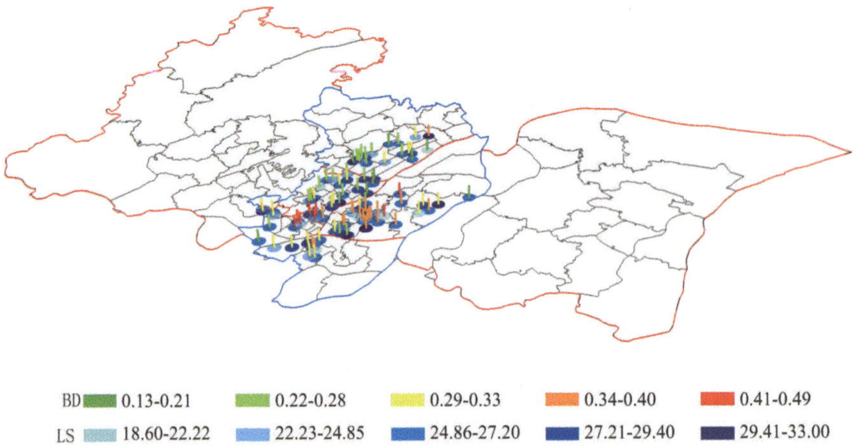


**Fig. 8.5** Geospatial relationship between community building density and average life satisfaction of residents. Source: Created by the authors

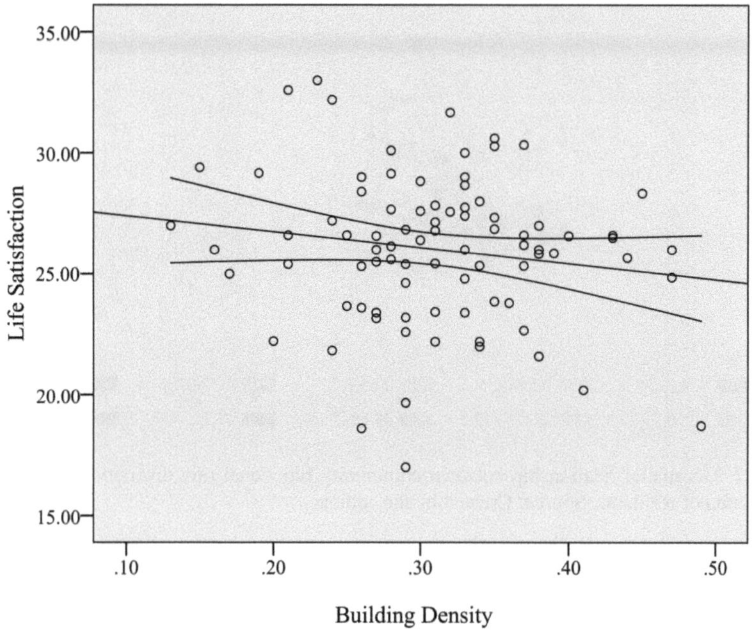


**Fig. 8.6** Fitting the trend of relationship between community building density and average life satisfaction of residents. ○ Average Life Satisfaction of survey community residents—Fitting the Trend line and confidence interval. Source: Created by the authors

### *8.1.2 Relationship Between Diversity and Life Satisfaction*

In empirical studies, land-use diversity and Functional Mix Diversity are often used as proxy variables for diversity. In this study, Functional Mix Diversity was used as a proxy variable for built environment diversity.[4] Figs. 8.7 and 8.8 show the geospatial relationship and data fit trends between Functional Mix Diversity and Life Satisfaction. The geographical spatial relationship between them shows that residents 'Life Satisfaction is higher in some communities with higher Functional Mix Diversity in the Yuexiu District and the Haizhu District. However, in some communities with low Functional Mix Diversity in Liwan District and Tianhe District, residents 'Life Satisfaction is not low.

Therefore, the relationship between community Functional Mix Diversity and community average Life Satisfaction is not clear. The trend graph of data fitting shows that there is a small correlation between community Functional Mix Diversity and Life Satisfaction. The Pearson correlation coefficient was −0.043, which indicated that the effect of Functional Mix Diversity on residents 'Life Satisfaction needs further study.

[4] Functional Mix Diversity (FMD) was calculated using the entropy index formula based on Point of Interest (POI) data obtained from the Amap API (2022), categorized into 10 primary types.

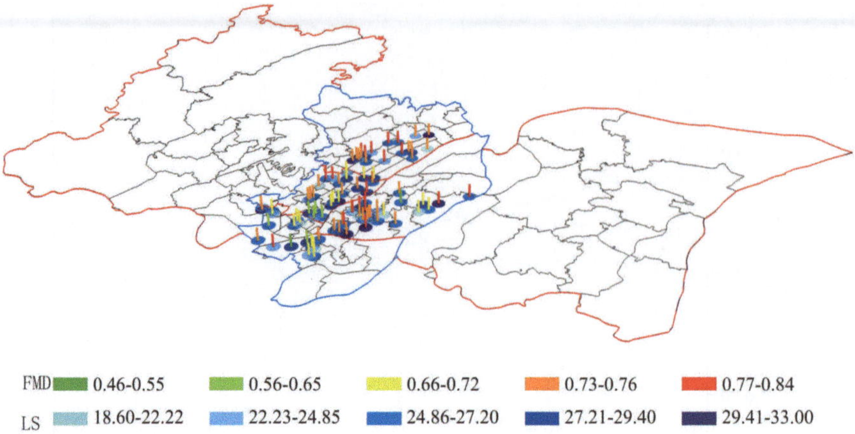


**Fig. 8.7** Geospatial relationship between community functional mix diversity and average life satisfaction of residents. Source: Created by the authors

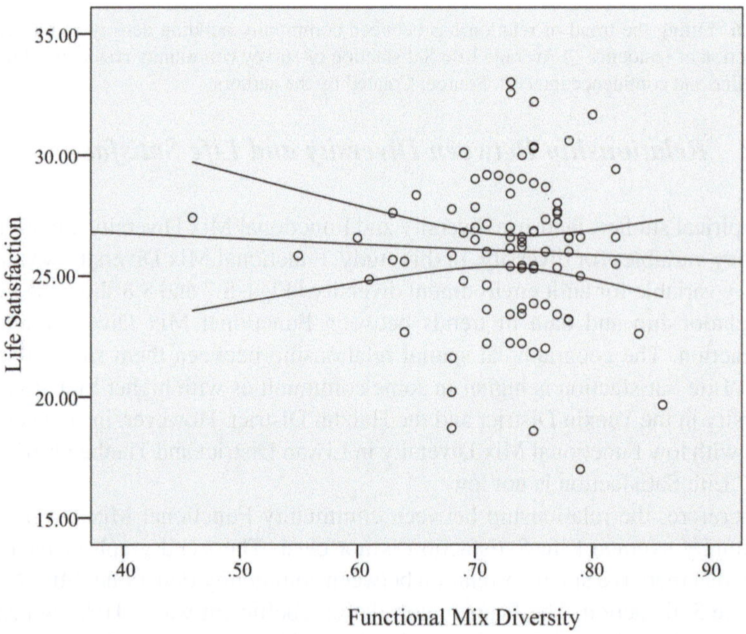


**Fig. 8.8** Fitting the trends of the relationship between community functional mix diversity and average life satisfaction of residents. ○ Average Life Satisfaction of survey community residents.— Fitting the Trend line and confidence interval. Source: Created by the authors

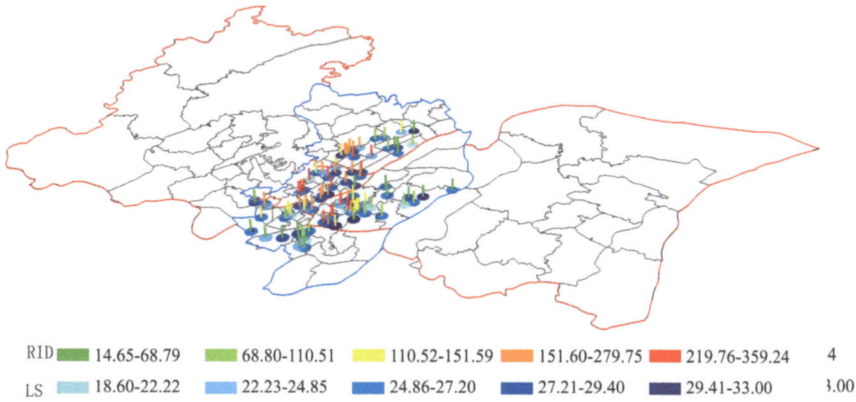


**Fig. 8.9** Geospatial relationship between community road intersection density and average life satisfaction of residents. Source: Created by the authors

## *8.1.3 Relationship Between Design and Life Satisfaction*

In empirical studies, design, i.e., street design, is usually measured by road network density and intersection density. This study uses road intersection density as a proxy variable. Figures 8.9 and 8.10 show the geospatial relationship and data Fitting the Trends between road intersection density and Life Satisfaction. The geographical spatial relationship between them shows that most communities with high road intersection density in Tianhe District, Yuexiu District and Haizhu District have low average Life Satisfaction. The average Life Satisfaction of residents in most communities with low road intersection density in the four districts is higher. However, in the four districts, there are also many communities with low road intersection density, and their average Life Satisfaction is not high. Therefore, the relationship between the two remains uncertain. The data Fitting the Trend chart shows that the road intersection density of the community has a small negative correlation with Life Satisfaction. The Pearson correlation coefficient was −0.029, which indicated that the effect of road intersection density on residents 'Life Satisfaction needs further study.

## *8.1.4 Distance to Transit and Life Satisfaction*

In empirical studies, Distance to Transit usually uses bus line density, bus stop density and bus stop distance as proxy variables. This study uses bus line density and bus stop density to measure.

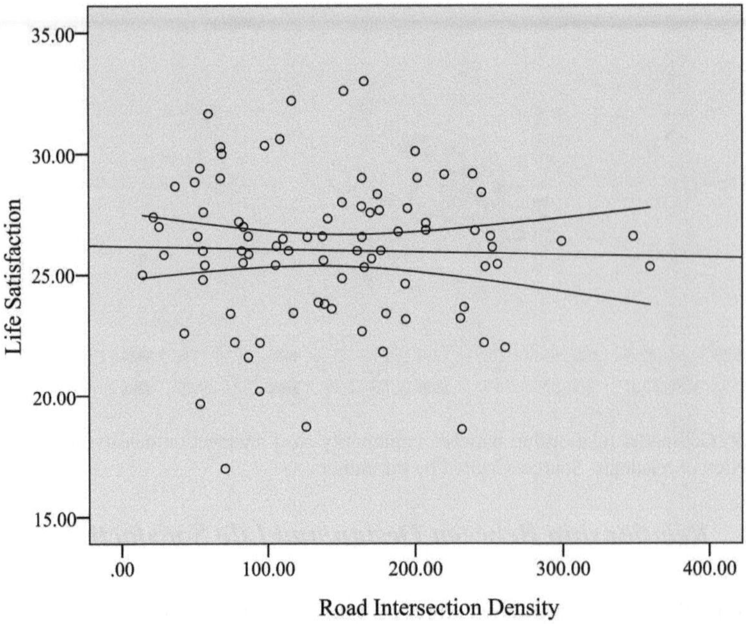


**Fig. 8.10** Fitting the trend of the relationship between road intersection density and average life satisfaction of residents. ○ Average Life Satisfaction of survey community residents.—Fitting the Trend line and confidence interval. Source: Created by the authors

1. Description of the Relationship between Bus Line Density and Life Satisfaction in the Community

Figures 8.11 and 8.12 show the geospatial relationship and data Fitting the Trend between community bus line density and Life Satisfaction. The geographical spatial relationship between them shows that in the four Main cities, the average Life Satisfaction of residents in some communities with high bus line density is high, while that of residents in some communities with low bus line density is not low. Therefore, the relationship between bus line density and community average Life Satisfaction is not clear. The data Fitting the Trend chart shows that the density of public transport routes and Life Satisfaction in the community have a small positive correlation; that is, with the increase of public transport route density in the community, Life Satisfaction will increase. Therefore, bus line density may be the built environment factor affecting residents 'Life Satisfaction, and the relationship between them needs to be discussed by more rigorous models.

2. Description of the Relationship between Bus Stop Density and Life Satisfaction in Community

Figures 8.13 and 8.14 show the geospatial relationship and data Fitting the Trend between community bus stop density and Life Satisfaction. The geographical spatial relationship between them shows that in the four Main cities, the average Life

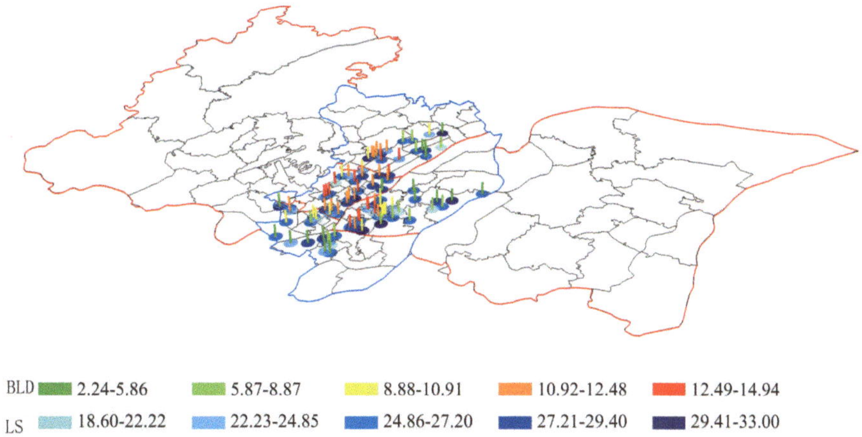


**Fig. 8.11** Geospatial relationship between community bus line density and average life satisfaction of residents. Source: Created by the authors

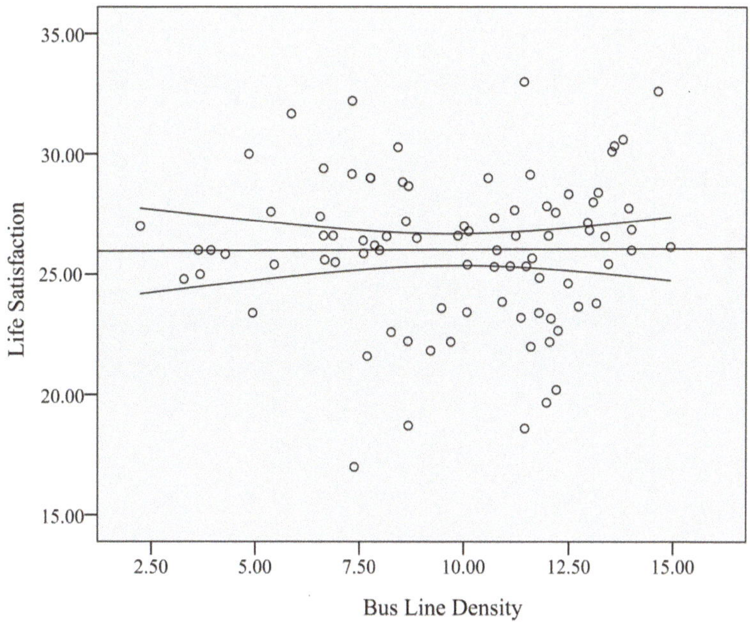


**Fig. 8.12** Fitting the trend of the relationship between bus line density and average life satisfaction of residents. O Average Life Satisfaction of survey community residents.—Fitting the Trend line and confidence interval. Source: Created by the authors

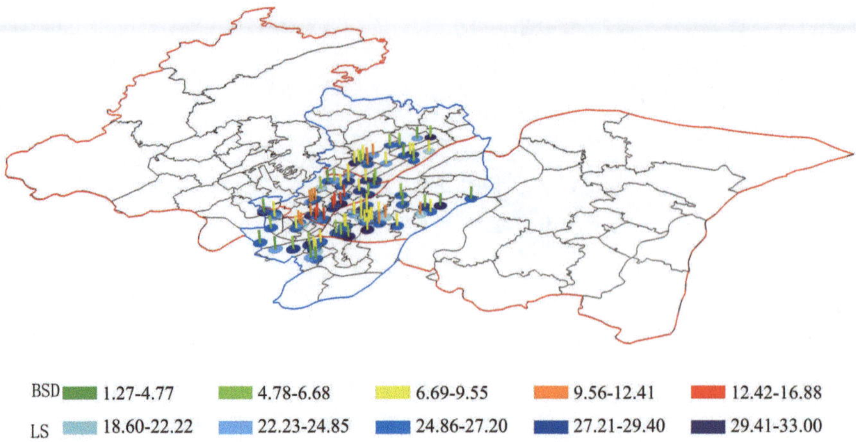


**Fig. 8.13** Geospatial relationship between community bus stop density and average life satisfaction of residents. Source: Created by the authors

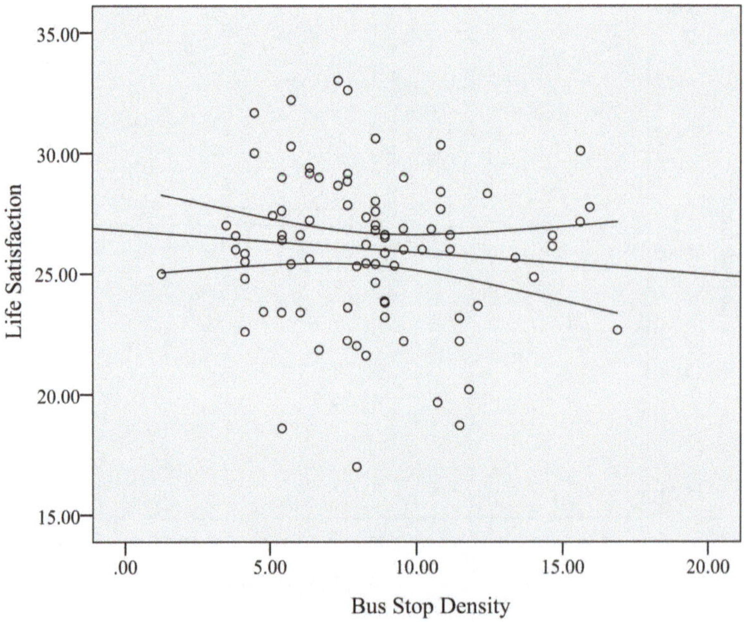


**Fig. 8.14** Fitting the trend of the relationship between bus stop density and average life satisfaction of residents. ○ Average Life Satisfaction of survey community residents.—Fitting the Trend line and confidence interval. Source: Created by the authors

Satisfaction of residents in most communities with high bus stop density is low, while the average Life Satisfaction of residents in most communities with low bus stop density is high. Therefore, there is a negative relationship between bus stop density and average Life Satisfaction. The data Fitting the Trend chart shows that the density of bus stops and Life Satisfaction are negatively correlated to some extent; that is, with the increase in bus stop density, Life Satisfaction will decrease. The Pearson correlation coefficient is −0.093, which indicates that bus stop density may be one of the potential built environment factors affecting residents' Life Satisfaction.

## 8.1.5 Destination Accessibility and Life Satisfaction

Destination accessibility generally refers to the convenience of the destination (Yin, 2020). It is usually measured by bus stop distance, city center distance, park distance, POI point density of facilities, etc.[5] Figs. 8.15 and 8.16 show the geospatial point relationships and data Fitting the Trends between POI point density and Life Satisfaction in communities. The geographical spatial relationship between them shows that in the four Main cities, most communities with higher POI density have relatively high average Life Satisfaction, but some communities have relatively low. Most of the communities with low POI density have low Life Satisfaction, but there are also many with high. Therefore, the relationship between POI density and average Life Satisfaction is not clear. The data Fitting the Trend chart shows that there is a

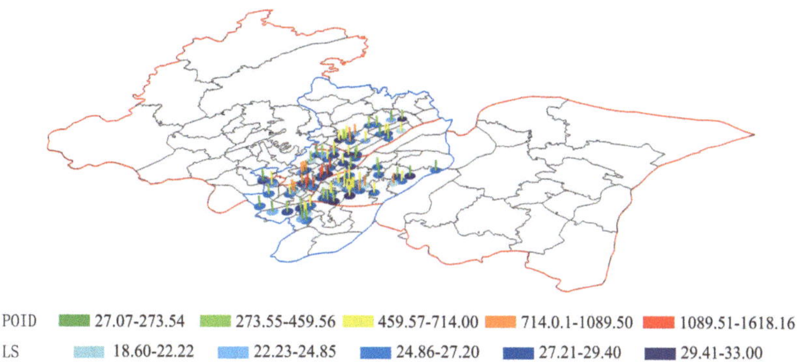


**Fig. 8.15** Geospatial relationship between POID and average life satisfaction of residents

[5] Using POI density as a proxy for destination accessibility emphasizes 'supply' over 'distance,' which can be more effective in high-density, mixed-use urban environments. See, for a discussion in the Chinese context, Wang and Zhou (2017). *Transportation Research Part D: Transport and Environment,* 52, 574–585.

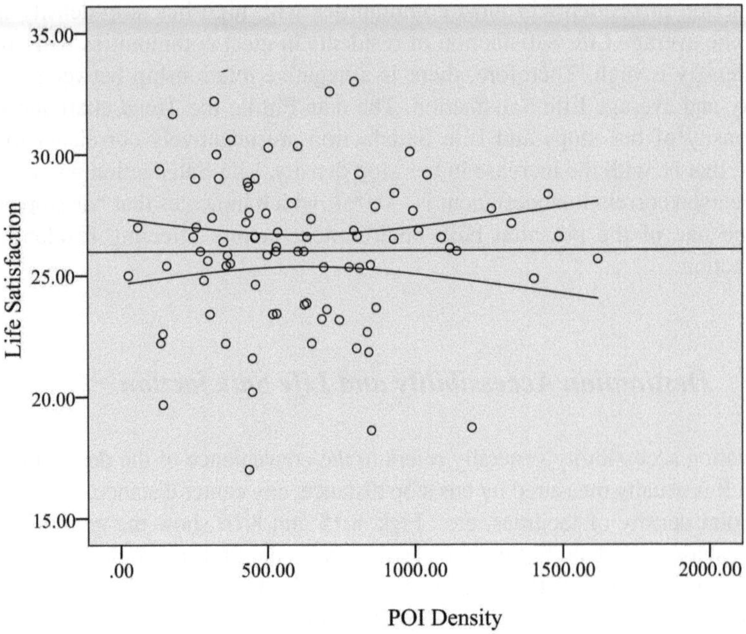


**Fig. 8.16** Fitting the trend of the relationship between POID and average life satisfaction of residents

small positive correlation between community POI point density and Life Satisfaction, that is, as community POI point density increases, Life Satisfaction will increase accordingly. Therefore, POI density may be a built environment factor affecting residents 'Life Satisfaction, which will be tested by more rigorous models later.

### *8.1.6 Relationship Between Perceived Environmental Quality and Life Satisfaction*

The relationship between Objective Built Environment factors and Life Satisfaction discussed above was found to be certain but not intuitively clear, underscoring the importance of complementing objective measures with perceptual data, as they capture different aspects influencing well-being (Cao & Ewing, 2022). In order to explore the relationship between the characteristics of the built environment and Life Satisfaction more directly, Fig. 8.17 plots the relationship between various convenience perceptions, traffic convenience perceptions, aesthetic perceptions. Figure 8.18 shows the correlation heat map between perceived built environment (divers convenience perceptions, traffic convenience perceptions, aesthetic perceptions) and Life Satisfaction. The results showed that the Perceived Environmental Quality of three types of environments was positively correlated with Life

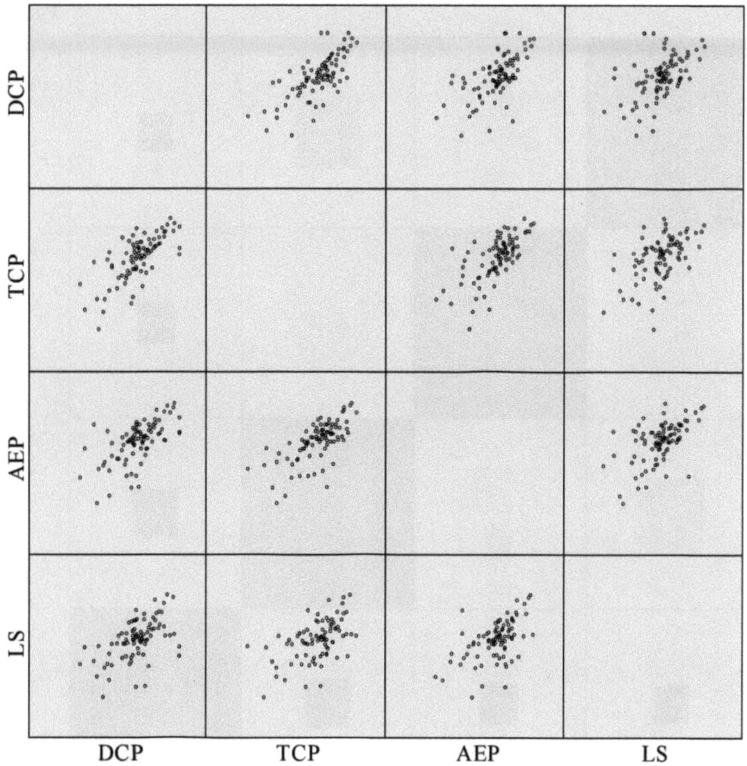


**Fig. 8.17** Matrix diagram of relationship between perceived built environment elements and average life satisfaction of residents. Source: Created by the authors

Satisfaction. Geospatial relationships and data fitting are discussed in detail below. These perceptions align with the urban design qualities identified in prior literature as crucial for walkability and livability (Ewing & Handy, 2009).

1. Description of the Relationship between Community Divers' Convenience Perception and Life Satisfaction

Figures 8.19 and 8.20 show the geospatial relationship and data Fitting the Trend between the perception of diverse convenience of the built environment and Life Satisfaction, respectively. The geographical spatial relationship between them shows that in the four Main cities, the average Life Satisfaction of most communities with high average diversity convenience perception will be better, while the average Life Satisfaction of most communities with low average diversity divers convenience perception will be lower. Therefore, the geospatial relationship indicates that the average diversity convenience perception of community residents may be positively correlated with Life Satisfaction. The data Fitting the Trend chart shows that there is a significant positive correlation between diverse convenience perceptions and Life Satisfaction, that is, as diverse convenience perceptions

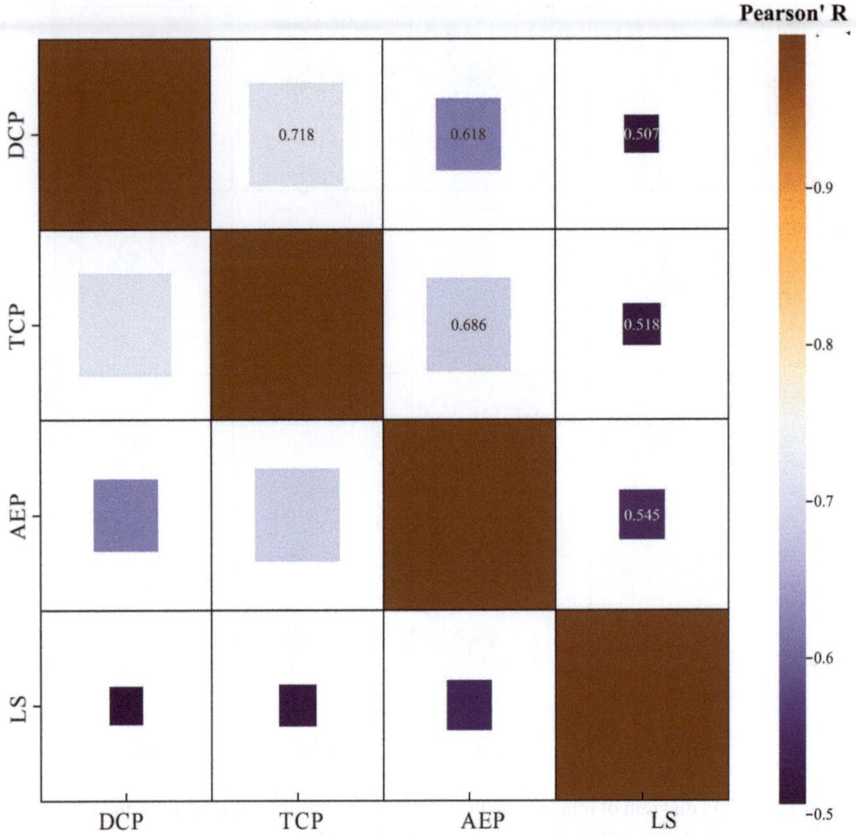


**Fig. 8.18** Heat map of correlation between perceived built environment factors and average life satisfaction of residents. Source: Created by the authors

increase, Life Satisfaction increases. The Pearson correlation coefficient was 0.507 ($P < 0.001$), which indicated that there was a significant positive correlation between them. Therefore, geospatial relationships and data fitting suggest that multiple convenience perceptions of the community built environment may be one of the more important factors affecting Life Satisfaction.

2. Description of the Relationship between Traffic Convenience Perception and Life Satisfaction in the Community Built Environment

Figures 8.21 and 8.22 show the geospatial relationship and data Fitting the Trend between traffic convenience perception and Life Satisfaction in the built environment respectively. The geospatial relationship between them shows that in the four Main Cities, the community residents with higher average traffic convenience perception will have better Life Satisfaction, while most communities with low average traffic convenience perception will have lower Life Satisfaction. Therefore, the geospatial relationship indicates that the average traffic convenience perception of

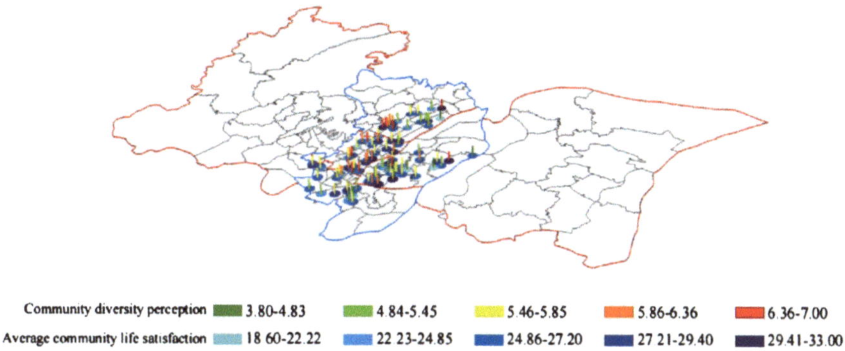


**Fig. 8.19** Geospatial relationship between divers convenience perception and average life satisfaction of residents. Source: Created by the authors

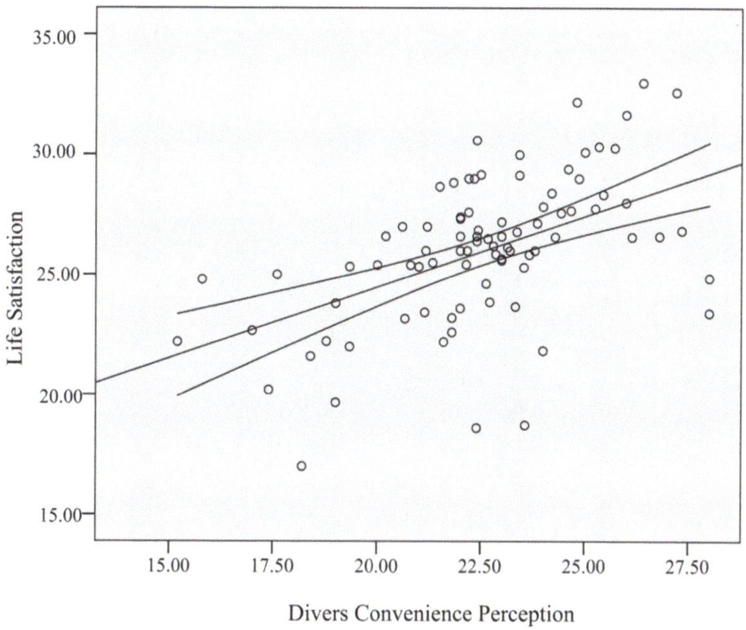


**Fig. 8.20** Fitting the trend of the relationship between divers convenience perception and average life satisfaction of residents. ○ Average Life Satisfaction of survey community residents.—Fitting the Trend line and confidence interval. Source: Created by the authors

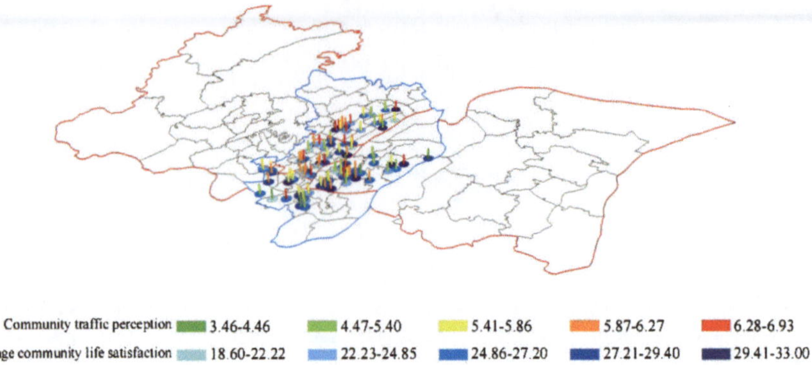


**Fig. 8.21** Geospatial relationship between traffic convenience perception and average life satisfaction of residents. Source: Created by the authors

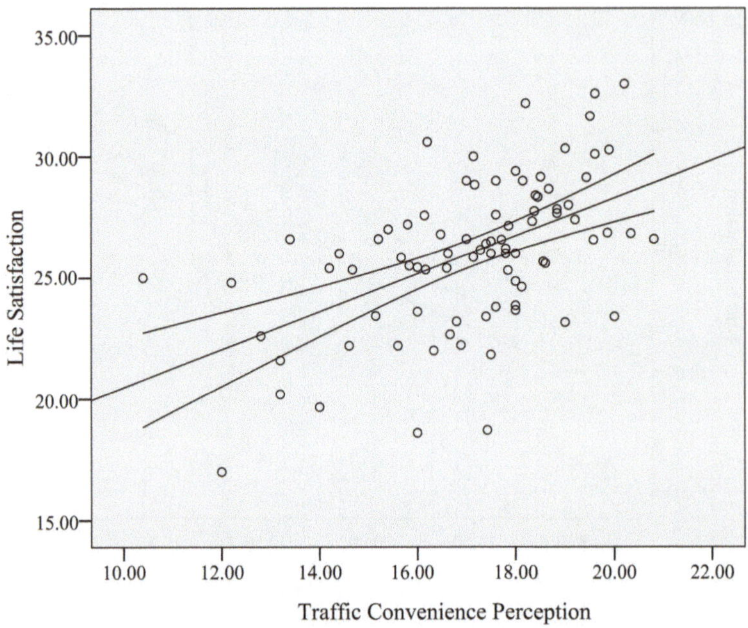


**Fig. 8.22** Fitting the trend of the Relationship between traffic convenience perception and average life satisfaction of residents. ○ Average Life Satisfaction of survey community residents.—Fitting the Trend line and confidence interval. Source: Created by the authors

community residents may be related to Life Satisfaction to some extent. The data Fitting the Trend chart shows that there is a positive correlation between traffic convenience perception and Life Satisfaction, that is, with the improvement of traffic convenience perception, Life Satisfaction increases. The Pearson correlation coefficient was 0.518 ($P < 0.001$), which indicated that there was a significant positive correlation between them. Therefore, geospatial relationships and data fitting indicate that the perception of traffic convenience in the community built environment may be an important factor closely related to Life Satisfaction.

3. Description of the Relationship between Aesthetic Perception of the Community Built Environment and Life Satisfaction

Figures 8.23 and 8.24 show the geospatial relationship and data Fitting the Trend between aesthetic perception of the built environment and Life Satisfaction, respectively. The geospatial relationship between them shows that in the four Main cities, the community with a higher average aesthetic perception will have better Life Satisfaction, while the community with a lower aesthetic perception will have worse Life Satisfaction. Therefore, the geospatial relationship indicates that the average aesthetic perception of community residents may be related to Life Satisfaction. The trend graph of data fitting shows that aesthetic perception is positively correlated with Life Satisfaction; that is, Life Satisfaction increases with the improvement of aesthetic perception. The Pearson correlation coefficient was 0.545 ($P < 0.001$), which indicated that there was a significant positive correlation between them. Therefore, the aesthetic perception of the community's built environment may be an important factor affecting Life Satisfaction.

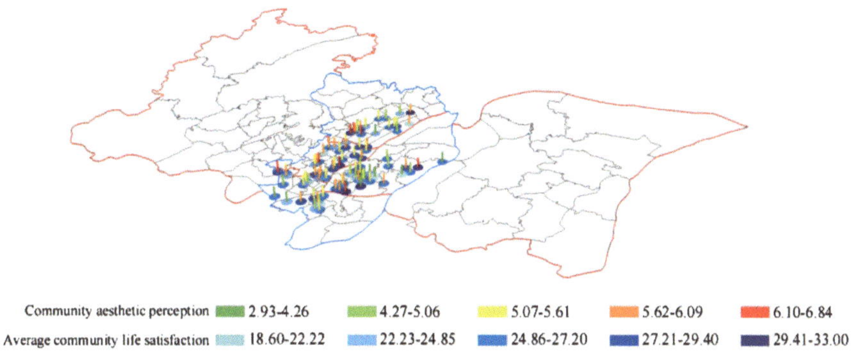


**Fig. 8.23** Geospatial relationship between aesthetic perception and average life satisfaction of residents. Source: Created by the authors

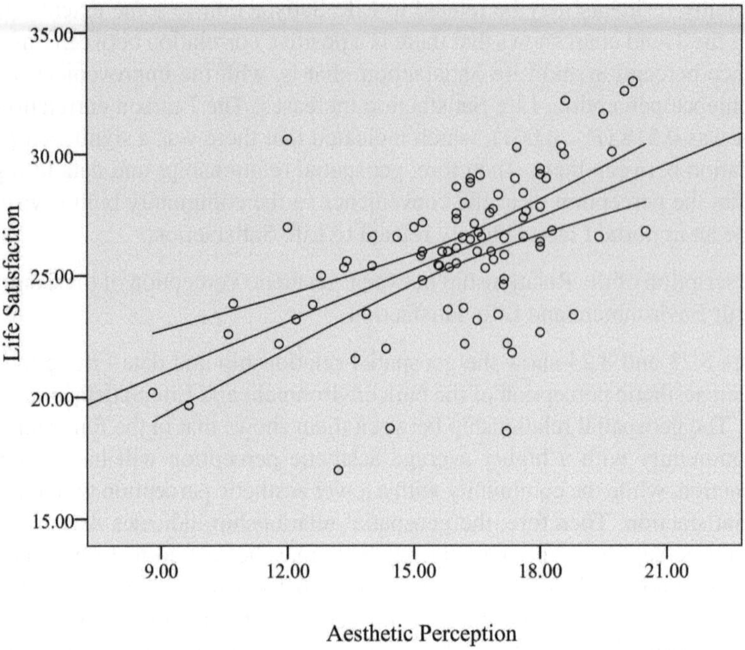


**Fig. 8.24** Fitting the trend of the Relationship between aesthetic perception and average life satisfaction of residents. ○ Average Life Satisfaction of survey community residents.—Fitting the Trend line and confidence interval. Source: Created by the authors

## *8.1.7 Summary of this Section*

In order to explore the difference in Life Satisfaction, explore the relationship between the built environment and Subjective Well-being, explore the geographical spatial relationship and data Fitting the Trend between Objective Built Environment elements and Life Satisfaction, perceived environment quality and Life Satisfaction, and find that there is spatial differentiation of Life Satisfaction among different built environments, however, the correlation between built environment and Life Satisfaction can only grasp the trend between the two, and can not effectively explain the theoretical relationship between the above simple factors and more complex Subjective Well-being, nor can it be inferred that the variables of built environment have an impact or causal relationship on Subjective Well-being. There are three reasons for this:

1. Relevant literature and Fig. 8.1 prove that high population density and high building density, which are the basic characteristics of high-density cities, are closely related to various elements of the built environment, such as higher building density, higher plot ratio, better accessible streets, more bus stops and subway stops distributed, and higher density of living service facilities. Therefore, the mixture degree and quantity of various facilities will increase. Correlation

only considers the relationship between two variables without considering the influence of other built environment factors, so that false correlation may occur.
2. Other personal socioeconomic attributes also influence the relationship between built environment factors and Life Satisfaction. Therefore, without considering the effect of other control variables besides the built environment, the effect of built environment factors and Life Satisfaction may not be rigorous and accurate enough.
3. This section describes the relationship between the built environment of a community and the average Life Satisfaction of its residents. Direct inference from the community-level relationship to the individual level may lead to an "ecological fallacy".

To sum up, the correlation discussed in this section may be insufficient, but three points of cognition can be obtained: firstly, the central Urban area of Guangzhou investigated is characterized by high building density, high Functional Mix Diversity, high road intersection density, many bus lines, many bus stops, large density of various facilities, and close correlation between built environment. A simple discussion of the relationship between the two factors may lead to misjudgment. Second, built environment factors and Life Satisfaction have positive and negative relations in different degrees, which may have different positive and negative effects on Life Satisfaction. Third, the relationship between the built environment and Life Satisfaction at the community level is not clear, which needs further research.

## 8.2 Impact of Objective Built Environment on Life Satisfaction

The former part discusses the relationship between the Objective Built Environment and average Life Satisfaction of community residents, and finds that different Life Satisfaction may accompany different built environment factors, but the relationship between them is still uncertain, so it is necessary to explore the impacts of Objective Built Environment factors on individual Life Satisfaction. In order to better study the impacts and more rigorously test the impacts of the Objective Built Environment. Based on the model constructed in Chap. 2 and the discussion on the relationship between the Objective Built Environment and Life Satisfaction, according to the 5D model, variables or proxy variables reflecting the characteristics of "5D" are included in the study to focus on the impact of the built environment on residents 'Life Satisfaction. This section will use multiple regression analysis to test the following assumptions:

H1: The density of the Objective Built Environment (OBE) has a significant impact on residents 'Life Satisfaction.
H1a: The Population Density (PD) of the Objective Built Environment (OBE) has a significant impact on residents 'Life Satisfaction.

H1b: The Building Density (BD) of the Objective Built Environment (OBE) has a significant impact on residents 'Life Satisfaction.

H2: The diversity of the Objective Built Environment (OBE) has a significant impact on residents 'Life Satisfaction.

H3: The design of the Objective Built Environment (OBE) has a significant impact on residents 'Life Satisfaction.

H4: The distance to transit in the Objective Built Environment (OBE) has a significant impact on residents 'Life Satisfaction.

H4a: The density of Bus Line Density (BLD) in the Objective Built Environment (OBE) has a significant impact on residents 'Life Satisfaction.

H4b: The density of bus stop density (BSD) in the Objective Built Environment (OBE) has a significant impact on residents 'Life Satisfaction.

H5: Destination accessibility in the Objective Built Environment (OBE) has a significant impact on residents 'Life Satisfaction.

The impact of the Objective Built Environment on Life Satisfaction is analysed below.

This study constructs a model to analyze and test the effect of the Objective Built Environment on Life Satisfaction. The regression model variables included density (population density, building density), diversity (Functional Mix Diversity), design (road intersection density), distance to transit (bus line density, bus stop density), destination accessibility (POI point density), and individual socioeconomic characteristics of the respondents within the defined community (i.e., buffer zone with a radius of 1000 meters from the community centroid); the dependent variable was the sum average of the five items of Life Satisfaction[6] (Fig. 8.25).

Prior to conducting regression analysis, comprehensive diagnostics were performed to ensure the model's adherence to the key assumptions of linear regression, including linearity, independence, normality, homoscedasticity, and the absence of multicollinearity.

Scatterplots with fitted lines suggested approximately linear relationships between the dependent and independent variables, satisfying the linearity assumption. The normality and zero-conditional-mean assumptions of the residuals were supported by a standardized normal P-P plot (Fig. 8.26), in which the observed values closely followed the expected diagonal line. The Durbin-Watson statistic ($d = 1.757$) fell within the acceptable range of 1.5 to 2.5, indicating no substantial autocorrelation in the residuals. Homoscedasticity was confirmed by examining the plot of standardized residuals against predicted values, which revealed no evident pattern. Furthermore, variance inflation factors (VIF) for all independent variables ranged from 1.049 to 3.463, well below the common threshold of 5, thus confirming the absence of multicollinearity.

[6] Including individual socioeconomic attributes as control variables in a model analyzing community-level built environment factors helps control for potential confounding effects of individual differences, a common practice in multi-level analysis.

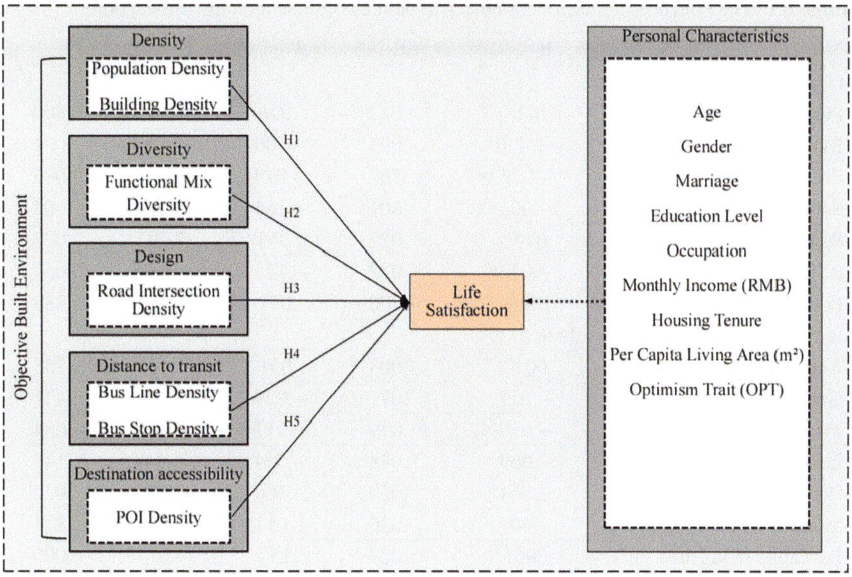


**Fig. 8.25** The effects of objective built environment factors on life satisfaction. Caption: Control variables (Individual Socioeconomic Attribute) are not shown. Source: Created by the authors

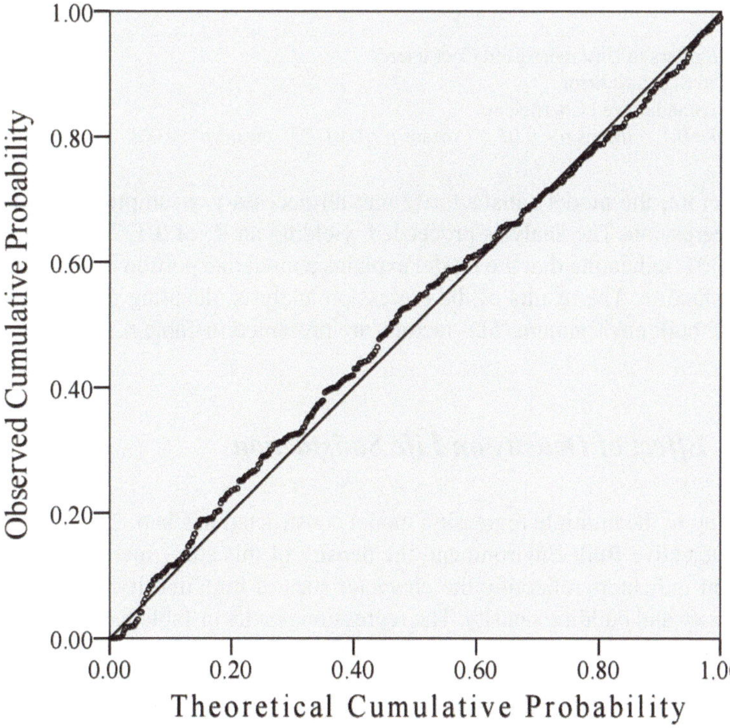


**Fig. 8.26** Normal P-P plot of life satisfaction. Source: Created by the authors

**Table 8.1** The effects of the factors of objective built environment on life satisfaction

| | B | SE | *β* | T | *P* |
|---|---|---|---|---|---|
| Objective built environment | | | | | |
| PD | .011 | .022 | .026 | .512 | .609 |
| BD | -1.430* | .693 | −.106 | −2.065 | .039 |
| FMD | −1.273# | .760 | −.075 | −1.674 | .095 |
| RID | −.002** | .001 | −.164 | −2.864 | .004 |
| BLD | .050* | .022 | .146 | 2.297 | .022 |
| BSD | −.042# | .022 | −.135 | −1.908 | .057 |
| POID | 1.804 | 1.700 | .077 | 1.061 | .289 |
| Individual socioeconomic attribute | | | | | |
| Age | .003 | .003 | .051 | 1.131 | .258 |
| Gender | −.057 | .071 | −.029 | −.804 | .421 |
| Marriage | −.020 | .049 | −.017 | −.408 | .684 |
| Education level | −.069 | .048 | −.064 | −1.444 | .149 |
| Occupation | −.064* | .029 | −.106 | −2.215 | .027 |
| Monthly income (RMB) | −.008 | .036 | −.011 | −.224 | .823 |
| Per capita living area ($m^2$) | .096*** | .024 | .157 | 4.017 | .000 |
| Housing tenure | .046 | .028 | .067 | 1.602 | .110 |
| Optimism trait | | | | | |
| OPT | .496*** | .036 | .537 | 13.750 | .000 |
| Constant | 3.943 | .771 | | 5.115 | .000 |
| $R^2$ | 0.322 | | | | |
| Adj $R^2$ | 0.301 | | | | |

Caption: *B* refers to Unstandardized Coefficient,
SE refers to Standard Error,
*β* refers to Standardized Coefficient.
# means $p < 0.1$, * means $p < 0.05$, ** means $p < 0.01$, *** means $p < 0.001$

Therefore, the model satisfactorily met all necessary assumptions for multiple linear regression. The analysis proceeded, yielding an $R^2$ of 0.322 and an adjusted $R^2$ of 0.301, indicating that the model explains a moderate portion of the variance in life satisfaction. The results of the regression analysis, detailing the impact of the objective built environment '5D' factors, are presented in Table 8.1."

## 8.2.1 *Effect of Density on Life Satisfaction*

According to the multiple regression model constructed in Chap. 2 and the analysis of the Objective Built Environment, the density of this study mainly includes two important indicators reflecting the characteristics of high density, namely population density and building density. The regression results in Table 8.1 show that there is a positive correlation between population density and Life Satisfaction to some extent (B = 0.011, $\beta$ = 0.026), that is, the increase of population density will be accompanied by the improvement of Life Satisfaction, but the significance test results show that the results of population density affecting Life Satisfaction do not reach statistical significance (P = 0.609). This conclusion is similar to that of some studies, such as a study in Shanghai that demonstrated a positive but not significant correlation between population density and mental health (Yin, 2020). For example, a study in the United States proved that population density was negatively correlated with residential satisfaction and not significant (Cao, 2016); a study in Guangzhou proved that population density and Life Satisfaction were negatively correlated and not significant within a 1000 m element buffer zone (Liu et al., 2020); a study in Shanghai proved that population density was negatively correlated with community satisfaction (Shen & Fu, 2019). However, a Beijing study demonstrated a positive but not significant correlation between population density and Life Satisfaction (Wang & Wang, 2020; Yang et al., 2021), so the effect of population density needs further verification and explanation.

The relationship between population density and Life Satisfaction is not significant. On the one hand, population density increases the opportunity to meet and interact with relatives and friends, which enhances the sense of belonging of residents (Leyden, 2003); on the other hand, the increase of population density, especially in the context of high density or ultra-high density, may lead to increased individual sensory and psychological pressure (Yin, 2020), resulting in residents hiding and becoming independent, reducing communication with people, and reducing sense of belonging and satisfaction (French et al., 2014; Lin, 2019). In addition, increased population density is usually accompanied by increased concentration and flow of people, which may bring noise to the environment, intrusion by strangers, increase the chances of contact between crime victims and offenders, and thus reduce the sense of security. Another Chinese study demonstrates that increasing population density often means that people of all shapes and colors gather, and property crime rates increase (Liu et al., 2012). Therefore, the relationship between population density and Life Satisfaction is complex, and the effect is not significant. This also shows that density is a complex concept that needs to be "disentangled" (Churchman, 1999). Therefore, it is necessary to establish a more rigorous measurement model for discussion.

When other variables are controlled, building density and Life Satisfaction are negatively correlated to some extent (B = −1.430, $\beta = -0.106$), that is, building density will be accompanied by a decrease in Life Satisfaction. The results of the significance test showed that building density had a significant effect on Life Satisfaction (P = 0.039). This conclusion is consistent with most studies. The possible explanation lies in that, on the one hand, the increase of building density squeezes the elasticity of public space, brings complicated road traffic, makes it difficult for residents to make convenient passage and convenient travel choice, and the inconvenience brought reduces Life Satisfaction; on the other hand, the increase of building density not only compresses the area of park green space and public space, but also affects the outdoor leisure and activity space of residents; In addition, the increase of building density, especially in the central Urban area, means that buildings of various ages and diversity are densely distributed. The combined effect of the above factors leads to the decline of the aesthetics of park green space, the disorder of street elevation and the diversity of building forms, which reduces the aesthetic perception level. The above factors combine to reduce residents 'Life Satisfaction. This conclusion is basically consistent with another study's conclusion in Guangzhou (Liu et al., 2020).

### *8.2.2 Effects of Diversity on Life Satisfaction*

The results in Table 8.1 show that, when other variables are controlled, the variable reflecting diversity characteristics, Functional Mix Diversity and Life Satisfaction are negatively correlated to a certain extent ($B = -1.273$, $\beta = -0.075$), that is, the improvement of community Functional Mix Diversity will lead to the decrease of Life Satisfaction. The results of the significance test showed that the relationship between them was not statistically significant but practically significant ($P = 0.095$). This conclusion is consistent with some studies. A study of Guangzhou residents 'instantaneous happiness demonstrated that the number of POI types was negatively and significantly correlated with happiness (Su et al., 2024). But it is also inconsistent with some research results. A Guangzhou study using land-use mix as an approximation of Functional Mix Diversity demonstrated that land-use mix significantly positively affects residents 'Life Satisfaction (Su et al., 2019). Another American study also demonstrated that land use mix affects residents 'satisfaction with housing (Cao, 2016). However, the effect of POI density in the built environment was not considered in the above two studies. In general, the density of public facilities in the Functional Mix Diversity has a significant negative effect on Life Satisfaction, while the density of commercial facilities related to daily life has a significant positive effect on Life Satisfaction (Yang et al., 2021). To sum up, due to the different variables in the study design, the influence of Functional Mix Diversity still exists in case studies (Song & Knaap, 2004), and more rigorous models are needed to explore the influence mechanism of Functional Mix Diversity on Life Satisfaction.

### *8.2.3 Effects of Design on Life Satisfaction*

In this study, intersection density representing road connectivity was used as a proxy variable to reflect the "design" dimension of the built environment. When other variables are controlled, intersection density is negatively correlated with Life Satisfaction ($B = -0.002$, $\beta = -0.164$). The results of the significance test showed that intersection density had a significant effect on Life Satisfaction ($P = 0.004$). This conclusion is consistent with most studies.

The possible reasons lie in two aspects: on the one hand, more road intersections may bring better road network connectivity and destination accessibility, improve the traffic environment around the community and increase people's travel convenience; On the other hand, more road intersections may represent the complexity of the traffic system, which means higher road network density, which will also bring increased traffic flow, traffic congestion, air pollution and noise exposure (Appleyard & Lintell, 1972), thus affecting people's Life Satisfaction (Marshall & Garrick, 2011).

### 8.2.4 *Impact of Public Transport Distance on Life Satisfaction*

In the related literature, distance from bus stop and bus stop density are often used to reflect the characteristics of bus travel. In this study, bus line density and bus stop density are included in the analysis.

When controlling for other variables, bus line density was positively correlated with Life Satisfaction (B = 0.050, β = 0.146), that is, the increase of bus line density would lead to an increase in Life Satisfaction. The results of the significance test showed that the relationship between them was statistically significant ($P = 0.022$). Bus Stop Density and Life Satisfaction are negatively correlated (B = -0.042, $\beta = -0.135$), that is, the increase of bus stop density will lead to the decrease of Life Satisfaction. The results of the significance test showed that the relationship between them was not statistically significant but practically significant ($P = 0.057$). Relevant literature research proves that transportation convenience and transportation center (Ma et al., 2018), Rail Transit (Cao, 2013), which in turn affects Life Satisfaction. Other studies have demonstrated that distance to bus stops is significantly negatively correlated with Life Satisfaction (Lin, 2019), i.e., Bus Stop Density is positively correlated with Life Satisfaction. However, the negative correlation found of this study are also in common with some of the conclusions of the study. Another study proved that in the built environment at different levels of the neighborhood committee's administrative jurisdiction, a 1000-meter buffer zone, and a 15-min reachable area, the relationship between bus stop density and Life Satisfaction is at the first two levels. There is a negative correlation between bus stop density and Life Satisfaction, and this conclusion is the same as the results of this study (Liu et al., 2020). However, it is also proven that there is a positive correlation at the latter level, which is different from the results of this study. Therefore, the relationship between bus stop density and Life Satisfaction is unstable at different study sites and different built environment levels (Liu et al., 2020).

After considering both bus line density and bus stop density, this model finds that bus line density is positively correlated with Life Satisfaction but negatively correlated with bus stop density. The possible explanation is that after controlling the density of bus routes, i.e., under the same bus route, the more bus stops, the more difficult it may be for people to choose travel or the timing is disordered, the traffic is scattered, and thus the Life Satisfaction is reduced.[7] This is consistent with some Western studies, where fewer bus stops mean less bus flow and lower traffic, which may lead to higher habitability for residents (Ewing & Handy, 2009).

### 8.2.5 Impact of Destination Accessibility on Life Satisfaction

POI density is used as a proxy variable of destination accessibility, reflecting the density of daily life service facilities. When controlling for other variables, POI Density was positively correlated with Life Satisfaction (B = 1.804, $\beta = 0.077$), i.e., increasing POI density would lead to an increase in Life Satisfaction. The results of the significance test showed that the relationship between them was not statistically significant ($P = 0.289$). This finding is partially consistent with previous research. Some studies have demonstrated a positive, though not always statistically significant, association between POI density and life satisfaction within a 1000 m buffer (Liu et al., 2020; Su et al., 2019), while others found a significant positive correlation with happiness after accounting for POI typology and density (Su et al., 2024).

The number of daily living facilities will also positively affect residents 'Life Satisfaction (Su et al., 2019). In addition, relevant studies have proved that the distance from the nearest commercial center is significantly negatively correlated with Life Satisfaction; that is, the closer to the commercial center, the higher the Life Satisfaction. The increase of POI density means that the density of accommodation, catering, shopping, leisure and entertainment, financial services and other facilities closely related to residents 'lives will increase, thus improving the diversified convenience level of residents (Litman, 2003). At the same time, the increase in POI density will attract more people to the streets, enhance the vitality of the streets, play the role of social supervision and natural monitoring (Jacobs, 1961; Litman, 2003), and improve people's sense of security and satisfaction. However, POI density may also have a negative impact on Life Satisfaction. The increase in people usually accompanies the increase in POI point density flow, which brings negative effects such as environmental noise and traffic congestion, and may also reduce the opportunity for residents to meet and interact at POI points when meeting their needs for daily life services and functions, thus affecting residents' Life Satisfaction.

[7] This nuanced finding has implications for transit planning: enhancing route coverage (positive effect) should be balanced with optimizing stop spacing to mitigate potential negative externalities like traffic scattering and noise, aligning with goals of improving operational efficiency within 'transit-first' strategies.

## 8.3 Summary of this Chapter

Based on the analysis of the characteristics of Objective Built Environment and Life Satisfaction, this chapter discusses the geographical spatial relationship and correlation between Objective Built Environment elements and Life Satisfaction, and finds that there are geographical differences in Life Satisfaction. Then, a multiple regression model is used to further explore the impact of Objective Built Environment factors on Life Satisfaction. There are different and close logical relationships between them.

The results show that building density, Functional Mix Diversity, intersection density and bus stop density have a significant negative effect on Life Satisfaction after controlling for individual socioeconomic variables, that is, Life Satisfaction decreases with the increase of these four factors. At the same time, bus line density has a positive and significant impact on Life Satisfaction; that is, Life Satisfaction increases with the increase in bus routes. Among the significant built environment factors mentioned above, the standardization coefficient shows that the density of road intersections is the most influential factor, followed by the density of bus lines, the density of bus stops, the density of buildings, and finally the degree of functional mixture. Therefore, the sample data support the effects of the five factors on Life Satisfaction, but there is no evidence that population density and POI density significantly affect Life Satisfaction.

Although the above studies found evidence that some elements of the Objective Built Environment had significant effects on Life Satisfaction, there were slight differences in their effects due to differences in the variables included in different studies. Therefore, multiple regression models alone cannot explain how the effects of built environment factors on Life Satisfaction occur for complex Life Satisfaction. In the following, we will examine the mechanism by which the Objective Built Environment affects Life Satisfaction through more rigorous structural equation models.

**Funding** This research was funded by

1. Guangdong Provincial Office of Philosophy and Social Sciences [Grant No. GD24XSH06]
2. Projects of PhDs' Start-up Research of GDUPT [Grant No. 2023bsqd1008].

## References

Appleyard, D., & Lintell, M. (1972). The environmental quality of city streets: The residents' viewpoint. *Journal of the American Planning Association, 38*(2), 84–101.

Cao, X. (2013). The influence of light rail transit on transit-oriented development in China. *Transportation Research Part A: Policy and Practice, 50*, 240–251.

Cao, X. (2016). How does neighborhood design affect life satisfaction? Evidence from twin cities. *Travel Behaviour and Society, 5*, 68–76.

Cao, X., & Ewing, R. (2022). How do perceived and objective measures of the built environment influence travel behavior? A meta-analysis. *Transportation, 49*(3), 787–816.

Cervero, R., & Kockelman, K. (1997). Travel demand and the 3Ds: Density, diversity, and design. *Transportation Research Part D: Transport and Environment, 2*(3), 199–219.

Churchman, A. (1999). Disentangling the concept of density. *Journal of Planning Literature, 13*(4), 389–411.

Ewing, R., & Cervero, R. (2010). Travel and the built environment. *Journal of the American Planning Association, 76*(3), 265–294.

Ewing, R., & Cervero, R. (2017). Does compact development make people drive less? *Journal of the American Planning Association, 83*(1), 7–18.

Ewing, R., & Handy, S. (2009). Measuring the unmeasurable: Urban design qualities related to walkability. *Journal of Urban Design, 14*(1), 65–84.

French, S., Wood, L., & Foster, S. A. (2014). The impact of crowding on mental health in urban environments. *Environment and Behavior, 46*(2), 189–210.

Jacobs, J. (1961). *The death and life of great American cities*. Random House.

Leyden, K. M. (2003). Social capital and the built environment: The importance of walkable neighborhoods. *American Journal of Public Health, 93*(9), 1546–1551.

Lin, T. (2019). The influence of transit accessibility on subjective Well-being: Evidence from Taipei. *Transportation Research Part D: Transport and Environment, 67*, 167–178.

Litman, T. (2003). Economic value of walkability. *Transportation Research Record: Journal of the Transportation Research Board, 1828*(1), 3–11.

Liu, Y., Wang, R., & Lu, Y. (2020). The nonlinear relationship between built environment and life satisfaction in Guangzhou. *Landscape and Urban Planning, 195*, 103725.

Liu, Z., Wang, Y., & Chen, S. (2012). The impact of population density on property crime in Chinese cities. *Journal of Criminal Justice, 40*(5), 423–431.

Ma, L., Li, Y., & Li, W. (2018). The impact of transit-oriented development on housing satisfaction in Chinese cities. *Transport Policy, 66*, 1–10.

Marshall, W. E., & Garrick, N. W. (2011). Does street network design affect traffic safety? *Accident Analysis & Prevention, 43*(3), 769–781.

Shen, Y., & Fu, X. (2019). The impact of population density on community satisfaction in urban China. *Habitat International, 85*, 1–10.

Song, Y., & Knaap, G. J. (2004). Measuring the effects of mixed land uses on housing values. *Regional Science and Urban Economics, 34*(6), 663–680.

Su, L., Chen, J., & Li, Y. (2019). The impact of land use mix on life satisfaction in Guangzhou. *Cities, 87*, 1–10.

Su, L., Zhang, H., & Wang, Y. (2024). POI diversity and happiness: A spatial analysis in high-density cities. *Applied Geography, 154*, 102345.

Wang, D., & Zhou, M. (2017). The built environment and travel behavior in urban China: A literature review. *Transportation Research Part D: Transport and Environment, 52*, 574–585.

Wang, F., & Wang, D. (2020). Changes in residential satisfaction after home relocation: A longitudinal study in Beijing, China. *Urban Studies, 57*(3), 583–605.

Yang, J., Li, X., & Zhang, W. (2021). The impact of built environment on life satisfaction in Beijing: A multi-level analysis. *Journal of Environmental Psychology, 75*, 101619.

Yin, L. (2020). The impact of built environment on mental health in high-density cities: A case study of Shanghai. *Journal of Urban Health, 97*(3), 345–360.

# Chapter 9
# Unveiling the Mechanisms: Mediating Pathways from the Objective Built Environment to Life Satisfaction

**Abstract** Extending prior findings that the objective built environment (OBE) significantly influences residents' life satisfaction (LS), this chapter aims to uncover the underlying mechanisms through which these effects occur. Specifically, it examines the mediating roles of Perceived Environmental Quality (PEQ), Perceived Safety (PES), Social Capital (SC), and Self-Rated Health (SRH) in the relationship between OBE and LS. Using a structural equation modeling (SEM) approach, the chapter first validates the measurement models for latent constructs and then tests a series of direct and indirect pathways based on proposed hypotheses. Section 9.1 presents confirmatory factor analysis, which confirms the reliability and validity of the measurement models for PEQ, SC, and LS. Section 9.2 details the structural model results, revealing significant effects of OBE factors—such as Population Density, Building Density, Functional Mix Diversity, Road Intersection Density, Bus Line Density, Bus Stop Density, POI Density—on the mediators, and subsequently on LS. Section 9.3 employs bootstrapping methods to test specific indirect effects, identifying multiple significant mediation pathways, including simple and serial mediations involving PEQ, PES, SC, and SRH. The findings demonstrate that OBE influences LS not only directly but also through intertwined perceptual and social mechanisms, highlighting the importance of mediating processes in urban well-being research. This chapter provides a nuanced understanding of how built environments shape subjective well-being, offering theoretical and practical insights for urban planning and policy-making.

**Keywords** Objective built environment · Life satisfaction · Perceived environmental quality · Perceived safety · Social capital · Self-rated health · Structural equation modeling · Mediation analysis · Urban Well-being · Guangzhou

Grounding the findings of Chap. 8, which identified significant associations between objective built environment (OBE) factors and residents' life satisfaction (LS) through multiple regression analysis, this chapter delves deeper into the underlying mechanisms through which these influences occur. While Chap. 8 established that



H. Li, G. Shen, *The Anatomy of Subjective Well-Being*, Resilient and Friendly Living Environments, https://doi.org/10.1007/978-981-95-4937-5_9

certain OBE elements directly affect LS, it also highlighted the need to explore the mediating roles of perceptual and social variables to fully understand the pathways of influence. Using a structural equation modeling (SEM) approach, Chap. 9 examines how OBE factors impact LS through four key mediators: perceived environmental quality (PEQ), perceived safety (PES), social capital (SC), and self-rated health (SRH). This chapter not only tests the significance of these mediating and chain-mediating effects but also provides a more nuanced understanding of how urban built environments shape subjective well-being, thereby bridging macro-level spatial analyses with micro-level perceptual and social processes.

The previous analysis examined the impact of the Objective Built Environment on Life Satisfaction. It was found that elements of the Objective Built Environment, such as Building Density, Functional Mix Diversity, road intersection density, and bus stop density, significantly reduce residents' Life Satisfaction, while bus line density significantly enhances it. How do these impacts of the built-environment elements occur and through which pathways? In the following, based on the structural equation model (SEM) of the relationships among the Objective Built Environment, perceived environment, self-rated health, public safety, Social Capital for public safety, and Life Satisfaction constructed in Chap. 2, we will further explore the impact of the Objective Built Environment on Life Satisfaction through mediators.

For a rigorous analysis, this chapter is mainly divided into three parts. First, the measurement model is tested. Second, the regression results of the paths through which built environment factors influence mediating variables and those through which mediating variables affect Life Satisfaction are reported and analyzed. Finally, the Bootstrap mediation effect test technique of the structural equation model is used to test the indirect effects statistically, and the significant paths are interpreted (Fang et al., 2011).

## 9.1 Measurement Model Analysis in the Structural Equation of the Impact of Objective Built Environment on Life Satisfaction

The structural equation model consists of two parts: the measurement model and the structural model (Hair et al., 2011; Wu, 2009). Confirmatory factor analysis (CFA) is an important part of SEM.[1] After verification and reliability and validity tests of the measurement models for each construct (Brown, 2015), it can be ensured that the measurement model correctly reflects the constructs of the study (Thompson, 2000), and then SEM is worth conducting (Brown, 2015; Kline, 2015).

[1] CFA is used to validate the structure of latent variables by testing how well measured variables represent underlying constructs. For a comprehensive guide on CFA applications in social sciences, see Brown (2015).

The research begins with the measurement model, estimating the measurement models of the main latent variables, namely Perceived Environmental Quality (PEQ), Social Capita (SC), and Life Satisfaction (LS).

Table 9.1 presents the goodness-of-fit indices for the measurement model of perceived environmental quality. The model demonstrated a robust fit to the data, with all incremental indices (TLI, CFI, IFI, RFI) exceeding .90 (Bentler & Bonett, 1980) and both absolute fit indices (RMSEA and SRMR) below the .08 cutoff (Hu & Hu & Bentler, 1999).[2] This construct, measured through three dimensions—a) perceived diversity and convenience (four items), b) perceived transportation convenience (three items), and c) perceived aesthetics (three items)—also exhibited strong internal consistency and convergent validity. The composite reliability (CR) was 0.874, and the average variance extracted (AVE) was 0.669 (Fornell & Larcker, 1981). As detailed in Table 9.4, all factor loadings were significant, with the majority exceeding 0.7; only item TCP3 showed a good but slightly lower loading (>0.600), collectively indicating acceptable to excellent indicator reliability (Hair et al., 2019). These results confirm that the latent construct is well-represented by its ten measurement items. The detailed factor structure is illustrated in Fig. 9.1.

The measurement model for social capital likewise achieved excellent fit (Table 9.2). All model fit indices met or surpassed the recommended standards: incremental indices >0.90 (Bentler & Bonett, 1980) and RMSEA/SRMR <0.08 (Hu & Bentler, 1999). The construct's reliability and validity were also confirmed. The composite reliability (CR) was strong at 0.931, and the average variance extracted (AVE) of 0.630 affirmed convergent validity (Fornell & Larcker, 1981). As shown in Table 9.4, all eight items loaded significantly ($p < .001$) onto their intended factor with standardized coefficients greater than 0.7, demonstrating strong indicator reliability (Hair et al., 2019). This provides conclusive evidence that the eight-item scale accurately measures the social capital construct. The factor structure is depicted in Fig. 9.2.

Finally, the life satisfaction measurement model also showed a strong fit to the data (Table 9.3), conforming to all established fit criteria (Bentler & Bonett, 1980; Hu & Bentler, 1999). The scale exhibited high internal consistency, with a composite reliability (CR) of 0.881. The average variance extracted (AVE) value of 0.598 confirmed its convergent validity (Fornell & Larcker, 1981). Furthermore, all five

**Table 9.1** Goodness-of-fit indices for the perceived environment quality measurement model

| Indicators | CMIN/DF | RMSEA | CFI | GFI | AGFI | IFI | NFI | SRMR |
|---|---|---|---|---|---|---|---|---|
| Recommended criterion | <3 | <0.08 | >0.9 | >0.9 | >0.9 | >0.9 | >0.9 | <0.08 |
| The model fit indicators | 2.789 | 0.056 | 0.981 | 0.969 | 0.956 | 0.981 | 0.970 | 0.031 |

Source: Created by the authors

[2] The cutoff of 0.90 for incremental fit indices is widely accepted, though some scholars advocate for stricter thresholds (e.g., CFI > 0.95) in complex models. See Hu & Hu and Bentler (1999) for a discussion on sensitivity and specificity of fit indices.

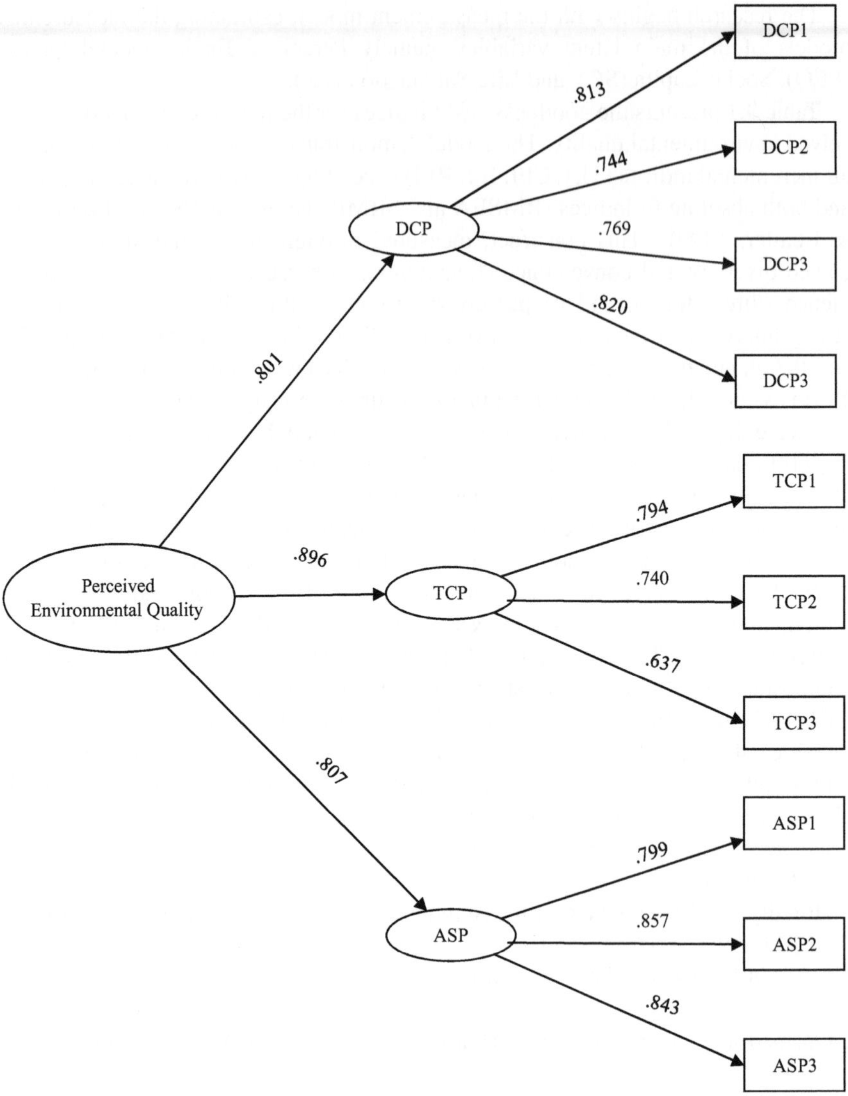


**Fig. 9.1** Confirmatory factor analysis results for the perceived environment quality (PEQ). Source: Created by the authors

factor loadings were highly significant ($p < .001$) and substantial (standardized coefficients >0.7), as reported in Table 9.4, which meets the threshold for strong indicator reliability (Hair et al., 2019). In summary, the results robustly support the five-item Life Satisfaction scale as a reliable and valid measure. The corresponding factor model is presented in Fig. 9.3.

Additionally, the Perceived Safety dimension was represented by two items following the deletion of one item during the pre-test. Due to the identification

**Table 9.2** Goodness-of-fit indices for the social capital measurement model

| Indicators | CMIN/DF | RMSEA | CFI | GFI | AGFI | IFI | NFI | SRMR |
|---|---|---|---|---|---|---|---|---|
| Recommended criterion | <3 | <0.08 | >0.9 | >0.9 | >0.9 | >0.9 | >0.9 | <0.08 |
| The model fit indicators | 1.560 | 0.031 | 0.997 | 0.991 | 0.956 | 0.997 | 0.995 | 0.061 |

Source: Created by the authors

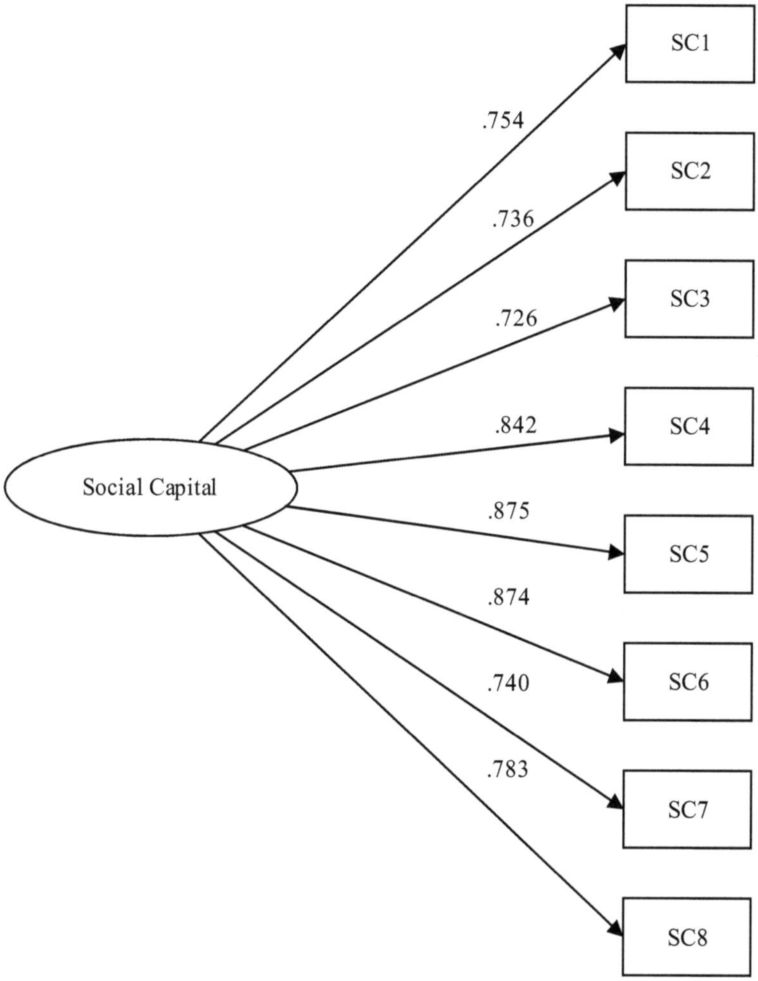


**Fig. 9.2** Confirmatory factor analysis results for the social capital (SC). Source: Created by the authors

constraints associated with latent variables measured by only two indicators, the scores of these two items were averaged to form a composite score, which was then treated as an observed variable in the model (Hayduk, 1987). This approach of using composite scores for parcels with few items is supported in the literature to enhance

**Table 9.3** Goodness-of-fit indices for the life satisfaction measurement model

| Indicators | CMIN/DF | RMSEA | CFI | GFI | AGFI | IFI | NFI | SRMR |
|---|---|---|---|---|---|---|---|---|
| Recommended criterion | <3 | <0.08 | >0.9 | >0.9 | >0.9 | >0.9 | >0.9 | <0.08 |
| The model fit indicators | 0.729 | 0.000 | 1.000 | 0.997 | 0.992 | 1.000 | 0.997 | 0.009 |

Source: Created by the authors

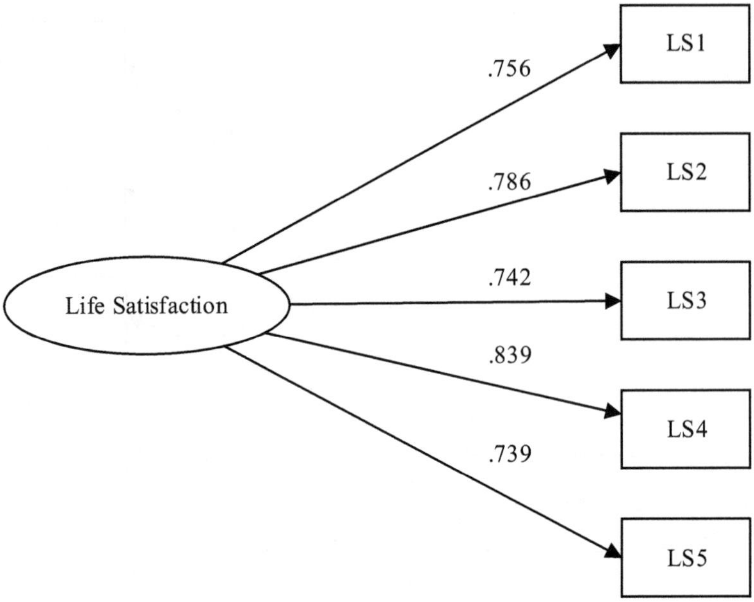


**Fig. 9.3** Confirmatory factor analysis results for the life satisfaction (LS). Source: Created by the authors

model stability (Little et al., 2013). Similarly, the single-item self-rated health measure was also included as an observed variable.

For the multi-item latent constructs in the model, a composite reliability (CR) value greater than 0.7 for each dimension indicates adequate internal consistency, with higher values indicating better reliability, meeting the recommended standards (Fornell & Larcker, 1981; Hair et al., 2022). An average variance extracted (AVE) value greater than 0.5 indicates that the latent variable has satisfactory explanatory power for its measurement variables, with higher values indicating better convergent validity (Fornell & Larcker, 1981).

In this specific model examining the influence of the Objective Built Environment on Subjective Well-being, the core independent variables (objective built environment factors) and the processed security composite score are observed variables. As observed variables, their measurement error is not estimated separately in the model specification, and thus, conventional reliability and AVE metrics are not applicable. The overall measurement model for any latent constructs demonstrated good properties as reported above. Therefore, based on the above results, the measurement

**Table 9.4** The statistics of latent variable, measurement model parameter estimation and convergent validity

| Latent Variable | Standardized Coefficient | Cronbach's α | CR | AVE |
|---|---|---|---|---|
| Perceived environment quality (PEQ) | | 0.871 | 0.874 | 0.669 |
| DCP | 0.801 | | | |
| TCP | 0.896 | | | |
| AEP | 0.807 | | | |
| Perceived safety (PES) | | 0.815 | 0.819 | 0.604 |
| PES1 | 0.674 | | | |
| PES2 | 0.866 | | | |
| PES3 | 0.779 | | | |
| Social capital(SC) | | 0.928 | 0.931 | 0.630 |
| SC1 | 0.754 | | | |
| SC2 | 0.736 | | | |
| SC3 | 0.726 | | | |
| SC4 | 0.842 | | | |
| SC5 | 0.875 | | | |
| SC6 | 0.874 | | | |
| SC7 | 0.740 | | | |
| SC8 | 0.783 | | | |
| Life satisfaction (LS) | | 0.879 | 0.881 | 0.598 |
| LS 1 | 0.756 | | | |
| LS 2 | 0.786 | | | |
| LS 3 | 0.742 | | | |
| LS 4 | 0.839 | | | |
| LS 5 | 0.739 | | | |

Caption: The factors of the objectively measured built environment are based on objective spatial data, without reporting psychometric properties like CR or AVE. (Ewing & Cervero, 2010) Since self-rated health was assessed using a single item, it is not standard practice to report its CR or AVE (Hair et al., 2022)
Source: Created by the authors

model exhibits satisfactory reliability and validity, allowing for the proceeding analysis of the structural equation model.

## 9.2 Analysis of the Structural Model of the Objective Built Environment on Life Satisfaction

In the previous section, the reliability and validity of the measurement model of the structural equation model were tested. According to the structural equation model of Objective Built Environment affecting Life Satisfaction through intermediary factors constructed in the previous section, this section tests the structural model of

Objective Built Environment affecting Life Satisfaction, focusing on intermediary effects, and tests the following hypothesis[3]:

H1: The Objective Built Environment affects Life Satisfaction.

H2: Objective Built Environment affects Life Satisfaction through Perceived Environmental Quality (OBE → PEQ → LS).

H3: Objective Built Environment affects Life Satisfaction through Perceived Safety (OBE → PES → LS).

H4: Objective Built Environment affects Life Satisfaction through Social Capital (OBE → SC → LS).

H5:Objective Built Environment affects Life Satisfaction through Self-Rated Health (OBE → SRH → LS).

H6: The Objective Built Environment affects Life Satisfaction through the chain-mediating roles of Perceived Environmental Quality and Perceived Safety (OBE → PEQ → PES → LS).

H7: The Objective Built Environment affects Life Satisfaction through the chain-mediating roles of Perceived Environmental Quality and Social Capital (OBE → PEQ → SC → LS).

H8: The Objective Built Environment affects Life Satisfaction through the chain-mediating roles of Perceived Environmental Quality and Self-Rated Health (OBE → PEQ → SRH → LS).

H9: The Objective Built Environment affects Life Satisfaction through the chain-mediating roles of Perceived Safety and Social Capital (OBE → PES → SC → LS).

H10: The Objective Built Environment affects Life Satisfaction through the chain-mediating roles of Perceived Safety and Self-Rated Health (OBE → PES → SRH → LS).

H11: The Objective Built Environment affects Life Satisfaction through the chain-mediating roles of Social Capital and Self-Rated Health (OBE → SC → SRH → LS).

H12: The Objective Built Environment affects Life Satisfaction through the chain-mediating roles of Perceived Environmental Quality, Perceived Safety, and Social Capital (OBE → PEQ → PES → SC → LS).

H13: The Objective Built Environment affects Life Satisfaction through the chain-mediating roles of Perceived Environmental Quality, Perceived Safety, and Self-Rated Health (OBE → PEQ → PES → SRH → LS).

H14: The Objective Built Environment affects Life Satisfaction through the chain-mediating roles of Perceived Environmental Quality, Social Capital, and Self-Rated Health (OBE → PEQ → SC → SRH → LS).

H15: The Objective Built Environment affects Life Satisfaction through the chain-mediating roles of Perceived Safety, Social Capital, and Self-Rated Health (OBE → PES → SC → SRH → LS).

---

[3] These hypotheses are based on a theoretical framework that links the built environment to well-being through the pathways of perceived and social capital and health, and are supported by previous work (e.g., Ewing & Cervero, 2010; Mouratidis, 2018).

H16: The Objective Built Environment affects Life Satisfaction through the chain-mediating roles of Perceived Environmental Quality, Perceived Safety, Social Capital, and Self-Rated Health (OBE → PEQ → PES → SC → SRH → LS).

In this structural equation model (SEM), the maximum likelihood estimation (MLE) method is used to estimate the parameters and test the comprehensive relationship among SEM variables. MLE is a full information estimation method for normally distributed, unbiased and asymptotically normal continuous variables (Hair, 2009; Kline, 2023). Although MLE has strong robustness and statistical validity when the data are not normally distributed, the mean error of parameter estimates may still be biased (Boomsma & Hoogland, 2001; Chou & Bentler, 1995). Therefore, a bootstrap procedure was applied to test the structural model in this study. In SEM analysis, the fitness index determines whether the study model is acceptable or not. If the degree of fit is acceptable, it means that the research model fits the sample data well, but it does not mean that the prediction ability of the model is good. When testing the fitness of the whole model, it is essential to examine the error variance term, standardization coefficient and standard error are tested first (Babin & Anderson, 2011; Wen et al., 2004). The results of model verification did not violate the estimation, thus allowing the model fitness to be tested (Wu, 2009).

Based on model fit metrics, we evaluated the degree of model fit using comparative fit index (CFI), goodness of fit index (GFI), adjusted goodness of fit index (AGFI), incremental fit index (IFI), normalized fit index (NFI), and standard root mean square residuals (SRMR) and root mean square approximation errors (RMSEA). The study reports the main goodness-of-fit indices of the structural equation model in Table 9.5 (Sivo et al., 2006). In the test of absolute goodness-of-fit indices, the chi-square ($\chi^2$) is the most primitive index in SEM, representing the difference between the sample covariance matrix and the model-implied covariance matrix. Generally, a smaller $\chi^2$ value is desirable, and a non-significant p-value ($P > 0.05$) suggests a good fit between the hypothesized model and the observed data (Wu, 2013). In this case, the $\chi^2$ is 342.130 and $p < 0.05$, suggesting a significant discrepancy between the sample data and the model, which may indicate a poor fit. The chi-square test is sensitive to the multivariate normality of data. Upon testing, the multivariate normal kurtosis was 66.323, indicating that the data do not meet the multivariate normality (Kline & Little, 2011). This violation can lead to an overestimation of the $\chi^2$ value, an underestimation of the standard error, and an increase in the CR value (CR = 23.388), resulting in a significant P value ($p < 0.001$). As noted in the literature, a CR value exceeding 5.00 typically indicates a departure from multivariate normality (Bentler & Wu, 2005; Savalei & Bentler, 2005). To discern whether the significant p-value stems from large sample size or genuine model misspecification (Bollen & Long, 1993; Golob, 2003; Jackson, 2003), we implemented the Bollen-Stine bootstrap correction (Bollen & Stine, 1992). Following 2000 bootstrap iterations, the Bollen-Stine, $p = .0005$ ($p < .05$) indicates that the misfit is attributable to sample size rather than model inadequacy. Accordingly, the model was adjusted following methodological recommendations (Byrne, 2009). After

**Table 9.5** Goodness-of-fit indices for the research model

| Fit Index | Ideal Criterion | Uncorrected Model | Bollen-Stine Corrected Model | Interpretation |
|---|---|---|---|---|
| $\chi^2$ | Smaller-is-better | 342.130 | 111.661 | Smaller-better |
| DF | Smaller-is-better | 103 | 103.000 | Smaller-better |
| $\chi^2$/DF | $\chi^2$/DF < 3 | 3.322 | 1.084 | Excellent |
| RMSEA | <0.080 | 0.064 | 0.012 | Excellent |
| SRMR | <0.080 | 0.050 | | Excellent |
| GFI | >0.800 acceptable >0.900 excellent | 0.953 | 0.980 | Excellent |
| AGFI | >0.800 acceptable >0.900 excellent | 0.873 | 1.000 | Excellent |
| TLI | >0.900 | 0.889 | 0.996 | Excellent |
| CFI | >0.900 | 0.955 | 0.998 | Excellent |
| IFI | >0.900 | 0.956 | 0.998 | Excellent |
| Hoelter's N(CN) | >200 | 233.000 | 524.757 | Excellent |
| RFI | >0.900 | 0.849 | 0.951 | Excellent |

Caption: The chi-square correction of the structural equation model was performed using the Bollen-Stine method (Bollen & Stine, 1990), and the model fit was recalculated. Bollen-Stine correction with 2000 bootstraps (Bollen & Long, 1993), $p = .0005$. All $\chi^2$ and df values satisfy the smaller-is-better criterion (Kline, 2023)

$\chi^2$/df = Chi-square to degrees of freedom ratio

Source: Created by the authors

correction, the mean $\chi^2$ was 111.661 with $\chi^2$/df = 1.084, which aligns with conventional standards (Kline, 2023).

Model fit indices were re-evaluated subsequently. Even before correction, the model showed an acceptable fit (RMSEA = 0.064, SRMR = 0.050). After applying the correction, the RMSEA improved to 0.012, indicating excellent fit (Hu & Hu & Bentler, 1999). Table 9.5 compares the fit indices before and after correction. All reported indices—GFI, AGFI, TLI, CFI, IFI, and RFI—exceeded the recommended threshold of 0.90 (Bentler & Bonett, 1980), collectively supporting the robustness of the model fit. Given satisfactory reliability, validity, and overall model fit, we proceeded to examine structural path coefficients. Table 9.6 presents regression results for relationships among exogenous variables (Objective Built Environment), mediators, and the endogenous variable (Life Satisfaction). Columns 1–4 in Table 9.6 report the effects of the Objective Built Environment on four mediators: Perceived Environmental Quality, Perceived Safety, Social Capital, and Self-Rated Health.

Table 9.6 presents the regression results from the structural equation model, illustrating the relationships among the Objective Built Environment (OBE), the mediating variables, and the dependent variable, Life Satisfaction (LS). Columns 1 through 4 report the effects of the Objective Built Environment on the four mediating factors: Perceived Environmental Quality, Perceived Safety, Social Capital, and Self-Rated Health, respectively. A column-by-column analysis is provided below.

### 9.2.1 *The Effects of the OBE Factors on the Mediating Variables*

In structural equation modeling, Squared Multiple Correlations (SMC) is a core indicator in structural equation models (SEM). To ensure the transparency of SEM results, it is necessary to report the SMC for each endogenous structure (Hancock & Mueller, 2011). SMC is used to measure the proportion of the variance of an endogenous variable (dependent variable) explained by the model, and it is essentially equivalent to the coefficient of determination ($R^2$) in multiple regression. The study uses the following method to calculate the squared multiple correlation coefficient for each endogenous variable (Kline, 2016), with the formula as follows:

$$\text{SMC} = 1 - \frac{\text{Disturbance variance}}{\text{Variance of endogenous variable})}$$

The results indicate that the squared multiple correlation (SMC) values are 0.407 for Perceived Environmental Quality (PEQ), 0.708 for Perceived Safety (PES), 0.440 for Social Capital (SC), 0.534 for Self-Rated Health (SRH), and 0.558 for the ultimate dependent variable, Life Satisfaction (LS). This means that the predictors collectively explain 40.7% of the variance in PEQ, 70.8% in PES, 44.0% in SC, 53.4% in SRH, and 55.8% in LS. These results suggest that the structural equation model exhibits strong explanatory power (Kline, 2023). The subsequent section will analyze the specific influencing effects on Perceived Environmental Quality (PEQ), Perceived Safety (PES), Social Capital (SC), Self-Rated Health (SRH), and Life Satisfaction (LS).

1. The Effect of Objective Built Environment Factors on the Perceived Environmental Quality (PEQ)

Column 1 of Table 9.6 reports the impact of the Objective Built Environment on Perceived Environmental Quality (PEQ). The Fig. 9.4 shows the effects of the factors of OBE on PEQ. The results demonstrate that, after controlling for other variables, Building Density (BD), Functional Mix Diversity (FMD), Bus Line Density (BLD), and POI Density (POID) have statistically significant impacts on PEQ ($p < 0.05$), while Bus Stop Density (BSD) has a marginally significant impact on PEQ ($p < 0.1$). Among them, Building Density ($\beta = -0.151$), Functional Mix Diversity ($\beta = -0.098$), and Bus Stop Density ($\beta = -0.140$) have significant negative impacts on PEQ. In contrast, Bus Line Density (BLD) ($\beta = 0.193$) and POI

**Table 9.6** The effects of OBE factors and mediating factors on life satisfaction

| | B (SE) | β | B | β | B | β | B | β | B | β |
|---|---|---|---|---|---|---|---|---|---|---|
| | PEQ | PEQ | SEC | SEC | SC | SC | SRH | SRH | LS | LS |
| PD | −.012<br>(.021) | −.032 | −.067**<br>(.021) | −.154 | .051*<br>(.023) | .118 | −.034<br>(.022) | −.074 | .021<br>(.021) | .049 |
| BD | −1.768**<br>(.660) | −.151 | 1.968**<br>(.686) | .145 | −.291<br>(.750) | −.021 | 2.244**<br>(.722) | .159 | −.945<br>(.703) | −.070 |
| FMD | −1.455*<br>(.723) | −.098 | 3.346***<br>(.748) | .195 | −.867<br>(.862) | −.050 | .693<br>(.828) | .039 | −.930<br>(.788) | −.054 |
| RID | .000<br>(.001) | −.029 | .001<br>(.001) | .042 | .000<br>(.001) | .007 | .001#<br>(.001) | .092 | −.002**<br>(.001) | −.161 |
| BLD | .057**<br>(.021) | .193 | −.013<br>(.022) | −.037 | −.062**<br>(.023) | −.180 | −.031<br>(.022) | −.086 | .030<br>(.021) | .087 |
| BSD | −.038#<br>(.021) | −.140 | .023<br>(.021) | .072 | .062**<br>(.022) | .197 | .029<br>(.022) | .088 | −.035#<br>(.021) | −.112 |
| POID | 5.604***<br>(1.559) | .271 | −2.898#<br>(1.634) | −.121 | −6.830***<br>(1.740) | −.285 | −3.733*<br>(1.744) | −.150 | .215<br>(1.679) | .009 |
| PEQ | | | .849***<br>(.072) | .735 | .665***<br>(.127) | .573 | .555***<br>(.134) | .460 | .354**<br>(.133) | .306 |
| SEC | | | | | .010<br>(.103) | .010 | −.030<br>(.099) | −.028 | .153#<br>(.093) | .153 |
| SC | | | | | | | .008<br>(.053) | .007 | .166***<br>(.050) | .167 |
| SRH | | | | | | | | | .070<br>(.054) | .073 |
| Age | −.003<br>(.003) | −.049 | −.001<br>(.003) | −.017 | .004<br>(.003) | .057 | .003<br>(.003) | .043 | .004<br>(.003) | .069 |

(continued)

**Table 9.6** (continued)

| | B (SE) | $\beta$ | B | $\beta$ | B | $\beta$ | B | $\beta$ | B | $\beta$ |
|---|---|---|---|---|---|---|---|---|---|---|
| | PEQ | PEQ | SEC | SEC | SC | SC | SRH | SRH | LS | LS |
| Gender | −.002 (.068) | −.001 | −.046 (.070) | −.023 | .135# (.073) | .068 | −.049 (.071) | −.024 | −.074 (.067) | −.037 |
| Marriage | −.039 (.046) | −.037 | −.255*** (.048) | −.210 | .062 (.056) | .050 | −.045 (.053) | −.035 | .036 (.051) | .029 |
| Education level | .032 (.046) | .034 | .069 (.048) | .063 | −.107* (.050) | −.098 | −.059 (.048) | −.052 | −.078# (.046) | −.072 |
| Occupation | −.007 (.027) | −.013 | .005 (.028) | .008 | −.021 (.029) | −.034 | −.057* (.028) | −.089 | −.053* (.027) | −.086 |
| Monthly income (RMB) | .003 (.034) | .005 | −.055 (.035) | −.075 | −.055 (.037) | −.074 | −.041 (.036) | −.053 | .016 (.034) | .022 |
| Housing tenure | −.025 (.026) | −.041 | −.046# (.027) | −.066 | .070* (.029) | .100 | −.145*** (.028) | −.200 | .070* (.028) | .101 |
| Per capita living area | .040# (.022) | .075 | .018 (.023) | .029 | .047* (.024) | .076 | .007 (.023) | .012 | .060** (.022) | .097 |
| OPT | .439*** (.037) | .543 | .066 (.046) | .071 | .134** (.046) | .143 | .263*** (.044) | .270 | .208*** (.044) | .222 |

Caption: *B* refers to Unstandardized Coefficient. SE refers to Standard Error. $\beta$ refers to Standardized Coefficient. # means $p < 0.1$, * means $p < 0.05$, ** means $p < 0.01$, *** means $p < 0.001$

Source: Created by the authors

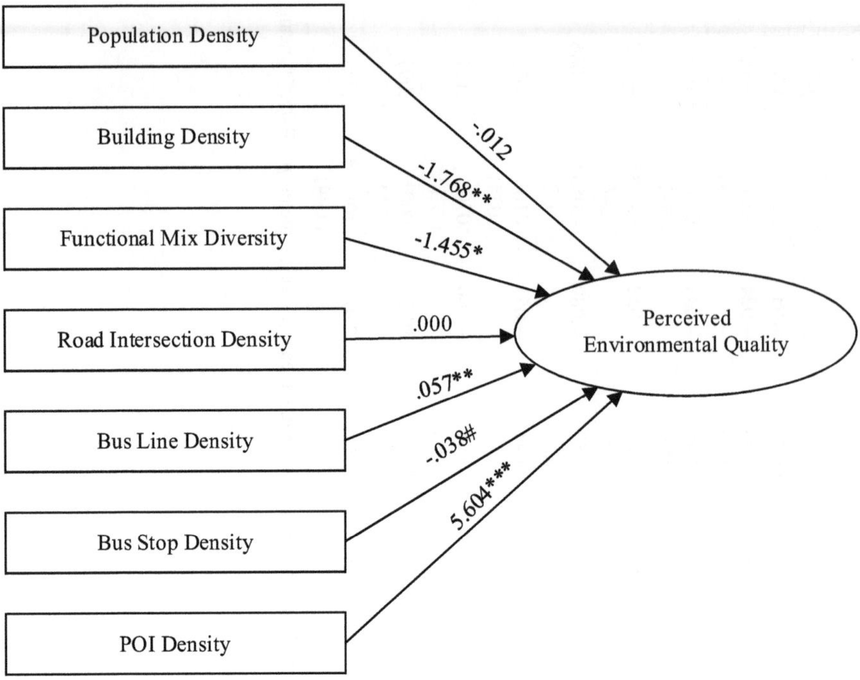


**Fig. 9.4** The effect of objective built environment factors on the perceived environmental quality (PEQ). Source: Created by the authors

Density ($\beta = 0.271$) have significant positive impacts on PEQ. Among the individual socioeconomic attributes, per capita housing area ($p < 0.1$) and optimistic personality ($p < 0.001$) both significantly influence the evaluation of Perceived Environmental Quality. Since this study focuses on the impact of Objective Built Environment factors on Perceived Environmental Quality, the effects of individual socioeconomic attributes that significantly affect Perceived Environmental Quality will not be discussed.

The standardized coefficients indicate that, among the three Objective Built Environment factors with significant negative effects on PEQ, the order of influence from largest to smallest is Building Density (BD) ($\beta = -0.151$), Bus Stop Density (BSD) ($\beta = -0.140$), and Functional Mix Diversity (FMD) ($\beta = -0.098$). A possible explanation is that increased building density may compress the flexibility of public spaces and lead to complex traffic conditions (Ewing & Handy, 2009), making it difficult for residents to travel conveniently or choose preferred modes of transportation. Furthermore, higher density can reduce the area of parks and green spaces and result in more diverse and complex building forms. These factors collectively contribute to a decline in aesthetic perception, with cluttered street facades and less cohesive urban scenery (Gehl, 2011). An increase in Functional Mix Diversity implies more varied land uses, but it can also lead to the compression of commercial, leisure, and public service facilities within the same space, thereby reducing the perceived level of diverse convenience (Leyden, 2003). However, properly planned land use mix has been shown to create positive externalities that enhance neighborhood attractiveness (Song & Knaap, 2004). Moreover, in line with the modernist architectural principle that "form follows function" (Wang, 2002), functional diversification often introduces varied building styles and appearances, which may adversely affect the aesthetic coherence of streets and buildings. The negative correlation between bus stop density and PEQ may be due to increased pedestrian flow density and directional convergence, which can overcrowd commercial and leisure facilities and cause inconvenience for residents (Litman, 2022).

The standardized coefficients also show that, among the two Objective Built Environment factors with significant positive effects on PEQ, the influence effect of POI Density is relatively large ($\beta$ = 0.271), followed by the Bus Line Density ($\beta$ = 0.193). A possible explanation is that, under the condition of controlling other variables, especially after controlling the bus stop density, an increase in bus lines means an increase in the number of bus trips at each bus stop, thereby increasing the convenience for residents to access more travel destinations and enhancing their perception of transportation convenience (Shen & Fu, 2019a, 2019b). An increase in POI Density implies an increase in the density of facilities closely related to residents' lives (Leyden, 2003), such as accommodation, catering, shopping, leisure and entertainment, and financial services, thus improving the level of diverse convenience for residents. From another perspective, higher POI density is often accompanied by better transportation infrastructure, which not only corroborates the planning efficacy of the transit-oriented development (TOD) model but also supports the feasibility of the service-oriented development (SOD) model (Cervero & Kockelman, 1997).

2. The Effect of Objective Built Environment Factors on Perceived Safety (PES)

Column 2 of Table 9.6 reports the impacts of Objective Built Environment (OBE) factors on Perceived Safety (PES). The Fig. 9.5 shows the effects of the factors of OBE on PES. After controlling for covariates, the results indicate that Population Density (PD), Building Density (BD), Functional Mix Diversity (FMD), and Perceived Environmental Quality (PEQ) all exhibit statistically significant effects on Perceived Safety (SEP) ($p < 0.05$). In contrast, POI Density (POID) shows a marginally significant association ($p < 0.1$). Among these, Population Density (PD) ($\beta = -0.154$) and POID ($\beta = -0.121$) exert significant negative effects on Perceived Safety, while Building Density (BD) ($\beta = 0.145$), Functional Mix Diversity (FMD) ($\beta = 0.195$), and Perceived Environmental Quality (PEQ) ($\beta = 0.735$) show significant positive effects. Regarding individual socioeconomic attributes, both marital status ($p < 0.001$) and Housing Tenure ($p < 0.1$) significantly influence evaluations of Perceived Safety. Since this study focuses on the impacts of OBE factors, the effects of socioeconomic attributes will not be discussed further.

After controlling for other variables, two OBE factors significantly and negatively affect Perceived Safety (PES). The standardized coefficients indicate that Population Density (PD) has a relatively strong influence ($\beta = -0.154$), followed by POID ($\beta = -0.121$). Higher population density may lead to a lower perceived sense

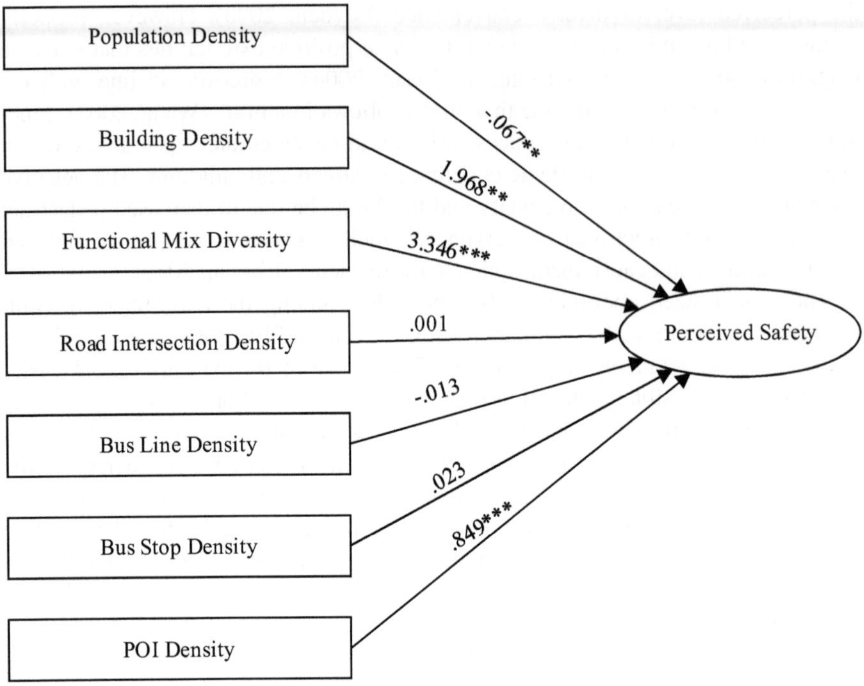


**Fig. 9.5** The effect of objective built environment factors on perceived safety (PES). Source: Created by the authors

of safety, a finding supported in previous literature. Routine Activity Theory (Cohen & Felson, 1979) suggests that the convergence of motivated offenders, suitable targets, and the absence of guardianship predicts criminal occurrence. Thus, when guardianship remains constant, higher population density can increase opportunities for criminal encounters, thereby reducing perceived safety (Browning et al., 2010). Additionally, an increase in leisure, commercial, and public service facilities may attract more visitors, further elevating population density and fostering environments where property crime rates rise (Liu et al., 2012).

Conversely, three factors show significant positive effects on Perceived Safety. The standardized coefficients reveal that Perceived Environmental Quality (PEQ) has the strongest influence ($\beta$ = 0.735), followed by Functional Mix Diversity (FMD) ($\beta$ = 0.195), and Building Density (BD) ($\beta$ = 0.145). An increase in functional diversity enhances perceived safety, corroborating earlier studies. Jacobs' concept of 'eyes on the street' (Jacobs, 1961) argues that mixed-use environments and public spaces facilitate natural surveillance, as diverse facilities and activities encourage continuous street presence, thereby improving safety (Gehl, 2011). This is supported by empirical evidence showing that neighborhood design characteristics significantly influence residents' fear of crime (Foster et al., 2010). Conversely, the Broken Windows Theory (Wilson & Kelling, 1982) posits that physical disorder—such as dilapidated public property and poor lighting—signals neglect, attracting criminal behavior and reducing perceived safety. Higher building density often indicates greater economic vitality and functional diversity, which may contribute to improved perceptions of safety.

Improved Perceived Environmental Quality—reflected in diverse convenience related to commercial, educational, medical, and leisure facilities—is often accompanied by enhanced public infrastructures such as street lighting and surveillance cameras. These improvements reduce perceived disorder and obstacles (Tao & Chai, 2020), thereby elevating residents' sense of safety. Similarly, the convenience of transportation stations usually correlates with well-maintained road networks and surveillance infrastructure. The presence of technologies such as facial recognition and AI monitoring systems, combined with higher pedestrian flow, can effectively deter crime and improve safety perceptions (Welsh & Farrington, 2009). According to the theory of environmental disorder (Sampson & Raudenbush, 2004), visible signs of neglect—such as abandoned buildings, graffiti, and litter—signal a lack of monitoring and may attract offenders (Jiang et al., 2018), reduce the perceived cost of crime, and diminish the sense of safety. Conversely, improved aesthetic quality can enhance public safety evaluations.

3. The Effect of the Objective Built Environment Factors on Social Capital (SC)

Column 3 of Table 9.6 reports the impact of the Objective Built Environment (OBE) on Social Capital (SC). The Fig. 9.6 shows the effects of the factors of OBE on SC. The results demonstrate that after controlling for other variables, Population Density (PD), Bus Line Density (BLD), Bus Stop Density (BSD), POI point density (POID), and Perceived Environmental Quality (PEQ) all have a strict significant impact on Life Satisfaction ($p < 0.05$). Among them, Population Density (PD), Bus Stop Density (BSD), and Perceived Environmental Quality (PEQ) all have a positive impact on Life Satisfaction, while bus line density (BLD) and POI point density (POID) both have a negative impact on Life Satisfaction. In individual socioeconomic attributes, the higher the education level ($p < 0.05$), the lower the community Social Capital. There is a strong correlation between Housing Tenure ($p < 0.05$) and Social Capital. Both per capita housing area ($p < 0.05$) and an optimistic personality ($p < 0.05$) significantly influence the formation of individual Social Capital. This study focuses on the impact of objective built-environment factors on perceived Social Capital, so the effects of individual socioeconomic attributes that significantly affect Social Capital will not be discussed in detail.

Specifically, after controlling for other variables, the Objective Built Environment (OBE) has a significant positive impact on two factors of Social Capital (SC). The standardized coefficients indicate that the Perceived Environmental Quality (PEQ) has the greatest influence ($\beta = 0.573$), followed by the Bus Stop Density (BSD, $\beta = 0.197$), and the Population Density (PD) has the smallest effect ($\beta = 0.118$). Among the two Objective Built Environment (OBE) factors that have a significant positive impact on Social Capital (SC), an increase in population density will improve Social Capital, and this result basically conforms to the expectations of a compact city. A possible explanation is that after controlling for other variables, especially the perceived safety, an increase in population density will increase the

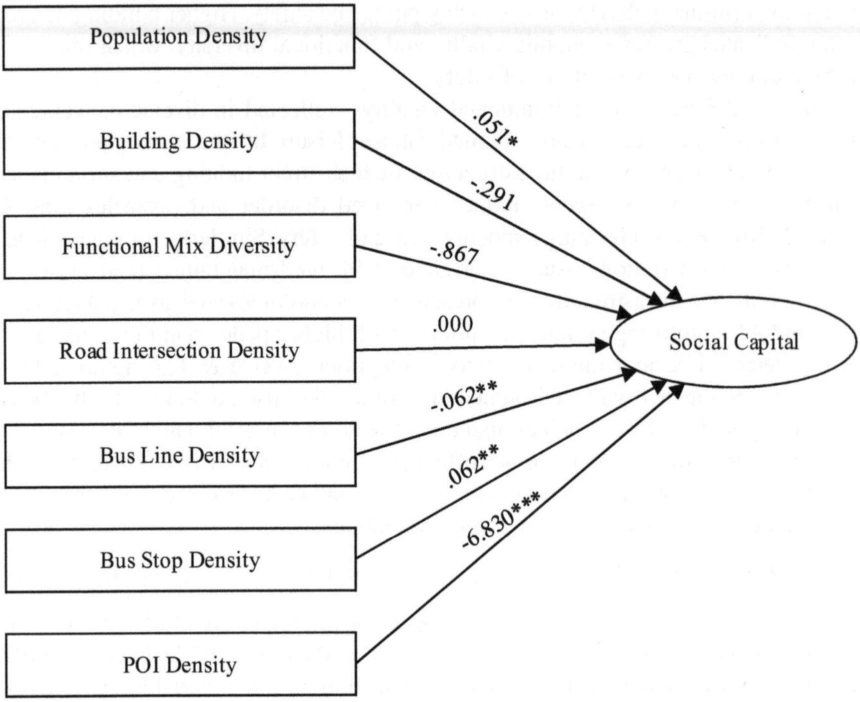


**Fig. 9.6** The effect of the objective built environment factors on social capital (SC). Source: Created by the authors

frequency of interaction and the network scale among community residents, thereby enhancing Social Capital. Although some studies suggest that population density has a significant negative relationship with residents' social interactions (French et al., 2014; Lin, 2019), an increase in population density may lead to the invisibility of residents and weaken social connections; it may also cause individual sensory overload and make individuals withdraw from social life (Yin, 2020). An increase in the bus stop density will attract residents to choose bus travel more often, increasing the opportunities for informal meetings during the journey from home to the bus stop and the chances of accidental encounters at the bus stop (Wood et al., 2010). Similarly, the website will increase the frequency of meetings, conversations, and interactions among residents due to bus travel, thereby enhancing Social Capital (Currie & Stanley, 2008). Meanwhile, after controlling for other variables, an improvement in the Perceived Environmental Quality will increase residents' Social Capital, which is consistent with the findings of relevant literature. On the one hand, the perception of diversity and convenience will increase the opportunities for residents to meet for daily services, enhance the communication among residents due to the satisfaction of living functions, and improve Social Capital. On the other hand, the perception of transportation convenience will affect residents' travel choices. More bus travel will increase the probability of residents meeting and interacting during travel, enhancing residents' Social Capital. Moreover, the maintenance and order of the clean, hygienic, and aesthetically pleasing appearance of the built environment, such as landscape greening, will attract more attention from residents and encourage outdoor activities, increasing the opportunities for residents to meet in parks, streets, and other public spaces, triggering different levels of communication and exchanges. This effectively expands the social network of residents within the community, increases the frequency of residents' communication, enhances the trust among residents and the sense of community belonging, reduces the perceived chaos, and thus affects residents' social interactions, improving the social relationships among neighbors (Tao & Chai, 2020).

Specifically, after controlling for other variables, the Objective Built Environment significantly and negatively affects two factors of Social Capital. The standardized coefficients indicate that the impact of POID is relatively large ($\beta = -0.285$), while that of BLD is relatively small ($\beta = -0.180$). Among the two Objective Built Environment factors that significantly and negatively affect Social Capital, an increase in POID will reduce residents' Social Capital, and this conclusion differs from the expectation of compact communities. However, it is similar to the conclusions of some relevant studies. There is a negative correlation between the accessibility density of public facilities and Social Capital (Yang et al., 2021), and a negative correlation between destination density and Social Capital (Wood et al., 2008). This may be because the increase in the density of diverse commercial, leisure, and public service facilities may reduce the opportunities for residents to communicate through daily life service behaviors such as clothing, food, housing, and transportation, thus affecting the formation of residents' Social Capital. After controlling for other variables, especially the density of bus stops, an increase in bus routes means an increase in the number of vehicles stopping at each bus stop. The expansion of the scale of bus routes and stops will reduce the contact and interaction among residents during bus-related behaviors such as waiting for buses and traveling together, which affects the formation of residents' Social Capital.

4. The Effect of the Objective Built Environment Factors on Self-Rated Health (SRH)

Column 4 of Table 9.6 presents the effects of the Objective Built Environment (OBE) on residents' Self-Rated Health (SRH). The Fig. 9.7 shows the effects of the factors of OBE on SRH. The results indicate that, after controlling for other variables, several objective built-environment factors significantly influence SRH. Specifically, Building Density (BD), Perceived Environmental Quality (PEQ), and POID show statistically significant effects on self-rated health ($p < 0.05$), while Road Intersection Density (RID) exhibits a marginally significant effect ($p < 0.1$). Among these factors, Building Density (BD, $\beta = 0.159$), Road Intersection Density (RID, $\beta = 0.092$), and Perceived Environmental Quality (PEQ, $\beta = 0.460$) exhibit positive effects on Self-Rated Health (SRH), whereas POID ($\beta = -0.150$) demonstrates a significant negative effect on SRH. Among the individual socioeconomic attributes, occupation ($p < 0.05$) and Housing Tenure ($p < 0.001$) are significantly associated with self-rated health. An optimistic personality ($p < 0.001$) also serves

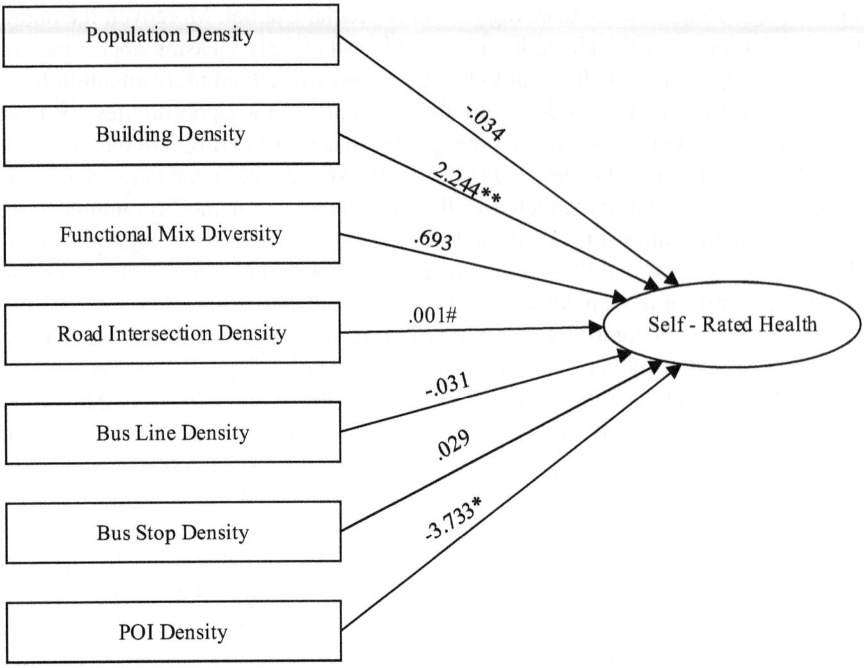


**Fig. 9.7** The effect of the objective built environment factors on self-rated health (SRH). Source: Created by the authors

as an important factor influencing Self-Rated Health. As this study focuses on the impact of objective built-environment factors on SRH, the effects of individual socioeconomic attributes will not be discussed in detail.

After controlling for other variables, the objective built environment shows a significant positive association with self-rated health. Standardized coefficients indicate that Building Density (BD) has a relatively substantial effect ($\beta = 0.159$), while Road Intersection Density (RID) exhibits a relatively modest influence ($\beta = 0.092$). One possible explanation is that increased building density may lead to the encroachment on public spaces and green areas, potentially limiting residents' opportunities for outdoor leisure activities, which could adversely affect health. Additionally, street network design characteristics, such as intersection density, have been shown to influence traffic safety outcomes (Marshall & Garrick, 2011), which may indirectly affect residents' health perceptions. On the other hand, higher building and road intersection densities might increase commuting distance and walking time, which has been associated with improved self-rated health (Owen et al., 2007). Furthermore, controlling for population density, an increase in building density could correlate with a lower floor area ratio, potentially mitigating negative psychological effects such as depression, loneliness, fear of crime, and low community spirit often associated with high-density high-rise residential buildings (Gifford, 2007). Conversely, an increase in POID density—particularly the density of restaurants, supermarkets, and shopping areas—is negatively associated with self-rated health.

A significant positive correlation is observed between Perceived Environmental Quality (PEQ) and SRH. Previous research suggests that better accessibility to bus stops and a more efficient and environmentally friendly public transportation system offer residents more travel options. Recreational activities during transit can contribute to improved self-rated health (Shen & Fu, 2019a, 2019b). Additionally, the aesthetic perception of the built environment can influence residents' evaluation of community walkability (Baum & Palmer, 2002; Sugiyama et al., 2008) and encourage recreational walking and outdoor activities (Sun et al., 2016), thereby enhancing SRH.

### 9.2.2 *The Influence of Mediating Factors on Life Satisfaction*

Column 5 of Table 9.6 presents the estimated effects of objective built-environment factors and mediating variables on Life Satisfaction. After controlling for other variables, among the built-environment factors, Road Intersection Density (RID), Perceived Environmental Quality (PEQ), and Perceived Safety (PES) exhibit statistically significant effects on Life Satisfaction ($p < 0.05$), while Bus Stop Density (BSD) shows a marginally significant effect ($p < 0.1$). Specifically, both Road Intersection Density (RID, $\beta = -0.161$) and Bus Stop Density (BSD, $\beta = -0.112$) are negatively associated with Life Satisfaction. A potential explanation is that higher road intersection density may reflect a more developed transportation system, and increased bus stop density could attract greater volumes of pedestrians and vehicles, potentially leading to traffic congestion, thereby adversely affecting residents' life satisfaction.

Among the individual socioeconomic attributes, education level, occupation ($p < 0.1$), Housing Tenure ($p < 0.05$), and per capita housing area ($p < 0.01$) are significantly associated with self-rated health. An optimistic personality ($p < 0.001$) also serves as an important factor influencing Self-Rated Health. As this study focuses on the impact of objective built-environment factors on SRH, the effects of individual socioeconomic attributes will not be discussed in detail. In addition to the objective built-environment factors, several mediating variables also significantly influence Life Satisfaction. This section focuses specifically on the effects of these mediating factors.

1. The Effect of Perceived Environmental Quality on Life Satisfaction

The results presented in Column 5 of Table 9.6 indicate that Perceived Environmental Quality (PEQ, $p < 0.01$), Perceived Safety (PES, $p < 0.1$), and Social Capital (SC, $p < 0.001$) are significantly associated with Life Satisfaction (LS). Among these mediating factors, PEQ and SC exhibit statistically significant effects on LS ($p < 0.05$), while PES shows a marginally significant effect ($p < 0.1$). All three factors demonstrate positive effects on Life Satisfaction. The standardized coefficients reveal that Perceived Environmental Quality has the strongest influence ($\beta = 0.306$), followed by Social Capital ($\beta = 0.167$), and Perceived Safety has the smallest effect ($\beta = 0.153$).

Specifically, higher levels of Perceived Environmental Quality (PEQ) are associated with greater Life Satisfaction (LS), a finding consistent with research on the determinants of subjective well-being across different cultural contexts (Diener & Seligman, 2004). For instance, satisfaction with public services (Basolo & Strong, 2002) and schools (Parkes et al., 2002) have been identified as significant predictors of community satisfaction. A study in the Netherlands demonstrated a positive correlation between perceived satisfaction with the diversity of shopping facilities and Life Satisfaction (Permentier et al., 2011). Similarly, research in Ohio, United States, indicated that perceived convenience of access to commercial and entertainment facilities is positively associated with Life Satisfaction (Hur & Morrow-Jones, 2008). A comprehensive study in the Twin Cities, United States, further confirmed that proximity to shopping malls and the convenience of commercial services significantly affect accessibility and thereby influence satisfaction (Cao, 2016).

Research in Beijing, China, has also shown that perceived quality of commercial, educational, medical, and sports facilities exerts a significant positive impact on Life Satisfaction (Chen & Ning, 2015). Additionally, studies indicate that satisfaction with transportation—including public transit convenience and access to transportation hubs (Ma et al., 2018) and rail transit (Cao, 2013)—is significantly related to residential satisfaction, which in turn affects Life Satisfaction. However, some scholars argue that lower traffic volumes enhance neighborhood livability (Appleyard, 1980; Bosselmann, 2012), whereas traffic exceeding certain thresholds can reduce resident satisfaction (e.g., Appleyard, 1980).

In the Chinese context, travel convenience has been confirmed as a key factor shaping environmental perceptions and influencing Subjective Well-Being (Chen & Ning, 2015). Commuting behavior—including time and mode choices (Wang & Wang, 2020; Wu, 2016)—has also been shown to affect individuals' Subjective Well-Being (Yin, 2020). Although one study noted that traffic congestion showed a non-significant negative correlation with community satisfaction in China (Shen & Fu, 2019a, 2019b), overall findings align with national conditions, indicating that positive perceptions of public transportation accessibility contribute positively to community satisfaction and subsequently to Life Satisfaction.

Furthermore, community aesthetics and appearance are identified as important variables influencing community satisfaction (Hur & Morrow-Jones, 2008; Parkes et al., 2002), and are among the most critical factors enhancing both community satisfaction and quality of life (Langdon, 1997; Sirgy & Cornwell, 2002). It should be noted, however, that the importance of aesthetic factors may vary across demographic contexts. For example, newer residents may prioritize physical appearance, whereas long-term residents may emphasize social factors (Potter & Cantarero, 2006). Domestic studies have corroborated that satisfaction with landscape and greenery significantly affects Subjective Well-Being (Chen & Ning, 2015), and overall aesthetic perception is positively correlated with community satisfaction (Shen & Fu, 2019a, 2019b).

2. The Impact of Perceived Safety on Life Satisfaction

A significant positive relationship exists between Perceived Safety (PES) and Life Satisfaction (LS). Data from the World Values Survey indicate that safety perceptions account for a notable portion of the variation in LS (Inglehart et al., 2003). A safe living environment not only enhances property values but also improves residents' satisfaction by fostering a greater sense of security and stability (Glaeser & Sacerdote, 1999). From a psychological perspective, environments with high PES help reduce fear and fulfill fundamental safety needs (Maslow, 1943), thereby directly improving LS (Ferriss, 2004; Pfeiffer & Cloutier, 2016). Furthermore, positive perceived safety encourages the use of public spaces—such as nighttime outings and community gatherings—which strengthens social cohesion and promotes a stronger sense of belonging and well-being (Jacobs, 1961; Janowitz, 1952). There is also evidence that PES may indirectly affect LS by facilitating leisure activities and physical exercise, as individuals are more likely to engage in outdoor activities when they feel secure (Harrison et al., 2007). Additionally, comparative urban studies suggest that the effect of PES on LS may be moderated by cultural and contextual factors. For instance, in high-density cities, the presence of "eyes on the street" and well-lit public areas significantly enhances perceived safety, particularly among vulnerable groups such as the elderly and women (Foster et al., 2013). Moreover, neighborhood watch programs and community policing have been shown to strengthen feelings of safety, which subsequently contribute to higher overall well-being (Lorenc et al., 2013).

3. The Impact of Social Capital on Life Satisfaction

A strong and significant positive relationship is observed between Social Capital (SC) and Life Satisfaction (LS), corroborating findings from numerous international studies (Kim & Kaplan, 2004; Hoogerbrugge & Burger, 2018). This aligns with sociological perspectives that emphasize the role of social connections in promoting well-being (Morris, 2011). SC serves as a crucial determinant of well-being, in which trust, reciprocity, and social networks collectively contribute to elevated LS (Putnam, 2000). At the community level, factors such as social cohesion (Liu et al., 2020) and neighborhood trust (Lin, 2019) are strongly associated with higher levels of life satisfaction. Moreover, SC acts as a significant mediator through which the built environment influences LS (Yang et al., 2021). On an individual level, strong social ties provide emotional support and alleviate psychological stress through the sharing of resources and burdens (Helliwell et al., 2020). Enhanced SC strengthens residents' sense of community belonging, which in turn raises LS (Sørensen et al., 2023). Additionally, communities characterized by high SC often exhibit better-maintained public spaces and lower perceived disorder, further reinforcing satisfaction and well-being (Sampson, 2012). It is also noteworthy that different forms of social capital—bridging, bonding, and linking—exert differential effects on well-being across diverse socioeconomic contexts (Szreter & Woolcock, 2004). For example, bonding social capital (strong ties among similar groups) may enhance emotional support, whereas bridging social capital (cross-group connections) facilitates access to opportunities and resources, thereby indirectly promoting LS (Aldrich & Meyer, 2015). Recent studies in East Asian contexts further emphasize the role of digital social networks in maintaining social capital in rapidly urbanizing settings (Li & Chen, 2022).

4. The Influence of Self-Rated Health on Life Satisfaction

Self-Rated Health (SRH) is positively associated with Life Satisfaction (LS), though this relationship was not statistically significant after controlling for other variables in the current model (Saelens & Handy, 2008). This result partially aligns with—yet does not fully replicate—existing literature. Extensive research has demonstrated that both physical and psychological health are strong predictors of LS (Dolan et al., 2008; Frey & Stutzer, 2010). Moreover, neurobiological studies suggest that urban living can affect stress processing mechanisms, which may indirectly influence self-rated health assessments (Lederbogen et al., 2011). Conversely, poor health conditions substantially diminish life satisfaction. In the Chinese context, multiple studies have confirmed significant positive correlations between SRH and LS (Lin, 2019), as well as between mental health and LS (Liu et al., 2020). Overall health perceptions also contribute notably to well-being (Shen & Fu, 2019a, 2019b). It is worth noting that the non-significance of SRH in this specific model may be attributable to contextual or mediating factors, such as adaptive coping mechanisms among individuals with chronic conditions or cultural variations in health perception (Baron-Epel et al., 2008). For instance, in collectivist societies, health perceptions may be less strongly tied to individual life satisfaction when social and familial support systems are strong (Kim et al., 2020). Moreover, methodological differences in measuring SRH—such as the use of single-item scales versus multidimensional health assessments—may also influence the observed effects (Jylhä, 2009). Further research incorporating longitudinal designs and culturally adapted instruments is needed to clarify these relationships.

### 9.2.3 Summary of this Section

This section presents an empirical study that employs a structural equation model to examine the influence of the Objective Built Environment on the mediating variables—namely, Perceived Environmental Quality (PEQ), Perceived Safety (PES), Self-Rated Health (SRH), and Social Capital (SC)—as well as the effects of these mediators on Life Satisfaction (LS).

As shown in Column 1 of Table 9.6, several Objective Built Environment factors, including building Density (BD), Functional Mix Diversity (FMD), Bus Line Density (BLD), Bus Stop Density (BLD), and POI Density (BLD), exert significant effects on Perceived Environmental Quality (PEQ). Column 2 indicates that Population Density (BLD), Building Density (BLD), Functional Mix Diversity (BLD), and POI Density (BLD) significantly influence Perceived Safety, with Perceived Environmental Quality (PEQ) also demonstrating a significant positive impact on Perceived Safety (PES). Column 3 reveals that Population Density (PD), Bus Line Density (BLD), Bus Stop Density (BSD), and POI Density (POID), significantly affect Social Capital (SC), and Perceived Environmental Quality (PEQ) likewise shows a significant effect. According to Column 4, Building Density (BD), Road Intersection Density (RID), and POI Density (POID) among the Objective Built Environment factors exhibit significant impacts on Self-Rated Health (SRH), and Perceived Environmental Quality also has a significant positive influence. Finally, Column 5 shows that Road Intersection Density (RID) and Bus Stop Density (BSD) significantly affect Life Satisfaction (LS), with Perceived Environmental Quality (PEQ), Perceived Safety (PES), and Social Capital (SC) all contributing significantly as well.

Based on these findings, Objective Built Environment factors may affect Life Satisfaction both directly and through mediating pathways. Specifically:

1. Objective Built Environment factors may influence Life Satisfaction (LS) through Perceived Environmental Quality (PEQ).
2. Objective Built Environment factors may affect residents' Life Satisfaction (LS) through Perceived Safety (PES).
3. Objective Built Environment factors may influence Life Satisfaction (LS) through Social Capital (SC).
4. Objective Built Environment factors may affect Life Satisfaction (LS) through Self-Rated Health (SRH).
5. Objective Built Environment factors may influence Life Satisfaction (LS) through chain mediation involving multiple factors such as Perceived Environmental Quality (PEQ), Perceived Safety (PES), Social Capital (SC), and Self-Rated Health (SRH).

It should be noted that Table 9.6 only examines the mediating factors through which the Objective Built Environment influences Life Satisfaction and the direct effects of these mediators. To further investigate whether the Objective Built Environment factors exert mediating effects on Life Satisfaction through Perceived Environmental Quality, Perceived Safety, Social Capital, and Self-Rated Health—as well as the magnitude and significance of these effects—the following section will present formal mediation tests.

## 9.3 Mediating Effect of Objective Built Environment Factors on Life Satisfaction

Columns 1–4 of Table 9.6 above report the mediating factors of Objective Built Environment impacts, and column 5 reports the mediating factors 'effects on Life Satisfaction. Fig. 9.8 shows how the Objective Built Environment may affect Life Satisfaction through four mediating factors: Perceived Environmental Quality, self-rated health, Perceived Safety and safety, and Social Capital. In order to further verify the mediating effect of Objective Built Environment on Life Satisfaction, and explore which paths Objective Built Environment has significant indirect effect on Life Satisfaction, this study reviewed relevant literature and combined with the actual subject (Hayes, 2009; Preacher & Hayes, 2008) Bootstrap mediating effect test was used to test the mediating effect of Objective Built Environment on Life Satisfaction, and bias-corrected non parametric percentile estimation method (Fang et al., 2011, 2012) was used to test the possible mediating effect in the above.

It is particularly emphasized that in order to understand the indirect effects of Objective Built Environment on Life Satisfaction through the above mediation more intuitively and conveniently, and effectively explain the influence mechanism of built environment on Life Satisfaction through mediation factors, the study will use the total-component-total form to verify the indirect effects (Yin Chun, 2020) and test the mediation effect of the path in Fig. 9.8.

To test the proposed hypotheses, a structural equation model (SEM) was estimated. The specific indirect effects for each hypothesized mediation pathway were calculated as the product of the corresponding path coefficients. The following paths were defined (Fig. 9.8):

$a_1$: The path from Objective Built Environment (OBE) to Perceived Environmental Quality (PEQ);
$a_2$: The path from OBE to Perceived Safety (PES);
$a_3$: The path from OBE to Social Capital (SC);
$a_4$: The path from OBE to Self-Rated Health (SRH);
$d_{21}$: The path from PEQ to PES;
$d_{31}$: The path from PEQ to SC;
$d_{32}$: The path from PES to SC;
$d_{41}$: The path from PEQ to SRH;
$d_{42}$: The path from PES to SRH;
$d_{43}$: The path from SC to SRH;
$b_1$: The path from PEQ to Life Satisfaction (LS);
$b_2$: The path from PES to LS;
$b_3$: The path from SC to LS;
$b_4$: The path from SRH to LS;
$c'$: The direct path from OBE to LS (controlling for all mediators).

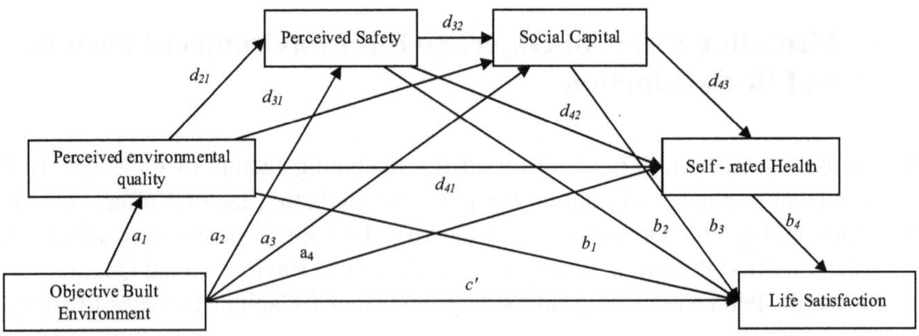


**Fig. 9.8** A schematic diagram of the path of Objective Built Environment influencing Life Satisfaction through mediation factors. Caption: Control variables are not shown in the structural equation models. Source: Created by the authors

Based on the research hypotheses developed in Sect. 9.2, the structural equation model illustrated in Fig. 9.8, and the definitions of these paths, the specific indirect effects for the hypothesized pathways are calculated as follows:

H1: The direct from OBE to LS ($c'$);
H2: Indirect effect via PEQ = $a_1 \times b_1$;
H3: Indirect effect via PES = $a_2 \times b_2$;
H4: Indirect effect via SC = $a_3 \times b_3$;
H5: Indirect effect via SRH = $a_4 \times b_4$;
H6: Indirect effect via PEQ → PES = $a_1 \times d_{21} \times b_2$;
H7: Indirect effect via PEQ → SC = $a_1 \times d_{31} \times b_3$;
H8: Indirect effect via PEQ → SRH = $a_1 \times d_{41} \times b_4$;
H9: Indirect effect via PES → SC = $a_2 \times d_{32} \times b_3$;
H10: Indirect effect via PES → SRH = $a_2 \times d_{42} \times b_4$;
H11: Indirect effect via SC → SRH = $a_3 \times d_{43} \times b_4$;
H12: Indirect effect via PEQ → PES → SC = $a_1 \times d_{21} \times d_{32} \times b_3$;
H13: Indirect effect via PEQ → PES → SRH = $a_1 \times d_{21} \times d_{42} \times b_4$;
H14: Indirect effect via PEQ → SC → SRH = $a_1 \times d_{31} \times d_{43} \times b_4$;
H15: Indirect effect via PES → SC → SRH = $a_2 \times d_{32} \times d_{43} \times b_4$;
H16: Indirect effect via PEQ → PES → SC → SRH = $a_1 \times d_{21} \times d_{32} \times d_{43} \times b_4$.

Based on the analysis methods in Sect. 6.6, the significance of the total, direct, and specific indirect effects was assessed using bootstrapping with 5000 resamples to generate bias-corrected confidence intervals (Hayes,2009; Preacher & Hayes, 2008).[4] The 95% and 90% confidence intervals of the bias-corrected non-parametric percentile estimation were reported to test the mediating effect. The following section presents the reporting and analysis of the results regarding the specific indirect effects, the total effects, the direct effects, and total direct effects, examining how the Objective Built Environment (OBE) influences Life Satisfaction (LS) through simple and serial mediation (also known as sequential or chain mediation) pathways.

### 9.3.1 *The Factors of Objective Built Environment Affect Life Satisfaction Through Perceived Environmental Quality (PEQ)*

Table 9.7 reports the indirect effect of seven factors of OBE on LS through PEQ. Figure 9.9 shows the effects of OBE on LS through this path. The three factors of OBE, e.g., BD, BLD, and POID, do not contain 0 in 95% confidence intervals,

[4] Bootstrapping is preferred over the traditional Sobel test because it does not assume normality of the sampling distribution of the indirect effect. A resample size of 5000 is recommended to ensure stable and accurate confidence interval estimates. The bias-corrected method is particularly effective in accounting for asymmetry in the sampling distribution. For detailed discussions, see Hayes (2009) and Preacher and Hayes (2008).

**Table 9.7** Objective built environment affects life satisfaction through perceived environment quality (OBE → PEQ → LS)

| | | | 95% CI | | 90% CI | |
|---|---|---|---|---|---|---|
| SIE | EST | SE | 2.5% | 97.5% | 5% | 95% |
| PD → PEQ → LS | −.004 | .008 | −.025 | .008 | −.022 | .005 |
| BD → PEQ → LS | −.627 | .433 | **−1.832** | **−.046** | **−1.629** | **−.121** |
| FMD → PEQ → LS | −.516 | .366 | −1.441 | .009 | **−1.270** | **−.055** |
| RID→PEQ → LS | .000 | .000 | −.001 | .001 | **−.001** | **−.000** |
| BLD → PEQ → LS | .020 | .012 | **.002** | **.052** | **.005** | **.046** |
| BSD → PEQ → LS | −.013 | .011 | −.046 | .000 | **−.040** | **−.001** |
| POID→PEQ → LS | 1.986 | 1.312 | **.172** | **5.563** | **.414** | **4.884** |

Caption: CI are estimated using the bias-corrected nonparametric percentile bootstrap
Boldface indicates statistical significance at the 95% and 90% confidence levels
SIE refers to specific indirect effect. EST refers to point estimate
SE refers to Standard Error
Source: Created by the authors

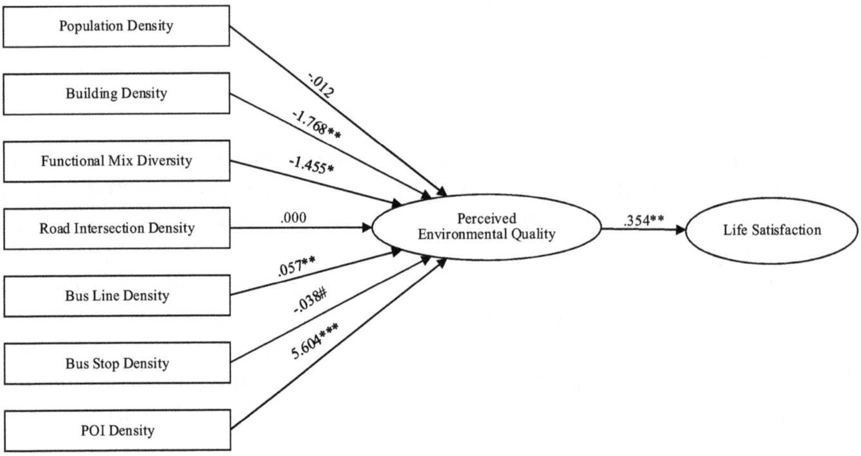


**Fig. 9.9** The effects of the OBE factors on LS through perceived environment quality (OBE → PEQ → LS). Caption: Control variables are omitted from the model for clarity. Source: Created by the authors

which were strict significance. The two factors of OBE, e.g., FMD, RID and BSD, do not contain 0 in 90% confidence intervals, which were marginal significant. The indirect effect of PD through this path is 0 in 95% and 90% confidence intervals, and the mediating effect of PD through Perceived Environmental Quality on Life Satisfaction is not significant. There are 6 significant indirect effects, among which the effects of BD, FMD, and BSD through this path are all negatively significant, while the effects of RID, BLD, and POID are all positive. Significant.

Specifically, for every 1 unit increase in BD, it decreases LS by 0.627 units through PEQ. After considering the factors of FMD and POID, the increase of BD means that the road traffic is more complicated, and it is difficult for residents to choose convenient roads to travel or pass through, which reduces the residents 'perception of diversity convenience and traffic convenience; the increase of BD means the compression of public space and the occupation of park green space, which affects the aesthetic perception evaluation of residents; At the same time, the increase of BD means that the diversity of buildings will also increase. The combination of the above factors leads to the decline of the aesthetics of park green space, the disorder of street facade and the diversity of building forms, which reduces the aesthetic perception level and further reduces Life Satisfaction. For every 1 unit increase in FMD, it decreases LS by 0.516 units through PEQ. The increase of FMD means the diversification of functions, which will affect residents 'perception of diverse convenience. At the same time, considering the effect of POID, the increase of Functional Mix Diversity means the decrease of functional POI points closely related to life, thus reducing residents' perception of diverse convenience. According to the principle "form follows function" advocated by Louis Sullivan, a master of modernist architecture in the Chicago School (Wang, 2002), diversification in building functions leads to a proliferation of architectural forms and styles. This stylistic heterogeneity negatively impacts the aesthetic harmony of buildings and streetscapes, thereby diminishing residential satisfaction.

Specifically, for every 1 unit increase in BSD, it decreases LS by 0.013 units through PEQ. After considering the influence of BSD, the increase of BSD means that the choice of travel route and travel destination of each bus stop is limited, which affects the Perceived Environmental Quality of residents and then affects the level of LS. The degree of RID also reduces LS by affecting tPEQ, mainly because the increase of RID means the complexity of the traffic system, which brings a lot of traffic flow and traffic congestion problems, which reduces the perception level of residents 'diversity convenience and traffic convenience, and then affects residents' LS. For every 1 unit increase in BLD, it increases LS by 0.020 units through PEQ. The increase of POID also has a positive impact on LS through PEQ. The possible explanation lies in that, after considering the effects of BSD and FMD, the increase of BLD means a large number of pedestrians gathering and various demands increasing, and the increase of POI point density means that the density of accommodation, catering, shopping, leisure entertainment, financial services and other facilities closely related to residents 'lives will increase, thus improving the diversified convenience level of residents (Wang & Wang, 2020).

### *9.3.2 The Factors of Objective Built Environment Affect Life Satisfaction Through Perceived Safety (PES)*

Table 9.8 reports the indirect effect of seven factors of OBE on LS through PES. Figure 9.10 shows the effect of OBE on LS through this path. The results show that the indirect effects of RID do not contain 0 in 95% confidence intervals, which were strict significance. The three factors of OBE, e.g., PD, BD, and FMD, do not

**Table 9.8** Objective built environment affects life satisfaction through perceived security (OBE → PES → LS)

| | | | 95% CI | | 90% CI | |
|---|---|---|---|---|---|---|
| SIE | EST | SE | 2.5% | 97.5% | 5% | 95% |
| PD → PES → LS | −.010 | .009 | −.033 | .002 | **−.029** | **−.000** |
| BD → PES → LS | .302 | .255 | −.053 | 1.041 | **.007** | **.856** |
| FMD → PES → LS | .513 | .402 | −.081 | 1.536 | **.010** | **1.345** |
| RID→PES → LS | .000 | .000 | **.000** | **.001** | **.000** | **.001** |
| BLD → PES → LS | −.002 | .004 | −.016 | .004 | −.013 | .002 |
| BSD → PES → LS | .003 | .006 | −.004 | .023 | −.002 | .019 |
| POID→PES → LS | −.444 | .491 | −2.037 | .104 | −1.684 | .022 |

Caption: CI are estimated using the bias-corrected nonparametric percentile bootstrap
Boldface indicates statistical significance at the 95% and 90% confidence levels
SIE refers to specific indirect effect. EST refers to point estimate
SE refers to Standard Error
Source: Created by the authors

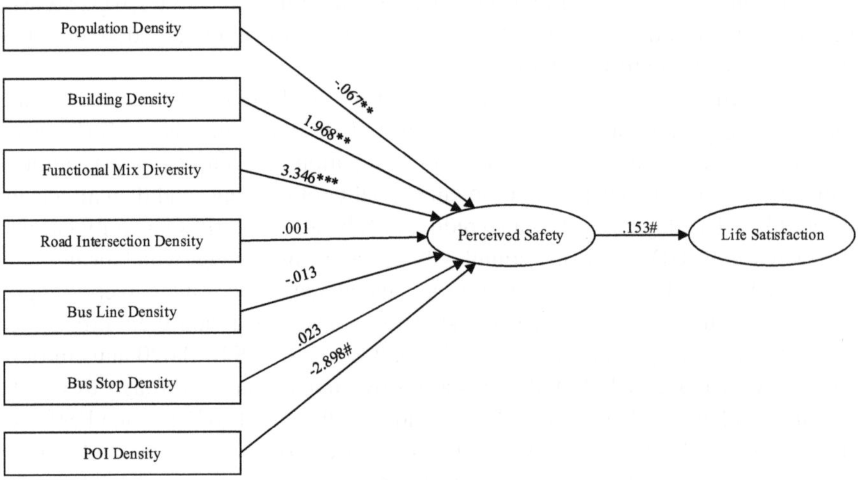


**Fig. 9.10** The effects of the OBE factors on LS through perceived safety (OBE → PES → LS). Caption: Control variables are omitted from the model for clarity. Source: Created by the authors

contain 0 in 90% confidence intervals, which were marginal significant. The indirect effect of PD was negatively significant, while the indirect effects of BD, FMD and RID were positively significant.

Specifically, for every 1 unit increase in PD, it decreases LS by 0.01 units through PES. For every 1 unit increase in BD, it increases 0.302 units of LS through PES. In the core Urban area of Guangzhou City, the increase of PD is accompanied by the increase of BD, which means the increase of PD. According to the routine activity theory, criminal targets, perpetrators, and preventive measures are all factors that influence the success of crime (Cohen & Felson, 1979)。 Thus, given similar precautions, increased population density increases the chances of contact between victims of crime and perpetrators, thereby reducing the sense of security. Relevant studies show that the increase of population density means the aggregation of various groups of people and the noise of the environment, which will increase the risk of Perceived Safety monitoring and the difficulty of preventive measures, increase the probability of contact between potential criminals and victims, and lead to an increase in property crime rate (Liu et al., 2012), resulting in a decrease in residents 'sense of Perceived Safety (Cohen & Felson, 1979; Liu et al., 2012). This, in turn, affects the Life Satisfaction of residents. The evaluation of crime and public order can significantly affect residents 'Subjective Well-being (Cohen & Felson, 1979).

For each additional unit of FMD, it increases 0.513 units of LS through PES. The increase of BD often means the improvement of FMD, especially the diversification of various types of living service facilities. An increase in FMD will increase the variety of commercial facilities, recreational facilities, public services, etc., attracting people to the streets, increasing street vitality, and various functional facilities can monitor street security in different ways "Therefore, the purpose of natural monitoring and social supervision is achieved (Zhang et al., 2020), which effectively ensures the safety of social security and crime and thus improves Life Satisfaction.

### *9.3.3 The Influence of Social Capital (SC) on Life Satisfaction*

Table 9.9 reports the indirect effect of seven factors of OBE on LS through SC. Figure 9.11 shows the effects of OBE on LS through this path. 1. The results show that five factors of OBE, e.g., PD, RID, BLD, BSD, and POID, do not contain 0 in 95% confidence intervals, which were strict different. The indirect effects of BD and FMD through this path all contain 0 in 95% and 90% confidence intervals, so the mediation effects of the two factors are not significant. The indirect effects of PD, RID, and BSD were significantly positive, while the indirect effects of BLD and POID were significantly negative.

Specifically, for every 1 unit increase in PD, it increases 0.009 units of LS through SC. PD increases LS indirectly through this path. The possible explanation is that increasing PD increases the frequency of interaction among residents, network size, Social Capital, and thus the Life Satisfaction of residents. RID also significantly increases LS through SC. It may be due to the increase in traffic convenience caused by the increase in road intersection density, which will drive the frequency of residents 'interaction and communication, thus improving residents' LS. For every 1 unit increase in BSD, it increases 0.01 units of LS through SC. Considering the emotion of bus density, the increase of BSD means that the website accessibility is better, which will attract more residents to choose bus travel, increase the chances of residents meeting, talking and interacting with each other due to bus travel, and further improve Social Capital and Life Satisfaction level.

For every 1 unit increase in BLD, it decreases 0.010 units of LS through SC. Considering the BSD, the increase of BLD reduces the chance of residents

**Table 9.9** Objective built environment affects life satisfaction through social capital (OBE → SC → LS)

| SIE | EST | SE | 95% CI | | 90% CI | |
|---|---|---|---|---|---|---|
| | | | 2.5% | 97.5% | 5% | 95% |
| PD → SC → LS | .009 | .006 | **.000** | **.023** | **.001** | **.021** |
| BD → SC → LS | −.048 | .127 | −.345 | .185 | −.292 | .128 |
| FMD → SC → LS | −.144 | .166 | −.598 | .095 | −.513 | .047 |
| RID→SC → LS | .000 | .000 | **.000** | **.000** | **.000** | **.000** |
| BLD → SC → LS | −.010 | .006 | **−.026** | **−.002** | **−.023** | **−.003** |
| BSD → SC → LS | .010 | .006 | **.002** | **.025** | **.003** | **.023** |
| POID→SC → LS | −1.134 | .522 | **−2.442** | **−.329** | **−2.203** | **−.446** |

Caption: CI are estimated using the bias-corrected nonparametric percentile bootstrap
Boldface indicates statistical significance at the 95% and 90% confidence levels
SIE refers to specific indirect effect. EST refers to point estimate
SE refers to Standard Error
Source: Created by the authors

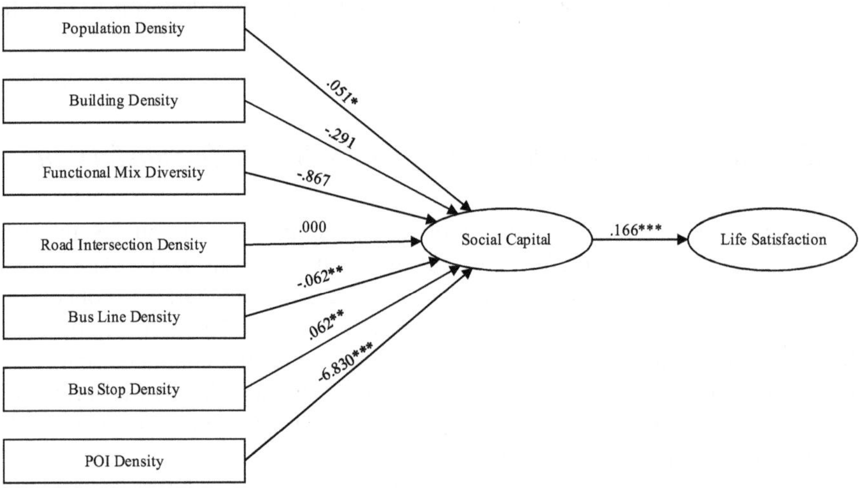


**Fig. 9.11** The effects of the OBE factors on LS through social capital (OBE → SC → LS). Caption: Control variables are omitted from the model for clarity. Source: Created by the authors

meeting for bus travel, and then affects Social Capital and reduces the Life Satisfaction level. The increase in POID has a significant negative impact on LS through SC. For every 1 unit increase in POID, it decreases 1.134 units of LS through SC. Increasing POI density reduces Social Capital, similar to a related study (Yang et al., 2021), where accessibility density of public facilities is negatively correlated with Social Capital. It is possible that the increase in POI density may reduce the opportunity and contact frequency of residents to meet and interact in functional services and demand satisfaction due to daily activities, affect the formation and evaluation of Social Capital, and then reduce Life Satisfaction.

### 9.3.4 The Influence of Self-Rated Health (SRH) on Life Satisfaction

Table 9.10 reports the mediating effect of the seven factors of OBE on LS through SRH. Figure 9.12 shows the effects of OBE on LS through this path. Among the seven factors of OBE, the indirect effect of RID through this path does not contain 0 in 95% confidence intervals, which were strict different. The other effects of the six factors of OBE contained 0 in 95%,90%confidence intervals, which were not significant.

**Table 9.10** Objective built environment affects life satisfaction through self-rated health (OBE → SRH → LS)

| | | | 95% CI | | 90% CI | |
|---|---|---|---|---|---|---|
| SIE | EST | SE | 2.5% | 97.5% | 5% | 95% |
| PD → SRH → LS | −.002 | .003 | −.013 | .001 | −.010 | .001 |
| BD → SRH → LS | .158 | .175 | −.104 | .625 | −.046 | .541 |
| FMD → SRH → LS | .049 | .117 | −.080 | .460 | −.044 | .375 |
| RID→SRH → LS | .000 | .000 | **.000** | **.001** | **.000** | **.000** |
| BLD → SRH → LS | −.002 | .003 | −.016 | .001 | −.013 | .000 |
| BSD → SRH → LS | .002 | .003 | −.001 | .015 | −.001 | .012 |
| POID→SRH → LS | −.262 | .336 | −1.289 | .154 | −1.080 | .065 |

Caption: CI are estimated using the bias-corrected nonparametric percentile bootstrap
Boldface indicates statistical significance at the 95% and 90% confidence levels
SIE refers to specific indirect effect. EST refers to point estimate
SE refers to Standard Error
Source: Created by the authors

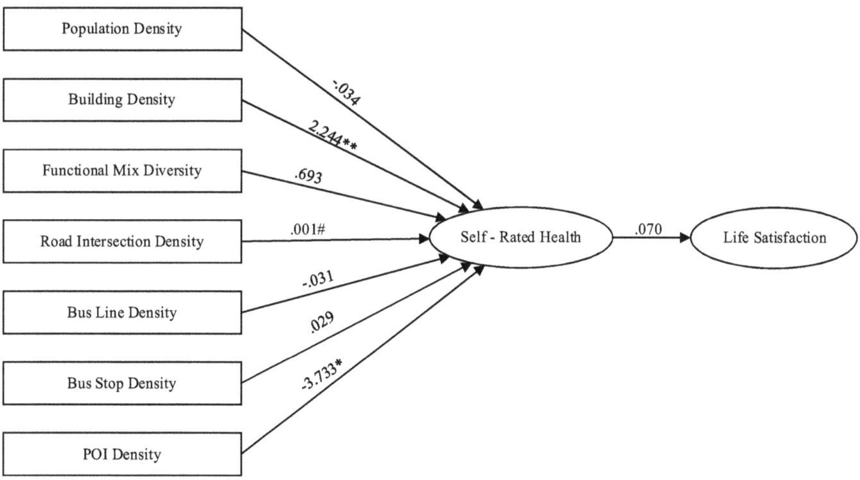


**Fig. 9.12** The effects of the OBE factors on LS through self-rated health (OBE → SRH → LS). Caption: Control variables are omitted from the model for clarity. Source: Created by the authors

Specifically, the increase of RID will improve residents 'self-rated health and thus enhance residents' Life Satisfaction. It may be that the increase in PD means that various facilities and resources per capita in the community will be reduced, and the resource limitation or long-term occupation will reduce the frequency and time of outdoor activities of community residents, affect the SRH level, and then reduce the LS. In the core urban areas of Guangzhou, especially in small housing communities where per capita public space, leisure and entertainment facilities, and fitness facilities are limited, a relatively high RID expands the activity space for residents to go out, encourages residents to engage in outdoor activities, and subsequently improves their self-rated health levels.

### 9.3.5 *The Effects on Life Satisfaction Through Serial Multiple Mediations*

The simple indirect effects of OBE through PEQ, PES, SC and SRH have been analyzed above. Since the relationship between the four mediators has been confirmed in some studies, that is, there is a series relationship between the mediation variables, which is different from the general multiple mediation with multiple mediation, called and serial mediation or chain multiple mediation (Liu and Ling 2009). For multi-mediation models, total mediation is not a prerequisite for analysis of individual mediation effects, so the analysis focuses on both testing individual mediation effects (Wen et al., 2005; Wen, 2012).

Table 9.11 reports the indirect effects of seven factors of OBE on LS through the chain mediation of PEQ affecting PES. Figure 9.13 shows the path results of these effects. The results show that the indirect effects of five factors of OBE, e.g., BD, RID, BLD, BSD and POID through the path contained 0 in 95% confidence

**Table 9.11** The objective built environment affects life satisfaction through the serial mediation of perceived environmental quality and perceived safety (OBE → PEQ → PES → LS)

| | | | 95% CI | | 90% CI | |
|---|---|---|---|---|---|---|
| SIE | EST | SE | 2.5% | 97.5% | 5% | 95% |
| PD → PEQ → PES → LS | −.002 | .003 | −.013 | .003 | −.010 | .002 |
| BD → PEQ → PES → LS | −.230 | .205 | −.869 | .028 | **−.714** | **−.009** |
| FMD → PEQ → PES → LS | −.189 | .181 | −.679 | .050 | −.581 | .009 |
| RID→PEQ → PES → LS | .000 | .000 | −.001 | .000 | **.000** | **.000** |
| BLD → PEQ → PES → LS | .007 | .006 | −.001 | .024 | **.000** | **.021** |
| BSD → PEQ → PES → LS | −.005 | .005 | −.021 | .001 | **−.018** | **−.000** |
| POID→PEQ → PES → LS | .729 | .605 | −.098 | 2.433 | **.045** | **2.126** |

Caption: CI are estimated using the bias-corrected nonparametric percentile bootstrap
Boldface indicates statistical significance at the 95% and 90% confidence levels
SIE refers to specific indirect effect. EST refers to point estimate
SE refers to Standard Error
Source: Created by the authors

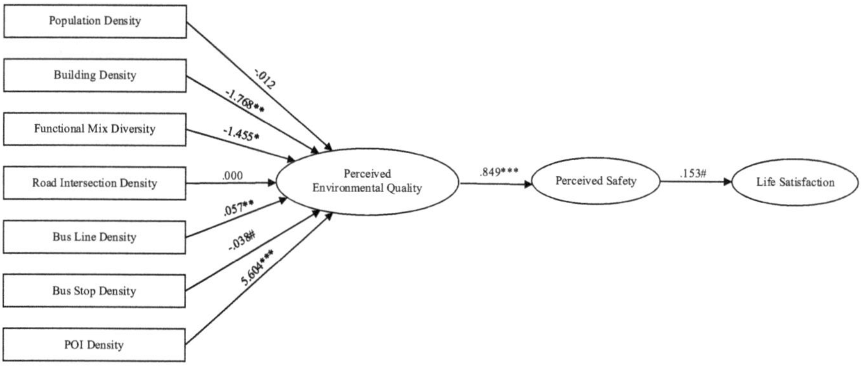


**Fig. 9.13** The effects of the OBE factors on LS through Serial Mediators of PEQ, PES (OBE → PEQ → PES → LS). Caption: Control variables are omitted from the model for clarity. Source: Created by the authors

intervals, which were not strict significant. However, these effects do not contain 0 in 90% confidence intervals, which were marginally significant. The indirect effects of BD and BSD were negative, while the indirect effects of RID, BLD and POID were positive.

Specifically, for every 1 unit increase in BD, it decreases 0.230 units of LS through PEQ, affecting PES. After considering the FMD and POID, the increase in building density means that the distribution of service facilities and resources related to life will be difficult to perceive convenience and convenience due to the complexity of roads, and at the same time, it will compress public space and occupy the green area of parks. In addition, buildings of different ages will also affect the street facade and architectural aesthetic evaluation. These combined effects not only affect the vitality of streets, limit the role of natural and social monitoring of streets, but also affect aesthetic evaluation and increase the perceived confusion and obstacles of residents (Tao & Chai, 2020). This reduces the Perceived Safety evaluation of residents and affects the level of Life Satisfaction. After considering the effect of FMD, POID significantly improves residents 'Life Satisfaction through this path. The increase of POID means that the convenience and convenience of commercial, leisure and entertainment, public services and other facilities closely related to life functions increase, which further improves the diversified convenience level of residents, is conducive to realizing the purpose of natural monitoring and social supervision, affects the PES of residents and further improves the level of Life Satisfaction.

The increase in RID means that the traffic system is developed, which brings residents a better perception of traffic convenience. At the same time, the high level of RID means that various types of electronic eyes, surveillance cameras, and a continuous flow of people will monitor the environment and enhance residents 'perception of security. The increase of BLD means the concentration of various demands caused by the large number of pedestrians, and the increase of service facilities closely related to life functions, thus improving the diversified convenience level of residents. The above-mentioned effects on PEQ will affect Life

Satisfaction by affecting PES. After considering the influence of BLD, the increase of BSD means that the choice of travel route and travel destination of each bus stop is limited, and the complex traffic stop system reduces residents 'convenience perception, thus affecting residents' Life Satisfaction.

Table 9.12 reports the indirect effects of seven factors of OBE on LS through the chain mediation of PEQ affecting SC. Figure 9.14 shows the impact of OBEt on LS through this path. The results showed that among the seven factors of OBE, except PD, all contained 0 in 95% and 90% confidence intervals, which were not significant. The six factors of OBE, namely BD, FMD, RID, BLD, and POID, are strict diffent at a 95% confidence level, while BSD is marginally significant at a 90% confidence level. Among the six significant mediating effects mentioned above, the indirect effects of BD, FMD, and BSD are negative, while the indirect effects of RID, BLD and POID are positive and significant.

**Table 9.12** The objective built environment affects life satisfaction through the serial mediation of perceived environmental quality and social capital (OBE → PEQ → SC → LS)

| | | | 95% CI | | 90% CI | |
|---|---|---|---|---|---|---|
| SIE | EST | SE | 2.5% | 97.5% | 5% | 95% |
| PD → PEQ → SC → LS | −.001 | .002 | −.008 | .003 | −.006 | .002 |
| BD → PEQ → SC → LS | −.195 | .115 | **−.528** | **−.041** | **−.449** | **−.062** |
| FMD → PEQ → SC → LS | −.161 | .113 | **−.466** | **−.010** | **−.410** | **−.031** |
| RID→PEQ → SC → LS | .000 | .000 | **.000** | **.000** | **.000** | **.000** |
| BLD → PEQ → SC → LS | .006 | .003 | **.002** | **.016** | **.002** | **.014** |
| BSD → PEQ → SC → LS | −.004 | .003 | −.013 | .000 | **−.011** | **−.001** |
| POID→PEQ → SC → LS | .619 | .315 | **.188** | **1.528** | **.254** | **1.342** |

Caption: CI are estimated using the bias-corrected nonparametric percentile bootstrap
Boldface indicates statistical significance at the 95% and 90% confidence levels
SIE refers to specific indirect effect. EST refers to point estimate
SE refers to Standard Error
Source: Created by the authors

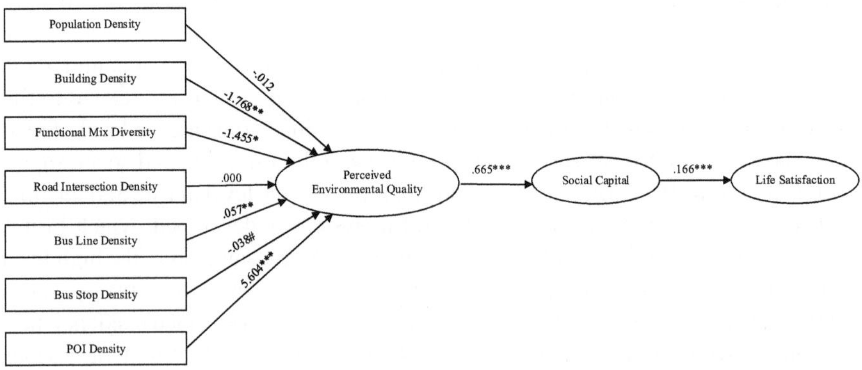


**Fig. 9.14** The effects of the OBE factors on LS through Serial Mediators of PEQ, SC (OBE → PEQ → SC → LS). Caption: Control variables are omitted from the model for clarity. Source: Created by the authors

Every unit increase in BD decreases LS by 0.195 units through PEQ, affecting SC. For each unit of Functional Mix Diversity, Life Satisfaction decreases by 0.161 units through PEQ, affecting SC. The increase in building density means that it is difficult to unify the facade of buildings of various ages, which means the complexity of road traffic, which affects the convenience of daily traffic and aesthetic evaluation of residents, reduces Social Capital and affects Life Satisfaction.

The increase of BLD means that public transport can go to more travel destinations and increase residents 'public travel choices, increase the frequency of neighborhood interaction due to bus travel, enhance Social Capital and increase residents' Life Satisfaction level. However, after considering the BLD, the high degree of BSD means that the choice of bus travel is more dispersed, and the probability of residents meeting due to bus travel will be reduced, which affects the formation of Social Capital and reduces Life Satisfaction. For every 1 unit increase in POID, it increases 0.619 units of LS through PEQ, affecting SC. The increase of POID not only brings about the improvement of diversified convenience perception level, but also increases the opportunity for residents to communicate with life service behaviors such as clothing, food, accommodation and transportation, affects residents' Social Capital and promotes the improvement of Life Satisfaction.

Table 9.13 reports the indirect effects of seven factors of OBE on LS through the chain mediation of PEQ affecting SRH. Figure 9.15 shows the path diagram of the OBE influence on LS through this path. The results prove that among the seven factors of OBE, RID does not contain 0 in the 95% confidence interval, thus reaching strict significance. The 90% confidence intervals of BLD and BSD do not contain 0, reaching marginal significant. Among the three significant mediating effects mentioned above, the indirect effects of RID and BLD are significantly positive, while the indirect effect of BSD is significantly negative.

Within the core urban area of Guangzhou, the density of road intersections implies an improvement in the perceived transportation convenience and diversity

**Table 9.13** The objective built environment affects life satisfaction through the serial mediation of perceived environmental quality and self-rated health (OBE → PEQ → SRH → LS)

| | | | 95% CI | | 90% CI | |
|---|---|---|---|---|---|---|
| SIE | EST | SE | 2.5% | 97.5% | 5% | 95% |
| PD → PEQ → SRH → LS | .000 | .001 | −.006 | .001 | −.005 | .000 |
| BD → PEQ → SRH → LS | −.069 | .094 | −.399 | .035 | −.315 | .012 |
| FMD → PEQ → SRH → LS | −.057 | .082 | −.328 | .030 | −.264 | .012 |
| RID→PEQ → SRH → LS | .000 | .000 | **.000** | **.000** | **.000** | **.000** |
| BLD → PEQ → SRH → LS | .002 | .003 | −.001 | .011 | **.000** | **.009** |
| BSD → PEQ → SRH → LS | −.001 | .002 | −.010 | .001 | **−.008** | **−.000** |
| POID→PEQ → SRH → LS | .219 | .281 | −.112 | 1.165 | −.040 | .950 |

Caption: CI are estimated using the bias-corrected nonparametric percentile bootstrap
Boldface indicates statistical significance at the 95% and 90% confidence levels
SIE refers to specific indirect effect. EST refers to point estimate
SE refers to Standard Error
Source: Created by the authors

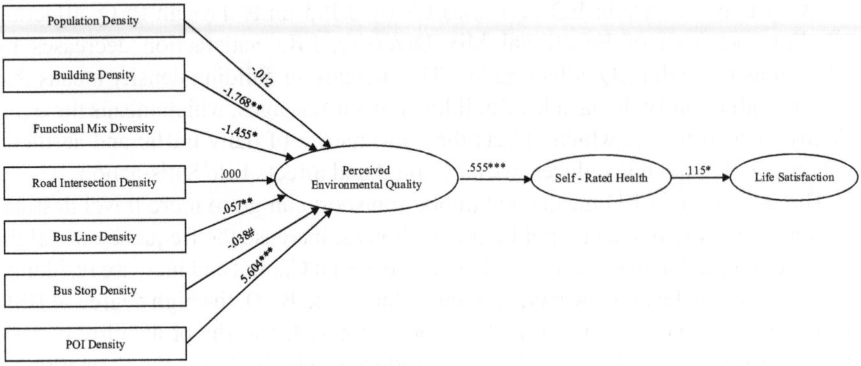


**Fig. 9.15** The effects of the OBE factors on LS through Serial Mediators of PEQ, SRH (OBE → PEQ → SRH → LS). Caption: Control variables are omitted from the model for clarity. Source: Created by the authors

among residents. Meanwhile, the surveillance cameras on the roads and the flow of people all contribute to environmental supervision. A high RID can increase the possibility of residents expanding their activity spaces outward, guiding them to engage in outdoor activities, and thus improving their self-rated health levels. An increase in the density of bus stops means that residents perceive greater travel convenience. However, this may lead to a reduction in the intensity or opportunity of health-related exercise during their commuting trips, thereby lowering their self-rated health levels.

Table 9.14 reports the indirect effects of seven factors of OBE on LS through the chain mediation of PES affecting SC. Figure 9.16 shows the path diagram of the OBE influence on LS through this path. The results show that RID does not include 0 in the 95% confidence interval, and its indirect effects are strict significant. The

**Table 9.14** The objective built environment affects life satisfaction through the serial mediation of perceived safety and social capital (OBE → PES → SC → LS)

| | | | 95% CI | | 90% CI | |
|---|---|---|---|---|---|---|
| SIE | EST | SE | 2.5% | 97.5% | 5% | 95% |
| PD → PES → SC → LS | .000 | .001 | −.004 | .002 | −.003 | .002 |
| BD → PES → SC → LS | .003 | .045 | −.076 | .115 | −.054 | .089 |
| FMD → PES → SC → LS | .005 | .074 | −.119 | .193 | −.087 | .153 |
| RID→PES → SC → LS | .000 | .000 | **.000** | **.000** | **.000** | **.000** |
| BLD → PES → SC → LS | .000 | .001 | −.002 | .001 | −.001 | .000 |
| BSD → PES → SC → LS | .000 | .001 | −.001 | .002 | −.001 | .002 |
| POID→PES → SC → LS | −.005 | .076 | −.211 | .114 | −.167 | .075 |

Caption: CI are estimated using the bias-corrected nonparametric percentile bootstrap
Boldface indicates statistical significance at the 95% and 90% confidence levels
SIE refers to specific indirect effect. EST refers to point estimate
SE refers to Standard Error
Source: Created by the authors

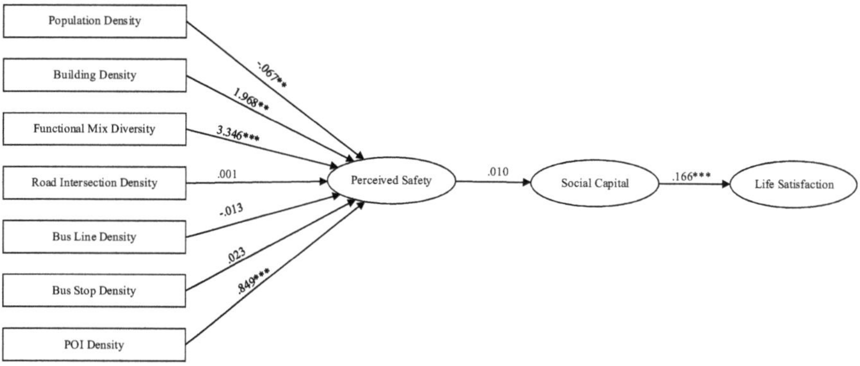


**Fig. 9.16** The effects of the OBE factors on LS through Serial Mediators of PES, SC (OBE → PES → SC → LS). Caption: Control variables are omitted from the model for clarity. Source: Created by the authors

95% and 90% confidence intervals of the other 6 factors of OBE all include 0, so their indirect effects are not significant. The indirect effect estimates prove that the indirect effect of intersection density on Life Satisfaction through the path is small but positively significant. The possible explanation is that the density of road intersections brings about a transportation system extending in all directions, which brings about an increase in the convenience of residents 'transportation, drives the frequency of residents' interaction and expands the social network, thus improving residents' Life Satisfaction.

Table 9.15 reports the indirect effects of seven factors of OBE on LS through the chain mediation of PES affecting SRH. Figure 9.17 shows the path diagram of the OBE influence on LS through this path. The results show that RID and BLD do not contain 0 in the 95% confidence interval, and their indirect effects are strict

**Table 9.15** The objective built environment affects life satisfaction through the serial mediation of perceived safety and self-rated health (OBE → PES → SRH → LS)

| | | | 95% CI | | 90% CI | |
|---|---|---|---|---|---|---|
| SIE | EST | SE | 2.5% | 97.5% | 5% | 95% |
| PD → PES → SRH → LS | .000 | .001 | −.001 | .004 | −.001 | .003 |
| BD → PES → SRH → LS | −.004 | .032 | −.117 | .029 | −.091 | .019 |
| FMD → PES → SRH → LS | −.007 | .049 | −.165 | .052 | −.119 | .034 |
| RID→PES → SRH → LS | .000 | .000 | **.000** | **.000** | **.000** | **.000** |
| BLD → PES → SRH → LS | .000 | .000 | **.000** | **.002** | **.000** | **.001** |
| BSD → PES → SRH → LS | .000 | .001 | −.003 | .000 | −.002 | .000 |
| POID→PES → SRH → LS | .006 | .063 | −.039 | .264 | −.024 | .188 |

Caption: CI are estimated using the bias-corrected nonparametric percentile bootstrap
Boldface indicates statistical significance at the 95% and 90% confidence levels
SIE refers to specific indirect effect. EST refers to point estimate
SE refers to Standard Error
Source: Created by the authors

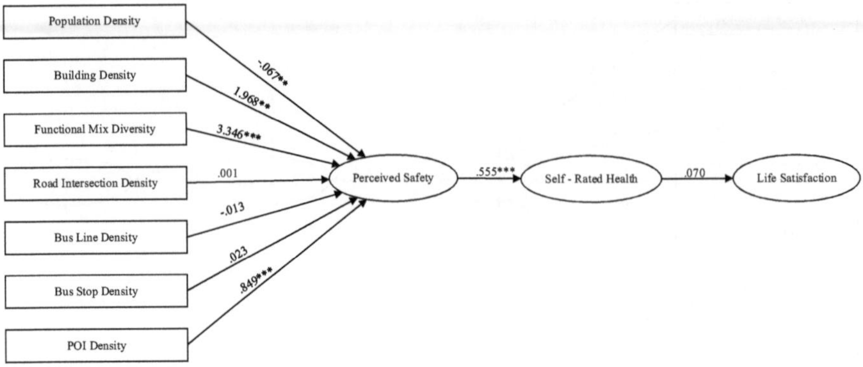


**Fig. 9.17** The effects of the OBE factors on LS through Serial Mediators of PES, SRH (OBE → PES → SRH → LS). Caption: Control variables are omitted from the model for clarity. Source: Created by the authors

significant. The 95% and 90% confidence intervals of the other five factors of OBE all contained 0, so their indirect effects were not significant. The indirect effect estimates prove that the indirect effects of RID and BLD on Life Satisfaction through the path are small but positively significant. It may be that the old residential area in the core Urban area of Guangzhou accounts for a relatively high proportion; its public space is crowded, leisure entertainment and fitness facilities are limited, etc. The higher road intersection density can attract residents to expand their activity space outward, guide residents to go out for activities, further improve the self-evaluation health level of residents, and further improve the Life Satisfaction level of residents. Under the condition of controlling the density of bus stops, the increase of bus line density means the concentration of bus stops, which brings about the increase of commuting distance from residents to bus stops, which is accompanied by commuting activities of residents, improves the self-rated health level of residents, and then improves the Life Satisfaction level of residents.

Table 9.16 reports the indirect effects of seven factors of OBE on LS through the chain mediation of SC affecting SRH. Figure 9.18 shows the path diagram of the OBE influence on LS through this path. The results show that PD, RID does not contain 0 in the 95% confidence interval, and its indirect effects are strict significant. The 90% confidence intervals of BSD do not contain 0, so their indirect effects are marginally significant. The indirect effect estimates prove that the indirect effects of PD, RID, and BSD on LS are small but positively significant. The indirect effect of PD is also significant, unlike the two paths mentioned above. It may be that the increase in population density will increase the frequency of interaction among residents, the scale of the network, enhance Social Capital, and then improve the Life Satisfaction of residents.

Table 9.17 reports the indirect effects of seven factors of OBE on LS through chain mediation of PEQ affecting PES and then affecting SC Fig. 9.19 shows the path diagram of the OBE influence on LS through this path. The results show that RID does not contain 0 in the 95% confidence interval, and its indirect effects are

**Table 9.16** The objective built environment affects life satisfaction through the serial mediation of social capital and self-rated health (OBE → SC → SRH → LS)

| SIE | EST | SE | 95% CI | | 90% CI | |
|---|---|---|---|---|---|---|
| | | | 2.5% | 97.5% | 5% | 95% |
| PD → SC → SRH → LS | .000 | .000 | **.000** | **.001** | **.000** | **.001** |
| BD → SC → SRH → LS | .000 | .005 | −.015 | .005 | −.011 | .003 |
| FMD → SC → SRH → LS | .000 | .007 | −.026 | .007 | −.019 | .004 |
| RID→SC → SRH → LS | .000 | .000 | **.000** | **.000** | **.000** | **.000** |
| BLD → SC → SRH → LS | .000 | .000 | −.001 | .001 | −.001 | .000 |
| BSD → SC → SRH → LS | .000 | .000 | −.001 | .001 | **.000** | **.001** |
| POID→SC → SRH → LS | −.004 | .043 | −.118 | .061 | −.093 | .040 |

Caption: CI are estimated using the bias-corrected nonparametric percentile bootstrap
Boldface indicates statistical significance at the 95% and 90% confidence levels
SIE refers to specific indirect effect. EST refers to point estimate
SE refers to Standard Error
Source: Created by the authors

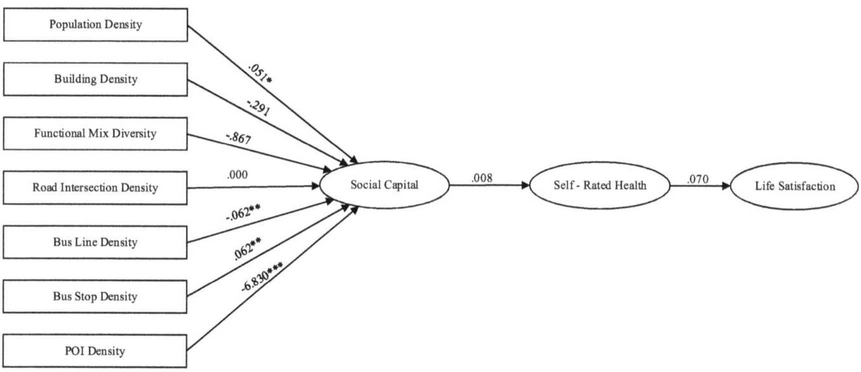


**Fig. 9.18** The effects of the OBE factors on LS through Serial Mediators of SC, SRH (OBE → SC → SRH → LS). Caption: Control variables are omitted from the model for clarity. Source: Created by the authors

strict significant. The 95% and 90% confidence intervals of the other 6 Objective Built Environment factors all contain 0, so their indirect effects are not significant. The indirect effect estimates prove that the indirect effect of RID on LS through the path is small but still positive.

Table 9.18 reports the indirect effects of seven factors of OBE on LS through the chain mediation of PEQ affecting PES and then affecting SRH. Figure 9.20 shows the path diagram of the OBE influence on LS through this path. The results show that PD, RID and BSD do not contain 0 in the 95% confidence interval, and their indirect effects are strict significant. The other four factors contained 0 in 95% and 90% confidence intervals, so their indirect effects were not significant. The indirect effect estimates prove that the indirect effects of PD, RID, and BSD on LS are small but positively significant.

**Table 9.17** The objective built environment affects life satisfaction through the serial mediation of perceived environmental quality, perceived safety and social capital (OBE → PEQ → PES → SC → LS)

| | | | 95% CI | | 90% CI | |
|---|---|---|---|---|---|---|
| SIE | EST | SE | 2.5% | 97.5% | 5% | 95% |
| PD → PEQ → PES → SC → LS | .000 | .000 | −.001 | .001 | −.001 | .000 |
| BD → PEQ → PES → SC → LS | −.002 | .037 | −.093 | .056 | −.074 | .040 |
| FMD → PEQ → PES → SC → LS | −.002 | .032 | −.076 | .048 | −.059 | .034 |
| RID→PEQ → PES → SC → LS | .000 | .000 | **.000** | **.000** | **.000** | **.000** |
| BLD → PEQ → PES → SC → LS | .000 | .001 | −.002 | .003 | −.001 | .002 |
| BSD → PEQ → PES → SC → LS | .000 | .001 | −.002 | .001 | −.002 | .001 |
| POID→PEQ → PES → SC → LS | .008 | .111 | −.175 | .275 | −.129 | .218 |

Caption: CI are estimated using the bias-corrected nonparametric percentile bootstrap
Boldface indicates statistical significance at the 95% and 90% confidence levels
SIE refers to specific indirect effect. EST refers to point estimate
SE refers to Standard Error
Source: Created by the authors

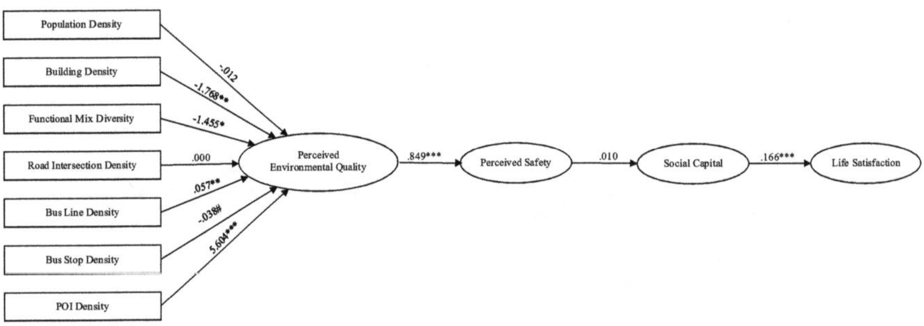


**Fig. 9.19** The effects of the OBE factors on LS through Serial Mediators of PEQ, PES, SC (OBE → PEQ → PES → SC → LS). Caption: Control variables are omitted from the model for clarity. Source: Created by the authors

**Table 9.18** The objective built environment affects life satisfaction through the serial mediation of perceived environmental quality, perceived safety and self-rated health (OBE → PEQ → PES → SRH → LS)

| | | | 95% CI | | 90% CI | |
|---|---|---|---|---|---|---|
| SIE | EST | SE | 2.5% | 97.5% | 5% | 95% |
| PD → PEQ → PES → SRH → LS | .000 | .000 | **.000** | **.002** | **.000** | **.001** |
| BD → PEQ → PES → SRH → LS | .003 | .029 | −.022 | .107 | −.014 | .079 |
| FMD → PEQ → PES → SRH → LS | .003 | .026 | −.018 | .087 | −.011 | .068 |
| RID→PEQ → PES → SRH → LS | .000 | .000 | **.000** | **.000** | **.000** | **.000** |
| BLD → PEQ → PES → SRH → LS | .000 | .001 | −.003 | .001 | −.002 | .001 |
| BSD → PEQ → PES → SRH → LS | .000 | .001 | **.000** | **.003** | **.000** | **.002** |
| POID→PEQ → PES → SRH → LS | −.010 | .088 | −.325 | .075 | −.235 | .049 |

Caption: CI are estimated using the bias-corrected nonparametric percentile bootstrap
Boldface indicates statistical significance at the 95% and 90% confidence levels
SIE refers to specific indirect effect. EST refers to point estimate
SE refers to Standard Error
Source: Created by the authors

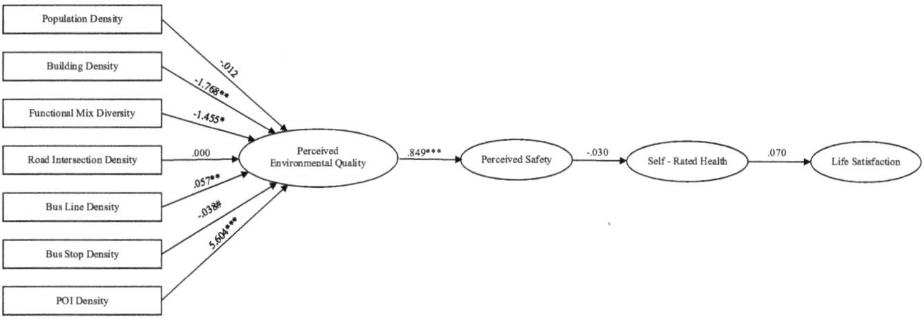


**Fig. 9.20** The effects of the OBE factors on LS through Serial Mediators of PEQ, PES, SRH (OBE → PEQ → PES → SRH → LS). Caption: Control variables are omitted from the model for clarity. Source: Created by the authors

**Table 9.19** The objective built environment affects life satisfaction through the serial mediation of perceived environmental quality, social capital and self-rated health (OBE → PEQ → SC → SRH → LS)

| | | | 95% CI | | 90% CI | |
|---|---|---|---|---|---|---|
| SIE | EST | SE | 2.5% | 97.5% | 5% | 95% |
| PD → PEQ → SC → SRH → LS | .000 | .000 | **.000** | **.000** | **.000** | **.000** |
| BD → PEQ → SC → SRH → LS | −.001 | .010 | −.024 | .010 | −.018 | .006 |
| FMD → PEQ → SC → SRH → LS | −.001 | .009 | −.020 | .009 | −.016 | .005 |
| RID→PEQ → SC → SRH → LS | .000 | .000 | **.000** | **.000** | **.000** | **.000** |
| BLD → PEQ → SC → SRH → LS | .000 | .000 | **.000** | **.001** | **.000** | **.001** |
| BSD → PEQ → SC → SRH → LS | .000 | .000 | −.001 | .000 | **.000** | **.000** |
| POID→PEQ → SC → SRH → LS | .002 | .028 | −.032 | .068 | −.020 | .053 |

Caption: CI are estimated using the bias-corrected nonparametric percentile bootstrap
Boldface indicates statistical significance at the 95% and 90% confidence levels
SIE refers to specific indirect effect. EST refers to point estimate
SE refers to Standard Error
Source: Created by the authors

Table 9.19 reports the indirect effects of seven factors of OBE on LS through the chain mediation of PEQ affecting SC and then affecting SRH. Figure 9.21 shows the path diagram of the OBE influence on LS through this path. The results show that PD, RID, and BLD do not contain 0 in the 95% confidence interval, and their indirect effects are strict significant. The other four factors contained 0 in 95% and 90% confidence intervals, so their indirect effects were not significant. The indirect effect estimates prove that the indirect effects of PD, RID, and BLD on LS through the path are small but positively significant.

Table 9.20 reports the indirect effects of seven factors of OBE on LS through the chain mediation of PES affecting SC and then affecting SRH. Figure 9.22 shows the path diagram of the OBE influence on LS through this path. The results show that PD, RID, BLD and BSD do not contain 0 in the 95% confidence interval, and their indirect effects are strict significant. The other three factors contained 0 in 95% and 90% confidence intervals, so their indirect effects were not significant. The indirect

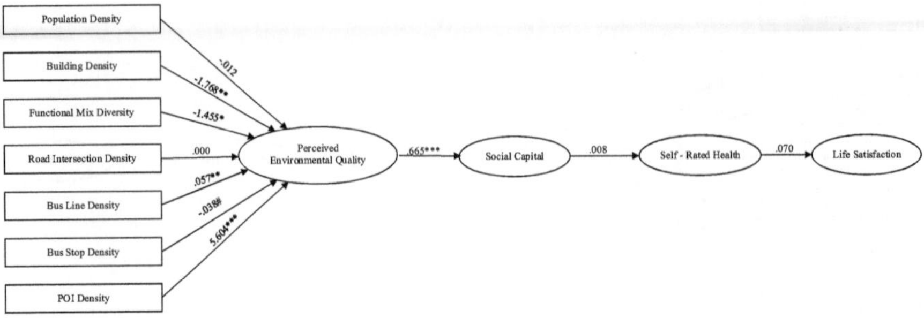


**Fig. 9.21** The effects of the OBE factors on LS through Serial Mediators of PEQ, SC, SRH (OBE → PEQ → SC → SRH → LS). Caption: Control variables are omitted from the model for clarity. Source: Created by the authors

**Table 9.20** The objective built environment affects life satisfaction through the serial mediation of perceived safety, social capital and self-rated health (OBE → PES → SC → SRH → LS)

| | | | 95% CI | | 90% CI | |
|---|---|---|---|---|---|---|
| SIE | EST | SE | 2.5% | 97.5% | 5% | 95% |
| PD | .000 | .000 | **.000** | **.000** | **.000** | **.000** |
| BD | .000 | .002 | −.002 | .004 | −.001 | .002 |
| FMD | .000 | .004 | −.004 | .006 | −.002 | .004 |
| RID | .000 | .000 | **.000** | **.000** | **.000** | **.000** |
| BLD | .000 | .000 | **.000** | **.000** | **.000** | **.000** |
| BSD | .000 | .000 | **.000** | **.000** | **.000** | **.000** |
| POID | .000 | .004 | −.007 | .003 | −.004 | .002 |

Caption: CI are estimated using the bias-corrected nonparametric percentile bootstrap
Boldface indicates statistical significance at the 95% and 90% confidence levels
SIE refers to specific indirect effect. EST refers to point estimate
SE refers to Standard Error
Source: Created by the authors

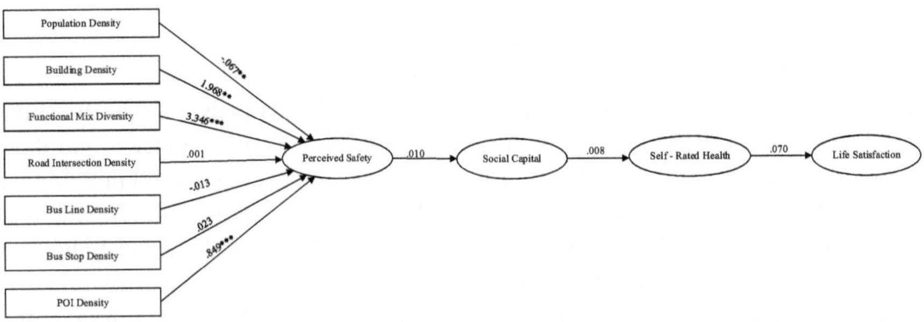


**Fig. 9.22** The effects of the OBE factors on LS through Serial Mediators of PES, SC, SRH (OBE → PES → SC → SRH → LS). Caption: Control variables are omitted from the model for clarity. Source: Created by the authors

effect estimates prove that PD, RID, BLD and BSD have small but positive effects on Life Satisfaction through the path.

Table 9.21 reports the indirect effects of seven factors of OBE on LS via serial mediation of PEQ affecting PES affecting SC, and then affecting SRH. Figure 9.23 shows the path diagram of the OBE influence on LS through this path. The results show that PD, RID, BLD and BSD do not contain 0 in the 95% confidence interval, and their indirect effects are strict significant. The other three factors contained 0 in 95% and 90% confidence intervals, so their indirect effects were not significant. The indirect effect estimates prove that PD, RID, BLD, and BSD have small but positive effects on Life Satisfaction through the path.

In this study, Bootstrap was used to test the mediating effect, and the possible action path of the factors of OBE affecting LS in the structural equation model was tested empirically, and some evidence supporting this hypothesis was found. The results show that the factors of OBE influence residents 'Life Satisfaction through intermediary factors. Specifically, the hypothesis test results are summarized as follows:

H1: Among the seven objective built environment factors affecting life satisfaction, all seven factors had 95% confidence intervals containing zero, indicating that none of their direct effects reached strict statistical significant. With the exception of RID, whose direct effect was marginally significant as evidenced by its 90% confidence interval excluding zero, no other factor showed a statistically meaningful direct influence.(OBE → LS). Thus, the hypothesis of significant direct effects is only partially supported, and only by RID at a marginal level.

H2: Among the seven factors of OBE on LS through PEQ, the three factors of OBE, e.g., BD, BLD, and POID, were statistically significant. The three factors of OBE, e.g., FMD, RID and BSD, were marginal significant. The indirect effect of PD through PEQ on LS is not significant. Therefore, this study supports the indirect effects of the above six factors of OBE through this path (OBE → PEQ → LS). This hypothesis is partially supported, with six out of seven factors exhibiting significant or marginally significant indirect effects.

H3: Among the seven factors of OBE on LS through PES, the factor of RID was statistical significant. The three factors of OBE, e.g., PD, BD, and FMD, were marginally significant. The indirect effect of the other factors of OBE, e.g., BLD, BSD and POID through PES on LS is not significant. Therefore, this study supports the indirect effects of the above four factors of OBE through this path (OBE → PES → LS). The hypothesis is partially supported, with four factors demonstrating significant or marginal indirect effects.

H4: Among the seven factors of OBE on LS through SC, the factors of OBE, e.g., PD, RID, BLD, BSD, and POID, are statistical significant. The indirect effect of the other factors of OBE, e.g., BD, FMD, through SC on LS is not significant. Therefore, this study supports the indirect effects of the above five factors of OBE through this path (OBE → SC → LS). This hypothesis is largely supported, with five significant mediators among the seven factors.

H5: Among the seven factors of OBE on LS through SRH, the factor of RID was statistical significant. The indirect effect of the other factors through SRH on LS is

**Table 9.21** The objective built environment affects life satisfaction through the serial mediation of perceived environmental quality, perceived safety, social capital and self-rated health (OBE → PEQ → PES → SC → SRH → LS)(OBE → PEQ → PES → SC → SRH → LS)

| SIE | EST | SE | 95% CI | | 90% CI | |
|---|---|---|---|---|---|---|
| | | | 2.5% | 97.5% | 5% | 95% |
| PD → PEQ → PES → SC → SRH → LS | .000 | .000 | **.000** | **.000** | **.000** | **.000** |
| BD → PEQ → PES → SC → SRH → LS | .000 | .002 | −.003 | .002 | −.002 | .001 |
| FMD → PEQ → PES → SC → SRH → LS | .000 | .003 | −.002 | .002 | −.002 | .001 |
| RID→PEQ → PES → SC → SRH → LS | .000 | .000 | **.000** | **.000** | **.000** | **.000** |
| BLD → PEQ → PES → SC → SRH → LS | .000 | .000 | **.000** | **.000** | **.000** | **.000** |
| BSD → PEQ → PES → SC → SRH → LS | .000 | .000 | **.000** | **.000** | **.000** | **.000** |
| POID→PEQ → PES → SC → SRH → LS | .000 | .007 | −.005 | .009 | −.004 | .006 |

Caption: CI are estimated using the bias-corrected nonparametric percentile bootstrap
Boldface indicates statistical significance at the 95% and 90% confidence levels
SIE refers to specific indirect effect. EST refers to point estimate
SE refers to Standard Error
Source: Created by the authors

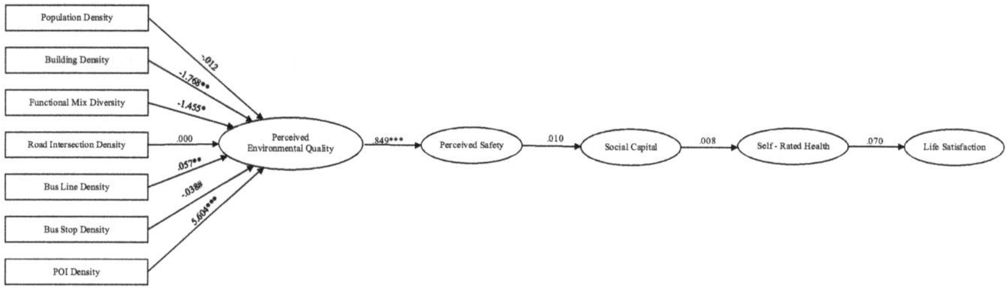


**Fig. 9.23** The effects of the OBE factors on LS through Serial Mediators of PEQ, PES, SC, SRH (OBE → PEQ → PES → SC → SRH → LS). Caption: Control variables are omitted from the model for clarity. Source: Created by the authors

not significant. Therefore, this study supports the indirect effects of the RID of OBE through this path (OBE → SRH → LS). The hypothesis is only partially supported, with a significant effect observed for RID exclusively.

H6: Among the seven factors of OBE on LS via serial mediators of PEQ affecting PES, the factors of OBE, e.g., BD, RID, BLD, BSD, POID, were marginally significant. The indirect effect of the other factors of OBE, e.g., PD and FMD, through this path is not significant. Therefore, this study supports the indirect effects of the above five factors of OBE through this path (OBE → PEQ → PES → LS). This hypothesis receives partial support, with five factors showing a marginally significant result.

H7 Among the seven factors of OBE on LS via serial mediators of PEQ affecting SC, the factors of OBE, e.g., BD, FMD, RID, BLD, and POID, were statistical significant. The indirect effect of BSD was marginally significant. Therefore, this study supports the indirect effects of the above six factors of OBE through this path (OBE → PEQ → SC → LS). The hypothesis is strongly supported, with all factors except one showing significant or marginal effects.

H8: Among the seven factors of OBE on LS via serial mediators of PEQ affecting SRH, the indirect effect of RID was statistically significant. The indirect effect of BLD, BSD was marginally significant. Therefore, this study supports the indirect effects of the above three factors of OBE through this path (OBE → PEQ → SRH → LS). This hypothesis is partially supported, with three factors showing significant or marginal indirect effects.

H9: Among the seven factors of OBE on LS through the chain mediation of PES affecting SC, the indirect effect of RID was statistically significant. The other factors of OBE were not significant. Therefore, this study supports the indirect effects of the above factor of OBE through this path (OBE → PES → SC → LS). The hypothesis is only minimally supported, with a significant effect for RID alone.

H10: Among the seven factors of OBE on LS via serial mediators of PES affecting SRH, the indirect effects of RID and BLD were statistically significant. The other factors of OBE were not significant. Therefore, this study supports the indirect effects of the above two factors of OBE through this path (OBE → PES → SRH → LS). This hypothesis is partially supported, with two factors exhibiting significant effects.

H11: Among the seven factors of OBE on LS through the chain mediation of SC affecting SRH, the indirect effects of PD and RID were statistically significant. The factor of BSD was marginally significant. Therefore, this study supports the indirect effects of the above three factors of OBE through this path (OBE → SC → SRH → LS). The hypothesis is partially supported, with three factors showing significant or marginal effects.

H12: Among the seven factors of OBE on LS via serial mediators of PEQ affecting PES, then affecting SC, the indirect effect of RID was statistically significant. The other factors of OBE were not significant. Therefore, this study supports the indirect effects of the above factor of OBE through this path (OBE → PEQ → PES → SC → LS). This hypothesis receives minimal support, with only one significant factor.

H13: Among the seven factors of OBE on LS via serial mediators of PEQ affecting PES, then affecting SRH, the indirect effect of the factors of OBE, e.g., PD, RID, and BSD, were statistically significant. The other factors of OBE were not significant. Therefore, this study supports the indirect effects of the above three factors of OBE through this path (OBE → PEQ → PES → SRH → LS). The hypothesis is partially supported, with three significant mediators.

H14: Among the seven factors of OBE on LS via serial mediators of PEQ affecting PES, then affecting SRH, the indirect effect of the factors of OBE, e.g., PD, RID, and BLD, were statistically significant. The indirect effect of BSD was marginally significant. The other factors of OBE were not significant. Therefore, this study supports the indirect effects of the above four factors of OBE through this path (OBE → PEQ → SC → SRH → LS). This hypothesis is partially supported, with four factors showing significant or marginal effects.

H15: Among the seven factors of OBE on LS via serial mediators of PES affecting SC, then affecting SRH, the indirect effect of the factors of OBE, e.g., PD, RID, BLD, and BSD, were statistically significant. The other factors of OBE were not significant. Therefore, this study supports the indirect effects of the above four factors of OBE through this path (OBE → PES → SC → SRH → LS). The hypothesis is partially supported, with four significant factors.

H16: Among the seven factors of OBE on LS via serial mediators of PEQ affecting PES affecting SC, then affecting SRH, the indirect effect of the factors of OBE, e.g., PD, RID, BLD, and BSD, were statistically significant. The other factors of OBE were not significant. Therefore, this study supports the indirect effects of the above four factors of OBE through this path (OBE → PEQ → PES → SC → SRH → LS). This hypothesis is partially supported, with four out of seven factors demonstrating significant indirect effects.

Table 9.22 also reports the total and direct effects of the Objective Built Environment on Life Satisfaction. The total effects show that for BD, RID, and BLD, the 95% confidence intervals do not contain 0, indicating strict significant (Preacher & Hayes, 2008), while for FMD, the 90% confidence interval does not contain 0, indicating marginal significant. Among them, the effects of BD, FMD, and RID are negatively significant, and the effect of BLD is positively significant. The results of the direct effects prove that RID reaches a marginal significant level

**Table 9.22** The test of the total effects, direct effects of the objective built environment on life satisfaction

| | | | 95% CI | | 90% CI | |
|---|---|---|---|---|---|---|
| SIE | EST | SE | 2.5% | 97.5% | 5% | 95% |
| Total effect | | | | | | |
| PD | .010 | .024 | −.039 | .058 | −.0320 | 0.050 |
| BD | −1.657 | .752 | −3.128 | −.200 | **−2.9** | **−0.46** |
| FMD | −1.437 | .816 | −3.015 | .217 | **−2.747** | **−0.107** |
| RID | −.002 | .001 | **−.004** | **−.001** | **−.003** | **−.001** |
| BLD | .052 | .024 | **.001** | **.098** | **0.01** | **0.091** |
| BSD | −.044 | .029 | −.100 | .014 | −.090 | .006 |
| POID | 1.925 | 1.967 | −2.055 | 5.702 | −1.418 | 5.086 |
| Direct effect | | | | | | |
| PD | .025 | .025 | −.025 | .075 | −.017 | .065 |
| BD | −.738 | .759 | −2.249 | .700 | −2.021 | .508 |
| FMD | −.770 | .827 | −2.451 | .806 | −2.153 | .542 |
| RID | −.002 | .001 | −.004 | .000 | **−.003** | **−.000** |
| BLD | .028 | .023 | −.016 | .073 | −.009 | .065 |
| BSD | −.030 | .024 | −.076 | .018 | −.070 | .009 |
| POID | .000 | .000 | −.001 | .000 | −.001 | .000 |
| Indirect effect | | | | | | |
| PD | −.011 | .015 | −.042 | .018 | −.037 | .014 |
| BD | −.711 | .599 | −1.987 | .392 | −1.747 | .233 |
| FMD | −.507 | .544 | −1.583 | .532 | −1.389 | .388 |
| RID | .000 | .001 | −.001 | .001 | −.001 | .001 |
| BLD | .022 | .015 | −.008 | .052 | −.002 | .047 |
| BSD | −.008 | .017 | −.042 | .024 | −.037 | .018 |
| POID | 1.710 | 1.623 | −1.199 | 5.334 | −.737 | 4.620 |

Caption: CI are estimated using the bias-corrected nonparametric percentile bootstrap
Boldface indicates statistical significance at the 95% and 90% confidence levels
SIE refers to specific indirect effect. EST refers to point estimate
SE refers to Standard Error
Source: Created by the authors

as its 90% confidence interval does not contain 0. The results indicate that the direct effect of road intersection density (RID) is marginally significant, as evidenced by the 90% confidence interval excluding zero. A possible explanation is that increased road intersection density may expose residents to ambient noise, traffic congestion, and daily inconveniences, thereby reducing their life satisfaction (Ewing & Cervero, 2010). The test of the total indirect effects shows that the total indirect effects of all objective built-environment elements on Life Satisfaction contain 0 within both the 95% and 90% confidence intervals, all being non-significant. This is mainly because there are both significantly positive and significantly negative specific indirect effects, which may have a suppressing effect (MacKinnon et al. 2000; Wen & Ye, 2014; Wen et al., 2004).[5]

## 9.4 Summary of this Chapter

Based on the previous exploration of the relationship between the objective built environment (OBE) and life satisfaction (LS), this chapter employs a structural equation modeling (SEM) approach to examine the mechanisms through which the OBE influences LS via multiple mediators: perceived environmental quality (PEQ), perceived safety (PES), social capital (SC), and self-rated health (SRH). The mediating effects of these variables are rigorously tested within a comprehensive theoretical framework.

The results indicate that, after controlling for key socioeconomic attributes, certain factors of the OBE exert significant effects on the proposed mediating variables. Notably, factors such as Building Density, road connectivity, and land use mix show particularly strong associations with perceived environmental quality and social capital (Litwin, 2020). In turn, the mediators—Perceived Environment Quality (PEQ), Perceived Safety (PES), and Social Capital (SC)—all demonstrate significant positive effects on life satisfaction, underscoring their roles as plausible mechanisms through which the built environment influences subjective well-being. Self-rated health also serves as a mediator, albeit for a narrower set of OBE factors, suggesting a more constrained pathway.

To empirically validate these pathways, mediation effect analyses were conducted using bootstrapping methods, which allow for robust inference of indirect effects (Hayes, 2018). The findings confirm that several OBE factors exert influence through both simple and serial mediation routes. For instance, green space availability significantly affects life satisfaction through perceived environmental quality and social capital sequentially, highlighting the interplay between environmental and social factors.

Figure 9.24 summarizes the significant mediating pathways, illustrating that the OBE influences life satisfaction not only directly but also through a network of mediating variables. These results reinforce the notion that improving the built environment can enhance residents' quality of life through both physical and perceptual channels (Leyden, 2003; Ewing & Cervero, 2010; Dong & Qin, 2017; Litman, 2022). However, it is also evident that not all OBE factors influence LS equally; rather, their effects are channeled through specific cognitive and social mechanisms. This study adds to a growing body of literature linking urban form to well-being through multiple mediating pathways (Park & Garcia, 2020).[6]

---

[5] Suppression effects occur when mediators have opposing signs, masking the total indirect effect. For methodological discussions, see Wen and Ye (2014) and MacKinnon et al. (2000).

[6] Future research could explore cross-cultural comparisons or longitudinal designs to validate these pathways. Limitations include the use of self-reported data and cross-sectional design.

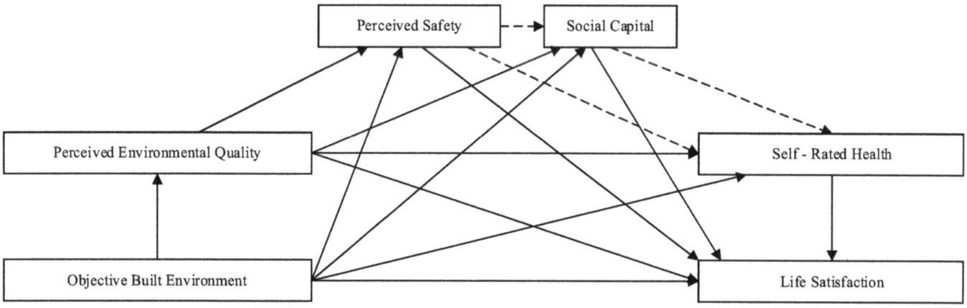


**Fig. 9.24** Pathways of the OBE factors effect Life Satisfaction. Caption: Solid lines with arrows indicate statistically significant paths, whereas dashed lines with arrows indicate paths that are not statistically significant Control variables are omitted from the model for clarity. Source: Created by the authors

**Funding** This research was funded by

1. Guangdong Provincial Office of Philosophy and Social Sciences [Grant No. GD24XSH06]
2. Projects of PhDs' Start-up Research of GDUPT [Grant No. 2023bsqd1008].

## References

Aldrich, D. P., & Meyer, M. A. (2015). Social capital and community resilience. *American Behavioral Scientist, 59*(2), 254–269.

Appleyard, D. (1980). Livable streets: Protected neighborhoods? *The Annals of the American Academy of Political and Social Science, 451*(1), 106–117.

Babin, B. J., & Anderson, R. E. (2011). *Multivariate data analysis: A global perspective* (7th ed.). Pearson.

Baron-Epel, O., Weinstein, R., Haviv-Mesika, A., Garty-Sandalon, N., & Green, M. S. (2008). Individual-level analysis of social capital and health: A comparison of Arab and Jewish Israelis. *Social Science & Medicine, 66*(4), 900–910.

Basolo, V., & Strong, D. (2002). Understanding the neighborhood: From residents' perceptions and needs to action. *Housing Policy Debate, 13*(1), 83–105.

Baum, F., & Palmer, C. (2002). Opportunity structures': Urban landscape, social capital and health promotion in Australia. *Health Promotion International, 17*(4), 351–361.

Bentler, P. M., & Bonett, D. G. (1980). Significance tests and goodness of fit in the analysis of covariance structures. *Psychological Bulletin, 88*(3), 588–606.

Bentler, P. M., & Wu, E. J. C. (2005). *EQS 6.1 for windows program manual*. Multivariate Software.

Bollen, K. A., & Long, J. S. (1993). *Testing structural equation models*. Sage.

Bollen, K. A., & Stine, R. A. (1990). Direct and indirect effects: Classical and bootstrap estimates of variability. *Sociological Methodology, 20*, 115–140.

Bollen, K. A., & Stine, R. A. (1992). Bootstraping goodness-of-fit measures in structural equation models. *Sociological Methods & Research, 21*(2), 205–229.

Boomsma, A., & Hoogland, J. J. (2001). The robustness of LISREL modeling revisited. In R. Cudeck, S. du Toit, & D. Sörbom (Eds.), *Structural equation models: Present and future* (pp. 139–168). Scientific Software International.

Bosselmann, P. (2012). *Urban transformation: Understanding city form and design*. Island Press.

Brown, T. A. (2015). Confirmatory factor analysis for applied research (2nd ed.). Guilford Press.

Browning, C. R., Byron, R. A., Calder, C. A., Krivo, L. J., Kwan, M.-P., Lee, J.-Y., & Peterson, R. D. (2010). Commercial density, residential concentration, and crime: Land use patterns and violence in neighborhood context. *Journal of Research in Crime and Delinquency, 47*(3), 329–357.

Byrne, B. M. (2009). *Structural equation modeling with AMOS: Basic concepts, applications, and programming* (2nd ed.). Routledge.

Cao, X. (2013). The influence of light rail transit on transit-oriented development in China. *Transportation Research Part A: Policy and Practice, 50*, 240–251.

Cao, X. (2016). How does neighborhood design affect life satisfaction? Evidence from twin cities. *Travel Behaviour and Society, 5*, 68–76.

Cervero, R., & Kockelman, K. (1997). Travel demand and the 3Ds: Density, diversity, and design. *Transportation Research Part D: Transport and Environment, 2*(3), 199–219.

Chen, J., & Ning, Z. (2015). The impact of perceived neighborhood environment on residents' mental health: A case study of Guangzhou. *Acta Geographica Sinica, 70*(4), 611–625.

Chou, C.-P., & Bentler, P. M. (1995). Estimates and tests in structural equation modeling. In R. H. Hoyle (Ed.), *Structural equation modeling: Concepts, issues, and applications* (pp. 37–55). Sage.

Cohen, L. E., & Felson, M. (1979). Social change and crime rate trends: A routine activity approach. *American Sociological Review, 44*(4), 588–608.

Currie, G., & Stanley, J. (2008). Investigating links between social capital and public transport. *Transport Reviews, 28*(4), 529–547.

Diener, E., & Seligman, M. E. P. (2004). Beyond money: Toward an economy of Well-being. *Psychological Science in the Public Interest, 5*(1), 1–31.

Dolan, P., Peasgood, T., & White, M. (2008). Do we really know what makes us happy? A review of the economic literature on the factors associated with subjective Well-being. *Journal of Economic Psychology, 29*(1), 94–122.

Dong, H., & Qin, B. (2017). Exploring the link between neighborhood environment and mental Well-being: A case study in Beijing, China. *Landscape and Urban Planning, 164*, 71–80.

Ewing, R., & Cervero, R. (2010). Travel and the built environment. *Journal of the American Planning Association, 76*(3), 265–294.

Ewing, R., & Handy, S. (2009). Measuring the unmeasurable: Urban design qualities related to walkability. *Journal of Urban Design, 14*(1), 65–84.

Fang, J., Wen, Z., & Hau, K.-T. (2011). A review of the development and application of mediation models. *Advances in Psychological Science, 19*(5), 765–774.

Fang, J., Zhang, M. Q., & Qiu, H. Z. (2012). Mediation analysis and effect size measurement: Retrospect and prospect. *Psychological Development and Education, 28*(1), 105–111.

Ferriss, A. L. (2004). The quality of life concept in sociology. *The American Sociologist, 35*(3), 37–51.

Fornell, C., & Larcker, D. F. (1981). Evaluating structural equation models with unobservable variables and measurement error. *Journal of Marketing Research, 18*(1), 39–50.

Foster, S., Giles-Corti, B., & Knuiman, M. (2010). Neighbourhood design and fear of crime: A social-ecological examination of the correlates of residents' fear in new suburban housing developments. *Health & Place, 16*(6), 1156–1165.

Foster, S., Giles-Corti, B., & Knuiman, M. (2013). Does fear of crime discourage walkers? A social-ecological exploration of the moderating role of neighborhood social capital. *Preventive Medicine, 56*(5), 390–393.

French, S., Wood, L., Foster, S. A., Giles-Corti, B., Frank, L., & Learnihan, V. (2014). Sense of community and its association with the neighborhood built environment. *Environment and Behavior, 46*(6), 677–697.

Frey, B. S., & Stutzer, A. (2010). *Happiness and economics: How the economy and institutions affect human Well-being*. Princeton University Press.

Gehl, J. (2011). *Life between buildings: Using public space*. Island Press.

Gifford, R. (2007). The consequences of living in high-rise buildings. *Architectural Science Review, 50*(1), 2–17.

Glaeser, E. L., & Sacerdote, B. (1999). Why is there more crime in cities? *Journal of Political Economy, 107*(S6), S225–S258.

Golob, T. F. (2003). Structural equation modeling for travel behavior research. *Transportation Research Part B: Methodological, 37*(1), 1–25.

Hair, J. F., Black, W. C., Babin, B. J., & Anderson, R. E. (2011). *Multivariate data analysis* (7th ed.). Pearson.

Hair, J. F., Black, W. C., Babin, B. J., & Anderson, R. E. (2019). *Multivariate data analysis* (8th ed.). Cengage Learning.

Hair, J. F., Hult, G. T. M., Ringle, C. M., & Sarstedt, M. (2022). *A primer on partial least squares structural equation modeling (PLS-SEM)* (3rd ed.). Sage.

Hancock, G. R., & Mueller, R. O. (2011). *Structural equation modeling: A second course* (2nd ed.). Information Age Publishing.

Harrison, R. A., Gemmell, I., & Heller, R. F. (2007). The population effect of crime and neighborhood on physical activity: An analysis of 15,461 adults. *Journal of Epidemiology and Community Health, 61*(1), 34–39.

Hayduk, L. A. (1987). *Structural equation modeling with LISREL: Essentials and advances*. Johns Hopkins University Press.

Hayes, A. F. (2009). Beyond Baron and Kenny: Statistical mediation analysis in the new millennium. *Communication Monographs, 76*(4), 408–420.

Hayes, A. F. (2018). *Introduction to mediation, moderation, and conditional process analysis: A regression-based approach* (2nd ed.). Guilford Press.

Helliwell, J. F., Layard, R., Sachs, J. D., & De Neve, J.-E. (2020). *World happiness report 2020*. Sustainable Development Solutions Network.

Hoogerbrugge, M. M., & Burger, M. J. (2018). Neighborhood-based social capital and life satisfaction: A case study in Rotterdam, The Netherlands. *Urban Geography, 39*(10), 1484–1509.

Hu, L.-t., & Bentler, P. M. (1999). Cutoff criteria for fit indexes in covariance structure analysis: Conventional criteria versus new alternatives. *Structural Equation Modeling: A Multidisciplinary Journal, 6*(1), 1–55.

Hur, M., & Morrow-Jones, H. (2008). Factors that influence residents' satisfaction with neighborhoods. *Environment and Behavior, 40*(5), 619–635.

Inglehart, R., Haerpfer, C., Moreno, A., Welzel, C., Kizilova, K., Diez-Medrano, J., & Jackson, D. L. (2003). Revisiting sample size and number of parameter estimates: Some support for the N: Q hypothesis. *Structural Equation Modeling, 10*(1), 128–141.

Jacobs, J. (1961). *The death and life of great American cities*. Random House.

Jackson, D. L. (2003). Revisiting sample size and number of parameter estimates: Some support for the N:q hypothesis. *Structural Equation Modeling, 10*(1), 128–141.

Janowitz, M. (1952). *The community press in an urban setting*. Free Press.

Jiang, B., Mak, C. N. S., Zhong, H., Larsen, L., & Webster, C. J. (2018). From broken windows to perceived routine activities: Examining the impacts of environmental cues on neighborhood safety. *Landscape and Urban Planning, 170*, 1–10.

Jylhä, M. (2009). What is self-rated health and why does it predict mortality? Towards a unified conceptual model. *Social Science & Medicine, 69*(3), 307–316.

Kim, E. S., Delaney, S. W., & Kubz ansky, L. D. (2020). Sense of purpose in life and cardiovascular disease: Underlying mechanisms and future directions. *Current Cardiology Reports, 22*(7), 49.

Kim, J., & Kaplan, R. (2004). Physical and psychological factors in sense of community: New urbanist Kentlands and nearby Orchard Village. *Environment and Behavior, 36*(3), 313–340.

Kline, R. B. (2015). *Principles and practice of structural equation modeling* (4th ed.). Guilford Press.

Kline, R. B. (2016). *Principles and practice of structural equation modeling* (4th ed.). Guilford Press.

Kline, R. B. (2023). *Principles and practice of structural equation modeling* (5th ed.). Guilford Press.

Kline, R. B., & Little, T. D. (Eds.). (2011). Principles and practice of structural equation modeling (3rd ed.). Guilford Press.

Langdon, P. (1997). *A better place to live: Reshaping the American suburb*. University of Massachusetts Press.

Lederbogen, F., Kirsch, P., Haddad, L., Streit, F., Tost, H., Schuch, P., Wüst, S., Pruessner, J. C., Rietschel, M., & de Meyer-Lindenberg, A. (2011). City living and urban upbringing affect neural social stress processing in humans. *Nature, 474*(7352), 498–501.

Leyden, K. M. (2003). Social capital and the built environment: The importance of walkable neighborhoods. *American Journal of Public Health, 93*(9), 1546–1551.

Li, W., & Chen, J. (2022). Digital social capital and subjective Well-being in urban China: The moderating role of migration status. *Cities, 120*, 103464.

Lin, T. (2019). The influence of transit accessibility on subjective Well-being: Evidence from Taipei. *Transportation Research Part D: Transport and Environment, 67*, 167–178.

Little, T. D., Rhemtulla, M., Gibson, K., & Schoemann, A. M. (2013). Why the items versus parcels controversy needn't be one. *Psychological Methods, 18*(3), 285–300.

Litwin, H. (2020). Social capital and Well-being in later life: A critical review. *The Journals of Gerontology, Series B: Psychological Sciences and Social Sciences, 75*(9), 2061–2071.

Litman, T. M. (2022). Evaluating transportation equity: Guidance for incorporating distributional impacts in transport planning. *Institute of Transportation Engineers. ITE Journal, 92*(4), 43–49.

Liu, H., & Ling, W. (2009). Serial mediation analysis in structural equation modeling. *Psychological Methods, 14*(3), 280–291.

Liu, Y., Wang, R., & Lu, Y. (2020). The nonlinear relationship between built environment and life satisfaction in Guangzhou. *Landscape and Urban Planning, 195*, 103725.

Liu, Z., Wang, Y., & Chen, S. (2012). The impact of population density on property crime in Chinese cities. *Journal of Criminal Justice, 40*(5), 423–431.

Lorenc, T., Clayton, S., Neary, D., Whitehead, M., Petticrew, M., Thomson, H., et al. (2013). Crime, fear of crime, environment, and mental health and Well-being: Mapping review of theories and causal pathways. *Health & Place, 18*, 757–765.

Ma, L., Li, Y., & Li, W. (2018). The impact of transit-oriented development on housing satisfaction in Chinese cities. *Transport Policy, 66*, 1–10.

MacKinnon, D. P., Krull, J. L., & Lockwood, C. M. (2000). Equivalence of the mediation, confounding, and suppression effect. *Prevention Science, 1*(4), 173–181.

Marshall, W. E., & Garrick, N. W. (2011). Does street network design affect traffic safety? *Accident Analysis & Prevention, 43*(3), 769–781.

Maslow, A. H. (1943). A theory of human motivation. *Psychological Review, 50*(4), 370–396.

Morris, E. W. (2011). *An introduction to the sociology of Well-being*. Routledge.

Mouratidis, K. (2018). Rethinking how built environments influence subjective well-being: A new conceptual framework. *Journal of Urbanism: International Research on Placemaking and Urban Sustainability, 11*(1), 24–40.

Owen, N., Cerin, E., Leslie, E., duToit, L., Coffee, N., Frank, L. D., et al. (2007). Neighborhood walkability and the walking behavior of Australian adults. *American Journal of Preventive Medicine, 33*(5), 387–395.

Park, S., & Garcia, M. (2020). The mediating role of social capital in the relationship between the built environment and mental health. *International Journal of Environmental Research and Public Health, 17*(17), 6113.

Parkes, A., Kearns, A., & Atkinson, R. (2002). What makes people dissatisfied with their neighbourhoods? *Urban Studies, 39*(13), 2413–2438.

Permentier, M., van Ham, M., & Bolt, G. (2011). Behavioural responses to neighbourhood reputations. *Journal of Housing and the Built Environment, 26*(2), 199–213.

Pfeiffer, D., & Cloutier, S. (2016). Planning for happy neighborhoods. *Journal of the American Planning Association, 82*(3), 267–279.

Potter, J., & Cantarero, R. (2006). How does increasing population density affect Well-being? An analysis of the mediating role of travel behavior. *Transportation Research Part D: Transport and Environment, 11*(5), 347–358.

Preacher, K. J., & Hayes, A. F. (2008). Asymptotic and resampling strategies for assessing and comparing indirect effects in multiple mediator models. *Behavior Research Methods, 40*(3), 879–891.

Putnam, R. D. (2000). *Bowling alone: The collapse and revival of American community*. Simon & Schuster.

Saelens, B. E., & Handy, S. L. (2008). Built environment correlates of walking: A review. *Medicine and Science in Sports and Exercise, 40*(7 Suppl), S550–S566.

Sampson, R. J. (2012). *Great American city: Chicago and the enduring neighborhood effect*. University of Chicago Press.

Sampson, R. J., & Raudenbush, S. W. (2004). Seeing disorder: Neighborhood stigma and the social construction of "broken windows". *Social Psychology Quarterly, 67*(4), 319–342.

Savalei, V., & Bentler, P. M. (2005). A statistically justified pairwise ML method for incomplete nonnormal data: A comparison with direct ML and pairwise ADF. *Structural Equation Modeling, 12*(2), 183–214.

Shen, Y., & Fu, X. (2019a). Impact of the built environment and perceived neighborhood characteristics on residents' satisfaction: Evidence from the Shanghai suburbs. *Progress in Geography, 38*(5), 686–697.

Shen, Y., & Fu, X. (2019b). The impact of population density on community satisfaction in urban China. *Habitat International, 85*, 1–10.

Sirgy, M. J., & Cornwell, T. (2002). How neighborhood features affect quality of life. *Social Indicators Research, 59*(1), 79–114.

Sivo, S. A., Fan, X., Witta, E. L., & Willse, J. T. (2006). The search for "optimal" cutoff properties: Fit index criteria in structural equation modeling. *The Journal of Experimental Education, 74*(3), 267–288.

Song, Y., & Knaap, G. J. (2004). Measuring the effects of mixed land uses on housing values. *Regional Science and Urban Economics, 34*(6), 663–680.

Sørensen, J. F. L., Davidsen, M., & Pedersen, M. H. (2023). The role of social capital in mitigating the mental health impact of the COVID-19 pandemic: A longitudinal study. *Social Science & Medicine, 317*, 115603.

Sugiyama, T., Leslie, E., Giles-Corti, B., & Owen, N. (2008). Associations of neighbourhood greenness with physical and mental health: Do walking, social coherence and local social interaction explain the relationships? *Journal of Epidemiology & Community Health, 62*(5), e9–e9.

Sun, G., Oreskovic, N. M., & Lin, H. (2016). How do changes to the built environment influence walking behaviors? A longitudinal study within a university campus in Hong Kong. *International Journal of Health Geographics, 15*(1), 1–13.

Szreter, S., & Woolcock, M. (2004). Health by association? Social capital, social theory, and the political economy of public health. *International Journal of Epidemiology, 33*(4), 650–667.

Tao, T., & Chai, Y. (2020). The mechanism of the built environment on residents' subjective Well-being: A case study of Beijing. *Acta Geographica Sinica, 75*(10), 2154–2173.

Thompson, B. (2000). Ten commandments of structural equation modeling. In L. G. Grimm & P. R. Yarnold (Eds.), *Reading and understanding more multivariate statistics* (pp. 261–283). American Psychological Association.

Wang, F., & Wang, D. (2020). Changes in residential satisfaction after home relocation: A longitudinal study in Beijing, China. *Urban Studies, 57*(3), 583–605.

Wang, Y. (2002). *A study on the form and function of modern architecture*. China Architecture & Building Press.

Welsh, B. C., & Farrington, D. P. (2009). Public area CCTV and crime prevention: An updated systematic review and meta-analysis. *Justice Quarterly, 26*(4), 716–745.

Wen, Z. (2012). The development and application of mediation models in psychological research. *Advances in Psychological Science, 20*(5), 731–745.

Wen, Z., Chang, L., Hau, K. T., & Liu, H. (2004). Testing and application of the mediating effects. *Acta Psychologica Sinica, 36*(5), 614–620.

Wen, Z., & Ye, B. (2014). Analyses of mediating effects: The development of methods and models. *Advances in Psychological Science, 22*(5), 731.

Wen, Z., Zhang, L., & Hou, J. (2005). Mediation effect analysis: A new method for developing and testing mediation models. *Advances in Psychological Science, 13*(5), 614–620.

Wilson, J. Q., & Kelling, G. L. (1982). Broken windows: The police and neighborhood safety. *Atlantic Monthly, 249*(3), 29–38.

Wood, L., Frank, L. D., & Giles-Corti, B. (2010). Sense of community and its relationship with walking and neighborhood design. *Social Science & Medicine, 70*(9), 1381–1390.

Wood, L., Shannon, T., Bulsara, M., Pikora, T., McCormack, G., & Giles-Corti, B. (2008). The anatomy of the safe and social suburb: An exploratory study of the built environment, social capital and residents' perceptions of safety. *Health & Place, 14*(1), 15–31.

Wu, C. (2016). The influence of commuting mode on subjective Well-being: Evidence from Beijing. *Transportation Research Part A: Policy and Practice, 94*, 669–678.

Wu, M. (2009). Structural equation modeling: Operation and application of AMOS (2nd). Chongqing University Press.

Wu, M. (2013). Structural equation modeling: AMOS operation and application (2nd). Chongqing University Press.

Yang, J., Li, X., & Zhang, W. (2021). The impact of built environment on life satisfaction in Beijing: A multi-level analysis. *Journal of Environmental Psychology, 75*, 101619.

Yin, C. (2020). The impact of built environment on mental health in high-density cities: A case study of Shanghai. *Journal of Urban Health, 97*(3), 345–360.

Zhang, H., Chen, B., Sun, Z., & Bao, Z. (2020). The impact of perceived neighborhood environment on residents' mental health: A case study of Guangzhou. *Landscape and Urban Planning, 203*, 103890.

# Part III
# Synthesis, Implications, and Future Directions

# Chapter 10
# Synthesis Discussion: Theoretical Integration and Research Implications

**Abstract** Against the backdrop of China's urbanization and the declining trend in residents' happiness, this chapter examines the complex relationship between the built environment and life satisfaction in high-density urban contexts. Focusing on the central urban area of Guangzhou, it aims to elucidate both the direct and indirect pathways through which objective built environment factors influence residents' overall cognitive evaluation of subjective well-being. Drawing on questionnaire survey data and statistical modeling, the study integrates a conceptual framework incorporating four key mediators: Perceived Environmental Quality (PEQ), Perceived Safety (PES), Social Capital (SC), and Self-Rated Health (SRH). Section 10.1 presents comprehensive research conclusions, detailing the direct effects of built environment factors such as Building Density, Functional Mix Diversity, and transportation infrastructure on life satisfaction, and further unveils complex mediation mechanisms through structural equation modeling. Section 10.2 discusses research implications and policy recommendations tailored for high-density urban areas, proposing a holistic policy framework that addresses density management, transit optimization, and mediating factor enhancement. Section 10.3 critically reflects on research limitations—including geographic specificity and methodological constraints—and suggests future directions for multi-scale and multi-context built environment analysis. This chapter contributes to the adaptation of compact city theory and environmental psychology within high-density Chinese urban settings, offering evidence-based insights for planning and policy interventions aimed at enhancing residents' quality of life.

**Keywords** Built environment · Life satisfaction · High-density cities · Mediation analysis · Perceived environmental quality · Urban policy · Subjective Well-being · Compact city theory · Guangzhou · Structural equation modeling

Synthesizing the empirical results from Chap. 9, which confirmed the significant direct and indirect pathways through which built environment attributes influence residents' well-being—particularly within high-density urban settings—this chapter both consolidates and expands upon these findings to derive broader theoretical



H. Li, G. Shen, *The Anatomy of Subjective Well-Being*, Resilient and Friendly Living Environments, https://doi.org/10.1007/978-981-95-4937-5_10

and practical implications. Against this backdrop, it is noteworthy that in recent years, China has repeatedly emphasized the goal of "continuously enhancing the people's sense of gain, happiness, and security," positioning the improvement of the living environment as a central task in urbanization and quality enhancement initiatives. Although China's urbanization level has risen substantially, with continuous urban expansion and increasing density leading to improved material living standards, the happiness index of Chinese residents has shown a declining trend since the release of the World Happiness Report. This declining sense of happiness amid rapid urbanization has drawn growing attention from various sectors of society, forming a salient contradiction that concerns multiple disciplines, including urban geography, planning, and management (Helliwell et al., 2020). While literature suggests that compact communities, as advocated by New Urbanism, contribute to healthy and happy cities and support the development of sustainable communities (Calthorpe, 2011; Duany et al., 2000), it remains unclear whether these conclusions apply to high-density cities within China's specific context. Previous studies on the built environment and subjective well-being have typically focused on either objective environmental factors or subjective evaluations, with limited integration of the two. Furthermore, although some studies have explored how the objective built environment affects life satisfaction through perceptual mediators, the dimensions of these perceptual variables are often ambiguously defined, and structural equation models rarely address the complex interrelationships among mediating factors (Huang et al., 2023; Mouratidis, 2018a, 2018b).

In response, this study takes the central urban area of Guangzhou as a high-density case example. Using statistical modeling based on questionnaire surveys and objective built-environment data, it examines the influence of the built environment on residents' life satisfaction—the cognitive core of subjective well-being—and its underlying pathways. This chapter summarizes the research conclusions, offers context-sensitive policy recommendations for enhancing life satisfaction, and critically reflects on the limitations of the study while proposing promising directions for future research.

## 10.1 Research Conclusions and Discussions

### *10.1.1 Comprehensive Research Conclusions*

Based on Campbell's framework regarding the impact of the built environment on Life Satisfaction and the new conceptual framework summarized by Mouratidis about the impact of the built environment on Subjective Well-being, this study integrated and constructed a comprehensive conceptual framework (Campbell et al., 1976; Mouratidis, 2018a, 2018b). In this model, there are four key mediating factors between the Objective Built Environment and Subjective Well-being: Perceived Environmental Quality (PEQ), Perceived Safety (PES), Social Capital (SC), and self-rated health (SRH). Figure 10.1 illustrates the conceptual framework showing the direct and mediated pathways through which built environment factors influence Life Satisfaction.

As illustrated in Fig. 10.1, the conceptual framework encompasses five core dimensions of the objective built environment—Density, Diversity, Design, Distance to transit, and Destination accessibility—along with their corresponding variables. These factors influence Life Satisfaction both directly and indirectly through four key mediators: Perceived Environmental Quality, Perceived Safety, Social Capital, and Self-Rated Health. Individual socioeconomic and attribute variables are controlled for in the model. This comprehensive framework captures the multifaceted pathways through which urban form impacts subjective well-being in high-density contexts.

1. Direct Effects of the Built Environment on Life Satisfaction

It is crucial to emphasize that the findings presented in this study are specifically derived from high-density urban core areas (central urban districts of Guangzhou with average building densities exceeding typical Western urban contexts). The applicability of these conclusions to other urban contexts, including suburban areas, new town developments, or lower-density urban environments, requires careful consideration and additional empirical validation (Liu et al., 2020; Mouratidis, 2019).

The built environment has a significant and multifaceted impact on residents' overall cognitive evaluation of Subjective Well-being, namely Life Satisfaction. This finding aligns with substantial international research demonstrating the critical

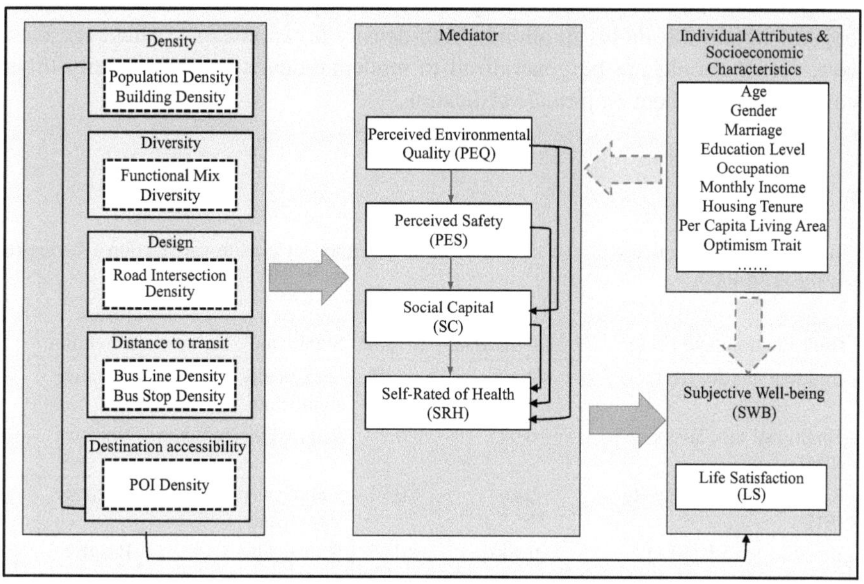


**Fig. 10.1** Conceptual framework of the impact of built environment on life satisfaction in high-density urban contexts. Source: Created by the authors

role of environmental factors in determining quality of life outcomes (Mouratidis, 2019; Sirgy & Cornwell, 2002). The present study reveals significant differences in Life Satisfaction at the community spatial level, which correlate substantially with residents' living environment and various built environment factors.

After controlling for individual socioeconomic attribute variables, several key built environment factors demonstrate significant effects on Life Satisfaction. Building Density (BD), Functional Mix Diversity (FMD), Road Intersection Density (RID), and Bus Stop Density (BSD) all show significant negative effects on Life Satisfaction, meaning that increases in these four factors reduce residents' Life Satisfaction levels. Conversely, Bus Line Density (BLD) shows a positive significant effect, indicating that increased Bus Line Density promotes improvements in residents' Life Satisfaction (Table 10.1). These significant negative relationships are further detailed in Table 10.1.

The multiple regression model reveals that these built environment factors explain 32.2% of the variance in Life Satisfaction (adjusted $R^2 = 0.301$), demonstrating substantial explanatory power. Among the significant factors, Road Intersection Density exhibits the strongest influence ($\beta = -0.164$), followed by Bus Line Density ($\beta = 0.146$) and Bus Stop Density ($\beta = -0.135$), while Building Density ($\beta = -0.106$) and Functional Mix Diversity ($\beta = -0.075$) show relatively smaller but still significant effects.

These findings present important implications specifically for high-density urban contexts and challenge some assumptions of compact city theories that originated in Western, lower-density environments (Ewing & Cervero, 2010). The negative effects of building density and functional mix diversity suggest that the benefits of compactness may have diminishing returns or even become counterproductive beyond certain thresholds in already high-density urban contexts. However, these conclusions should not be generalized to moderate-density or low-density urban environments without empirical verification.

**Table 10.1** Direct effects of the factors of built environment on life satisfaction (Regression Analysis Results)

| Built Environment Factor | β Coefficient | p-value | Significant | Effect Direction |
|---|---|---|---|---|
| Building density (BD) | −0.106 | < 0.05 | Statistically significant | Negative |
| Functional mix diversity (FMD) | −0.075 | < 0.1 | Sarginally significant | Negative |
| Road intersection density (RID) | −0.164 | < 0.01 | Statistically significant | Negative |
| Bus line density (BLD) | +0.146 | < 0.05 | Statistically significant | Positive |
| Bus stop density (BSD) | −0.135 | < 0.1 | Marginally significant | Negative |

Source: Created by the authors

2. Indirect Effects Through Mediating Pathways

The Objective Built Environment exerts indirect impacts on Life Satisfaction through four key mediating factors: Perceived Environmental Quality, Perceived Safety, Social Capital, and self-rated health. This mediation framework reflects the complex psychological and social processes through which physical environment characteristics influence subjective well-being outcomes (Evans & McCoy, 1998; Stoknes, 2015).

The structural equation model analysis reveals multiple significant mediation pathways. Simple mediation effects are observed through individual mediating factors (OBE → PEQ → LS, OBE → PES → LS, OBE → SC → LS, OBE → SRH → LS). More complex chain mediation relationships among the four mediators also significantly affect Life Satisfaction through various pathways, including:

- OBE → PEQ → SP → LS (Environmental Quality affecting Perceived Safety)
- OBE → PEQ → SC → LS (Environmental Quality affecting Social Capital)
- OBE → PEQ → SRH → LS (Environmental Quality affecting Health)
- OBE → PES → SC → LS (Perceived Safety affecting Social Capital)
- OBE → PES → SRH → LS (Perceived Safety affecting Health)
- OBE → SC → SRH → LS (Social Capital affecting Health)

Additionally, three-way and four-way chain mediation paths demonstrate significant effects, indicating the interconnected nature of these psychological and social mediating processes. Table 10.2 provides a comprehensive summary of these specific mediation pathways, detailing which built environment variables significantly influence Life Satisfaction through each type of mediating mechanism.

The Squared Multiple Correlations (SMC) values demonstrate strong explanatory power: 40.7% for Perceived Environmental Quality, 70.8% for Perceived Safety, 44.0% for Social Capital, 53.4% for Self-Rated Health, and 55.8% for Life Satisfaction.

3. Perceived Environmental Quality as the Primary Mediator

Path analysis confirms that Perceived Environmental Quality emerges as the most important mediator among the four intermediate factors, with the largest standardized coefficient ($\beta = 0.306$). This finding aligns with environmental psychology research emphasizing the primacy of environmental perception in determining behavioral and affective responses to built environments (Mehta, 2013; Gehl, 2010).

The mediation tests reveal that simple mediation through Perceived Environmental Quality (OBE → PEQ → LS) is significant for six of seven built environment factors. Furthermore, chain mediation analyses show that the Objective Built Environment affects Perceived Safety, Social Capital, and self-rated health through Perceived Environmental Quality, which then influences Life Satisfaction. This suggests that residents' subjective evaluation of their environment serves as a critical gateway through which objective environmental characteristics translate into well-being outcomes.

**Table 10.2** Summary of significant mediation pathways

| Types of Mediation Effects | Significant Variables | Significant Variables |
|---|---|---|
| | 95% CI (statistically significant) | 90%CI (marginally significant) |
| Simple mediation | | |
| OBE → PEQ → LS | BD, BLD, POID | FMD, RID, BSD |
| OBE → PES → LS | RID | PD, BD, FMD |
| OBE → SC → LS | PD, RID, BLD, BSD, POID | |
| OBE → SRH → LS | RID | |
| Two-way chain mediation | | |
| OBE → PEQ → PES → LS | BD, RID, BLD, BSD, POID | |
| OBE → PEQ → SC → LS | BD, FMD, RID, BLD, POID | BSD |
| OBE → PEQ → SRH → LS | RID | BLD, BSD |
| OBE → PES → SC → LS | RID | |
| OBE → PES → SRH → LS | RID, BLD | |
| OBE → SC → SRH → LS | PD, RID | BSD |
| Three way chain mediation | | |
| OBE → PEQ → PES → SC → LS | RID | |
| OBE → PEQ → PES → SRH → LS | PD, RID, BSD | |
| OBE → PEQ → SC → SRH → LS | PD, RID, BLD | BSD |
| OBE → PES → SC → SRH → LS | PD, RID, BLD, BSD | |
| Four-way chain mediation | | |
| OBE → PEQ → PES → SC → SRH → LS | PD, RID, BLD, BSD | |

Source: Created by the authors

This finding has important theoretical implications, as it bridges objective environmental measurements with subjective well-being outcomes through perceptual processes. It suggests that environmental interventions should consider not only objective improvements but also how these improvements are perceived and interpreted by residents (Gifford, 2007; Bell et al., 2001).

### *10.1.2 Detailed Analysis of Built Environment Effects*

1. Building Density Effects

Building density reflecting density characteristics consistently reduces residents' Life Satisfaction across multiple analytical approaches. This conclusion specifically applies to high-density urban core areas and should be interpreted cautiously when considering moderate-density residential areas or suburban contexts where different density-satisfaction relationships may apply (Bramley & Power, 2009; Liu et al., 2020). The negative impact operates through several mechanisms:

First, increased building density compresses the elasticity of public spaces, reducing park green space and public space areas. This compression intensifies competition for public facility usage (green spaces and fitness areas), reduces accessibility, and affects residents' outdoor leisure and activity spaces (Newman & Kenworthy, 2015). The spatial constraints imposed by high building density create a cascade of effects that limit residents' opportunities for restorative activities and social interaction.

Second, high building density creates more complex road networks, making convenient travel and transportation choices difficult for residents, resulting in mobility-related stress that reduces Life Satisfaction. The navigation complexity in high-density environments can lead to way-finding difficulties and increased cognitive load during daily activities (Lynch, 1960).

Third, high density creates spatial oppression effects on residents, potentially increasing anxiety levels and inducing stress responses. Research in environmental psychology demonstrates that spatial crowding can trigger physiological stress responses and reduce psychological well-being (Evans & McCoy, 1998). High-density high-rise housing also weakens neighborhood interactions and reduces social support networks, as documented in numerous studies of social effects in high-rise environments (Gifford, 2007).

Fourth, building density often correlates with buildings of various ages, types, and styles, creating disorderly street facades and reduced aesthetics. This visual disorder affects residents' Perceived Environmental Quality evaluation and contributes to environmental stress (Nasar, 1998).

The mediation analysis reveals that building density affects Life Satisfaction primarily through Perceived Environmental Quality ($\beta = -0.151$), confirming that the negative effects operate substantially through residents' subjective environmental evaluations rather than purely objective conditions.

2. Functional Mix Diversity Paradox

Functional Mix Diversity, reflecting land use diversity characteristics, shows a negative relationship with Life Satisfaction in high-density contexts, presenting findings that diverge from some compact city expectations but align with other high-density urban research. This finding reveals important nuances in how functional diversity affects well-being, specifically in high-density environments and may not apply to lower-density mixed-use developments (Cao, 2016; Su et al., 2024).

A study of residents' instantaneous happiness in Guangzhou demonstrated that POI type numbers negatively correlate with happiness (Su et al., 2024), supporting the present findings. Similarly, research on American cities proved that land use mixture negatively correlates with Life Satisfaction in certain contexts (Cao, 2016). However, contradictory findings exist, with some Guangzhou studies showing that land-use mix significantly positively affects residents' Life Satisfaction (Su et al., 2019a, 2019b). Relevant studies in Beijing have proved that land use mix significantly improves residents' life satisfaction at both residence and workplace (Wu et al., 2022). American research demonstrates its positive effects on housing satisfaction (Cao, 2016). The inconsistency in findings suggests that the effect of functional mix diversity may be context-dependent and potentially follows a non-linear relationship (Cao, 2016; Mouratidis, 2021). In high-density environments, the hypothesis regarding the relationship between functional mixed diversity and life satisfaction might follow an inverted U-shaped curve. While optimal diversity levels can maximize satisfaction, excessive diversity may have adverse effects (Grant & Perrott, 2011).

In high-density environments, excessive functional mixing may create several negative effects:

First, high functional diversity in constrained spaces can lead to conflicts between different land uses, creating noise, congestion, and environmental disruption that reduces residential satisfaction (Jacobs, 1961). While Jane Jacobs advocated for mixed-use neighborhoods, her prescriptions may require modification in extremely high-density contexts.

Second, functional diversity can contribute to visual and environmental chaos when not carefully managed, affecting aesthetic perception and environmental coherence (Alexander et al., 1977). The principle of "form follows function" suggests that functional diversification brings different architectural styles and forms, potentially creating visual discord when not integrated thoughtfully.

Third, in high-density contexts, functional mixing may increase pedestrian and vehicle traffic to levels that exceed residents' comfort thresholds, creating crowding and congestion effects that outweigh convenience benefits (Appleyard, 1981).

The mediation analysis shows that functional mix diversity affects Life Satisfaction through Perceived Environmental Quality ($\beta = -0.098$), suggesting that residents perceive excessive functional mixing as environmental degradation rather than enhancement in high-density contexts.

3. Transportation Infrastructure Complexity

Road Intersection Density, representing design characteristics and connectivity, demonstrates significant negative effects on Life Satisfaction ($\beta = -0.164$), representing the strongest individual effect among built environment factors. This finding reveals important insights about transportation infrastructure, specifically in high-density contexts, and may not apply to lower-density areas where connectivity benefits may outweigh congestion costs (Ewing & Cervero, 2010).

The negative effects operate through multiple pathways. While increased intersection density may improve road network connectivity and destination accessibility, these benefits appear to be outweighed by negative consequences in high-density environments. More intersections create increased complexity in the transportation system, leading to higher traffic volumes, increased congestion probability, and more frequent traffic accidents (Ewing & Cervero, 2010).

Environmental health impacts constitute a significant pathway through which road intersection density negatively affects life satisfaction. The operation of vehicles at intersections—characterized by frequent acceleration, deceleration, and idling—significantly elevates localized concentrations of air pollutants, particularly PM2.5 (Tong et al., 2016). Concurrently, traffic noise peaks in these high-conflict zones. Chronic exposure to this compounded environmental stressor of poor air quality and noise pollution has been consistently linked to physiological stress responses, including elevated cortisol levels, which detrimentally impact resident well-being (Babisch, 2008; Wallas et al., 2018). Critically, the relationship between intersection density and well-being is not linear but is hypothesized to follow an inverted U-shaped curve (Ewing & Cervero, 2010; Mouratidis, 2021). This suggests that while moderate connectivity improves accessibility and satisfaction, an overabundance of intersections leads to diminishing returns. The inflection point occurs where the benefits of connectivity are eclipsed by the externalities of congestion, pollution, and noise. Our scatter plot trend analysis confirms this pattern, showing life satisfaction first increasing, then decreasing with rising intersection density. Furthermore, the mediation analysis confirms that intersection density affects life satisfaction through both a strong direct pathway ($\beta = -0.161$) and multiple indirect pathways, primarily by degrading Perceived Environmental Quality, impairing Perceived Safety, and eroding Social Capital. This multi-pathway effect underscores that transportation infrastructure complexity has cascading negative effects across multiple domains of the resident experience.

4. Public Transit Accessibility Paradox

The analysis of public transportation factors reveals a complex and nuanced relationship between transit accessibility and Life Satisfaction, specifically within high-density urban core areas. Bus Line Density shows positive effects ($\beta = 0.146$), while Bus Stop Density demonstrates negative effects ($\beta = -0.135$), creating an apparent paradox that requires careful interpretation within the context of already well-served high-density environments (Cervero, 2013; Litman, 2017).

The apparent paradox stems from a fundamental conceptual difference between these two metrics, as delineated in Fig. 10.2. Bus Line Density and Bus Stop Density measure distinct aspects of transit service provision, which consequently exert opposing influences on resident well-being in high-density contexts.

Bus Line Density represents the variety and coverage of transit routes available to residents. Increased bus line density enhances residents' ability to reach diverse destinations without private vehicle ownership, supporting sustainable transportation behaviors and providing mobility options that enhance Life Satisfaction (Cervero, 2013). The positive effect aligns with transit-oriented development principles and research showing that transit accessibility improves quality of life (Litman, 2017).

Conversely, Bus Stop Density represents the frequency of stops along transit routes. After controlling for bus line density, increased stop density appears to create negative effects on Life Satisfaction in high-density contexts where transit service is already abundant. This counterintuitive finding can be explained through several mechanisms:

First, higher stop density may indicate inefficient transit operations, with frequent stops increasing travel times and reducing service quality for existing users (Vuchic, 2017). When stops are too closely spaced, the transit system becomes slower and less competitive with other transportation modes.

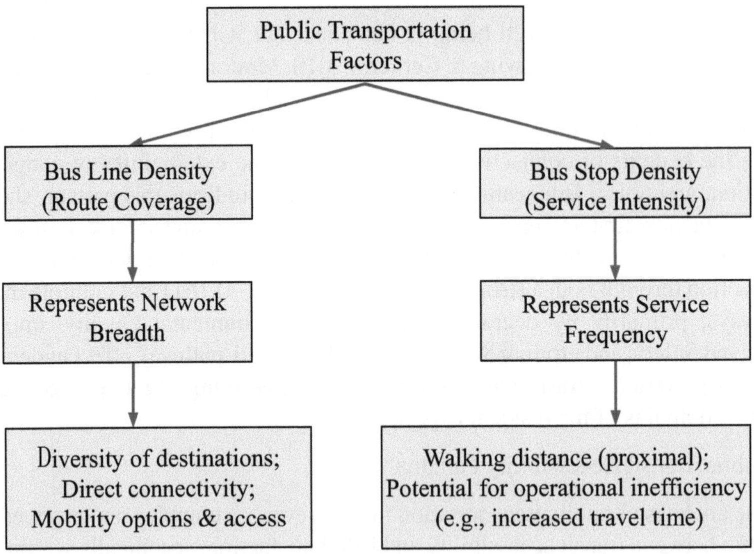


**Fig. 10.2** Conceptual Differentiation: Bus Line Density vs. Bus Stop Density. Caption: This flowchart delineates the fundamental conceptual distinction between the two key independent variables. Bus Line Density reflects the extent and variety of route coverage, directly enabling access to diverse destinations. Bus Stop Density reflects the intensity of service provision along routes, representing a trade-off between walking distance and operational efficiency, with potential negative externalities in already saturated environments. Source: Created by the authors

Second, increased bus stop density creates higher concentrations of pedestrian traffic, noise, and congestion around residential areas. These localized impacts can reduce residential environmental quality and create nuisance effects that outweigh accessibility benefits (Appleyard, 1981).

Third, the relationship between stop density and accessibility follows a curve of diminishing returns, where optimal stop spacing provides the best balance between accessibility and service efficiency. Excessive stop density can actually reduce effective accessibility by degrading service quality (Murray, 2001).

To visually synthesize these opposing mechanisms, Fig. 10.3 and Fig. 10.4 below juxtapose the nuanced impacts of bus stop density and bus line density on Life Satisfaction. This contrast clarifies the apparent paradox: enhancing network breadth (Bus Line Density) provides unambiguously positive benefits by expanding access and destination choices, aligning with transit-oriented development principles (Cervero, 2013; Litman, 2017). Conversely, intensifying network intensity (Bus Stop Density) in already well-served urban cores creates significant trade-offs between highly localized accessibility and broader systemic efficiency and environmental quality, often resulting in net negative outcomes (Vuchic, 2017; Appleyard, 1981). This understanding of the underlying mechanisms directly informs the following trend analysis, which seeks to identify the optimal balance between these factors.

The trend analysis reveals that bus stop density effects vary across density ranges, with negative effects most pronounced in the range affecting approximately 63.2% of sample communities in the high-density context studied. This suggests that optimal transit planning should consider density thresholds and context-specific factors rather than simply maximizing stop coverage.

5. Point of Interest (POI) Density Effects

POI Density, representing destination accessibility and functional facility availability, shows positive but non-significant direct effects on Life Satisfaction within the high-density urban context examined. However, the mediation analysis reveals that POI density operates through complex indirect pathways that sometimes conflict with each other (Montgomery, 2013; Putnam, 2000).

POI density positively affects Life Satisfaction through Perceived Environmental Quality ($\beta$ = 0.271), reflecting that residents value access to diverse commercial, recreational, and service facilities. Increased facility density improves residents' perception of environmental convenience and supports daily life activities (Montgomery, 2013).

However, POI density negatively affects Life Satisfaction through Social Capital ($\beta = -0.285$) and self-rated health ($\beta = -0.150$). These negative indirect effects suggest that while facility density provides functional benefits, it may reduce opportunities for social interaction and healthy behaviors in high-density contexts where facility abundance is already high. High facility density might enable more individualized consumption patterns that reduce communal activities and social bonding (Putnam, 2000).

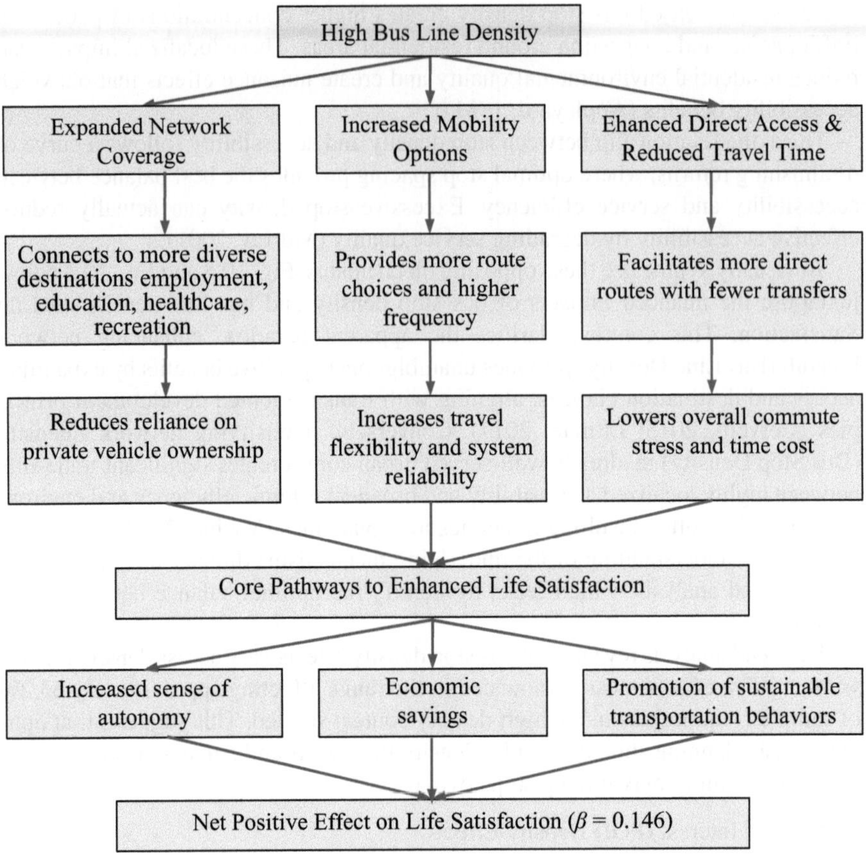


**Fig. 10.3** Mechanisms of positive impact from Bus Line Density (β = 0.146) on Life Satisfaction in High-Density urban cores. Caption: This flowchart delineates the primary mechanisms through which excessive bus stop density negatively influences Life Satisfaction after controlling for bus line density. It illustrates the pathway from inefficient operations and increased travel times (Vuchic, 2017), to localized nuisances such as pedestrian traffic, noise, and congestion that degrade environmental quality (Appleyard, 1981), and finally to the principle of diminishing returns in accessibility benefits (Murray, 2001), collectively resulting in a net reduction of Life Satisfaction. Source: Created by the authors

The health effects may relate to increased commercial density as demonstrated in research linking land use to public health outcomes (Su et al., 2016, 2019a, 2019b), creating environments that promote sedentary behaviors, unhealthy consumption, or environmental stressors such as noise and air pollution from increased commercial activity (Sallis et al., 2016).

These conflicting indirect effects help explain why the total effect of POI density is non-significant despite strong individual pathways, demonstrating the importance of mediation analysis in understanding complex built environment relationships,

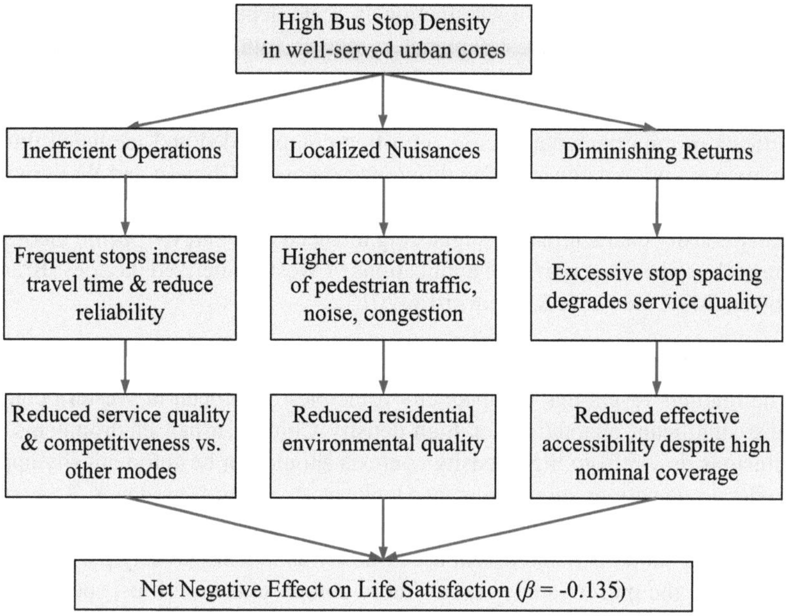


**Fig. 10.4** Mechanisms of negative impact from Bus Stop Density (β = −0.135) on Life Satisfaction in High-Density urban cores. Caption: This flowchart delineates the primary mechanisms through which excessive bus stop density negatively influences Life Satisfaction after controlling for bus line density. It illustrates the pathway from inefficient operations and increased travel times (Vuchic, 2017), to localized nuisances such as pedestrian traffic, noise, and congestion that degrade environmental quality (Appleyard, 1981), and finally to the principle of diminishing returns in accessibility benefits (Murray, 2001), collectively resulting in a net reduction of Life Satisfaction. Source: Created by the authors

particularly in high-density urban environments where multiple effects operate simultaneously.

### *10.1.3 Theoretical Framework Validation and Integration*

The empirical research results on the mechanism by which high-density urban community built environments affect the overall cognitive evaluation of Subjective Well-being (Life Satisfaction) validate the theoretical conceptual framework constructed in this study specifically for high-density urban core contexts. The framework proves both reasonable and effective in enriching theoretical understanding of the built environment's effects on well-being within these specific environmental conditions.

This study demonstrates both similarities and important differences compared with research results on built environment impacts in low-density Western cities. Supporting this perspective, Li et al. (2025) demonstrated that the built environment's impact on life satisfaction is scale-dependent, with housing-scale factors exhibiting the strongest explanatory power, followed by neighborhood and community-scale factors. This hierarchical pattern underscores the need for contextually adapted theoretical frameworks, particularly in high-density settings where the effects of environmental factors are often compounded and mediated through complex perceptual pathways. The differences between Chinese and Western studies appear to emerge from different national conditions, cultural contexts, and particularly density characteristics, suggesting the need for context-specific theoretical frameworks rather than universal applications of Western-derived theories (Bramley & Power, 2009; Newman & Kenworthy, 2015).

1. Contextual Adaptations of Compact City Theory

The findings reveal that compact city principles developed in Western contexts require significant adaptation for high-density Chinese urban environments, and conclusions drawn from high-density contexts should not be automatically applied to moderate-density or suburban areas (Jenks et al., 1996; Newman & Kenworthy, 2015). While Western compact city theory generally promotes higher density, mixed-use development, and transit accessibility as unambiguously positive (Jenks et al., 1996), the present study reveals more complex relationships, specifically in high-density contexts.

First, density effects show threshold behaviors where benefits transform into costs beyond certain levels specific to high-density environments. The built environment significantly affects residents' Life Satisfaction, but in high-density contexts, increased building density, functional mix diversity, and intersection density create negative effects rather than the positive effects predicted by compact city theory (Newman & Kenworthy, 2015). These threshold effects likely differ substantially in moderate-density and low-density contexts.

Second, the mediating factors (Perceived Environmental Quality, Perceived Safety, Social Capital, and Self-Rated Health) serve as critical pathways through which built environment factors affect Life Satisfaction. Since direct effects of built environment factors on Life Satisfaction are often non-significant, the environment influences well-being primarily through these psychological and social mediating processes rather than through direct environmental impacts.

Third, Perceived Environmental Quality emerges as the most important mediating factor, providing a critical bridge between Objective Built Environment characteristics and Life Satisfaction outcomes. This finding emphasizes that environmental interventions must consider perceptual and psychological processes rather than focusing solely on objective environmental improvements (Gifford, 2007).

2. Integration with Environmental Psychology Theory

The mediating role of Perceived Environmental Quality aligns with environmental psychology theories emphasizing the importance of person-environment fit and environmental preference in determining behavioral and affective responses (Mehta, 2013). The finding that objective environmental characteristics influence well-being primarily through subjective environmental evaluations supports theoretical frameworks that emphasize cognitive appraisal processes in environmental psychology (Lazarus & Folkman, 1984).

The chain mediation relationships among Perceived Environmental Quality, Perceived Safety, Social Capital, and Self-Rated Health reflect the interconnected nature of environmental experience. These relationships align with ecological systems theory, which emphasizes the multiple, interconnected systems that influence human development and well-being (Bronfenbrenner, 1979). The built environment operates as part of the microsystem that directly affects residents, but its effects cascade through multiple psychological and social processes.

3. Cultural and Contextual Considerations

The differences between this study's findings and Western research reflect important cultural and contextual factors that influence built environment relationships, specifically in high-density Chinese urban environments. These findings should be interpreted cautiously when considering different cultural contexts or urban development patterns (Leaf, 1998; Altman & Chemers, 1980). Chinese urban development patterns, cultural preferences, social structures, and lifestyle patterns may create different optimal configurations for built environment design (Leaf, 1998).

High-density contexts may require different balance points between accessibility, diversity, and environmental quality compared to lower-density contexts, and these balance points may not apply to suburban or new town development contexts. The negative effects of functional mix diversity and intersection density in high-density contexts suggest that optimization principles developed in lower-density environments may not transfer directly to high-density contexts.

Furthermore, cultural factors such as preferences for environmental harmony, community social structures, and lifestyle patterns may influence how built environment characteristics are perceived and experienced by residents (Altman & Chemers, 1980). Future research should consider these cultural and contextual factors when developing built environment theories and policies, particularly when extending findings across different density contexts.

## 10.2 Research Implications and Policy Recommendations

This study demonstrates that Objective Built Environment factors have significant impacts on Life Satisfaction, specifically in high-density urban contexts. The policy implications and recommendations presented here are primarily applicable to high-density urban core areas and should be adapted carefully when applied to suburban areas, new towns, or moderate-density urban environments (Liu et al., 2020; Mouratidis, 2019). Against the backdrop of China's national initiative to "continuously enhance people's well-being,"[1] research on the relationship between the built environment and subjective well-being provides crucial evidence from high-density Chinese cities for the fields of healthy cities, happy cities, and urban planning practice.

The findings can guide the design and development of healthy and happy city planning, specifically in high-density Chinese urban contexts, aiming to continuously improve residents' satisfaction with housing and living conditions, enhance residents' happiness index, and offer policy implications for promoting the construction of a happy China.

### *10.2.1 Comprehensive Policy Framework for High-Density Contexts*

Based on the research findings, we propose a comprehensive policy framework (Fig. 10.5) that integrates key intervention strategies for high-density urban environments. This framework encompasses density management, transportation optimization, functional facility planning, and mediating factor enhancement, providing a holistic approach to improving life satisfaction in high-density contexts.

Overall, urban planning, design, construction, and management practice in high-density urban areas should adopt a two-pronged approach specifically calibrated for high-density conditions. First, reasonable and scientific planning of objective built-environment factors should be implemented to promote Life Satisfaction and enhance residents' Subjective Well-being within the constraints and opportunities presented by high-density environments. Second, urban planning and construction processes should actively create conditions enabling mediating factors to effectively exert their indirect effects, promoting Life Satisfaction improvements and strengthening the optimization effects of the built environment on well-being outcomes.

This comprehensive approach recognizes that built environment effects on well-being operate through complex psychological and social processes that require attention to both objective environmental characteristics and subjective experiential factors, with particular attention to the unique conditions present in high-density urban environments. Policy interventions should therefore address both physical environmental improvements and the social and psychological conditions that mediate environmental effects on well-being within high-density contexts.

1. Building Density Management in High-Density Contexts

Building density represents one of the most significant factors negatively affecting Life Satisfaction, specifically in high-density contexts. Policy interventions

---

[1]The State Council of the People's Republic of China. (2017). *The 13th Five-Year Plan for Economic and Social Development of the People's Republic of China (2016–2020)*. Beijing: State Council.

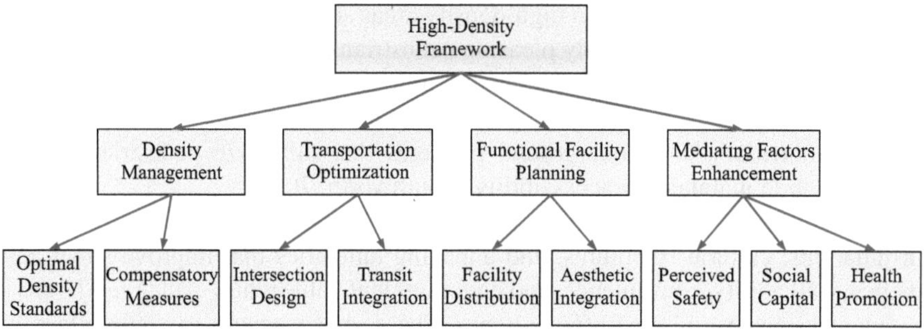


**Fig. 10.5** Comprehensive policy framework for High-Density urban environments. Source: Created by the authors

should focus on optimizing density levels rather than simply minimizing or maximizing density, with particular attention to the unique challenges and constraints of high-density urban core areas (Newman & Kenworthy, 2015).

First, urban planning in high-density areas should establish scientifically-based density standards that account for local context, infrastructure capacity, and resident well-being outcomes specific to high-density environments. Strict control should be exercised over building density and floor area ratios, with development guidance encouraging urban expansion toward peripheral areas where appropriate infrastructure and environmental capacity exist (Newman & Kenworthy, 2015). However, density management strategies appropriate for high-density core areas may not be suitable for suburban or new town contexts.

Second, in existing high-density urban areas where density reduction is impractical, compensatory measures should be implemented to mitigate negative density effects specific to high-density conditions. These measures include enhancing Perceived Safety through improved lighting and surveillance, increasing leisure and sports facilities to compensate for reduced per-capita space, and optimizing public space design to maximize functionality and aesthetic quality within spatial constraints (Gehl, 2010).

Third, density management in high-density contexts should consider the relationship between building density and other environmental factors such as green space availability, public facility access, and transportation infrastructure capacity. Integrated planning approaches should ensure that density increases are accompanied by proportional improvements in supporting infrastructure and amenities calibrated for high-density conditions.

2. Road Network and Intersection Design Enhancement for High-Density Areas

Road intersection density shows the strongest negative effect on Life Satisfaction among all built environment factors, specifically in high-density contexts, requiring careful attention in transportation planning and urban design for high-density areas (Southworth, 2005; Ewing & Cervero, 2010).

First, road network design in high-density areas should optimize walkability and create comfortable, aesthetically pleasing pedestrian environments while managing the complexity inherent in high-density transportation systems. This includes appropriate landscape greening integration, optimized construction of crosswalks, overpass corridors, and underground passages that minimize pedestrian-vehicle conflicts while maintaining accessibility (Southworth, 2005).

Second, intersection design in high-density contexts should incorporate art installations, cultural sculptures, and engaging amenities that improve street aesthetics, walkability, convenience, and entertainment value while managing the environmental impacts of high traffic volumes. These improvements can help offset the negative effects of intersection complexity by creating positive environmental experiences (Jacobs, 1961).

Third, intersection management in high-density areas should employ intelligent transportation systems and traffic optimization technologies specifically designed for high-volume, complex urban environments to minimize congestion, reduce pollution, and improve safety outcomes. Smart traffic signal systems, real-time traffic monitoring, and adaptive intersection management can help maximize connectivity benefits while minimizing negative environmental impacts (Ewing & Cervero, 2010).

3. Transportation System Integration for High-Density Contexts

The contrasting effects of bus line density and bus stop density reveal the importance of integrated transportation planning that optimizes both route coverage and stop spacing specifically for high-density urban environments where transit demand is high and service frequency is critical (Vuchic, 2017; Cervero, 2013).

Bus route planning in high-density areas should be conducted scientifically to balance the relationship between route density and stop density while considering the unique travel patterns and service demands of high-density environments. This requires a comprehensive analysis of travel demand patterns, service coverage objectives, and operational efficiency considerations specific to high-density contexts. Route networks should be designed to maximize accessibility while maintaining service quality and operational efficiency (Vuchic, 2017).

Transportation planning in high-density contexts should enhance residents' travel convenience, improve bus travel accessibility, and encourage residents to choose public transportation while managing the environmental impacts of high-frequency service. Well-designed transit systems can increase interaction opportunities during travel and enhance Social Capital formation, thereby improving Life Satisfaction levels (Cervero, 2013).

Integration with other transportation modes should be considered specifically for high-density environments, including pedestrian and cycling infrastructure that connects seamlessly with transit services while managing the spatial constraints of high-density areas. Multi-modal transportation planning can help optimize the overall transportation system while minimizing the negative environmental impacts of individual transportation components (Litman, 2017).

4. Functional Facility Planning and Management for High-Density Areas

The negative effects of functional mix diversity, specifically in high-density contexts, require sophisticated approaches to facility planning and management that recognize the unique conditions and constraints of high-density environments (Montgomery, 2013; Nasar, 1998).

Urban design and management in high-density areas should reasonably plan and configure various functional facility types and quantities, with particular attention to service facility distribution for residents' daily lives, while avoiding the negative effects of excessive functional mixing observed in high-density contexts. This planning should consider both functional needs and environmental quality impacts, ensuring that facility density does not exceed environmental capacity or resident comfort thresholds specific to high-density conditions (Montgomery, 2013).

Big data applications should be utilized to provide guidance on location selection and business format optimization for commercial operators, specifically in high-density environments where facility impacts are amplified. Data-driven approaches can help identify optimal facility distributions that maximize convenience benefits while minimizing negative environmental impacts such as congestion, noise, and visual disorder within high-density constraints (Batty, 2013).

Facility design and management in high-density contexts should emphasize environmental integration and aesthetic quality while managing the cumulative impacts of multiple facilities in close proximity. Commercial and service facilities should be designed to complement rather than disrupt residential environmental quality, with attention to building design, signage, operational hours, and environmental impacts (Nasar, 1998).

### *10.2.2 Mediating Factor Enhancement Strategies for High-Density Environments*

Given the significant relationships between Perceived Environmental Quality, Perceived Safety, Social Capital, self-rated health, and Life Satisfaction, specifically in high-density contexts, creating conditions conducive to these mediating factors represents a crucial strategy for optimizing built environment effects on well-being within high-density urban environments (Gifford, 2007; Gehl, 2010).

1. Perceived Environmental Quality Enhancement in High-Density Contexts

Perceived Environmental Quality emerges as the most important mediator between built environment and Life Satisfaction, specifically in high-density contexts, requiring comprehensive attention to how residents perceive and evaluate their environmental conditions within the unique constraints and opportunities of high-density urban environments (Gifford, 2007; Nasar, 1998).

Environmental design in high-density areas should prioritize resident perceptions and preferences rather than focusing solely on objective environmental characteristics, with particular attention to how high-density conditions influence environmental perception and preference. This requires understanding how different demographic groups, cultural backgrounds, and lifestyle patterns influence environmental perception and preference within high-density contexts (Gifford, 2007).

Aesthetic quality improvements should be integrated throughout high-density built environments, including landscape design, architectural quality, public art, and environmental maintenance specifically calibrated for high-density conditions. These improvements should reflect local cultural preferences and create coherent, harmonious environmental experiences that enhance residents' aesthetic satisfaction despite the visual complexity inherent in high-density environments (Nasar, 1998).

Resident engagement in environmental planning and design processes can help ensure that environmental improvements align with resident perceptions and preferences, specifically within high-density contexts. Participatory planning approaches can identify resident priorities and concerns that may not be apparent through objective environmental assessment alone (Sanoff, 2000).

2. Perceived Safety Enhancement Strategies for High-Density Areas

Perceived Safety significantly influences Life Satisfaction and mediates built environment effects on well-being outcomes, specifically in high-density contexts where security concerns may be amplified by density conditions (Newman, 1972; Crowe, 2000). Security enhancement requires both objective security improvements and attention to residents' subjective Perceived Safety within high-density environments.

Community monitoring and supervision systems should be strengthened to enhance security management capabilities and promote residents' sense of safety, specifically in high-density environments where security challenges may be more complex. This includes both technological solutions such as surveillance systems, and social solutions, such as community watch programs and neighborhood organization (Newman, 1972).

Environmental design in high-density contexts should incorporate Crime Prevention Through Environmental Design (CPTED) principles that use environmental design to reduce crime opportunities and enhance Perceived Safety while addressing the unique challenges of high-density environments. These principles include natural surveillance, access control, territorial reinforcement, and maintenance strategies that create environments that feel safe and secure (Crowe, 2000).

Community engagement and social cohesion programs can enhance Perceived Safety through strengthening social networks and community organization, specifically in high-density environments where social anonymity may be higher. Strong communities with active resident participation tend to experience both lower crime rates and higher Perceived Safety among residents (Sampson et al., 1997).

3. Social Capital Development in High-Density Environments

Social Capital serves as an important mediator through which the built environment affects Life Satisfaction, specifically in high-density contexts where social interaction patterns may differ from lower-density environments (Putnam, 2000; Flora & Flora, 2013). Environmental design and community programming should actively promote social interaction and community formation while addressing the unique social dynamics of high-density living.

Community spaces in high-density areas should be designed to facilitate social interaction and community activities while managing the higher usage intensity typical of high-density environments. This includes providing diverse spaces for different types of social activities, ensuring accessibility and comfort, and creating environments that encourage spontaneous social encounters within the spatial constraints of high-density development (Oldenburg, 1999).

Community programming and activities should be organized through multiple platforms and channels to promote communication and interaction among residents, increase a sense of community belonging, strengthen social relationships, and facilitate Social Capital formation specifically in high-density contexts where traditional community formation patterns may not apply. This includes cultural events, recreational activities, educational programs, and volunteer opportunities that bring residents together around common interests and goals (Putnam, 2000).

Government and community organizations should collaborate to create comprehensive support systems for community development, specifically adapted to high-density urban contexts. This includes funding for community programs, professional support for community organizations, and policy frameworks that enable and encourage community participation within high-density living conditions (Flora & Flora, 2013).

4. Health and Well-being Promotion in High-Density Contexts

Self-Rated Health influences Life Satisfaction and mediates some built environment effects, specifically in high-density environments, where health-promoting opportunities may be more limited due to spatial constraints. Environmental design and community programming should actively promote physical and mental health outcomes within high-density contexts (Jackson, 2003; Sallis et al., 2016; Nieuwenhuijsen, 2016).

Active living environments should be created through providing recreational facilities, walkable neighborhoods, accessible green spaces (Wang et al., 2020), and infrastructure that supports physical activity while addressing the space limitations typical of high-density areas. These environmental features can encourage healthy behaviors and improve residents' physical health outcomes (Sallis et al., 2016).

Mental health support systems should be integrated into community planning and programming specifically for high-density environments where mental health challenges may be amplified by density-related stressors. This includes providing spaces for relaxation and stress reduction, programming that addresses social isolation and mental health concerns, and access to mental health services and support (Jackson, 2003).

Environmental health considerations should be integrated into planning and design processes with particular attention to the cumulative health impacts common in high-density environments. This includes attention to air quality, noise levels, access to nature, and other environmental factors that influence physical and mental health outcomes within high-density constraints (Frumkin et al., 2017).

### 10.2.3 *Integrated Community Development Approach for High-Density Contexts*

The research findings suggest that effective built environment interventions in high-density contexts require integrated approaches that address multiple environmental factors and mediating processes simultaneously while recognizing the unique conditions and constraints of high-density urban environments.

1. Personality and Mental Health Development in High-Density Settings

The formation and cultivation of residents' positive personalities represents an important long-term strategy, specifically in high-density environments, where environmental stressors may be higher. Personality development is related not only to family education but also to school cultivation, social environment, and community conditions within high-density contexts (Saxena et al., 2007; Seligman, 2011).

Multi-dimensional support systems should be established involving "government-community-school-family" collaboration to guide residents in continuously improving their mental health levels and promoting positive, optimistic personality formation, specifically within high-density urban environments. These systems should provide mental health resources, educational opportunities, and social support that enable residents to develop resilience and well-being (Seligman, 2011).

Mental health service networks should be integrated into community planning and service delivery with particular attention to the mental health challenges that may be amplified in high-density living conditions. This includes providing accessible mental health services, peer support programs, and preventive mental health programming that addresses community mental health needs (Saxena et al., 2007).

Community environments in high-density areas should be designed and managed to support positive mental health outcomes despite the inherent stressors of high-density living. This includes creating environments that reduce stress, provide opportunities for restoration and relaxation, and support social connection and community engagement within the constraints of high-density development (Kaplan & Kaplan, 1989).

2. Comprehensive Environmental Management for High-Density Areas

Environmental management in high-density contexts should adopt holistic approaches that address the complex interactions among built environment factors and their effects on resident well-being while recognizing the unique management challenges of high-density environments (Holling, 1978; Beatley, 2000).

Integrated planning approaches should consider the relationships among different built environment factors rather than optimizing individual factors in isolation, with particular attention to how factor interactions may be amplified in high-density contexts. This requires comprehensive planning frameworks that can account for complex interactions and trade-offs among different environmental characteristics (Beatley, 2000).

Adaptive management approaches should be implemented to monitor environmental conditions and resident well-being outcomes over time, allowing for continuous improvement and adaptation of environmental interventions based on empirical evidence and resident feedback, specifically within high-density contexts. This includes recognition that optimal environmental configurations may change as density levels and resident needs evolve (Holling, 1978).

Cross-sector collaboration should be fostered among government agencies, community organizations, private sector actors, and residents to ensure comprehensive and coordinated approaches to environmental improvement and community development specifically calibrated for high-density urban contexts. This collaboration should recognize the unique challenges and opportunities present in high-density environments (Healey, 1997).

## 10.3 Research Limitations and Future Directions

### *10.3.1 Comprehensive Assessment of Research Limitations*

1. Sample and Data Limitations

This study focuses specifically on the central urban area of Guangzhou to investigate built environment-Life Satisfaction relationships within high-density contexts. While Guangzhou provides an excellent case study of high-density Chinese urban development, the geographic and contextual specificity of the findings raises important questions about generalizability to other high-density cities in China and internationally, and particularly about applicability to moderate-density or suburban contexts (Liu et al., 2020; Menard, 2002).

The conclusions derived from Guangzhou's high-density urban core may reflect city-specific and density-specific factors, including local culture, urban development history, policy context, economic conditions, and demographic characteristics that may not apply to other Chinese cities, international contexts, or different density environments. Future research should conduct comparative studies across multiple high-density cities in different regions of China to test the generalizability and identify context-specific factors that moderate built environment effects on well-being. Additionally, research should explicitly examine whether these relationships hold in moderate-density and suburban contexts before generalizing conclusions beyond high-density urban environments.

The study utilizes cross-sectional rather than longitudinal panel data, limiting the ability to establish causal relationships between built environment factors and Life Satisfaction outcomes. While cross-sectional analysis can identify associations and test theoretical relationships, it cannot definitively establish temporal precedence or rule out reverse causation (Menard, 2002).

Life Satisfaction evaluation exhibits characteristics of subjectivity, comprehensiveness, and relative stability within certain time periods, as demonstrated in psychological research (Diener, 1984; Diener et al., 1985). Similarly, built environment elements in central urban areas remain relatively stable over time periods relevant to individual well-being assessment. However, the absence of longitudinal data prevents analysis of how changes in built environment characteristics influence changes in Life Satisfaction over time.

Obtaining longitudinal data presents significant practical challenges, including the difficulty of tracking panel samples over extended time periods, the limited availability of historical built environment data, and resource constraints for longitudinal survey research. Future research should gradually adopt follow-up survey approaches to establish longitudinal panel data for in-depth analysis of causal relationships and temporal dynamics in built environment-well-being relationships across different density contexts.

2. Methodological Limitations

The study employs regression analysis and structural equation modeling techniques that are widely used and well-established in built environment and well-being research. However, these methods have inherent limitations that should be acknowledged and addressed in future research.

Recent methodological advances in built environment research include machine learning and deep learning approaches that can identify non-linear relationships, analyze complex environmental patterns, and integrate multiple data sources. Some studies have begun using convolutional neural networks to analyze street view images, quantifying environmental quality through PSPNet and DeepLabv3+ semantic segmentation models, and employing random forest, gradient boosting, or multi-layer perceptron approaches to explore non-linear built environment effects on well-being (Wang et al., 2024).

Multi-modal deep learning technologies enable integrated analysis by fusing street view images, text data (social media emotions), and spatio-temporal trajectory data (GPS). These approaches can reveal micro-environmental mechanisms and support precise urban design interventions that would not be possible with traditional analytical methods (Kang et al., 2020).

However, machine learning and deep learning approaches also present challenges, including interpretability limitations and generalization difficulties. The logical progressive relationship between research questions in this study suggests that traditional linear regression and structural equation models may be more conducive to the interpretation of action pathways and theoretical understanding, particularly when results need to inform policy decisions across different urban contexts. Future research should attempt to integrate multi-source data with machine learning techniques while maintaining theoretical interpretability and practical applicability.

3. Causal Inference Challenges

This study employs regression models and structural equation models to investigate built environment Subjective Well-being relationships, but the relationships explored are correlational rather than strictly causal. Several factors limit causal inference capabilities:

First, cross-sectional data cannot establish the temporal precedence necessary for causal inference. Built environment effects on Life Satisfaction may exhibit time-lag characteristics, where environmental changes influence well-being outcomes over extended periods rather than immediately. Additionally, reciprocal relationships may exist where Life Satisfaction influences environmental choices and behaviors, creating bidirectional causation that cannot be disentangled with cross-sectional analysis.

Second, this study demonstrates that health status improves Life Satisfaction levels, but research also shows that Life Satisfaction improvements can influence health outcomes, suggesting close reciprocal causal relationships (Lyubomirsky et al., 2005). Similarly, some studies indicate that self-rated health can affect Social Capital, demonstrating complex mutual influence relationships among potential mediating variables that require longitudinal analysis to understand fully.

Third, omitted variable bias represents a persistent challenge in observational research.[2] Despite controlling for numerous individual socioeconomic characteristics, unmeasured factors such as personality traits, genetic predispositions, life experiences, and environmental preferences may influence both built environment exposure and Life Satisfaction outcomes, creating spurious associations that cannot be ruled out with observational data alone.

Future research should employ quasi-experimental designs, natural experiments, and longitudinal approaches to strengthen causal inference, with particular attention to whether causal relationships vary across different density contexts and urban environments. Policy interventions and urban development projects provide opportunities for natural experiments that can help establish causal relationships through comparing outcomes before and after environmental changes or between treatment and control areas.

4. Hierarchical and Spatial Structure Limitations

The model constructed in this study involves variables at multiple levels, creating potential hierarchical data structure issues that require careful methodological consideration. This limitation may be particularly important when extending findings to different spatial scales or urban contexts (Raudenbush & Bryk, 2002). Objective built environment variables represent community-level characteristics,

[2] Wooldridge, J. M. (2010). *Econometric Analysis of Cross Section and Panel Data* (2nd ed.). Cambridge, MA: MIT Press.

while Perceived Environmental Quality, Perceived Safety, Social Capital, self-rated health, and socioeconomic attributes represent individual-level variables.[3]

Individuals are nested within communities, and communities are nested within cities, potentially creating data dependency that violates independence assumptions in traditional regression analysis. Some studies address these issues using hierarchical regression models or geographically weighted regression approaches (Raudenbush & Bryk, 2002).

However, hierarchical modeling applications in mediation analysis remain limited, and few studies have successfully implemented multi-level structural equation models that can simultaneously address hierarchical data structures and complex mediation relationships. The complexity of the theoretical model and interpretive challenges led this study to employ traditional analytical approaches while acknowledging these limitations.

The study focuses on four administrative districts within Guangzhou's central urban area, providing some control for city-level factors while maintaining sufficient variation in community-level built environment characteristics. However, this geographic specificity limits generalizability to other urban contexts and density levels. Future research should consider broader geographic scales and more sophisticated analytical approaches to address hierarchical data structures while maintaining theoretical interpretability and practical relevance across different urban development contexts.

5. Variable Selection and Measurement Limitations

Built environment variable selection in this study follows the "5D" model framework developed by Ewing and Cervero (2010), which builds on earlier work by Cervero and Kockelman (1997), representing the most widely recognized and applied framework in academic research. However, Ewing and Cervero acknowledge that the 5D model remains insufficiently mature and rigorous, with unclear dimensional boundaries and frequent overlaps between dimensions during analysis (Ewing & Cervero, 2010).

Current literature lacks more scientific, rigorous, effective, and clearly defined alternative frameworks for built environment measurement that can account for contextual differences across different density levels and urban development types (Yu & Hu, 2017). Most studies on the built environment and Subjective Well-being adopt "3D" or "5D" model frameworks, typically including 3–6 built environment variables. While including more variables could improve model estimation rigor and accuracy, unlimited variable inclusion creates model complexity and interpretive challenges.

This study employs the 5D framework while focusing on Objective Built Environment characteristics that represent core research interests within high-density contexts. Variables outside the primary research focus are excluded to

[3]Raudenbush, S. W., & Bryk, A. S. (2002). *Hierarchical Linear Models: Applications and Data Analysis Methods* (2nd ed.). Thousand Oaks, CA: Sage Publications.

maintain analytical focus and interpretive clarity. Consequently, other factors affecting Life Satisfaction beyond control variables and built environment factors are included in model residuals, potentially limiting explanatory completeness.

The study excludes community management and governance factors characteristic of Chinese contexts, such as neighborhood committees, homeowners' associations, property management companies, and volunteer services. These factors may significantly influence resident well-being in Chinese urban contexts but fall outside the theoretical framework emphasizing physical built environment characteristics, and their importance may vary across different density levels and urban contexts (Leaf, 1998).

Mediating variable selection includes perceived environment quality, public safety perception, Social Capital, and self-rated health, but excludes leisure activities and emotions proposed in Mouratidis's framework. This selective inclusion may limit understanding of the complete mediation pathways through which the built environment affects well-being outcomes, and the importance of different mediating pathways may vary across urban contexts.

Future research should develop more comprehensive theoretical frameworks that integrate physical built environment characteristics with social, cultural, and governance factors relevant to specific national and cultural contexts, with particular attention to how these factors may operate differently across different density levels and urban development types. Additionally, expanded mediation frameworks should include broader ranges of psychological, social, and behavioral mediating variables to capture the full complexity of built environment effects on well-being.

### *10.3.2 Future Research Directions and Opportunities*

1. Multi-Scale and Multi-Context Built Environment Analysis

This study examines built environment-Life Satisfaction relationships within 1000-meter buffer zones around community centroids in high-density urban contexts, representing residents' approximate daily activity ranges within these specific environments. However, this approach may not accurately represent individual activity spaces, which vary by demographic characteristics, lifestyle patterns, and transportation behaviors, and may operate quite differently in moderate-density or suburban contexts (Liu et al., 2020; Oshan et al., 2019).

Future research should explore built environment impacts at multiple spatial scales and across different urban density contexts, including administrative boundaries, walkable distances, and individualized activity spaces defined by GPS tracking or activity surveys. In this regard, the multi-scale analytical approach adopted by Li et al. (2025) offers a valuable methodological precedent, demonstrating how hierarchical regression can effectively parse the distinct contributions of housing, neighborhood, and community-scale environmental variables to overall well-being. Different spatial scales may reveal different built environment relationships, as demonstrated in studies finding heterogeneous effects across buffer zone sizes (Liu et al., 2020). Importantly, research should explicitly examine whether spatial scale effects vary across different urban density contexts.

Research should also investigate how built environment relationships vary across different community types, neighborhood characteristics, and urban density levels. While some literature explores community type effects, the interaction between community characteristics and urban density context may significantly influence built environment-well-being relationships and requires systematic investigation.

Advanced spatial analysis techniques, including Multi-scale Geographically Weighted Regression (MGWR), can address spatial heterogeneity and modifiable areal unit problems while identifying optimal spatial scales for different built environment relationships across different urban contexts (Oshan et al., 2019). MGWR bandwidth parameters can reflect spatial scales of different environmental processes, enabling a more sophisticated understanding of scale-dependent built environment effects and their variation across urban development types.

2. Advanced Methodological Approaches for Multi-Context Research

Future research should integrate multiple methodological approaches to address current limitations and enhance understanding of built environment-well-being relationships across different urban contexts and density levels. Non-linear estimation techniques can explore threshold effects, interaction relationships, and complex functional forms that may better represent built environment relationships than traditional linear models, with particular attention to how these relationships may vary across urban contexts.

Machine learning and deep learning approaches offer opportunities to integrate multi-source data and identify complex patterns not detectable through traditional statistical methods. Convolutional neural networks can analyze street view imagery to quantify environmental characteristics such as visual quality, green space visibility, and architectural diversity. Natural language processing can analyze social media data to understand residents environmental perceptions and experiences (Kang et al., 2020). These techniques may be particularly valuable for identifying context-specific patterns that vary across urban development types.

However, methodological sophistication should be balanced with theoretical interpretability and practical relevance. Advanced techniques should complement rather than replace theory-driven approaches, ensuring that methodological innovation contributes to theoretical understanding and policy application rather than purely technical advancement. This balance is particularly important when research findings need to inform policy decisions across diverse urban contexts.

Longitudinal and quasi-experimental designs represent critical methodological priorities for future research, with particular attention to how causal relationships may vary across different urban density contexts and development types. Natural experiments created by policy interventions, development projects, and environmental changes provide opportunities to strengthen causal inference while maintaining real-world relevance and external validity.

3. Causal Inference Enhancement across Urban Contexts

Establishing causal relationships between built environment characteristics and well-being outcomes represents a fundamental challenge requiring innovative research designs and analytical approaches, with particular attention to how causal relationships may vary across different urban contexts and density levels (Angrist & Pischke, 2009; Cook & Campbell, 1979).

Longitudinal panel studies can track individuals over time to observe how environmental changes influence well-being changes, providing evidence for temporal precedence and dynamic relationships. However, longitudinal studies should include diverse urban contexts to understand how causal relationships may vary across density levels and development types. These studies require substantial resources and face challenges, including sample attrition, measurement consistency, and changing environmental contexts.

Natural experiments and quasi-experimental designs can leverage policy interventions, development projects, and environmental changes to create treatment and control conditions that approximate experimental research designs across different urban contexts. These approaches can provide stronger causal evidence while maintaining real-world relevance and external validity (Cook & Campbell, 1979). Particularly valuable would be natural experiments that allow comparison of effects across different urban density contexts.

Instrumental variable approaches can address endogeneity concerns by identifying variables that influence built environment characteristics but affect well-being only through environmental pathways. Geographic, historical, and policy instruments may provide opportunities for causal identification in built environment research (Angrist & Pischke, 2009). These approaches should consider how instrumental validity may vary across different urban contexts.

Mixed-methods approaches combining quantitative analysis with qualitative research can provide a deeper understanding of causal mechanisms and processes across different urban contexts. In-depth interviews, focus groups, and ethnographic studies can illuminate how residents experience and interpret built environment characteristics, providing insights into causal pathways that quantitative analysis alone cannot reveal.

4. Expanded Variable and Context Consideration

Future research should expand theoretical and empirical frameworks to include broader ranges of environmental, social, and cultural factors that influence built environment-well-being relationships, with explicit attention to how these factors may operate differently across urban density contexts and development types (Bray, 2006; Oldenburg, 1999).

Community governance and management factors characteristic of Chinese contexts deserve systematic investigation across different urban density levels. Neighborhood committees, property management organizations, and community social programs may significantly influence resident well-being through pathways not captured in physical built environment frameworks developed in Western contexts (Bray, 2006). The importance and operation of these factors may vary significantly between high-density urban cores and lower-density suburban areas.

Third place concepts, including commercial, leisure, and entertainment spaces beyond residential and workplace environments, represent important environmental factors for understanding contemporary urban life across different urban contexts. The influence and mechanisms of third places on Life Satisfaction require systematic investigation as urban lifestyles increasingly depend on commercial and recreational environments, with particular attention to how third place functions may differ across urban density contexts (Oldenburg, 1999).

POI categorization should be refined to distinguish different types of facilities and services with potentially different effects on well-being across urban contexts. Commercial facilities, recreational amenities, and public services may have distinct effects on different aspects of well-being through different mechanisms, and these effects may vary across urban density levels and development types.

Cultural and demographic factors should receive greater attention as moderating variables that influence built environment relationships across different urban contexts. Age, gender, socioeconomic status, cultural background, and lifestyle preferences may significantly influence how individuals experience and respond to built environment characteristics (Altman & Chemers, 1980). These moderating effects may operate differently in different urban density contexts.

5. Alternative Mediation Pathways across Urban Contexts

Current research focuses on four mediating factors (Perceived Environmental Quality, Perceived Safety, Social Capital, and self-rated health), but the built environment may influence well-being through additional pathways requiring investigation across different urban contexts (Kaplan & Kaplan, 1989; Ulrich, 1984).

Leisure activities and recreational behaviors represent important mediating pathways through which built environment characteristics influence well-being outcomes, but these pathways may operate differently across urban density contexts. Environmental support for different types of leisure activities may significantly influence Life Satisfaction through promoting physical activity, social interaction, and psychological restoration (Kaplan & Kaplan, 1989). The availability and importance of different leisure opportunities may vary substantially between high-density urban cores and suburban environments.

Emotional pathways, including stress, restoration, and mood regulation represent important mechanisms through which environmental characteristics influence well-being across urban contexts. Environmental stressors such as noise, crowding, and visual disorder can directly influence psychological well-being, while restorative environments can promote psychological recovery and positive emotions (Ulrich, 1984). These emotional responses may be particularly important in high-density contexts but operate differently in lower-density environments.

Behavioral pathways, including transportation behavior, consumption patterns, and social interaction behaviors, may mediate built environment effects on well-being differently across urban contexts. Environmental characteristics that promote or constrain different behaviors can indirectly influence well-being through supporting or hindering goal achievement and lifestyle preferences (Sallis et al., 2016). The importance of different behavioral pathways may vary significantly across urban density levels.

Chain mediation relationships among different mediating pathways require more sophisticated investigation across urban contexts. The current study demonstrates complex relationships among mediating factors specifically in high-density contexts, but more comprehensive frameworks should include broader ranges of mediating variables and their interrelationships across different urban development types.

## 10.4 Theoretical and Practical Contributions

### *10.4.1 Theoretical Contributions*

This research makes several significant theoretical contributions to understanding built environment effects on subjective well-being, specifically in high-density urban contexts, while highlighting the need for context-specific theoretical development.

1. Contextual Adaptation of Environmental Psychology Theory

The study demonstrates that environmental psychology theories developed in Western, lower-density contexts require substantial adaptation for high-density urban environments, and conclusions derived from high-density contexts should not be automatically applied to other urban development types. The negative effects of building density, functional mix diversity, and intersection density challenge universal application of compact city theories and suggest context-dependent optimization principles that may vary significantly across urban density levels (Bramley & Power, 2009; Newman & Kenworthy, 2015).

These findings contribute to environmental psychology theory by demonstrating that person-environment fit and environmental preference relationships may vary significantly across different urban contexts and density levels. The threshold effects observed in several built environment relationships suggest non-linear theoretical models may better represent environmental psychology relationships than linear models commonly assumed in existing theory, but these threshold effects may occur at different levels across urban contexts.

2. Mediation Framework Development for High-Density Contexts

The comprehensive mediation framework integrating Perceived Environmental Quality, Perceived Safety, Social Capital, and self-rated health represents a significant theoretical contribution to understanding built environment-well-being relationships, specifically in high-density contexts. The demonstration that Perceived Environmental Quality serves as the primary mediator with cascading effects on other mediating factors provides important insights into the psychological processes through which environmental characteristics influence well-being within high-density environments.

The chain mediation relationships identified in this study demonstrate the complex, interconnected nature of environmental experience and well-being outcomes, specifically in high-density contexts. These findings contribute to systems theories of human-environment relationships by demonstrating how environmental effects cascade through multiple psychological and social processes rather than operating through simple direct pathways, but the specific patterns observed may be unique to high-density urban environments.

3. Cultural and Contextual Theory Development

The differences between this study's findings and Western research contribute to developing culturally and contextually appropriate theories of built environment effects on well-being, with particular emphasis on the need for density-specific theoretical frameworks. The research demonstrates that environmental preferences, optimal environmental configurations, and mediation pathways may vary across cultural and contextual conditions, with urban density representing a particularly important contextual factor.

These contributions suggest the need for theoretical frameworks that can accommodate both cultural variation and contextual adaptation across different urban development types rather than assuming universal environmental effects. Future theory development should consider cultural factors, including social structures, lifestyle patterns, environmental preferences, and value systems that influence built environment relationships, while also accounting for how these factors may operate differently across urban density contexts.

### *10.4.2 Practical and Policy Contributions*

1. Evidence-Based Planning Guidance for High-Density Contexts

This research provides empirical evidence for built environment planning and design decisions specifically in high-density urban contexts, with explicit recognition that these findings may not apply to moderate-density or suburban planning contexts. The specific findings about optimal ranges and threshold effects for different built environment characteristics can inform planning standards and design guidelines specifically for high-density urban development.

The demonstration that built environment effects operate primarily through mediating factors rather than direct pathways provides important guidance for environmental intervention strategies in high-density contexts. Policy interventions should consider both objective environmental improvements and the psychological and social processes through which environmental changes influence well-being outcomes within the specific conditions of high-density urban environments.

2. Integrated Policy Framework for High-Density Urban Development

The research supports integrated policy approaches that address multiple built environment factors and mediating processes simultaneously rather than optimizing individual environmental characteristics in isolation, with specific application to high-density urban contexts. The complex interaction effects among built environment factors suggest that holistic planning approaches may be more effective than sectoral interventions particularly in high-density environments where factor interactions may be amplified.

The emphasis on mediating factors suggests that environmental interventions in high-density contexts should be accompanied by social and community development programs that enhance residents' environmental perceptions, security feelings, social connections, and health behaviors within the specific conditions and constraints of high-density urban living.

3. Context-Specific Planning Standards

The research contributes to developing planning standards and design guidelines appropriate specifically for high-density Chinese urban contexts rather than adapting Western standards without consideration of contextual and density differences. The specific threshold effects and optimal ranges identified in this research can inform density standards, facility planning guidelines, and transportation infrastructure design for high-density urban development, while recognizing that different standards may be appropriate for other urban development types.

Future policy development should explicitly consider urban density context when developing planning guidelines, recognizing that optimal built environment configurations may vary significantly between high-density urban cores, moderate-density neighborhoods, and suburban areas.

### *10.4.3 Summary*

This comprehensive investigation of built environment effects on Life Satisfaction, specifically in high-density urban contexts, provides important theoretical insights and practical guidance for urban planning and community development within similar high-density environments. The research demonstrates that built environment characteristics significantly influence residents' subjective well-being through complex mediation pathways involving environmental perception, security evaluation, social relationships, and health outcomes, with these relationships being specifically observed and validated within high-density urban core conditions.

The findings reveal that compact city theories developed in Western contexts require substantial adaptation for high-density Chinese urban environments, and these adaptations should not be assumed to apply to moderate-density or suburban contexts without empirical verification. Building density, functional mix diversity, and transportation infrastructure complexity can create negative effects on Life Satisfaction when they exceed optimal thresholds, specifically in high-density contexts, challenging assumptions about the universal benefits of urban compactness.

The mediating role of Perceived Environmental Quality as the primary pathway through which the built environment affects well-being emphasizes the importance of residents' subjective environmental experiences rather than purely objective environmental characteristics within high-density contexts. This finding suggests that environmental interventions in high-density areas should consider perceptual and experiential factors alongside physical environmental improvements.

The research provides practical guidance for urban planning and policy development, specifically in high-density contexts, emphasizing the need for integrated approaches that address both objective built environment optimization and the social and psychological conditions that mediate environmental effects on well-being within high-density urban environments. The specific findings about optimal ranges and threshold effects for different built environment characteristics can inform evidence-based planning standards and design guidelines appropriate specifically for high-density urban contexts.

Future research should address the limitations identified in this study through longitudinal designs, multi-scale analysis, advanced methodological approaches, and expanded theoretical frameworks that include broader ranges of environmental, social, and cultural factors. Critically, future research should explicitly examine whether the relationships identified in this high-density context apply to other urban development types, including moderate-density neighborhoods and suburban areas. The ultimate goal of this research program is to contribute to the development of built environments that effectively support human well-being and quality of life in increasingly urbanized societies across diverse urban density contexts and development types.

The policy implications and theoretical conclusions presented in this study should be applied cautiously when considering urban contexts other than high-density urban cores, and additional research is needed to develop appropriate frameworks for moderate-density and suburban environments. The context-specific nature of these findings underscores the importance of developing nuanced, contextually appropriate approaches to built environment planning and design rather than applying universal principles across diverse urban development contexts.

**Funding** This research was funded by

1. Guangdong Provincial Office of Philosophy and Social Sciences [Grant No. GD24XSH06]
2. Projects of PhDs' Start-up Research of GDUPT [Grant No. 2023bsqd1008].

## References

Alexander, C., Ishikawa, S., & Silverstein, M. (1977). *A pattern language: Towns, buildings, construction.* Oxford University Press.

Altman, I., & Chemers, M. M. (1980). *Culture and environment.* Cambridge University Press.

Angrist, J. D., & Pischke, J. S. (2009). *Mostly harmless econometrics: An empiricist's companion*. Princeton University Press.
Appleyard, D. (1981). *Livable streets*. University of California Press.
Babisch, W. (2008). Road traffic noise and cardiovascular risk. *Noise and Health, 10*(38), 27–33.
Batty, M. (2013). *The new science of cities*. MIT Press.
Beatley, T. (2000). *Green urbanism: Learning from European cities*. Island Press.
Bell, P. A., Greene, T. C., Fisher, J. D., & Baum, A. (2001). *Environmental psychology* (5th ed.). Harcourt College Publishers.
Bramley, G., & Power, S. (2009). Urban form and social sustainability: The role of density and housing tenure. *Environment and Planning B: Planning & Design, 36*(1), 30–48.
Bray, D. (2006). Building 'community': New strategies of governance in urban China. *Economy and Society, 35*(4), 530–549.
Bronfenbrenner, U. (1979). *The ecology of human development: Experiments by nature and design*. Harvard University Press.
Calthorpe, P. (2011). *Urbanism in the age of climate change*. Island Press.
Campbell, A., Converse, P. E., & Rodgers, W. L. (1976). *The quality of American life: Perceptions, evaluations, and satisfactions*. Russell Sage Foundation.
Cao, J. (2016). How does neighborhood design affect life satisfaction? Evidence from twin cities. *Travel Behaviour and Society, 5*, 68–76.
Cervero, R. (2013). *Transport infrastructure and global competitiveness: Balancing mobility and livability*. Routledge.
Cervero, R., & Kockelman, K. (1997). Travel demand and the 3Ds: Density, diversity, and design. *Transportation Research Part D: Transport and Environment, 2*(3), 199–219.
Cook, T. D., & Campbell, D. T. (1979). *Quasi-experimentation: Design and analysis issues for field settings*. Houghton Mifflin.
Crowe, T. (2000). *Crime prevention through environmental design* (2nd ed.). Butterworth-Heinemann.
Diener, E. (1984). Subjective Well-being. *Psychological Bulletin, 95*(3), 542–575.
Diener, E., Emmons, R. A., Larsen, R. J., & Griffin, S. (1985). The satisfaction with life scale. *Journal of Personality Assessment, 49*(1), 71–75.
Duany, A., Plater-Zyberk, E., & Speck, J. (2000). *Suburban nation: The rise of sprawl and the decline of the American dream*. North Point Press.
Evans, G. W., & McCoy, J. M. (1998). When buildings don't work: The role of architecture in human health. *Journal of Environmental Psychology, 18*(1), 85–94.
Ewing, R., & Cervero, R. (2010). Travel and the built environment: A meta-analysis. *Journal of the American Planning Association, 76*(3), 265–294.
Flora, C. B., & Flora, J. L. (2013). *Rural communities: Legacy and change* (4th ed.). Westview Press.
Frumkin, H., Bratman, G. N., Breslow, S. J., Cochran, B., Kahn, P. H., Jr., Lawler, J. J., Levin, P. S., Tandon, P. S., Varanasi, U., Wolf, K. L., & Wood, S. A. (2017). Nature contact and human health: A research agenda. *Environmental Health Perspectives, 125*(7), 075001.
Gehl, J. (2010). *Cities for people*. Island Press.
Gifford, R. (2007). The consequences of living in high-rise buildings. *Architectural Science Review, 50*(1), 2–17.
Grant, J., & Perrott, K. (2011). Where is the café? The challenge of making retail uses viable in mixed-use subdivisions. *Journal of Urban Design, 16*(1), 77–95.
Healey, P. (1997). *Collaborative planning: Shaping places in fragmented societies*. UBC Press.
Helliwell, J., Layard, R., & Sachs, J. (2020). *World happiness report 2020*. UN Sustainable Development Solutions Network.
Holling, C. S. (1978). *Adaptive environmental assessment and management*. John Wiley & Sons.
Huang, L., Yang, W., & Liang, Z. (2023). How does the built environment affect residents' subjective Well-being? A systematic review and meta-analysis. *Cities, 131*, 103957.
Jackson, R. J. (2003). The impact of the built environment on health: An emerging field. *American Journal of Public Health, 93*(9), 1382–1384.
Jacobs, J. (1961). *The death and life of great American cities*. Random House.

Jenks, M., Burton, E., & Williams, K. (1996). *The compact city: A sustainable urban form?* E & FN Spon.
Kang, Y., Zhang, F., Gao, S., Lin, H., & Liu, Y. (2020). A review of urban physical environment sensing using street view imagery in public health studies. *Annals of GIS, 26*(3), 261–275.
Kaplan, R., & Kaplan, S. (1989). *The experience of nature: A psychological perspective*. Cambridge University Press.
Lazarus, R. S., & Folkman, S. (1984). *Stress, appraisal, and coping*. Springer.
Leaf, M. (1998). Urban planning and urban reality under Chinese economic reforms. *Journal of Planning Education and Research, 18*(2), 145–153.
Litman, T. (2017). *Well measured: Developing indicators for sustainable and livable transport planning*. Victoria Transport Policy Institute.
Li, H., Pan, C., Qiu, N., Guo, J., & Wu, J. (2025). Does the multi-scale built environment impact on residents' subjective Well-being? *Buildings, 15*(18), 3311.
Liu, Y., Wang, R., Grekousis, G., Liu, Y., Yuan, Y., & Li, Z. (2020). Neighbourhood greenness and mental Well-being in Guangzhou, China: What are the pathways? *Landscape and Urban Planning, 190*, 103602.
Lynch, K. (1960). *The image of the city*. MIT Press.
Lyubomirsky, S., King, L., & Diener, E. (2005). The benefits of frequent positive affect: Does happiness lead to success? *Psychological Bulletin, 131*(6), 803–855.
Mehta, V. (2013). *The street: A quintessential social public space*. Routledge.
Menard, S. (2002). *Longitudinal research* (2nd ed.). Sage Publications.
Montgomery, C. (2013). *Happy city: Transforming our lives through urban design*. Farrar, Straus and Giroux.
Mouratidis, K. (2018a). Built environment and leisure satisfaction: The role of commute time, social interaction, and active travel. *Journal of Transport Geography, 66*, 325–337.
Mouratidis, K. (2018b). Is compact city livable? The impact of compact versus sprawled neighbourhoods on neighbourhood satisfaction. *Urban Studies, 55*(11), 2408–2430.
Mouratidis, K. (2019). Compact city, urban sprawl, and subjective Well-being. *Cities, 92*, 261–272.
Mouratidis, K. (2021). Urban planning and quality of life: A review of pathways linking the built environment to subjective Well-being. *Cities, 115*, 103229.
Murray, A. T. (2001). Strategic analysis of public transport coverage. *socioeconomic Planning Sciences, 35*(3), 175–188.
Nasar, J. L. (1998). *The evaluative image of the city*. Sage Publications.
Newman, O. (1972). *Defensible space*. Macmillan.
Newman, P., & Kenworthy, J. (2015). *The end of automobile dependence: How cities are moving beyond car-based planning*. Island Press.
Nieuwenhuijsen, M. J. (2016). Urban and transport planning, environmental exposures and health-new concepts, methods and tools to improve health in cities. *Environmental Health, 15*(Suppl 1), S38.
Oldenburg, R. (1999). *The great good place: Cafes, coffee shops, bookstores, bars, hair salons, and other hangouts at the heart of a community*. Marlowe & Company.
Oshan, T. M., Li, Z., Kang, W., Wolf, L. J., & Fotheringham, A. S. (2019). Mgwr: A python implementation of multiscale geographically weighted regression for investigating process spatial heterogeneity and scale. *ISPRS International Journal of Geo-Information, 8*(6), 269.
Putnam, R. D. (2000). *Bowling alone: The collapse and revival of American community*. Simon & Schuster.
Raudenbush, S. W., & Bryk, A. S. (2002). *Hierarchical linear models: Applications and data analysis methods* (2nd ed.). Sage Publications.
Sallis, J. F., Bull, F., Guthold, R., Heath, G. W., Inoue, S., Kelly, P., et al. (2016). Progress in physical activity over the Olympic quadrennium. *The Lancet, 388*(10051), 1325–1336.
Sampson, R. J., Raudenbush, S. W., & Earls, F. (1997). Neighborhoods and violent crime: A multilevel study of collective efficacy. *Science, 277*(5328), 918–924.
Sanoff, H. (2000). *Community participation methods in design and planning*. John Wiley & Sons.

Saxena, S., Thornicroft, G., Knapp, M., & Whiteford, H. (2007). Resources for mental health: Scarcity, inequity, and inefficiency. *The Lancet, 370*(9590), 878–889.

Seligman, M. E. (2011). *Flourish: A visionary new understanding of happiness and Well-being*. Free Press.

Sirgy, M. J., & Cornwell, T. (2002). How neighborhood features affect quality of life. *Social Indicators Research, 59*(1), 79–114.

Southworth, M. (2005). Designing the walkable city. *Journal of Urban Planning and Development, 131*(4), 246–257.

Stoknes, P. E. (2015). *What we think about when we try not to think about global warming: Toward a new psychology of climate action*. Chelsea Green Publishing.

Su, S., Pi, J., Xie, H., Cai, Z., & Weng, M. (2019a). Community deprivation, walkability, and public health: Highlighting the social inequalities in land use planning for health promotion. *Land Use Policy, 67*, 315–326.

Su, S., Zhang, H., Wang, M., Weng, M., & Kang, M. (2024). Multi-scale analysis of spatially varying relationships between neighborhood disadvantage and life satisfaction. *Applied Geography, 156*, 102965.

Su, S., Zhang, Q., Pi, J., Wan, C., & Weng, M. (2016). Public health in linkage to land use: Theoretical framework, empirical evidence, and critical implications for reconnecting health promotion to land use policy. *Land Use Policy, 57*, 605–618.

Su, S., Zhou, H., Xu, M., Ru, H., Wang, W., & Weng, M. (2019b). Auditing street walkability and associated social inequalities for planning implications. *Journal of Transport Geography, 74*, 62–76.

The State Council of the People's Republic of China. (2017). *The 13th five-year plan for economic and social development of the People's Republic of China (2016–2020)*. State Council.

Tong, Z., Baldauf, R. W., Isakov, V., Deshmukh, P., & Zhang, K. M. (2016). Roadside vegetation barrier designs to mitigate near-road air pollution impacts. *Science of the Total Environment, 541*, 920–927.

Ulrich, R. (1984). View through a window may influence recovery from surgery. *Science, 224*(4647), 420–421.

Vuchic, V. R. (2017). *Urban transit systems and technology*. John Wiley & Sons.

Wallas, A., Eriksson, C., Gruzieva, O., Lind, T., Pyko, A., Sjöström, M., et al. (2018). Road traffic noise and determinants of saliva cortisol levels among adolescents. *International Journal of Hygiene and Environmental Health, 221*(2), 276–282.

Wang, R., Feng, Z., Liu, Y., & Lu, Y. (2020). Relationship between neighbourhood green space and mental Well-being in older adults. *Social Science & Medicine, 222*, 112–120.

Wang, Y., Chen, Y., Liu, H., & Zhang, L. (2024). Deep learning applications in urban environment analysis: A comprehensive review. *Computers, Environment and Urban Systems, 98*, 101892.

Wooldridge, J. M. (2010). *Econometric analysis of cross section and panel data* (2nd ed.). MIT Press.

Wu, W., Chen, W. Y., Yun, Y., Wang, F., & Gong, Z. (2022). Urban greenness, mixed land-use, and life satisfaction: Evidence from residential locations and workplace settings in Beijing. *Landscape and Urban Planning, 224*, 104428.

Yu, B., & Hu, Z. (2017). Research progress and prospect of built environment and health. *Progress in Geography, 36*(10), 1205–1217.

# Chapter 11
# Conclusion, Policy Implications and Future Directions

**Abstract** Synthesizing the empirical findings presented in previous chapters regarding the relationship between the built environment and life satisfaction in high-density urban contexts, this concluding chapter evaluates the study's contributions while critically reflecting on its limitations and proposing future research trajectories. It aims to provide a comprehensive evaluation of the research scope, methodological constraints, and policy relevance, particularly within the Chinese urban setting. Section 11.1 revisits and expands upon the study's key limitations, delving into issues of geographical specificity (with a focus on Guangzhou's central urban area), cross-sectional data constraints, methodological linearity, challenges in establishing causal inference, and variable selection within hierarchical data structures. Section 11.2 translates empirical results into actionable policy recommendations across national, provincial, and municipal levels, emphasizing the optimization of objective built environment elements and the enhancement of key mediating factors such as Perceived Environmental Quality and Social Capital. Finally, Section 11.3 outlines future research directions, advocating for multi-scale and multi-city analyses, methodological innovations—including machine learning and longitudinal designs—and theory development tailored to China's unique urban and institutional contexts. This chapter not only contextualizes the study's findings within broader urban policy frameworks such as China's "people-centered urbanization" but also highlights the imperative for culturally and contextually sensitive approaches in urban well-being research.

**Keywords** Built environment · Life satisfaction · High-density cities · Policy implications · Research limitations · Future research directions · Guangzhou · China · Urban planning · Subjective Well-being

Synthesizing the comprehensive theoretical integration and empirical findings presented in Chap. 10, which elucidate the complex mediation pathways through which the built environment influences residents' life satisfaction in high-density urban contexts. Concluding the volume, this chapter proceeds to synthesize the study's broader implications. It critically assesses research limitations, translates key



H. Li, G. Shen, *The Anatomy of Subjective Well-Being*, Resilient and Friendly Living Environments, https://doi.org/10.1007/978-981-95-4937-5_11

findings into actionable policy recommendations across governance levels, and outlines promising avenues for future scholarly inquiry. By embedding the study within China's urbanization context, it bridges academic insight with practical application and highlights the need for culturally and contextually adaptive approaches in urban well-being research.

## 11.1 Research Limitations

### *11.1.1 Limitations of Research Samples and Data*

Against the backdrop of high-density cities, this study focused on the central urban area of Guangzhou to investigate the relationship between the built environment and Life Satisfaction. This geographical specificity presents both strengths and limitations for the research findings. While Guangzhou serves as an exemplary case of high-density urban development in China, representing many characteristics shared by other major Chinese cities, the unique socioeconomic, cultural, and urban planning contexts of Guangzhou may limit the generalizability of findings to other high-density cities in China (Lin, 2019; Liu et al., 2019, 2020a, 2020b).

The Pearl River Delta region, where Guangzhou is located, has experienced distinctive urbanization patterns compared to other major urban agglomerations such as the Yangtze River Delta or the Beijing-Tianjin-Hebei region.[1] These regional differences in economic development trajectories, governance structures, and cultural contexts may influence the relationship between built environments and life satisfaction in ways that are not captured in this study. Whether the conclusions of this study are similar to or different from those of other high-density cities in China remains to be further explored, so the applicability of the conclusions needs to be verified across different urban contexts, climate zones, and regional development levels.

In future research, comparative studies can be conducted on high-density cities in different regions of China, including northern cities like Beijing and Tianjin, eastern cities like Shanghai and Nanjing, and western cities like Chengdu and Chongqing. Such comparative analysis would help identify universal patterns versus region-specific relationships between built environments and subjective well-being. Additionally, the inclusion of medium-density cities and rapidly urbanizing smaller cities would provide a more comprehensive understanding of how density itself mediates the relationship between built environments and life satisfaction.

The data collection methodology presents another significant limitation. The data in this study are cross-sectional data from the survey of the built environment

[1]For a comparative analysis of China's major urban agglomerations, see Fang and Yu (2017). Urban agglomeration: An evolving concept of an emerging phenomenon. *Landscape and Urban Planning*, 162, 126–136.

and Life Satisfaction, rather than panel data.[2] Therefore, it is difficult to confirm causal relationships between variables. According to literature research, the evaluation of Life Satisfaction mainly measures individuals' overall satisfaction with life, which has the evaluation characteristics of subjectivity, integrity, and relative stability (Diener, 1984; Diener et al., 1985). However, life satisfaction can also fluctuate due to life events, seasonal variations, and changing life circumstances (Lucas & Donnellan, 2012).

The temporal dimension is particularly important in the context of rapidly changing urban environments in China. The built-environment elements of variables in the study area, namely the central urban area, are also basically stable within a certain time, but this assumption may not hold over longer periods, given the rapid pace of urban development in Chinese cities. Since it is difficult to obtain panel data of objective built-environment data, there may be a problem of non-independent samples, and it is difficult to track and survey samples of Subjective Well-being within a limited time frame while controlling for the numerous confounding variables that could influence both built environment exposure and life satisfaction outcomes.

The timing of data collection (specific dates should be mentioned) may also introduce temporal bias, as life satisfaction can vary seasonally and in response to major events or policy changes. Therefore, based on practical considerations, cross-sectional data from the survey of the relationship between the built environment and Subjective Well-being are used for the study. In the future, follow-up surveys can be gradually adopted to establish longitudinal panel data for in-depth research on the relationship and mechanism between the two, ideally spanning multiple years to capture both short-term fluctuations and long-term trends.

### 11.1.2 Limitations of Research Methods

Based on the current relevant research literature, the vast majority of studies have employed regression analysis techniques to explore the impact of the built environment on physical health, mental health, and Subjective Well-being (Mouratidis, 2018; Nieuwenhuijsen, 2020). Meanwhile, a considerable number of studies have also used structural equation models to investigate underlying mechanisms (Ettema & Schekkerman, 2016). This study adopts similar statistical analysis methods, which, while well-established and widely accepted, may not capture the full complexity of the relationships under investigation.

The limitation of linear modeling approaches becomes apparent when considering the potentially non-linear relationships between built environment

[2] The challenges of obtaining longitudinal data in urban studies, especially panel data on subjective well-being, are discussed in detail in: Oishi and Kesebir (2015). Income inequality explains why economic growth does not always translate to an increase in happiness. *Psychological Science, 26*(10), 1630–1638.

characteristics and life satisfaction (Yang et al., 2021).[3] There are individual studies that use the three-factor theory and non-linear estimation to explore the relationship between the built environment and Subjective Well-being (Cao et al., 2018b, 2020), but comprehensive methodological frameworks for exploring complex mechanisms remain underdeveloped. The assumption of linear relationships may oversimplify the reality where threshold effects, tipping points, and interaction effects between different built environment characteristics may be more important than additive effects.

In the past 2 years, technological advances have opened new methodological possibilities. Some studies have used convolutional neural network technology to identify street view images with the help of PSPNet and DeepLabv3+ semantic segmentation models, quantify environmental quality more objectively (Yin & Wang, 2016), and use random forest (RF), gradient boosting trees (XGBoost), or multi-layer perceptrons to explore the non-linear impact of the built environment on subjective well-being (Zhang et al., 2018). These approaches can capture complex spatial patterns and interactions that traditional regression methods might miss.

Furthermore, there are individual studies that use multi-modal deep learning technology to conduct integrated research by fusing street view images, text data (social media emotions), and spatio-temporal trajectory data (GPS) (Wang et al., 2021). Machine learning and deep learning are reshaping the research paradigm of the built environment and well-being through automated environmental perception, mining complex mechanisms, fusing multi-source data, and simulating individualized interventions. They show irreplaceable value in revealing micro-environmental mechanisms and supporting precise urban design interventions.

However, these advanced methods also face significant challenges, such as interpretability, generalization across different contexts, data quality requirements, and the need for substantial computational resources. The "black box" nature of many machine learning approaches makes it difficult to understand the underlying mechanisms, which is crucial for policy development and urban planning applications. Therefore, based on the logical progressive relationship between the research questions posed in this study, using the currently popular linear regression and structural equation models for statistical analysis may be more conducive to the interpretation of action pathways and mechanism understanding. Subsequent studies can attempt to integrate multi-source data with machine learning and deep learning techniques for more comprehensive analytical approaches. However, it is worth noting that traditional regression approaches can still yield valuable insights when applied within a multi-scale framework. For example, Li et al. (2025) employed a hierarchical regression model to systematically examine the effects of built environments at housing, neighborhood, and community scales on life satisfaction, demonstrating a clear gradient of explanatory power from proximate to distal scales. This

[3] For a foundational discussion on the limitations of linear models and the importance of considering nonlinearity, see: Wooldridge, J. M. (2010). *Econometric Analysis of Cross Section and Panel Data (2nd ed.)*. Cambridge, MA: MIT Press.

underscores the importance of scale-sensitive analysis even within conventional methodological paradigms.

### 11.1.3 The Issue of Causal Effects in Research Conclusions

This paper conducts an empirical study on the relationship between the built environment and Subjective Well-being using regression models and structural equation models. Strictly speaking, the relationships between variables explored in this study are correlational rather than causal relationships, which represents a fundamental limitation in drawing policy implications and theoretical conclusions. The establishment of causality requires meeting several criteria, including temporal precedence, covariation, and the elimination of alternative explanations (Hill, 1965).

On one hand, the study uses cross-sectional data rather than longitudinal panel data, making it difficult to prove the chronological relationship of the data and establish temporal precedence. The simultaneity problem is particularly acute in studies of built environments and well-being, as people may self-select into neighborhoods that match their preferences, creating endogeneity issues that can bias estimates of built environment effects (Cao, Handy, & Mokhtarian, 2018a; Mokhtarian & Cao, 2008; Van Wee & Cao, 2022).

On the other hand, the impact of the built environment on Life Satisfaction may have complex time-lag effects that vary across different types of built environment characteristics. For instance, the effects of green space access might be immediate, while the effects of neighborhood walkability might only become apparent after residents have time to change their activity patterns and develop new social connections. Therefore, the influence and mechanism of the built environment on Life Satisfaction, explored through multiple regression and structural model analysis using cross-sectional data, cannot establish strictly causal relationships.

The issue of reverse causation further complicates causal inference. This study demonstrates that health status may improve the level of Life Satisfaction. However, relevant studies have also shown that improvements in Life Satisfaction can affect health outcomes, suggesting a close reciprocal causal relationship (Boehm & Kubzansky, 2012). This bidirectional relationship makes it challenging to determine which factor is the cause and which is the effect.

In addition, some studies have shown that self-rated health can affect Social Capital formation and vice versa, indicating close relationships of mutual influence among potential mediating variables (Kim et al., 2011). The interconnectedness of mediating variables creates complex feedback loops that are difficult to disentangle using cross-sectional data and traditional statistical approaches.

### *11.1.4 The Issue of Stratification and Multi-Level Analysis*

In the model constructed in this study, variables at multiple levels are involved, creating a hierarchical data structure that requires careful methodological consideration. Among the studied variables, the objective built-environment elements belong to community-level variables, while the Perceived Environmental Quality represents residents' evaluation of the Objective Built Environment and belongs to individual-level variables, highlighting the distinction between measured and perceived attributes of the environment (Ma et al., 2018). In addition, variables such as Perceived Safety, Social Capital, self-rated health, and socioeconomic attributes involved in the model all belong to individual-level variables.

Strictly speaking, the Objective Built Environment characteristics are shared among individuals within the same community but vary across different communities.[4] Individuals are nested within communities, and communities are nested within cities and regions, which creates a hierarchical data structure that may lead to statistical dependency issues and violation of the independence assumption underlying standard regression techniques (Raudenbush & Bryk, 2002).

The intra-class correlation (ICC) in hierarchical data can lead to underestimation of standard errors and inflation of significance levels if not properly addressed. Some studies have used hierarchical regression models (also known as multi-level models or mixed-effects models) to handle data nesting problems or adopted geographically weighted regression for spatially-aware analysis (Fotheringham et al., 2009). These approaches can account for the clustering of observations and provide more accurate estimates of both individual-level and community-level effects.

Some literature has used hierarchical regression techniques to confirm that the intra-class correlation coefficient of the Life Satisfaction index is less than 0.06 (Bliese, 2000), indicating that the data do not significantly violate the independent assumption of samples in their specific context (Kreft & de Leeuw, 1998). However, this threshold may vary across different studies and contexts, and the decision to use or not use hierarchical models should be based on both statistical criteria and theoretical considerations about the nature of the phenomena under study.

Additionally, when it comes to the problem of mediating mechanisms, few studies have successfully implemented hierarchical mediation models due to their complexity. The cross-level interactions between community-level built environment characteristics and individual-level mediating variables add another layer of complexity that is difficult to model and interpret using traditional approaches.

Therefore, referring to relevant literature, this study explores both the direct impact of the built environment on Life Satisfaction and the mechanisms through which the built environment affects Life Satisfaction via mediating variables. Considering the complexity of multi-level mediation models and the need for clear

[4] The conceptual basis for hierarchical data structures is elaborated in: Raudenbush, S. W., & Bryk, A. S. (2002). *Hierarchical Linear Models: Applications and Data Analysis Methods (2nd ed.).* Thousand Oaks, CA: Sage Publications.

interpretation of results for policy purposes, the hierarchical model approach was not used for empirical research. However, this decision represents a limitation that should be addressed in future research.

Moreover, the data used in this study are all from four administrative districts in the central urban area of Guangzhou. Compared with cross-city scale studies, the built-environment elements at the broader urban scale are relatively similar across the study area, so the impact of city-level built-environment elements is not explicitly considered. This limitation means that the findings may not capture broader urban-level effects that could influence the relationship between neighborhood-level built environments and life satisfaction.

### *11.1.5 Constraints on Variable Selection and Measurement*

In terms of variable selection, the conceptual framework for built environment variables in this study is based on the "5D" model commonly recognized and applied in the academic community, as constructed by Ewing and Cervero (2010). This model includes Density, Diversity, Design, Distance to transit, and Destination accessibility. However, Ewing and Cervero themselves acknowledge that the "5D" model is still not sufficiently mature and rigorous, with unclear boundaries between dimensions and potential overlaps when applied in empirical analysis.

When used for analysis, there are often overlaps between dimensions that can create multicollinearity issues and make it difficult to isolate the effects of specific built environment characteristics. For instance, areas with higher density often also have greater diversity and better transit access, making it challenging to determine which specific aspect of the built environment is driving observed effects on life satisfaction.

In terms of research progress, there is no more scientific, rigorous, effective, and universally accepted alternative model and variable framework (Yu & Hu, 2017). In current literature research, most studies on the built environment and Subjective Well-being adopt the "3D" or "5D" model frameworks, and the number of variables included in built environment analyses typically ranges from 3 to 6 variables. This limitation in theoretical framework development constrains the ability to capture the full complexity of built environment influences on well-being.

Although including more variables can potentially make model estimation more rigorous and accurate by capturing additional aspects of the built environment, unrestricted inclusion of variables can make models complex, cumbersome, and difficult to interpret for policy purposes. There is also the risk of overfitting, particularly with the relatively limited sample size typical of community-based studies.

This study uses the "5D" model framework as a foundation while making practical adaptations. Considering the progressive relationship of the research questions and for the sake of research focus and ease of interpretation, only variables reflecting the key characteristics of the Objective Built Environment are included in the model, while variables not central to the core research questions are not explicitly considered in the statistical models.

Therefore, the influence of other potentially important factors on Life Satisfaction, except for the control variables (i.e., individual socioeconomic attribute characteristics) and the built environment variables, is absorbed into the residuals. This approach has a relatively limited impact on the validity of the existing model results but may miss important contextual factors that could enhance understanding of the mechanisms under investigation.

The study enriches empirical research results in the Chinese context and verifies the effectiveness and applicability of relevant Western theories in the context of high-density cities in China, while supplementing the theoretical foundation for healthy cities and happy cities research. However, it does not include community management and governance factors with Chinese characteristics, such as the management services of organizations like sub-district offices, neighborhood committees, homeowners' associations, and property management companies, as well as volunteer services and community-based organizations.

These institutional factors are particularly important in the Chinese context, where community governance structures play a significant role in residents' daily lives and potentially influence their life satisfaction through mechanisms not captured in Western theoretical frameworks. The Chinese urban governance system includes unique institutions and practices that may moderate the relationship between built environments and well-being in ways that differ from Western contexts.

In terms of the selection of mediating variables, the existing literature predominantly uses single mediators, such as perceived environment quality, public safety perceptions, self-rated health, and Social Capital, to study the relationship between the Objective Built Environment and Life Satisfaction. Although this study considers the impact of the Objective Built Environment on Life Satisfaction through multiple mediators and chain mediation relationships, including perceived environment quality, public safety perceptions, Social Capital, and self-rated health, it does not include all potential mediating variables suggested in comprehensive theoretical frameworks.

Specifically, the study does not involve leisure activities and emotional states as mediating variables, which were included in the framework proposed by Mouratidis (2018). These omitted mediating variables may represent important pathways through which built environments influence life satisfaction, and their exclusion may lead to an incomplete understanding of the mechanisms under investigation.

### *11.1.6 Summary*

In summary, this study is subject to five principal sets of limitations pertaining to geographical specificity, cross-sectional data constraints, methodological linearity, causal inference challenges, and variable selection alongside hierarchical data structures. It is crucial to emphasize that these limitations do not diminish the validity or value of the core findings within the high-density urban context from which they were derived. Rather, they serve to precisely delineate the scope and conditions under which these conclusions are most applicable. Furthermore, they provide a critical and constructive framework that charts essential pathways for future research, guiding inquiries toward more geographically diverse, methodologically sophisticated, longitudinally designed, and theoretically comprehensive examinations of the complex relationships between the built environment and subjective well-being.

## 11.2 Policy Implications and Recommendations

### *11.2.1 National Policy Alignment and Integration*

The findings of this study have significant implications for national urban development policies in China. The research results align closely with the national strategic objectives outlined in China's 14th Five-Year Plan (2021–2025), which emphasizes "people-centered urbanization" and the goal of "continuously enhancing people's sense of gain, happiness, and security" (National Development and Reform Commission, 2021).[5] The empirical evidence from this study provides scientific support for policy directions aimed at improving urban living environments as a pathway to enhancing residents' subjective well-being, aligning with subsequent guidelines on promoting high-quality urban development (State Council, 2022).

The study's findings are particularly relevant to the implementation of the National New Urbanization Plan (2021–2035), which calls for high-quality urban development that prioritizes livability and sustainability over mere economic growth (Ministry of Housing and Urban-Rural Development, 2021). The research demonstrates that in high-density urban contexts like Guangzhou, traditional compact city principles derived from Western contexts may not directly apply, necessitating adapted approaches that account for Chinese urban characteristics and residents' preferences.

The negative relationship found between building density and life satisfaction has important implications for China's urban density policies. While the national government has promoted compact urban development to improve land use efficiency and reduce urban sprawl, this study suggests that excessive density can negatively impact residents' well-being. This finding supports the recent policy shift

[5] The full text and key objectives of the 14th Five-Year Plan can be accessed via: Central People's Government of the People's Republic of China. (2021). *The 14th Five-Year Plan for National Economic and Social Development of the People's Republic of China and the Outline of Long-Range Objectives for 2035*. Retrieved from http://www.gov.cn

toward "intensive urban development" that emphasizes quality over quantity in urban densification efforts.[6]

The study's results also inform the implementation of the "Beautiful China" initiative and the construction of "livable cities". The identification of Perceived Environmental Quality as the most important mediating factor between built environments and life satisfaction provides empirical support for environmental improvement policies and suggests that investments in environmental quality enhancement can yield significant benefits for residents' well-being.

### *11.2.2 Provincial and Municipal Policy Applications*

At the provincial level, the findings are highly relevant to Guangdong Province's urban development strategies. Guangdong's 14th Five-Year Plan emphasizes the development of the Guangdong-Hong Kong-Macao Greater Bay Area as a world-class urban agglomeration that prioritizes livability and sustainability (Guangdong Provincial Government, 2021).[7] The research provides evidence-based guidance for achieving these livability objectives in high-density urban environments typical of the Pearl River Delta region.

The negative effects of excessive intersection density and bus stop density found in this study have direct implications for Guangdong's transportation infrastructure planning. The province's massive investments in public transportation systems, including metro networks and bus rapid transit systems, should be designed with careful attention to optimizing accessibility while minimizing negative impacts on neighborhood livability (Guangdong Department of Transportation, 2022).

At the municipal level, the findings directly inform Guangzhou's urban planning and development policies. Guangzhou's Master Plan (2017–2035) aims to build a "globally influential and livable city", and this study provides specific guidance on how built environment characteristics can support or undermine these livability objectives (Guangzhou Municipal Government, 2018).[8]

The research supports Guangzhou's recent policy emphasis on "micro-renewal" of old urban areas rather than wholesale redevelopment. The findings suggest that careful attention to building density, functional mix, and public space quality in

[6] *Guiding Opinions of the Central Committee of the Communist Party of China and the State Council on Promoting High-Quality Urban Develop*ment ( in Chinese). Issued by the Central Committee of the Communist Party of China and the State Council, August 15, 2025. Retrieved from: https://www.gov.cn/zhengce/202508/content_7038144.

[7] *The 14th Five-Year Plan for National Economic and Social Development of Guangdong Province and the Long-Range Objectives Through the Year 2035*. Guangdong Provincial People's Government. (2021). Retrieved from https://www.gd.gov.cn/zwgk/wjk/qbwj/yf/content/post_3268751.html

[8] Guangzhou City Master Plan (2017–2035). Promulgated by the Guangzhou Municipal People's Government (2018). Retrieved from: https://ghzyj.gz.gov.cn

renewal projects can significantly impact residents' life satisfaction. This aligns with the city's "Three Old Renovation" policy, which emphasizes improving living conditions in existing urban areas (Guangzhou Urban Renewal Bureau, 2020).

Guangzhou's "15-minute community life circle" planning initiative can also benefit from this research. The study's findings about the complex relationship between functional mix diversity and life satisfaction suggest that simply increasing the variety of services and facilities within walking distance may not automatically improve residents' well-being. Instead, careful planning is needed to optimize the types, locations, and density of facilities to maximize positive impacts while minimizing negative effects such as noise, congestion, and overcrowding.

### *11.2.3 Specific Policy Recommendations*

1. Objective Built Environment Optimization

Based on the research findings, several specific policy recommendations emerge for optimizing objective built environment characteristics:

(1) Building Density Management: The negative relationship between building density and life satisfaction suggests that current approaches to urban densification need refinement. Rather than pursuing maximum density, planners should seek optimal density levels that balance land use efficiency with livability. This could involve:

- Establishing more nuanced density guidelines that consider local context, infrastructure capacity, and community characteristics
- Implementing "density bonuses" that are contingent on providing enhanced public amenities, green space, and community facilities
- Developing design standards that mitigate the negative effects of density through improved building spacing, height variations, and public space integration

(2) Functional Mix Optimization: The complex relationship between functional mix diversity and life satisfaction indicates that simply maximizing the variety of land uses and facilities may not improve residents' well-being. Policy recommendations include:

- Developing a more sophisticated understanding of which types of functional mix contribute positively to life satisfaction, versus those that may have negative effects
- Using big data analytics to guide location selection and business format decisions for commercial operators, as suggested in the study
- Implementing zoning policies that encourage beneficial types of mixed-use development while controlling potentially problematic combinations

(3) Transportation Infrastructure Planning: The contrasting effects of bus line density (positive) and bus stop density (negative) on life satisfaction provide important guidance for public transportation planning:

- Optimizing bus route planning to maximize network coverage while minimizing the number of stops needed to serve residents effectively, This approach is supported by research showing that neighborhood design significantly influences satisfaction with access to public transportation (Cao, 2013).
- Implementing smart transportation technologies to improve service efficiency and reduce the need for excessive stop density
- Considering the full range of impacts (accessibility, congestion, noise, safety) when planning transportation infrastructure improvements

(4) Road Network Design: The negative relationship between intersection density and life satisfaction suggests the need for more careful road network planning:

- Balancing connectivity benefits with livability concerns in road network design, including the need to audit and improve street walkability for all social groups (Su et al., 2019).
- Implementing traffic calming measures and improved intersection design to mitigate the negative effects of high intersection density
- Using complete streets design principles that consider pedestrians, cyclists, and community life in addition to vehicle movement[9]

2. Mediating Factor Enhancement Strategies

The identification of four key mediating factors provides clear guidance for policy interventions:

(1) Perceived Environmental Quality Enhancement: As the most important mediating factor, perceived environmental quality deserves priority attention in policy development:

- Implementing comprehensive urban greening programs that go beyond simple tree planting to create high-quality green spaces that residents perceive as beneficial, which can also contribute to mitigating environmental stressors and promoting mental health (Huang et al., 2023).
- Establishing urban design standards that prioritize aesthetic quality, cleanliness, and maintenance of public spaces
- Developing participatory planning processes that incorporate residents' perceptions and preferences into environmental improvement projects
- Public Safety and Security Improvement: The importance of Perceived Safety suggests that both actual safety and residents' feelings of safety need attention:

[9] This people-centered approach to street design echoes the foundational ideas of "livable streets" that prioritize community well-being over mere traffic movement (Appleyard, 1980).

- Enhancing community policing programs and neighborhood watch initiatives
- Improving lighting, sight lines, and natural surveillance in public spaces through environmental design to enhance safety and reduce environmental stressors such as noise, which has been linked to adverse health outcomes (Wallas et al., 2018)
- Establishing rapid response systems for addressing safety concerns and maintaining community confidence

(2) Social Capital Development: The role of social capital in mediating built environment effects on life satisfaction suggests the need for policies that actively promote community social connections:

- Creating and programming community spaces that encourage resident interaction and relationship building
- Supporting community organizations, resident associations, and volunteer programs
- Implementing community development programs that strengthen neighborhood social networks

(3) Health Promotion: The mediating role of self-rated health indicates that built environment policies should explicitly consider health impacts (Su et al., 2016):

- Integrating Health Impact Assessment into urban planning processes
- Developing active transportation infrastructure that encourages physical activity
- Ensuring adequate access to healthcare facilities and health-promoting amenities

### 11.2.4 Implementation Framework and Governance Mechanisms

1. Multi-Level Governance Coordination

Effective implementation of these policy recommendations requires coordination across multiple levels of government and different policy domains. A comprehensive implementation framework should include coordinated policies at multiple levels.

(1) National Level Coordination: The central government should provide policy guidance, funding mechanisms, and performance evaluation criteria that support local implementation of evidence-based urban planning approaches. This could include:

- Updating national urban planning standards and guidelines to reflect research findings about optimal built environment characteristics

- Establishing funding programs that incentivize local governments to prioritize residents' well-being in urban development projects
- Creating national monitoring systems to track progress on urban livability indicators

1. Provincial Level Integration: Provincial governments should coordinate across municipalities and provide technical support for policy implementation:

   - Developing regional coordination mechanisms for transportation, environmental, and development policies that affect multiple municipalities
   - Establishing provincial-level research and technical assistance programs to support evidence-based local planning
   - Creating inter-municipal learning and best practice sharing programs

(2) Municipal Level Implementation: Local governments are the key implementers of built environment policies and should develop comprehensive implementation strategies:

- Integrating well-being considerations into all urban planning processes and development approval procedures
- Establishing cross-departmental coordination mechanisms that align transportation, housing, environmental, and social service policies
- Developing community engagement processes that ensure resident voices are incorporated into planning decisions

2. Community Governance and Participation

The Chinese context includes unique community governance institutions that can play important roles in implementing built environment improvements:

(1) Neighborhood Committee Enhancement: Neighborhood committees can serve as important intermediaries between residents and government agencies:

- Training neighborhood committee members to understand the connections between built environments and resident well-being
- Providing neighborhood committees with resources and authority to address local environmental quality issues
- Establishing feedback mechanisms that allow neighborhood committees to communicate resident concerns to planning authorities

(2) Property Management Integration: Property management companies are increasingly important actors in Chinese urban communities and can contribute to built environment improvements:

- Developing professional standards and training programs for property managers that include an understanding of environment well-being relationships
- Creating incentive systems that reward property management companies for improvements in resident satisfaction and community environmental quality

- Establishing coordination mechanisms between property managers and municipal agencies for addressing infrastructure and service issues

(3) Homeowners' Association Empowerment: Homeowners' associations represent an important mechanism for resident participation in community improvement:

- Providing legal and technical support for homeowners' associations to engage effectively in community planning processes
- Creating funding mechanisms that allow homeowners' associations to implement small-scale environmental improvements
- Establishing dialogue forums between homeowners' associations and planning authorities

## 11.3 Prospects for Future Research

### *11.3.1 Expanding Geographic and Scale Analysis*

This study focused on 87 communities within four districts of Guangzhou's central urban area, using 1000-meter buffer zones centered on community centroids to define the built environment exposure areas. While this approach provides valuable insights into neighborhood-level relationships between built environments and life satisfaction, several opportunities exist for expanding the geographic scope and scale of analysis.

1. Multi-city Comparative Studies: Future research should extend beyond Guangzhou to include other high-density cities in China with different characteristics. Comparative studies across cities with different climate conditions, economic development levels, cultural backgrounds, and governance structures would help identify which findings from Guangzhou are generalizable versus context-specific. Priority cities for comparison might include:
   - Northern cities like Beijing and Tianjin, which have different climate conditions and urban development histories
   - Eastern coastal cities like Shanghai and Nanjing, which have different economic structures and population compositions
   - Western cities like Chengdu and Chongqing, which represent different geographic and cultural contexts
   - Rapidly developing secondary cities, which may show different patterns due to their development stage
2. Multi-scale Environmental Assessment: Current research predominantly constructs buffer zones around residential centroids to define built environment exposure, but this approach has limitations. Most residents' daily activity spaces extend beyond simple circular buffers and vary based on individual characteris-

tics, transportation modes, and activity patterns (Kwan, 2012).[10] Future research should explore:

- Activity-based exposure assessment using GPS tracking or activity diaries to capture actual environmental exposure patterns
- Multiple buffer distances to understand how the built environment effects vary with spatial scale
- Integration of the workplace and other frequently visited locations into the environmental exposure assessment
- Consideration of commuting routes and transportation-based environmental exposures

3. Administrative and Planning Unit Analysis: Different administrative and planning scales may reveal different relationships between built environments and well-being. Future studies should examine:

    - Administrative jurisdiction levels (district, sub-district, neighborhood committee areas)
    - Planning units (such as 15-minute community life circles, transit-oriented development areas)
    - Natural geographic units (watersheds, topographic areas)
    - Economic units (central business districts, industrial zones, mixed-use areas)

### *11.3.2 Methodological Innovation and Integration*

The rapid advancement of data collection technologies, analytical methods, and computational capabilities creates numerous opportunities for methodological innovation in built environment and well-being research.

1. Advanced Spatial Analysis Techniques: Traditional regression approaches assume spatial independence and may not adequately account for spatial relationships in the data. Future research should explore:

    - Multi-scale Geographically Weighted Regression (MGWR) to address spatial heterogeneity and examine how relationships vary across geographic space
    - Spatial lag and spatial error models to account for spatial dependence in both built environment characteristics and well-being outcomes
    - Geographically and temporally weighted regression (GTWR) to examine how relationships vary across both space and time
    - Machine learning approaches that can capture complex spatial patterns and non-linear relationships

[10] The concept of the "uncertain geographic context problem" (UGCoP) and the limitations of static residential buffers are critically discussed in: Kwan, M.-P. (2012). The uncertain geographic context problem. *Annals of the Association of American Geographers*, 102(5), 958–968.

2. Big Data and Multi-source Data Integration: The availability of large-scale datasets from multiple sources creates opportunities for more comprehensive analysis:

   - Social media data analysis to capture real-time sentiment and activity patterns related to specific places and built environment characteristics
   - Mobile phone location data to understand actual activity patterns and environmental exposure
   - Street view imagery analysis using computer vision techniques to assess environmental quality more objectively and comprehensively
   - Satellite imagery analysis to monitor environmental changes over time and assess their impacts on well-being

3. Machine Learning and Artificial Intelligence Applications: Recent advances in AI and machine learning offer new possibilities for understanding complex relationships:

   - Deep learning models that can integrate multiple data types (imagery, text, numerical data) to provide a more comprehensive environmental assessment
   - Natural language processing of resident feedback and social media content to identify specific environmental quality issues
   - Predictive modeling to forecast how built environment changes might affect resident well-being
   - Personalized recommendation systems that could suggest optimal built environment characteristics for different population groups

4. Longitudinal and Panel Data Analysis: Moving beyond cross-sectional analysis is crucial for establishing causal relationships and understanding dynamic processes:

   - Multi-wave panel surveys that follow the same residents over time to observe how built environment changes affect well-being trajectories
   - Natural experiment approaches that take advantage of policy changes, infrastructure investments, or development projects to identify causal effects
   - Difference-in-difference analyses comparing areas that experience built environment changes with similar areas that do not
   - Event study approaches examining how specific built environment interventions affect resident well-being over time

### *11.3.3 Theoretical Framework Development and Testing*

While this study builds on existing theoretical frameworks, particularly the integration of Campbell's framework and Mouratidis's conceptual model, there are significant opportunities for theoretical advancement.

1. Chinese Context-Specific Theory Development: Existing theories of built environment-well-being relationships are primarily derived from Western contexts and may not fully capture the unique aspects of Chinese urban development and social organization:
   - Development of theoretical frameworks that explicitly incorporate Chinese institutional contexts, including the role of work units, neighborhood committees, and other uniquely Chinese urban institutions
   - Investigation of how Confucian cultural values and collectivistic orientations might moderate relationships between built environments and individual well-being
   - Examination of how rapid urbanization and social change in China create different dynamics compared to more stable Western urban contexts, while also drawing insights from international perspectives on urban transformation (Bosselmann, 1988).
   - Integration of concepts from Chinese urban planning traditions and contemporary "people-centered" development philosophies

2. Expanded Mediating Variable Analysis: This study examined four mediating factors (Perceived Environmental Quality, Perceived Safety, Social Capital, and Self-rated Health), but additional mediating pathways likely exist, as highlighted in recent comprehensive reviews of the pathways linking urban form to subjective well-being (Mouratidis, 2021):
   - Leisure activities and recreation opportunities as mediators of built environment effects
   - Environmental stress and restoration as psychological mechanisms linking environments to well-being
   - Social identity and place attachment as mediators connecting built environments to life satisfaction
   - Economic opportunities and accessibility to employment as mediators of built environment effects

3. Dynamic and Interactive Relationships: Current theoretical models tend to assume linear and additive relationships, but reality likely involves more complex dynamics:
   - Investigation of threshold effects and non-linear relationships between built environment characteristics and well-being outcomes
   - Analysis of interaction effects between different built environment characteristics
   - Examination of how individual characteristics moderate built environment effects
   - Development of feedback loop models that account for how well-being outcomes might influence environmental preferences and residential choices

### 11.3.4 Policy-Relevant Research Directions

Given the policy implications of this research, future studies should be designed with explicit attention to policy relevance and application.

1. Policy Intervention Evaluation: Research should move beyond correlation analysis to evaluate the effectiveness of specific policy interventions:
   - Quasi-experimental evaluation of urban renewal projects, transportation investments, and other built environment interventions
   - Cost-benefit analysis of different types of built environment improvements in terms of their impacts on resident well-being
   - Comparative evaluation of different policy approaches (regulatory vs. incentive-based vs. participatory) for achieving built environment improvements
   - Long-term follow-up studies to assess the sustainability of policy intervention effects
2. Implementation Science Applications: Understanding how research findings can be effectively translated into policy and practice:
   - Investigation of barriers and facilitators to evidence-based urban planning practice
   - Analysis of how different stakeholder groups interpret and use built environment-well-being research
   - Development of decision support tools and planning guidelines based on research findings
   - Training and capacity building programs for planners and policymakers
3. Equity and Environmental Justice Analysis: Future research should explicitly address questions of equity in built environment-well-being relationships:
   - Analysis of how the built environment affects vary across different socioeconomic, demographic, and cultural groups
   - Investigation of environmental justice issues related to the distribution of beneficial vs. harmful built environment characteristics
   - Examination of how built environment improvements can reduce or exacerbate existing inequalities
   - Development of equity-focused metrics and evaluation criteria for built environment policies

### 11.3.5 Interdisciplinary Integration and Collaboration

The complexity of built environment-well-being relationships requires interdisciplinary approaches that integrate insights from multiple fields.

1. Health and Medical Sciences Integration: Collaboration with health researchers can strengthen the health-related aspects of built environment research:
   - Integration of objective health measures (biomarkers, physical fitness, clinical assessments) with subjective well-being measures
   - Investigation of how built environments affect specific health conditions and disease outcomes
   - Development of health-focused built environment interventions and evaluation of their effectiveness
   - Integration of medical and public health expertise in research design and interpretation
2. Psychology and Behavioral Science Applications: Deeper integration with psychological research can enhance understanding of mechanisms:
   - Investigation of cognitive and emotional processes that mediate built environment effects on well-being
   - Analysis of how environmental preferences develop and change over time
   - Examination of how built environments interact with personality characteristics and individual differences
   - Application of environmental psychology theories to Chinese urban contexts
3. Economics and Social Policy Analysis: Integration with economic and social policy research can strengthen policy relevance:
   - Cost-effectiveness analysis of different built environment interventions
   - Investigation of how built environment improvements affect property values, economic development, and social mobility
   - Analysis of financing mechanisms and governance structures for built environment improvements
   - Development of social impact assessment tools for urban planning projects
4. Technology and Data Science Collaboration: Partnership with computer scientists and data specialists can advance methodological capabilities:
   - Development of automated systems for monitoring built environment quality and resident well-being
   - Creation of real-time feedback systems that connect resident experiences to planning decisions
   - Application of advanced analytical techniques from machine learning and artificial intelligence
   - Development of simulation and modeling tools for predicting built environment intervention effects

**Funding** This research was funded by

1. Guangdong Provincial Office of Philosophy and Social Sciences [Grant No. GD24XSH06]
2. Projects of PhDs' Start-up Research of GDUPT [Grant No. 2023bsqd1008].

# References


Appleyard, D. (1980). *Livable streets*. University of California Press.

Bliese, P. D. (2000). Within-group agreement, non-independence, and reliability: Implications for data aggregation and analysis. In K. J. Klein & S. W. J. Kozlowski (Eds.), *Multilevel theory, research, and methods in organizations* (pp. 349–381). Jossey-Bass.

Boehm, J. K., & Kubzansky, L. D. (2012). The heart's content: The association between positive psychological Well-being and cardiovascular health. *Psychological Bulletin, 138*(4), 655–691.

Bosselmann, P. (1988). *Urban transformation*. Island Press.

Cao, X. (2013). The influence of light rail transit on transit-oriented development in China. *Transportation Research Part A: Policy and Practice, 50*, 240–251.

Cao, X. J., Ettema, D., & Næss, P. (2020). Residential self-selection in the relationships between the built environment and travel behaviour: Introduction to the special issue. *Journal of Transport Geography, 65*, 102–106.

Cao, X. J., Handy, S. L., & Mokhtarian, P. L. (2018a). The influences of the built environment and residential self-selection on pedestrian behavior: Evidence from Austin, TX. *Transportation, 33*(1), 1–20.

Cao, X., Wu, X., & Yuan, Y. (2018b). Examining built environmental correlates of neighborhood satisfaction: A focus on analysis approaches. *Journal of Planning Literature, 33*(4), 419–432.

Guiding Opinions of the Central Committee of the Communist Party of China and the State Council on Promoting High-Quality Urban Development (in Chinese). (2015, August 15). Issued by the Central Committee of the Communist Party of China and the State Council. Retrieved from: https://www.gov.cn/zhengce/202508/content_7038144

Diener, E. (1984). Subjective Well-being. *Psychological Bulletin, 95*(3), 542–575.

Diener, E., Emmons, R. A., Larsen, R. J., & Griffin, S. (1985). The satisfaction with life scale. *Journal of Personality Assessment, 49*(1), 71–75.

Ettema, D., & Schekkerman, M. (2016). How do spatial characteristics influence Well-being and mental health? Comparing the effect of objective and subjective characteristics at different spatial scales. *Travel Behaviour and Society, 5*, 56–67.

Ewing, R., & Cervero, R. (2010). Travel and the built environment: A meta-analysis. *Journal of the American Planning Association, 76*(3), 265–294.

Fang, C., & Yu, D. (2017). Urban agglomeration: An evolving concept of an emerging phenomenon. *Landscape and Urban Planning, 162*, 126–136.

Fotheringham, A. S., Brunsdon, C., & Charlton, M. (2009). Geographically weighted regression. In *The sage handbook of spatial analysis* (Vol. 1, pp. 243–254). Wiley.

Guangdong Department of Transportation. (2022). *Guangdong Province transportation Development plan (2021–2025)*. Guangdong Provincial Government Press.

Guangdong Provincial Government. (2021). *Guangdong Province 14th five-year plan for economic and social Development and 2035 vision*. Guangdong Provincial Government Press.

Guangzhou Municipal Government. (2018). *Guangzhou City master plan (2017–2035)*. Guangzhou Municipal Government Press.

Guangzhou Urban Renewal Bureau. (2020). Implementation guidelines for three old renovation projects in Guangzhou. *Guangzhou Municipal Government.*

Hill, A. B. (1965). The environment and disease: Association or causation? *Proceedings of the Royal Society of Medicine, 58*(5), 295–300.

Huang, L., Chen, K., & Zhou, M. (2023). Climate change and mental health: Risks, impacts and priority actions. *International Journal of Mental Health Systems, 17*(1), 28.

Kim, D., Subramanian, S. V., & Kawachi, I. (2011). Social capital and physical health: A systematic review of the literature. In I. Kawachi, S. V. Subramanian, & D. Kim (Eds.), *Social capital and health* (pp. 139–190). Springer.

Kreft, I., & de Leeuw, J. (1998). *Introducing multilevel modeling*. Sage.

Kwan, M. P. (2012). The uncertain geographic context problem. *Annals of the Association of American Geographers, 102*(5), 958–968.

Li, H., Pan, C., Qiu, N., Guo, J., & Wu, J. (2025). Does the multi-scale built environment impact on residents' subjective Well-being? *Buildings, 15*(18), 3311.

Lin, L. (2019). Association between built environment and life satisfaction: Evidence from Beijing. *Urban Studies, 56*(7), 1332–1346.

Liu, Y., Wang, R., Grekousis, G., Liu, Y., Yuan, Y., & Li, Z. (2019). Neighbourhood greenness and mental Well-being in Guangzhou, China: What are the pathways? *Landscape and Urban Planning, 190*, 103602.

Liu, Y., Wang, R., Lu, Y., Li, Z., Chen, H., Cao, M., Zhang, Y., & Song, Y. (2020a). Natural outdoor environment, neighbourhood social cohesion and mental health: Using multi-level structural equation modelling, streetscape and remote-sensing metrics. *Urban Forestry & Urban Greening, 48*, 126576.

Liu, Y., Xiao, T., Liu, Y., Qiu, Y., Liu, Y., & Li, Z. (2020b). Impacts of urban built environments on residents' subjective Well-being: An analysis based on 15-minute walking distance. *Progress in Geography, 39*(8), 1270–1282.

Lucas, R. E., & Donnellan, M. B. (2012). Personality development across the life span: Longitudinal analyses with a national sample from Germany. *Journal of Personality and Social Psychology, 101*(4), 847–861.

Ma, L., Dill, J., & Mohr, C. (2018). The objective versus the perceived environment: What matters for bicycling? *Transportation, 41*(6), 1135–1152.

Ministry of Housing and Urban-Rural Development. (2021). *National new Urbanization Plan (2021–2035)*. China Architecture & Building Press.

Mokhtarian, P. L., & Cao, X. (2008). Examining the impacts of residential self-selection on travel behavior: A focus on methodologies. *Transportation Research Part B: Methodological, 42*(3), 204–228.

Mouratidis, K. (2018). Built environment and social Well-being: How does urban form affect social life and personal relationships? *Cities, 74*, 7–20.

Mouratidis, K. (2021). Urban planning and quality of life: A review of pathways linking the built environment to subjective Well-being. *Cities, 115*, 103229.

National Development and Reform Commission. (2021). *The 14th five-year plan for National Economic and social Development and the long-range objectives through the year 2035*. People's Publishing House.

Nieuwenhuijsen, M. J. (2020). Green infrastructure and health. *Annual Review of Public Health, 41*, 317–328.

Oishi, S., & Kesebir, S. (2015). Income inequality explains why economic growth does not always translate to an increase in happiness. *Psychological Science, 26*(10), 1630–1638.

Raudenbush, S. W., & Bryk, A. S. (2002). *Hierarchical linear models: Applications and data analysis methods* (2nd ed.). Sage.

State Council. (2022). *Guidelines on promoting high-quality urban development*. State Council of the People's Republic of China.

Su, S., Zhou, H., Xu, M., Ru, H., Wang, W., & Weng, M. (2019). Auditing street walkability and associated social inequalities for planning implications. *Journal of Transport Geography, 74*, 62–76.

Su, S., Zhang, Q., Pi, J., Wan, C., Weng, M., Cai, Z., Mei, L., & Li, M. (2016). Public health in linkage to land use: Theoretical framework, empirical evidence, and critical implications for reconnecting health promotion to land use policy. *Land Use Policy, 57*, 605–618.

Van Wee, B., & Cao, X. J. (2022). Residential self-selection in the relationship between the built environment and travel behavior: A literature review and research agenda. *Advances in Transport Policy and Planning, 9*, 75–94.

Wallas, A., Eriksson, C., Gruzieva, O., Lind, T., Pyko, A., Sjöström, M., Ögren, M., & Pershagen, G. (2018). Road traffic noise and determinants of saliva cortisol levels among adolescents. *International Journal of Hygiene and Environmental Health, 221*(2), 276–282.

Wang, R., Feng, Z., Pearce, J., Yao, Y., Li, X., & Liu, Y. (2021). The distribution of greenspace quantity and quality and their association with neighbourhood socioeconomic conditions in

Guangzhou, China: A new approach using deep learning method and street view images. *Sustainable Cities and Society, 66*, 102664.
Yang, L., Ao, Y., Ke, J., Lu, Y., & Liang, Y. (2021). To walk or not to walk? Examining non-linear effects of streetscape greenery on walking propensity of older adults. *Journal of Transport Geography, 94*, 103099.
Yin, L., & Wang, Z. (2016). Measuring visual enclosure for street walkability: Using machine learning algorithms and Google street view imagery. *Applied Geography, 76*, 147–153.
Yu, B., & Hu, Z. (2017). *Urban planning and design methods*. China Architecture & Building Press.
Zhang, F., Zhou, B., Liu, L., Liu, Y., Fung, H. H., Lin, H., & Ratti, C. (2018). Measuring human perceptions of a large-scale urban region using machine learning. *Landscape and Urban Planning, 180*, 148–160.